Intelligent Systems Reference Library

Volume 62

Series editors

Janusz Kacprzyk, Polish Academy of Sciences, Warsaw, Poland
e-mail: kacprzyk@ibspan.waw.pl

Lakhmi C. Jain, University of Canberra, Canberra, Australia
e-mail: Lakhmi.Jain@unisa.edu.au

For further volumes:
http://www.springer.com/series/8578

About this Series

The aim of this series is to publish a Reference Library, including novel advances and developments in all aspects of Intelligent Systems in an easily accessible and well structured form. The series includes reference works, handbooks, compendia, textbooks, well-structured monographs, dictionaries, and encyclopedias. It contains well integrated knowledge and current information in the field of Intelligent Systems. The series covers the theory, applications, and design methods of Intelligent Systems. Virtually all disciplines such as engineering, computer science, avionics, business, e-commerce, environment, healthcare, physics and life science are included.

Bo Xing · Wen-Jing Gao

Innovative Computational Intelligence: A Rough Guide to 134 Clever Algorithms

Springer

Bo Xing
Department of Mechanical and Aeronautical Engineering
University of Pretoria
Pretoria
South Africa

Wen-Jing Gao
Department of New Product Development
Meiyuan Mould Design and Manufacturing Co., Ltd
Xianghe
People's Republic of China

ISSN 1868-4394 ISSN 1868-4408 (electronic)
ISBN 978-3-319-34930-5 ISBN 978-3-319-03404-1 (eBook)
DOI 10.1007/978-3-319-03404-1
Springer Cham Heidelberg New York Dordrecht London

Printed on acid-free paper

Springer is part of Springer Science+Business Media (www.springer.com)

Foreword

Computational intelligence (CI) is a relatively new discipline, and accordingly, there is little agreement about its precise definition. Nevertheless, most academicians and practitioners would include techniques such as artificial neural network, fuzzy systems, many versions of evolutionary algorithms (e.g. evolution strategies, genetic algorithm, genetic programming, differential evolution), as well as ant colony optimization, artificial immune systems, multi-agent systems, particle swarm optimization, and the hybridization versions of these, under the umbrella of CI.

In contrast to this common trend, Bo and Wen-Jing offer us a brand new perspective in the field of CI research through their book entitled *Innovative Computational Intelligence: A Rough Guide to 134 Clever Algorithms*. This book is unique because it contains in one source an overview of a wide range of newly developed CI algorithms that are normally found in scattered resources. The authors succeed in identifying this vast amount of novel CI algorithms and grouping them into four large classes, namely, biology-, physics-, chemistry-, and mathematics-based CI algorithms. Furthermore, the organization of the book is such that each algorithm covered in the book contains the corresponding core working principles and some preliminary performance evaluations. This style would, no doubt, lead to the further development of these fascinating algorithms.

This book will be beneficial to a broad audience: First, university students, particularly those pursuing their postgraduate studies in advanced subjects; Second, the algorithms introduced in this book can serve as foundations for researchers to build bodies of knowledge in the fast growing area of CI research; Finally, practitioners can also use the algorithms presented in this book to solve and analyze specific real-world problems. Overall, this book makes a worthwhile read and is a welcome edition to the CI literature.

Adelaide, Australia, September 2013 Zbigniew Michalewicz

Foreword

Computational intelligence (CI) is a fast evolving area in which many novel algorithms, stemmed from various inspiring sources, were developed during the past decade. Nevertheless, many of them are dispersed in different research directions and their true potential is thus not fully utilized yet. Therefore, there is an urgent need to have these newly developed CI algorithms compiled into one single reference source.

Through over 1,630 non-repetitive supporting references, Bo and Wen-Jing have made great efforts to respond to this requirement. In their book entitled *Innovative Computational Intelligence: A Rough Guide to 134 Clever Algorithms*, the readers will enjoy their readings of a vast amount of novel CI algorithms which have been carefully classified by Bo and Wen-Jing into four main groups, i.e., biology-, physics-, chemistry-, and mathematics-based CI algorithms. The four parts of the book dedicated to these four groups of algorithms, respectively, are independent of each other which also makes it an easy-to-use reference handbook.

The broad spectrum of articles collected in this monograph is a tribute to the richness of the huge tree of CI research, which undoubtedly will continue to bear fruit, develop offshoots, and shape new research directions in the near future. Thus this book, to be published by Springer Intelligent Systems Reference Library Series, should have a great appeal to graduate students, researchers, and practitioners.

Birmingham, UK October 2013 Xin Yao

Preface

During the past decade, a number of new computational intelligence (CI) algorithms have been proposed. Unfortunately, they spread in a number of unrelated publishing directions which may hamper the use of such published resources. These provide us with motivation to analyze the existing research for categorizing and synthesizing it in a meaningful manner. The mission of this book is really important since those algorithms are going to be a new revolution in computer science. We hope it will stimulate the readers to make novel contributions or to even start a new paradigm based on nature phenomena. This book introduces 134 innovative CI algorithms. The book consists of 28 chapters which are organized as five parts. Each part can be reviewed in any order and a brief description of each individual chapter is provided as follows:

Part I Introduction

Chapter 1: In this chapter, we introduce some general knowledge relative to the realm of CI. The desirable merits of these intelligent algorithms and their initial successes in many domains have inspired researchers (from various backgrounds) to continuously develop their successors. Such truly interdisciplinary environment of the research and development provides more and more rewarding opportunities for scientific breakthrough and technology innovation. We first introduce some historical information regarding CI in Sect. 1.1. Then, the organizational structures are detailed in Sect. 1.2. Finally, Sect. 1.3 summarizes this chapter.

Part II Biology-based CI Algorithms

Chapter 2: In this chapter, we present a set of algorithms that are inspired by the different bacteria behavioral patterns, i.e., bacterial foraging algorithm (BFA), bacterial colony chemotaxis (BCC) algorithm, superbug algorithm (SuA), bacterial colony optimization (BCO) algorithm, and viral system (VS) algorithm. We first describe the general knowledge of bacteria foraging behavior in Sect. 2.1. Then, the fundamentals and performance of BFA, BCC algorithm, SuA, BCO algorithm, and VS algorithm are introduced in Sects. 2.2 and 2.3, respectively. Finally, Sect. 2.4 summarizes this chapter.

Chapter 3: In this chapter, we present two algorithms that are inspired by the behaviors of bats, i.e., bat algorithm (BaA) and bat intelligence (BI) algorithm. We first describe the general knowledge of the foraging behavior of bats in Sect. 3.1. Then, the fundamentals and performance of the BaA and BI algorithm are introduced in Sects. 3.2 and 3.3, respectively. Finally, Sect. 3.4 summarizes this chapter.

Chapter 4: In this chapter, we present a set of algorithms that are inspired by different honeybees behavioral patterns, i.e., artificial bee colony (ABC) algorithm, honeybees mating optimization (HBMO) algorithm, artificial beehive algorithm (ABHA), bee colony optimization (BCO) algorithm, bee colony inspired algorithm (BCiA), bee swarm optimization (BSO) algorithm, bee system (BS) algorithm, BeeHive algorithm, bees algorithm (BeA), bees life algorithm (BLA), bumblebees algorithm, honeybee social foraging (HBSF) algorithm, OptBees algorithm, simulated bee colony (SBC) algorithm, virtual bees algorithm (VBA), and wasp swarm optimization (WSO) algorithm. We first describe the general knowledge about honeybees in Sect. 4.1. Then, the fundamentals and performance of these algorithms are introduced in Sects. 4.2–4.4, respectively. Finally, Sect. 4.5 summarizes this chapter.

Chapter 5: In this chapter, we introduce a novel optimization algorithm called biogeography-based optimization (BBO) which is inspired by the science of biogeography. We first describe the general knowledge about the science of biogeography in Sect. 5.1. Then, the fundamentals and performance of BBO are introduced in Sect. 5.2. Finally, Sect. 5.3 summarizes this chapter.

Chapter 6: In this chapter, we present a new population-based method, called cat swarm optimization (CSO) algorithm, which imitates the natural behavior of cats. We first describe the general knowledge about the behavior of cats in Sect. 6.1. Then, the fundamentals and performance of CSO are introduced in Sect. 6.2. Next, some selected variations of CSO are explained in Sect. 6.3. Right after this, Sect. 6.4 presents a representative CSO application. Finally, Sect. 6.5 summarizes this chapter.

Chapter 7: In this chapter, a set of cuckoo-inspired optimization algorithms, i.e., cuckoo search (CA) algorithm and cuckoo optimization algorithm (COA) are introduced. We first, in Sect. 7.1, describe the general knowledge about cuckoos. Then, the fundamentals and performance of CS are introduced in Sect. 7.2. Next, the selected variants of CS are outlined in Sect. 7.3 which is followed by a presentation of representative CS application in Sect. 7.4. Right after this, Sect. 7.5 introduces an emerging algorithm, i.e., COA, which also falls within this category. Finally, Sect. 7.6 draws the conclusions of this chapter.

Chapter 8: In this chapter, we present three algorithms that are inspired by the flashing behavior of luminous insects, i.e., firefly algorithm (FA), glowworm swarm optimization (GlSO) algorithm, and bioluminescent swarm optimization

(BiSO) algorithm. We first describe the general knowledge of the luminous insects in Sect. 8.1. Then, the fundamentals, performances, and selected applications of FA, GlSO algorithm, and BiSO algorithm are introduced in Sects. 8.2–8.4, respectively. Finally, Sect. 8.5 summarises this chapter.

Chapter 9: In this chapter, we present several fish algorithms that are inspired by some key features of the fish school/swarm, namely, artificial fish school algorithm (AFSA), fish school search (FSS), group escaping algorithm (GEA), and shark-search algorithm (SSA). We first provide a short introduction in Sect. 9.1. Then, the detailed descriptions regarding AFSA and FSS can be found in Sects. 9.2 and 9.3, respectively. Next, Sect. 9.4 briefs two emerging fish inspired algorithms, i.e., GEA and SSA. Finally, Sect. 9.5 summarizes this chapter.

Chapter 10: In this chapter, we present two frog-inspired CI algorithms, namely, shuffled frog leaping algorithm (SFLA) and frog calling algorithm (FCA). We first provide a brief introduction in Sect. 10.1. Then, the fundamentals and performance of SFLA are introduced in Sect. 10.2. Next, Sect. 10.3 outlines some core working principles and preliminary experimental studies relative to FCA. Finally, Sect. 10.4 summarizes this chapter.

Chapter 11: In this chapter, we present a novel optimization algorithm called fruit fly optimization algorithm (FFOA) which is inspired by the behavior of fruit flies. We first describe the general knowledge about the foraging behavior of fruit flies in Sect. 11.1. Then, the fundamentals and performance of FFOA are introduced in Sect. 11.2. Finally, Sect. 11.3 summarizes this chapter.

Chapter 12: In this chapter, we introduced a new optimization algorithm called group search optimizer (GrSO) which is inspired from the relationship of group foraging behaviors, i.e., producer-scrounger paradigm. We first describe the general knowledge about the producer-scrounger model in Sect. 12.1. Then, the fundamentals and performance of GrSO are introduced in Sect. 12.2. Finally, Sect. 12.3 summarizes this chapter.

Chapter 13: In this chapter, we present an interesting algorithm called invasive weed optimization (IWO) which is inspired from colonizing weeds. We first describe the general knowledge of the biological invasion in Sect. 13.1. Then, the fundamentals and performance of IWO are introduced in Sect. 13.2. Finally, Sect. 13.3 summarises this chapter.

Chapter 14: In this chapter, we introduce a set of music inspired algorithms, namely harmony search (HS), melody search (MeS) algorithm, and method of musical composition (MMC) algorithm. We first describe the general knowledge about harmony in Sect. 14.1. Then, the fundamentals and performances of HS, MeS algorithm, and MMC algorithm are introduced in Sects. 14.2 and 14.3, respectively. Finally, 14.4 summarizes this chapter.

Chapter 15: In this chapter, we present a new optimization algorithm called imperialist competitive algorithm (ICA) which is inspired by the human socio-political evolution process. We first describe the general knowledge about the imperialism in Sect. 15.1. Then, the fundamentals and performance of ICA are introduced in Sect. 15.2. Finally, Sect. 15.3 summarizes this chapter.

Chapter 16: In this chapter, we present an interesting algorithm called teaching–learning-based optimization (TLBO) which is inspired by the teaching and learning behavior. We first describe the general knowledge about the teacher-student relationships in Sect. 16.1. Then, the fundamentals and performance of TLBO algorithm are introduced in Sect. 16.2. Finally, Sect. 16.3 summarizes this chapter.

Chapter 17: In this chapter, a group of (more specifically 56 in total) emerging biology-based CI algorithms are introduced. We first, in Sect. 17.1, describe the organizational structure of this chapter. Then, from Sect. 17.2 to 17.57, each section is dedicated to a specific algorithm which falls within this category. The fundamentals of each algorithm and their corresponding performances compared with other CI algorithms can be found in each associated section. Finally, the conclusions drawn in Sect. 17.58 closes this chapter.

Part III Physics-based CI Algorithms

Chapter 18: In this chapter, the big bang–big crunch (BB–BC), a global optimization method inspired from one of the cosmological theories known as closed universe, is introduced. We first, in Sect. 18.1, describe the background knowledge regarding the big bang and big crunch. Then, Sect. 18.2 details the fundamentals of BB–BC, the selected variants of BB–BC, and the representative BB–BC application, respectively. Finally, Sect. 18.3 draws the conclusions of this chapter.

Chapter 19: In this chapter, we introduce a new deterministic multidimensional search algorithm called central force optimization (CFO), which is based on the metaphor of gravitational kinematics. We first, in Sect. 19.1, describe the general knowledge about the gravitational force. Then, in Sect. 19.2, the fundamentals and performance of CFO are detailed. Finally, Sect. 19.3 draws the conclusions of this chapter.

Chapter 20: In this chapter, we introduce a novel algorithm called charged system search (CSS) algorithm which is inspired by the coulomb's law and laws of motion. We fist describe the general knowledge of the coulomb's law and laws of motion in Sect. 20.1. Then, the fundamentals and performance of CSS are introduced in Sect. 20.2. Finally, Sect. 20.3 summarizes this chapter.

Chapter 21: In this chapter, we present an electromagnetism-like mechanism (EM) algorithm which is inspired by the theory of electromagnetism. We first describe the general knowledge about the electromagnetism field theory in Sect. 21.1. Then, the fundamentals and performance of EM are introduced in Sect. 21.2. Finally, Sect. 21.3 summarizes this chapter.

Chapter 22: In this chapter, we present a gravitational search algorithm (GSA) which is based on the law of gravity. We first describe the general information about the science of gravity and the definition of mass in Sect. 22.1, respectively. Then, the fundamentals and performance of GSA are introduced in Sect. 22.2. Finally, Sect. 22.3 summarizes this chapter.

Chapter 23: In this chapter, an intelligent water drops (IWD) algorithm is introduced. We first, in Sect. 23.1, describe the general knowledge about nature water drops and the Newton's law of gravity, respectively. Then, the fundamentals of IWD, the selected variant of IWD, and the representative IWD application are detailed in Sect. 23.2, respectively. Finally, Sect. 23.3 draws the conclusions of this chapter.

Chapter 24: In this chapter, a set of (more specifically 22 in total) emerging physics-based CI algorithms are introduced. We first, in Sect. 24.1, describe the organizational structure of this chapter. Then, from Sect. 24.2 to 24.23, each section is dedicated to a specific algorithm which falls within this category. The fundamentals of each algorithm and their corresponding performances compared with other CI algorithms can be found in each associated section. Finally, the conclusions drawn in Sect. 24.24 closes this chapter.

Part IV Chemistry-based CI Algorithms

Chapter 25: In this chapter, we present a novel optimization approach named chemical-reaction optimization (CRO) algorithm. The main idea behind CRO is that a simulation of the molecules' movements and their resultant chemical reactions. We first describe the general knowledge about the chemical reaction in Sect. 25.1. Then, the fundamentals and performance of CRO are introduced in Sect. 25.2. Next, a selected variation of CRO is explained in Sect. 25.3. Right after this, Sect. 25.4 presents a representative CRO application. Finally, Sect. 25.5 summarizes this chapter.

Chapter 26: In this chapter, a set of emerging chemistry-based CI algorithms are introduced. We first, in Sect. 26.1, describe the organizational structure of this chapter. Then, from Sect. 26.2 to 26.5 , each section is dedicated to a specific algorithm which falls within this category. The fundamentals of each algorithm and their corresponding performances compared with other CI algorithms can be found in each associated section. Finally, the conclusions drawn in Sect. 26.6 closes this chapter.

Part V Mathematics-based CI Algorithms

Chapter 27: In this chapter, the base optimization algorithm (BaOA), a global optimization method inspired from mathematics research, is introduced. We first, in Sect. 27.1, describe the background knowledge about mathematics. Then, the fundamentals and performance of BaOA are detailed in Sect. 27.2. Finally, Sect. 27.3 draws the conclusions of this chapter.

Chapter 28: In this chapter, an emerging mathematics-based CI category called matheuristics is introduced. We first, in Sect. 28.1, describe the background knowledge regarding the metaheuritics. Then, the fundamentals and representative application of matheuristics are briefed in Sect. 28.2. Finally, Sect. 28.3 draws the conclusions of this chapter.

Target Audience of this Book

This book will be useful to multidisciplinary students including those in aeronautic engineering, mechanical engineering, industrial engineering, electrical and electronic engineering, chemical engineering, computer science, applied mathematics, physics, economy, biology, and social science, and particularly those pursuing postgraduate studies in advanced subjects.

Moreover, the algorithms introduced in this book can motivate researchers to further develop more efficient and effective algorithms in dealing with many cutting-edge challenges that may sit on the periphery of their present fields of interest.

Finally, practitioners can also use the models presented in this book as a starting point to solve and analyze specific real-world problems. The book is carefully written to achieve a good balance between the theoretical depth and the comprehensiveness of the innovative CI paradigms.

Pretoria, South Africa, September 2013

Bo Xing
Wen-Jing Gao

Acknowledgments

The authors owe their gratitude to all colleagues and practitioners who have collaborated directly or indirectly in writing this manuscript. In particular, the first author of this book would like to thank the Department of Mechanical and Aeronautical Engineering, University of Pretoria, for providing a joyful research environment during the writing of this book; the second author of this book would like to thank Mr. Chang-Song Liu, the president of the Mei Yuan Mould Design and Manufacturing Co., Ltd, P. R. China, for his endless support during this project.

Also, we would like to thank our supervisors from Tianjin, P. R. China; Kassel, Germany; and Durban, Pretoria, and Johannesburg, South Africa, respectively, who had played pivotal roles in our education.

In addition, we also want to thank Springer International Publishing for their commitment to publish and stimulate innovative ideas.

Finally, this book is dedicated to both authors' families: Mr. Tan Xing, Mrs. Qiu-Lan Ma, Mr. Ming-Sheng Gao, and Mrs. Fan Wang, for their unconditional love and support for making this book a reality.

Pretoria, South Africa
September 2013

Bo Xing
Wen-Jing Gao

Contents

Abbreviations

ABC	Artificial Bee Colony
ABHA	Artificial Beehive Algorithm
ACMO	Atmosphere Clouds Model Optimization
ACO	Ant Colony Optimization
ACROA	Artificial Chemical Reaction Optimization Algorithm
AFSA	Artificial Fish School Algorithm
AOA	Amoeboid Organism Algorithm
ASSA	Artificial Searching Swarm Algorithm
BaA	Bat Algorithm
BaOA	Base Optimization Algorithm
BB–BC	Big Bang–Big Crunch
BBO	Biogeography-based Optimization
BCC	Bacterial Colony Chemotaxis
BCiA	Bee Colony-Inspired Algorithm
BCO	Bee Colony Optimization
BEA	Bats Echolocation Algorithm
BeA	Bees Algorithm
BeOA	Bean Optimization Algorithm
BeSO	Bee Swarm Optimization
BFA	Bacterial Foraging Algorithm
BI	Bat Intelligence
BiSO	Bioluminescent Swarm Optimization
BLA	Bees Life Algorithm
BNMR	Blind, Naked Mole-Rats
BO	Bionic Optimization
BS	Bee System
BSA	Backtracking Search Algorithm
BSOA	Brain Storm Optimization Algorithm
BSs	Bar Systems
CA	Culture Algorithm
CFO	Central Force Optimization

ChOA	Chaos Optimization Algorithm
CI	Computational Intelligence
CMBA	Cloud Model-based Algorithm
CMDE	Cloud Model-based Differential Evolution
CRA	Chemical Reaction Algorithm
CRO	Chemical Reaction Optimization
CS	Cuckoo Search
CSA	Clonal Selection Algorithm
CSO	Cat Swarm Optimization
CSOA	Cockroach Swarm Optimization Algorithm
CSS	Charged System Search
CuOA	Cuckoo Optimization Algorithm
DE	Differential Evolution
DS	Differential Search
DSO	Dove Swarm Optimization
EA	Evolutionary Algorithm
EBB–BC	Exponential Big Bang–Big Crunch
EM	Electromagnetism-like Mechanism
EO	Extremal Optimization
ES	Eagle Strategy
FA	Firefly Algorithm
FBA	Flocking-based Algorithm
FCA	Frog Calling Algorithm
FFOA	Fruit Fly Optimization Algorithm
FOA	Fireworks Optimization Algorithm
FPA	Flower Pollinating Algorithm
FSA	Fish Search Algorithm
FSS	Fish School Search
GA	Genetic Algorithm
GBMO	Gases Brownian Motion Optimization
GbSA	Galaxy-based Search Algorithm
GCA	Gravitational Clustering Algorithm
GDA	Great Deluge Algorithm
GEA	Group Escaping Algorithm
GFA	Gravitation Field Algorithm
GIO	Gravitational Interactions Optimization
GLOA	Group Leaders Optimization Algorithm
GlSO	Glowworm Swarm Optimization
GOA	Goose Optimization Algorithm
GrSO	Group Search Optimizer
GSA	Gravitational Search Algorithm
HBB–BC	Hybrid Big Bang–Big Crunch

HBMOA	Honey Bees Mating Optimization Algorithm
HBSF	Honey Bee Social Foraging
HEA	Harmony Elements Algorithm
HGF	Human Group Formation
HO	Hysteretic Optimization
HS	Harmony Search
HuS	Hunting Search
ICA	Imperialist Competitive Algorithm
IRO	Integrated Radiation Optimization
IWD	Intelligent Water Drops
IWO	Invasive Weed Optimization
KH	Krill Herd
LCA	League Championship Algorithm
LRO	Light Ray Optimization
MA	Membrane Algorithm
MBB–BC	Modified Big Bang–Big Crunch
MBO	Migrating Birds Optimization
MeS	Melody Search
MMC	Method of Musical Composition
MOA	Magnetic Optimization Algorithm
MOCSO	Multiobjective Cat Swarm Optimization
MSA	Monkey Search Algorithm
OSA	Oriented Search Algorithm
PA	Photosynthetic Algorithm
PCA	Particle Collision Algorithm
PCSO	Parallel Cat Swarm Optimization
PFA	Paddy Field Algorithm
PSO	Particle Swarm Optimization
RFD	River Formation Dynamics
RO	Ray Optimization
SA	Simulated Annealing
SBC	Simulated Bee Colony
SCA	Society and Civilization Algorithm
SCOA	Stem Cells Optimization Algorithm
SEOA	Social Emotion Optimization Algorithm
SeOA	Seeker Optimization Algorithm
SFHM	Sheep Flock Heredity Model
SFLA	Shuffled Frog Leaping Algorithm
SFS	Stochastic Focusing Search Algorithm
SGO	Space Gravitational Optimization
SGuA	Saplings Growing Up Algorithm
SMA	Slime Mold Algorithm

SOM	Sub-Optimization Mechanism
SOMA	Self-Organizing Migrating Algorithm
SpOA	Spiral Optimization Algorithm
SPOT	Simple Optimization
SSA	Shark-Search Algorithm
SSO	Swallow Swarm Optimization
SSOA	Social Spider Optimization Algorithm
SuA	Superbug Algorithm
ThA	Termite-hill Algorithm
TS	Tabu search
UBB–CBC	Uniform Big Bang–Chaotic Big Crunch
UBS	Upper Bound Strategy
US	Unconscious Search
VBA	Virtual Bees Algorithm
VS	Viral System
WCA	Wolf Colony Algorithm
WCOA	Water Cycle Optimization Algorithm
WFA	Water Flow Algorithm
WFlA	Water Flow-like Algorithm
WoAC	Wisdom of Artificial Crowds
WPS	Wolf Pack Search
WSO	Wasp Swarm Optimization

Objective and Mission

Introduction to the Subject Area

About 20 years ago, the term "Computational Intelligence" (CI), coined by Bezdek, triggered the development of a new field dedicated to computer-based intelligence which can be regarded as a timely intent to avoid some tough issues. In principle, CI consists of any science-supported approaches and technologies for analyzing, creating, and developing intelligent systems. The broad usage of this term was formalized by the IEEE Neural Network Council and the IEEE World Congress on Computational Intelligence in Orlando, Florida in the summer of 1994. With the advances of many advanced theories and methodologies, many obstacles that may have previously hindered the development of CI research have now been overcome. During the past two decades, as evidenced by the promising results of numerous researches, CI has enjoyed a wide acceptance and an unprecedented popularity. By applying it in various application settings, CI has opened many brand new dimensions for scientific research.

Objective and Mission

Traditional CI primarily concentrates on artificial neural network (ANN), fuzzy logic (FL), multi-agent system (MAS), evolutionary algorithms (EA) (e.g., genetic algorithm (GA), genetic programming (GP), evolutionary programming (EP), and evolutionary strategy (ES)), artificial immune systems (AIS), simulated annealing (SA), Tabu search (TS), as well as two variants of swarm intelligence (SI), i.e., ant colony optimization (ACO) and particle swarm optimization (PSO). In the literature, there are thousands of (if not more) books, conference proceedings, and edited monographs devoted to CI and its corresponding vast amount of applications.

Innovative CI, unlike its counterparts, i.e., highly developed and refined traditional CI, is a new CI category introduced by the authors of this book. Although most innovative CI algorithms introduced in this book hold considerable promise, a majority of them are still in their infancy. There are currently very few books specifically dedicated to these novel CI paradigms. Therefore it is the authors' hope that this book will inspire other far better qualified researchers to bring these new CI family members to their full potential. This serves as the main objective and mission of this book.

Unique Characteristics

The first notable feature of this book is its **innovation**: Computational intelligence (CI), a fast evolving area, is currently attracting a lot of researchers' attention in dealing with many complex problems. At present, there are quite a lot of competing books existing in the market. Nevertheless, the present book is markedly different from the existing books in that it presents **new paradigms of CI** that have rarely mentioned before, as opposed to the traditional CI techniques or methodologies employed in other books. During the past decade, a number of new CI algorithms are proposed. Unfortunately, they spread in a number of unrelated publishing directions which may hamper the use of such published resources. These provide us with motivation to analyze the existing research for categorizing and synthesizing it in a meaningful manner. The mission of this book is really important since those algorithms are going to be a new revolution in computer science. We hope it will stimulate the readers to make novel contributions or even start a new paradigm based on nature phenomena. Although structured as a textbook, the book's straightforward, self-contained style will also appeal to a wide audience of professionals, researchers, and independent learners. We believe that the book will be instrumental in initiating an integrated approach to complex problems by allowing cross-fertilization of design principles from different design philosophies.

The second feature of this book is its **comprehensiveness**: Through an extensive literature research, there are 134 innovative CI algorithms covered in this book.

Prospective Audience

This book will be useful to multidisciplinary students including those in aeronautic engineering, mechanical engineering, industrial engineering, electrical and electronic engineering, chemical engineering, computer science, applied mathematics, physics, economy, biology, and social science, and particularly those pursuing postgraduate studies in advanced subjects.

Moreover, the algorithms introduced in this book can motivate researchers to further develop more efficient and effective algorithms in dealing with many cutting-edge challenges that may sit on the periphery of their present fields of interest.

Finally, practitioners can also use the models presented in this book as a starting point to solve and analyse specific real-world problems. The book is carefully written to achieve a good balance between the theoretical depth and the comprehensiveness of the innovative CI paradigms.

Bo Xing
Wen-Jing Gao

Part I
Introduction

Chapter 1
Introduction to Computational Intelligence

Abstract In this chapter, we introduce some general knowledge relative to the realm of computational intelligence (CI). The desirable merits of these intelligent algorithms and their initial successes in many domains have inspired researchers (from various backgrounds) to continuously develop their successors. Such truly interdisciplinary environment of the research and development provides more and more rewarding opportunities for scientific breakthrough and technology innovation. We first introduce some historical information regarding CI in Sect. 1.1. Then, the organizational structures are detailed in Sect. 1.2. Finally, Sect. 1.3 summarises in this chapter.

1.1 Introduction

About 20 years ago, the term "Computational Intelligence" (CI), coined by Bezdek (1992, 1994), triggered the development of a new field dedicated to computer-based intelligence which can be regarded as a timely intent to avoid some of the tough issues. In principle, CI consists of any science-supported approaches and technologies for analyzing, creating, and developing intelligent systems. The broad usage of this term was formalized by the IEEE Neural Network Council and the IEEE World Congress on Computational Intelligence in Orlando, Florida in the summer of 1994. With the advances of many advanced theories and methodologies, many obstacles that may previously hindered the development of CI research have now been overcome. During the past two decades, as evidenced by the promising results of numerous researches, CI has enjoyed a widely acceptance and an unprecedented popularity. By applying it in various application settings, CI has opened many brand new dimensions for scientific research.

B. Xing and W.-J. Gao, *Innovative Computational Intelligence: A Rough Guide to 134 Clever Algorithms*, Intelligent Systems Reference Library 62,
DOI: 10.1007/978-3-319-03404-1_1,

1.1.1 Traditional CI

Traditional CI primarily concentrates on artificial neural network (ANN), fuzzy logic (FL), multi-agent system (MAS), evolutionary algorithms (EA) [e.g., genetic algorithm (GA), genetic programming (GP), evolutionary programming (EP), and evolutionary strategy (ES)], artificial immune systems (AIS), simulated annealing (SA), Tabu search (TS), as well as two variants of swarm intelligence (SI), i.e., ant colony optimization (ACO) and particle swarm optimization (PSO). In the literature, there are thousands of (if not more) books, conference proceedings, and edited monographs are devoted to CI and its corresponding vast amount of applications, e.g., (Wang and Kusiak 2001; Engelbrecht 2007; Fink and Rothlauf 2008; Marwala 2009, 2010, 2012; Rutkowski 2008; Sumathi and Paneerselvam 2010; Marwala and Lagazio 2011; Xing and Gao 2014; Chatterjee and Siarry 2013; Yang 2008, 2010a; Eberhart and Shi 2007; Fulcher and Jain 2008; Mumford and Jain 2009; Michalewicz 1996), to name just a few.

1.1.2 Innovative CI

Innovative CI, unlike its counterparts, i.e., highly developed and refined traditional CI, is a new CI category introduced by the authors of this book. Although most innovative CI algorithms introduced in this book hold considerable promise, majority of them are still in their infancy. There are currently very few books specifically dedicated to these novel CI paradigms. Therefore it is the authors' hope that this book will inspire other far better qualified researchers to bring these new CI family members to their full potential.

1.2 Organization of the Book

In this book, we will cover a vast amount of innovative algorithms (more specifically, 134 in total) that all involving some aspect of CI, but each one taking a somewhat pragmatic view. In order to present a clear picture over these approaches, we have organized them into four main classes, namely, biology-, physics-, chemistry-, and mathematics-based CI algorithms.

1.2.1 Biology-based CI Algorithms

Briefly, biology can be defined as a comprehensive science concerning all functions of living systems (Glaser 2012). From an evolutionary process point of view, biological systems possess many appealing characteristics such as sophistication, robustness, and adaptability (Floreano and Mattiussi 2008). These features represent a strong motivation for imitating the mechanisms of natural evolution in an attempt to create CI algorithms with merits comparable to those of biological systems.

In this class, we have covered 99 novel biology-based CI algorithms which are outlined as follows. Each algorithm's original reference has been attached for readers' convenience to trace their origination.

- Amoeboid Organism Algorithm (Zhang et al. 2013).
- Artificial Bee Colony (Karaboga and Basturk 2007).
- Artificial Beehive Algorithm (Muñoz et al. 2009).
- Artificial Fish Swarm Algorithm (Li 2003).
- Artificial Searching Swarm Algorithm (Chen 2009).
- Artificial Tribe Algorithm (Chen et al. 2012).
- Backtracking Search Algorithm (Civicioglu 2013).
- Bacterial Colony Chemotaxis (Müller et al. 2002).
- Bacterial Colony Optimization (Niu and Wang 2012).
- Bacterial Foraging Algorithm (Passino 2002).
- Bar Systems (Acebo and Rosa 2008).
- Bat Algorithm (Yang 2010b).
- Bat Intelligence (Malakooti et al. 2012).
- Bean Optimization Algorithm (Zhang et al. 2010).
- Bee Colony Optimization (Teodorović and Dell'Orco 2005).
- Bee Colony-inspired Algorithm (Häckel and Dippold 2009).
- Bee Swarm Optimization (Akbari et al. 2009).
- Bee System (Sato and Hagiwara 1997).
- BeeHive (Wedde et al. 2004).
- Bees Algorithm (Pham et al. 2006).
- Bees Life Algorithm (Bitam and Mellouk 2013).
- Biogeography-based optimization (Simon 2008).
- Bioluminescent Swarm Optimization (Oliveira et al. 2011).
- Bionic Optimization (Steinbuch 2011).
- Blind, Naked Mole-Rats (Taherdangkoo et al. 2012).
- Brain Storm Optimization (Shi 2011).
- Bumblebees Algorithm (Comellas and Martínez-Navarro 2009).
- Cat Swarm Optimization (Chu and Tsai 2007).
- Clonal Selection Algorithm (Castro and Zuben 2000).
- Cockroach Swarm Optimization (Chen and Tang 2010).
- Collective Animal Behaviour (Cuevas et al. 2013b).
- Cuckoo Optimization Algorithm (Rajabioun 2011).
- Cuckoo Search (Yang and Deb 2009).
- Cultural Algorithm (Reynolds 1994).
- Differential Search (Civicioglu 2012).
- Dove Swarm Optimization (Su et al. 2009).

- Eagle Strategy (Yang and Deb 2010).
- Firefly Algorithm (Łukasik and Żak 2009).
- Fireworks Algorithm (Tan and Zhu 2010).
- Fish School Search (Bastos-Filho et al. 2008).
- FlockbyLeader (Bellaachia and Bari 2012).
- Flocking-based Algorithm (Cui et al. 2006).
- Flower Pollinating Algorithm (Yang 2012).
- Frog Calling Algorithm (Mutazono et al. 2012).
- Fruit Fly Optimization Algorithm (Pan 2012).
- Glowworm Swarm Optimization (Krishnanand and Ghose 2005).
- Goose Optimization Algorithm (Sun and Lei 2009).
- Great Deluge Algorithm (Dueck 1993).
- Grenade Explosion Algorithm (Ahrari et al. 2009).
- Group Escaping Algorithm (Min and Wang 2010).
- Group Leaders Optimization Algorithm (Daskin and Kais 2011).
- Group Search Optimizer (He et al. 2006).
- Harmony Elements Algorithm (Cui et al. 2008).
- Harmony Search (Geem et al. 2001).
- Honeybee Social Foraging (Quijano and Passino 2010).
- Honeybees Mating Optimization (Abbass 2001).
- Human Group Formation (Thammano and Moolwong 2010).
- Hunting Search (Oftadeh et al. 2010).
- Imperialist Competition Algorithm (Atashpaz-Gargari and Lucas 2007).
- Invasive Weed Optimization (Mehrabian and Lucas 2006).
- Krill Herd (Gandomi and Alavi 2012).
- League Championship Algorithm (Kashan 2009).
- Melody Search (Ashrafi and Dariane 2011).
- Membrane Algorithm (Nishida 2005).
- Method of Musical Composition (Mora-Gutiérrez et al. 2012).
- Migrating Birds Optimization (Duman et al. 2012).
- Mine Blast Algorithm (Sadollah et al. 2012).
- Monkey Search (Mucherino and Seref 2007).
- Mosquito Host-Seeking Algorithm (Feng et al. 2009).
- OptBees (Maia et al. 2012).
- Oriented Search Algorithm (Zhang et al. 2008).
- Paddy Field Algorithm (Premaratne et al. 2009).
- Photosynthetic Algorithm (Murase 2000).
- Population Migration Algorithm (Zhang et al. 2009).
- Roach Infestation Optimization (Havens et al. 2008).
- Saplings Growing Up Algorithm (Karci and Alatas 2006).
- Seeker Optimization Algorithm (Dai et al. 2007).
- Self-Organizing Migrating Algorithm (Davendra et al. 2013).
- Shark-Search Algorithm (Hersovici et al. 1998).
- Sheep Flock Heredity Model (Nara et al. 1999).

- Shuffled Frog Leaping Algorithm (Eusuff and Lansey 2003).
- Simple Optimization (Hasançebi and Azad 2012).
- Simulated Bee Colony (McCaffrey and Dierking 2009).
- Slime Mould Algorithm (Shann 2008).
- Social Emotional Optimization Algorithm (Wei et al. 2010).
- Social Spider Optimization Algorithm (Cuevas et al. 2013a).
- Society and Civilization Algorithm (Ray and Liew 2003).
- Stem Cells Algorithm (Taherdangkoo et al. 2011).
- Stochastic Focusing Search (Zheng et al. 2009)
- Superbug Algorithm (Anandaraman et al. 2012).
- Swallow Swarm Optimization (Neshat et al. 2013).
- Teaching–learning-based Optimization (Rao et al. 2011).
- Termite-hill Algorithm (Zungeru et al. 2012).
- Unconscious Search (Ardjmand and Amin-Naseri 2012).
- Viral System (Cortés et al. 2008).
- Virtual Bees Algorithm (Yang 2005).
- Wasp Swarm Optimization (Theraulaz et al. 1991).
- Wisdom of Artificial Crowds (Ashby and Yampolskiy 2011).
- Wolf Colony Algorithm (Liu et al. 2011).
- Wolf Pack Search (Yang et al. 2007).

1.2.2 Physics-based CI Algorithms

The word physics is derived from the Greek word physika, which means "natural things" (Holzner 2011). As the most fundamental science, physics is concerned with the basic principles of the universe. It is therefore the foundation of many other sciences such as biology, chemistry, and geology. In physics, just a small number of concepts and models can dramatically alter and expand our view of the world around us. Typically, the research of physics can be classified into the following areas such as classical mechanics, relativity, thermodynamics, electromagnetism, optics, and quantum mechanics (Serway and Jewett 2014). The simplicity of all these fundamental principles is not only the real beauty of physics, but also the main momentum in developing innovative CI algorithms.

In this class, we have included 28 novel physics-based CI algorithms which are listed as follows. Each algorithm's original reference has also been attached for readers' convenience to trace their origination.

- Artificial Physics Optimization (Xie and Zeng 2009).
- Atmosphere Clouds Model Optimization (Yan and Hao 2012).
- Big Bang-Big Crunch (Erol and Eksin 2006).
- Central Force Optimization (Formato 2007).
- Chaos Optimization Algorithm (Li and Jiang 1998).
- Charged System Search (Kaveh and Talatahari 2010).
- Cloud Model-based Algorithm (Zhu and Ni 2012).

- Electromagnetism-like Mechanism (Birbil and Fang 2003).
- Extremal Optimization (Boettcher and Percus 2000).
- Galaxy-based Search Algorithm (Shah-Hosseini 2011).
- Gravitation Field Algorithm (Zheng et al. 2010).
- Gravitational Clustering Algorithm (Kundu 1999).
- Gravitational Emulation Local Search (Barzegar et al. 2009).
- Gravitational Interactions Optimization (Flores et al. 2011).
- Gravitational Search Algorithm (Rashedi et al. 2009).
- Hysteretic Optimization (Zaránd et al. 2002).
- Integrated Radiation Optimization (Chuang and Jiang 2007).
- Intelligent Water Drops (Shah-Hosseini 2007).
- Light Ray Optimization (Shen and Li 2009).
- Magnetic Optimization Algorithm (Tayarani et al. 2008).
- Particle Collision Algorithm (Sacco and Oliveira 2005).
- Ray Optimization (Kaveh and Khayatazad 2012).
- River Formation Dynamics Algorithm (Rabanal et al. 2007).
- Space Gravitational Optimization (Hsiao et al. 2005).
- Spiral Optimization Algorithm (Jin and Tran 2010).
- Water Cycle Algorithm (Eskandar et al. 2012).
- Water Flow Algorithm (Basu et al. 2007).
- Water Flow-like Algorithm (Yang and Wang 2007).

1.2.3 Chemistry-based CI Algorithms

Chemistry can be usually viewed as a branch of physical science, but it is distinct from physics. In fact, the chemistry can be defined as a molecular view of matter. The major concern of chemistry is about the matters' properties, the changes that matter undergoes, and the energy changes that accompany those processes (Whitten et al. 2014). In other words, to understand living systems fully, an important question to consider at this point is which factors that control and affect the chemical behaviours, such as photochemical reactions, oxidation–reduction reactions, combination reactions, decomposition reactions, displacement reactions, gas-formation reactions, and metathesis reactions. The analyzing of all these types of chemical reactions is not only the real beauty of chemistry, but also the main momentum in developing innovative CI algorithms.

In this class, we have included 5 novel chemistry-based CI algorithms which are listed as follows. Each algorithm's original reference has also been attached for readers' convenience to trace their origination.

- Artificial Chemical Process (Irizarry 2005).
- Artificial Chemical Reaction Optimization Algorithm (Alatas 2011).
- Chemical Reaction Algorithm (Melin et al. 2013).
- Chemical-Reaction Optimization Algorithm (Lam and Li 2010).
- Gases Brownian Motion Optimization (Abdechiri et al. 2013).

1.2.4 Mathematics-based CI Algorithms

During the past decades, we have witnessed a proliferation of personal computers, smart phones, high-speed Internet, to name a few. The rapid development of various technologies has reduce the necessity for human beings to perform manual tasks which are either tedious or dangerous in nature, as computers may now accomplish most of them. As one of the most important building blocks, mathematics plays a crucial role in realizing all these technologies. The history of mathematics is no doubt tremendous long. According to Anglin (1994), Aristotle thought that is the priests in Egypt who actually started mathematics since the priestly class was allowed leisure. Whereas, Herodotus, believed that geometry was created to re-determine land boundaries due to the annual flooding of the Nile. The accurate beginning of mathematics is of course out of the scope of this book, but the widely employed mathematical modelling approaches indeed help us to gain insight and make reasonable accurate predictions towards the targeted problems (Yang 2013). Apart from that, the real beauty of mathematics also forms the main thrust in developing innovative CI algorithms.

In this class, we have included 2 novel mathematics-based CI algorithms which are listed as follows. Each algorithm's original reference has also been attached for readers' convenience to trace their origination.

- Base Optimization Algorithm (Salem 2012).
- Matheuristics (Maniezzo et al. 2009).

1.3 Conclusions

Our natural world conceals many characteristics of different creatures, and all of them have some unique behaviour or features to keep them survive. In this chapter, a brief background of CI (both in terms of traditional and innovative perspectives) has been discussed from an introductory perspective. The organizational structure of this book has also been explained. Interested readers are referred to them as a starting point for a further exploration and exploitation of any of these 134 algorithms that may draw their attention.

References

Abbass, H. A. (2001, May 27–30). MBO: marriage in honey bees optimization. A Haplometrosis Polygynous swarming approach. In *2001 Congress on Evolutionary Computation (CEC), Seoul, South Korea* (pp. 207–214). IEEE.

Abdechiri, M., Meybodi, M. R. & Bahrami, H. (2013). Gases Brownian motion optimization: an algorithm for optimization (GBMO). *Applied Soft Computing*. http://dx.doi.org/10.1016/j.asoc.2012.03.068.

Acebo, E. D. & Rosa, J. L. D. L. (2008, April 1–4). Introducing bar systems: a class of swarm intelligence optimization algorithms. In *AISB 2008 Symposium on Swarm Intelligence Algorithms and Applications*, University of Aberdeen (pp. 18–23). The Society for the Study of Artificial Intelligence and Simulation of Behaviour.

Ahrari, A., Shariat-Panahi, M., & Atai, A. A. (2009). GEM: a novel evolutionary optimization method with improved neighborhood search. *Applied Mathematics and Computation, 210*, 379–386.

Akbari, R., Mohammadi, A. & Ziarati, K. (2009). A powerful bee swarm optimization algorithm. In *13th International Multitopic Conference* (*INMIC*), pp. 1–6. IEEE.

Alatas, B. (2011). ACROA: artificial chemical reaction optimization algorithm for global optimization. *Expert Systems with Applications, 38*, 13170–13180.

Anandaraman, C., Sankar, A. V. M., & Natarajan, R. (2012). A new evolutionary algorithm based on bacterial evolution and its applications for scheduling a flexible manufacturing system. *Jurnal Teknik Industri, 14*, 1–12.

Anglin, W. S. (1994). *Mathematics: a concise history and philosophy.* New York: Springer. ISBN 0-378-94280-7.

Ardjmand, E. & Amin-Naseri, M. R. (2012). Unconscious search: A new structured search algorithm for solving continuous engineering optimization problems based on the theory of psychoanalysis. In Y. Tan, Y. Shi, & Z. Ji, (Eds.), *ICSI 2012, Part I*, LNCS (Vol. 7331, pp. 233–242). Berlin: Springer.

Ashby, L. H., & Yampolskiy, R. V. (2011). Genetic algorithm and wisdom of artificial Crowds algorithm applied to light up. In *IEEE 16th International Conference on Computer Games* (*CGAMES 2011*) (pp. 27–32).

Ashrafi, S. M., & Dariane, A. B. (2011, December 5–8) A novel and effective algorithm for numerical optimization: melody search (MS). In *11th International Conference on Hybrid Intelligent Systems* (*HIS*), *Melacca* (pp 109–114). IEEE.

Atashpaz-Gargari, E., & Lucas, C. (2007). Imperialist competitive algorithm: an algorithm for optimization inspired by imperialistic competition. In *IEEE Congress on Evolutionary Computation* (*CEC 2007*) (pp. 4661–4667). IEEE.

Barzegar, B., Rahmani, A. M., & Zamanifar, K. (2009). Gravitational emulation local search algorithm for advanced reservation and scheduling in grid systems. In *First Asian Himalayas International Conference on Internet* (*AH-ICI*) (pp. 1–5). IEEE.

Bastos-Filho, C. J. A., LIMA-NETO, F. B. D., LINS, A. J. C. C., Nascimento, A. I. S., & Lima, M. P. (2008). A novel search algorithm based on fish school behavior. In *IEEE International Conference on Systems, Man and Cybernetics* (*SMC*) (pp. 2646–2651). IEEE.

Basu, S., Chaudhuri, C., Kundu, M., Nasipuri, M., & Basu, D. K. (2007). Text line extraction from multi-skewed handwritten documents. *Pattern Recognition, 40*, 1825–1839.

Bellaachia, A., & Bari, A. (2012). Flock by leader: a novel machine learning biologically inspired clustering algorithm. In: Y. Tan, Y. Shi, & Z. Ji (Eds.). *ICSI 2012, Part I, LNCS* (Vol. 7332, pp. 117–126). Berlin: Springer.

Bezdek, J. C. (1992). On the relationship between neural networks, pattern recognition and intelligence. *International Journal of Approximate Reasoning, 6*, 85–107.

Bezdek, J. C. (1994). What is computational intelligence? In J. M. Zurada, R. J. Marks, & C. J. Robinson (Eds.), *Computational intelligence imitating life* (pp. 1–12). Los Alamitos: IEEE Press.

Birbil, Şİ., & Fang, S.-C. (2003). An electromagnetism-like mechanism for global optimization. *Journal of Global Optimization, 25*, 263–282.

Bitam, S., & Mellouk, A. (2013). Bee life-based multi constraints multicast routing optimization for vehicular ad hoc networks. *Journal of Network and Computer Applications, 36*, 981–991.

Boettcher, S., & Percus, A. (2000). Nature's way of optimizing. *Artificial Intelligence, 119*, 275–286.

Castro, L. N. D., & Zuben, F. J. V. (2000, July). The clonal selecton algorithm with engineering applications. Workshop on Artificial Immune Systems and Their Applications, Las Vegas, USA (pp. 1–7).

Chatterjee, A., & Siarry, P. (Eds.). (2013). *Computational intelligence in image processing*. Berlin: Springer. ISBN 978-3-642-30620-4.

Chen, T. (2009). A simulative bionic intelligent optimization algorithm: artificial searching swarm algorithm and its performance analysis. In *International Joint Conference on Computational Sciences and Optimization (CSO)* (pp. 864–866). IEEE.

Chen, T., Wang, Y., & Li, J. (2012). Artificial tribe algorithm and its performance analysis. *Journal of Software, 7*, 651–656.

Chen, Z., & Tang, H. (2010). Cockroach swarm optimization. In *2nd International Conference on Computer Engineering and Technology (ICCET)* (pp. 652–655). IEEE.

Chu, S.-C., & Tsai, P.-W. (2007). Computational intelligence based on the behavior of cats. *International Journal of Innovative Computing, Information and Control, 3*, 163–173.

Chuang, C.-L., & Jiang, J.-A. (2007, September 25–28). Integrated radiation optimization: inspired by the gravitational radiation in the curvature of space-time. In *IEEE Congress on Evolutionary Computation (CEC), Singapore* (pp. 3157–3164). IEEE.

Civicioglu, P. (2012). Transforming geocentric Cartesian coordinates to geodetic coordinates by using differential search algorithm. *Computers and Geosciences, 46*, 229–247.

Civicioglu, P. (2013). Backtracking search optimization algorithm for numerical optimization problems. *Applied Mathematics and Computation, 219*, 8121–8144.

Comellas, F., & Martínez-Navarro, J. (2009). Bumblebees: a multiagent combinatorial optimization algorithm inspired by social insect behaviour. In: *First ACM/SIGEVO Summit on Genetic and Evolutionary Computation (GEC)* (pp. 811–814). New york: ACM.

Cortés, P., García, J. M., Muñuzuri, J., & Onieva, L. (2008). Viral systems: a new bio-inspired optimisation approach. *Computers and Operations Research, 35*, 2840–2860.

Cuevas, E., Cienfuegos, M., Zaldívar, D., & Pérez-Cisneros, M. (2013a). A swarm optimization algorithm inspired in the behavior of the social-spider. *Expert Systems with Applications*.http://dx.doi.org/10.1016/j.eswa.2013.05.041.

Cuevas, E., Zaldívar, D., & Pérez-Cisneros, M. (2013b). A swarm optimization algorithm for multimodal functions and its application in multicircle detection. *Mathematical Problems in Engineering, 2013*, 1–22.

Cui, X., Gao, J., & Potok, T. E. (2006). A flocking based algorithm for document clustering analysis. *Journal of Systems Architecture, 52*, 505–515.

Cui, Y. H., Guo, R., Rao, R. V., & Savsani, V. J. (2008, December 15–17). Harmony element algorithm: A naive initial searching range. In *International Conference on Advances in Mechanical Engineering*, (pp. 1–6). Gujarat: S. V. National Institute of Technology.

Dai, C., Zhu, Y., & Chen, W. (2007). Seeker optimization algorithm. In: Y. Wang, Cheung, Y., & Liu, H. (Eds.). *CIS 2006, LNAI*. (Vol. 4456, pp. 167–176). Berlin: Springer.

Daskin, A., & Kais, S. (2011). Group leaders optimization algorithm. *Molecular Physics, 109*, 761–772.

Davendra, D., Zelinka, I., Bialic-Davendra, M., Senkerik, R., & Jasek, R. (2013). Discrete self-organising migrating algorithm for flow-shop scheduling with no-wait makespan. *Mathematical and Computer Modelling, 57*, 100–110.

Dueck, G. (1993). New optimization heuristics: the great deluge algorithm and the record-to-record travel. *Journal of Computational Physics, 104*, 86–92.

Duman, E., Uysal, M., & Alkaya, A. F. (2012). Migrating birds optimization: a new metaheuristic approach and its performance on quadratic assignment problem. *Information Sciences, 217*, 65–77.

Eberhart, R. C., & Shi, Y. (2007). *Computational intelligence: concepts to implementations*. Los Altos: Morgan Kaufmann. ISBN 1558607595.

Engelbrecht, A. P. (2007). *Computational intelligence: an introduction*. West Sussex: Wiley. ISBN 978-0-470-03561-0.

Erol, O. K., & Eksin, I. (2006). A new optimization method: Big Bang–Big Crunch. *Advances in Engineering Software, 37*, 106–111.
Eskandar, H., Sadollah, A., Bahreininejad, A., & Hamdi, M. (2012). Water cycle algorithm: A novel metaheuristic optimization for solving constrained engineering optimization problems. *Computers and Structures, 110–111*, 151–166.
Eusuff, M. M., & Lansey, K. E. (2003). Optimization of water distribution network design using the shuffled frog leaping algorithm. *Journal of Water Resources Planning and Management, 129*, 210–225.
Feng, X., Lau, F. C. M., & Gao, D. (2009). A new bio-inspired approach to the traveling salesman problem. In: J. Zhou. (Ed.). *Complex 2009, Part II, LNICST*, (Vol. *5*, pp. 1310–1321). Institute for Computer Sciences, Social Informatics and Telecommunications Engineering.
Fink, A., & Rothlauf, F. (Eds.). (2008). *Advances in computational intelligence in transport, logistics, and supply chain management*. Berlin: Springer. ISBN 978-3-540-69024-5.
Floreano, D., & Mattiussi, C. (2008). *Bio-inspired artificial intelligence: theories, methods, and technologies*. Cambridge: The MIT Press. ISBN 978-0-262-06271-8.
Flores, J. J., López, R., & Barrera, J. (2011). Gravitational interactions optimization. *Learning and Intelligent Optimization*, (pp. 226–237). Berlin: Springer.
Formato, R. A. (2007). Central force optimization: a new metaheuristic with applications in applied electromagnetics. *Progress in Electromagnetics Research, PIER, 77*, 425–491.
Fulcher, J., & Jain, L. C. (Eds.). (2008). *Computational intelligence: a compendium*. Berlin: Springer. ISBN 978-3-540-78292-6.
Gandomi, A. H., & Alavi, A. H. (2012). Krill herd: a new bio-inspired optimization algorithm. *Communications in Nonlinear Science and Numerical Simulation, 17*, 4831–4845.
Geem, Z. W., Kim, J. H., & Loganathan, G. V. (2001). A new heuristic optimization algorithm: harmony search. *Simulation, 76*, 60–68.
Glaser, R. (2012). *Biophysics: an introduction*. Berlin: Springer. ISBN 978-3-642-25211-2.
Häckel, S., & Dippold, P. (2009, July 8–12). The bee colony-inspired algorithm (BCiA): A two-stage approach for solving the vehicle routing problem with time windows. GECCO'09 (pp. 25–32). Nontréal, Québec, Canada.
Hasançebi, O., & Azad, S. K. (2012). An efficient metaheuristic algorithm for engineering optimization: SPOT. *International Journal of Optimization in Civil Engineering, 2*, 479–487.
Havens, T. C., Spain, C. J., Salmon, N. G., & Keller, J. M. (2008, September 21–23). Roach infestation optimization. In *IEEE Swarm Intelligence Symposium* (pp. 1–7). St. Louis MO USA. IEEE.
He, S., Wu, Q. H., & Saunders, J. R. (2006, July 16–21). A novel group search optimizer inspired by animal behavioural ecology. In *IEEE Congress on Evolutionary Computation* (*CEC*) (pp. 1272–1278). Vancouver: Sheraton Vancouver Wall Centre Hotel. IEEE.
Hersovici, M., Jacovi, M., Maarek, Y. S., Pelleg, D., Shtalhaim, M., & Ur, S. (1998). The shark-search algorithm. An application: tailored Web site mapping. *Computer Networks and ISDN Systems, 30*, 317–326.
Holzner, S. (2011). *Physics I for dummies*. River Street: Wiley. ISBN 978-0-470-90324-7.
Hsiao, Y.-T., Chuang, C.-L., Jiang, J.-A., & Chien, C.-C. (2005, October 10–12). A novel optimization algorithm: space gravitational optimization. *IEEE International Conference on Systems, Man and Cybernetics* (*SMC*) (pp. 2323–2328). IEEE.
Irizarry, R. (2005). A generalized framework for solving dynamic optimization problems using the artificial chemical process paradigm: applications to particulate processes and discrete dynamic systems. *Chemical Engineering Science, 60*, 5663–5681.
Jin, G.-G., & Tran, T.-D. (2010, August 18–21). A nature-inspired evolutionary algorithm based on spiral movements. In *SICE Annual Conference* (pp. 1643–1647). The Grand Hotel: Taipei. IEEE.
Karaboga, D., & Basturk, B. (2007). A powerful and efficient algorithm for numerical function optimization: artificial bee colony (ABC) algorithm. *Journal of Global Optimization, 39*, 459–471.

Karci, A., & Alatas, B. (2006). Thinking capability of saplings growing up algorithm. In *Intelligent Data Engineering and Automated Learning (IDEAL 2006), LNCS* (Vol. 4224, pp. 386–393). Berlin: Springer.

Kashan, A. H. (2009). League championship algorithm: a new algorithm for numerical function optimization. In *International Conference of Soft Computing and Pattern Recognition* (SoCPAR) (pp. 43–48). IEEE.

Kaveh, A., & Khayatazad, M. (2012). A new meta-heuristic method: ray optimization. *Computers and Structures, 112–113*, 283–294.

Kaveh, A., & Talatahari, S. (2010). A novel heuristic optimization method: charged system search. *Acta Mechanica, 213*, 267–289.

Krishnanand, K. N., & Ghose, D. (2005). Detection of multiple source locations using a glowworm metaphor with applications to collective robotics. *IEEE Swarm Intelligence Symposium (SIS)* (pp. 84–91). IEEE.

Kundu, S. (1999). Gravitational clustering: a new approach based on the spatial distribution of the points. *Pattern Recognition, 32*, 1149–1160.

L.Mumford, C., & JAIN, L. C. (Eds.). (2009). *Computational intelligence: collaboration, fusion and emergence*. Berlin: Springer. ISBN 978-3-642-01798-8.

Lam, A. Y. S., & Li, V. O. K. (2010). Chemical-reaction-inspired metaheuristic for optimization. *IEEE Transactions on Evolutionary Computation, 14*, 381–399.

Li, B., & Jiang, W. (1998). Optimizing complex functions by chaos search. *Cybernetics and Systems: An International, 29*, 409–419.

Li, X.-L. (2003). *A new intelligent optimization method: Artificial fish school algorithm (in Chinese with English abstract)*. Unpublished Doctoral Thesis, Zhejiang University.

Liu, C., Yan, X., Liu, C., & Wu, H. (2011). The wolf colony algorithm and its application. *Chinese Journal of Electronics, 20*, 212–216.

Łukasik, S., & Żak, S. (2009). Firefly algorithm for continuous constrained optimization tasks. In *Computational Collective Intelligence. Semantic Web, Social Networks and Multiagent Systems, LNCS,* (Vol. 5796, pp. 97–106). Berlin: Spinger.

Maia, R. D., Castro, L. N. D., & Caminhas, W. M. (2012, June 10–15). Bee colonies as model for multimodal continuous optimization: the OptBees algorithm. In *IEEE World Congress on Computational Intelligence (WCCI)* (pp. 1–8). Brisbane, Australia. IEEE.

Malakooti, B., Sheikh, S., Al-Najjar, C., & Kim, H. (2013). Multi-objective energy aware multiprocessor scheduling using bat intelligence. *Journal of Intelligent Manufacturing, 24*, 805–819. doi: 10.1007/s10845-012-0629-6.

Maniezzo, V., Stützle, T. & VOß, S. (Eds.). (2009). *Matheuristics: hybridizing metaheuristics and mathematical programming.* New York: Springer. ISBN 978-1-4419-1305-0.

Marwala, T. (2009). *Computational intelligence for missing data imputation, estimation and management: knowledge optimization techniques*. New York: IGI Global. ISBN 978-1-60566-336-4.

Marwala, T. (2010). *Finite-element-model updating using computational intelligence techniques: applications to structural dynamics*. London: Springer. ISBN 978-1-84996-322-0.

Marwala, T. (2012). *Condition monitoring using computational intelligence methods: applications in mechanical and electrical systems*. London: Springer. ISBN 978-1-4471-2379-8.

Marwala, T., & Lagazio, M. (2011). *Militarized conflict modeling using computational intelligence*. London: Springer. ISBN 978-0-85729-789-1.

Mccaffrey, J. D., & Dierking, H. (2009). An empirical study of unsupervised rule set extraction of clustered categorical data using a simulated bee colony algorithm. In G. Governatori, Hall, J., & Paschke, A. (Eds.). *RuleML 2009, LNCS,* (Vol. *5858*, pp. 182–193). Berlin: Springer.

Mehrabian, A. R., & Lucas, C. (2006). A novel numerical optimization algorithm inspired from weed colonization. *Ecological Informatics, 1*, 355–366.

Melin, P., Astudillo, L., Castillo, O., Valdez, F., & Valdez, F. (2013). Optimal design of type-2 and type-1 fuzzy tracking controllers for autonomous mobile robots under perturbed torques using a new chemical optimization paradigm. *Expert Systems with Applications, 40*, 3185–3195.

Michalewicz, Z. (1996). *Genetic algorithms + data structures = evolution programs*. Berlin: Springer. ISBN 3-540-60676-9.

Min, H., & Wang, Z. (2010, December 14–18). Group escape behavior of multiple mobile robot system by mimicking fish schools. In *IEEE International Conference on Robotics and Biomimetics* (*ROBIO*) (pp. 320–326). Tianjin, China. IEEE.

Mora-Gutiérrez, R. A., Ramírez-Rodríguez, J., & Rincón-García, E. A. (2012). An optimization algorithm inspired by musical composition. *Artificial Intelligence Review*. doi: 10.1007/s10462-011-9309-8.

Mucherino, A., & Seref, O. (2007). Monkey search: a novel metaheuristic search for global optimization. *AIP Conference Proceedings, 953*, 162–173.

Müller, S. D., Marchetto, J., Airaghi, S., & Koumoutsakos, P. (2002). Optimization based on bacterial chemotaxis. *IEEE Transactions on Evolutionary Computation, 6*, 16–29.

Muñoz, M. A., López, J. A., & Caicedo, E. (2009). An artificial beehive algorithm for continuous optimization. *International Journal of Intelligent Systems, 24*, 1080–1093.

Murase, H. (2000). Finite element inverse analysis using a photosynthetic algorithm. *Computers and Electronics in Agriculture, 29*, 115–123.

Mutazono, A., Sugano, M., & Murata, M. (2012). Energy efficient self-organizing control for wireless sensor networks inspired by calling behavior of frogs. *Computer Communications, 35*, 661–669.

Nara, K., Takeyama, T., & Kim, H. (1999). A new evolutionary algorithm based on sheep flocks heredity model and its application to scheduling problem. In *IEEE International Conference on Systems, Man, and Cybernetics* (*SMC*) (pp. VI-503–VI-508). IEEE.

Neshat, M., Sepidnam, G., & Sargolzaei, M. (2013). Swallow swarm optimization algorithm: a new method to optimization. *Neural Computing and Application*. doi:10.1007/s00521-012-0939-9.

Nishida, T. Y. (2005, 18–21 July). Membrane algorithm: an approximate algorithm for NP-complete optimization problems exploiting P-systems. In: R. Freund, G. Lojka, M. Oswald, & G. Păun, (Eds.). *6th International workshop on membrane computing* (*WMC*) (pp. 26–43). Vienna, Austria. Institute of Computer Languages, Faculty of Informatics, Vienna University of Technology.

Niu, B., & Wang, H. (2012). Bacterial colony optimization. *Discrete Dynamics in Nature and Society, 2012*, 1–28.

Oftadeh, R., Mahjoob, M. J., & Shariatpanahi, M. (2010). A novel meta-heuristic optimization algorithm inspired by group hunting of animals: hunting search. *Computers and Mathematics with Applications, 60*, 2087–2098.

Oliveira, D. R. D., Parpinelli, R. S., & Lopes, H. S. (2011). Bioluminescent swarm optimization algorithm. In *Evolutionary Algorithms, Chapter 5* (pp. 71–84). Eisuke Kita: InTech.

Pan, W.-T. (2012). A new fruit fly optimization algorithm: Taking the financial distress model as an example. *Knowledge-Based Systems, 26*, 69–74.

Passino, K. M. (2002). Biomimicry of bacterial foraging for distributed optimization and control. *IEEE Control System Management, 22*, 52–67.

Pham, D. T., Ghanbarzadeh, A., Koç, E., Otri, S., Rahim, S., & Zaidi, M. (2006) The bees algorithm: A novel tool for complex optimisation problems. In *Second International Virtual Conference on Intelligent production machines and systems* (*IPROMS*) (pp. 454–459). Oxford: Elsevier.

Premaratne, U., Samarabandu, J., & Sidhu, T. (2009, December 28–31). A new biologically inspired optimization algorithm. In *Fourth International Conference on Industrial and Information Systems* (*ICIIS*) (pp. 279–284). Sri Lanka. IEEE.

Quijano, N., & Passino, K. M. (2010). Honey bee social foraging algorithms for resource allocation: theory and application. *Engineering Applications of Artificial Intelligence, 23*, 845–861.

Rabanal, P., Rodríguez, I., & Rubio, F. (2007. Using river formation dynamics to design heuristic algorithms. In: S. G. Akl, C. S. C., M.J. Dinneen, G. Rozenber, H.T. Wareham (Eds.). *UC 2007, LNCS*, Vol. 4618, (pp. 163–177). Berlin: Springer.

Rajabioun, R. (2011). Cuckoo optimization algorithm. *Applied Soft Computing, 11*, 5508–5518.

Rao, R. V., Savsani, V. J., & Vakharia, D. P. (2011). Teaching–learning-based optimization: a novel method for constrained mechanical design optimization problems. *Computer-Aided Design, 43*, 303–315.

Rashedi, E., Nezamabadi-Pour, H., & Saryazdi, S. (2009). GSA: a gravitational search algorithm. *Information Sciences, 179*, 2232–2248.

Ray, T., & Liew, K. M. (2003). Society and civilization: an optimization algorithm based on the simulation of social behavior. *IEEE Transactions on Evolutionary Computation, 7*, 386–396.

Reynolds, R. G. (1994). An introduction to cultural algorithms. In Sebald, A. V., & Fogel, L. J., (Eds.). *The 3rd Annual Conference on Evolutionary Programming* (pp. 131–139). World Scientific Publishing.

Rutkowski, L. (2008). *Computational intelligence: methods and techniques*. Berlin: Springer. ISBN 978-3-540-76287-4.

Sacco, W. F., & Oliveira, C. R. E. D. (2005, 30 May–03 June) A new stochastic optimization algorithm based on a particle collision metaheuristic. In *6th World Congresses of Structural and Multidisciplinary Optimization* (pp. 1–6). Rio de Janeiro, Brazil.

Sadollah, A., Bahreininejad, A., Eskandar, H., & Hamdi, M. (2012). Mine blast algorithm for optimization of truss structures with discrete variables. *Computers and Structures, 102–103*, 49–63.

Salem, S. A. Boa. (2012, October 10–11). A novel optimization algorithm. In *International Conference on Engineering and Technology* (*ICET*) (pp. 1–5). Cairo, Egypt. IEEE.

Sato, T., & Hagiwara, M. (1997). Bee system: finding solution by a concentrated search. In *IEEE International Conference on Systems, Man, and Cybernetics* (*SMC*) (pp. 3954–3959). IEEE.

Serway, R. A., & Jewett, J. W. (2014). *Physics for scientists and engineers with modern physics*. Boston, MA, USA: Brooks/Cole CENAGE Learning. ISBN 978-1-133-95405-7.

Shah-Hosseini, H. (2007, September 25–28). Problem solving by intelligent water drops. In *IEEE Congress on Evolutionary Computation* (*CEC*) (pp. 3226–3231). IEEE.

Shah-Hosseini, H. (2011). Otsu's criterion-based multilevel thresholding by a nature-inspired metaheuristic called galaxy-based search algorithm. *Third World Congress on Nature and Biologically Inspired Computing* (*NaBIC*) (pp. 383–388). IEEE.

Shann, M. (2008). *Emergent behavior in a simulated robot inspired by the slime mold*. Unpublished Bachelor Thesis, University of Zurich.

Shen, J., & Li, Y. (2009, April 24–26). Light ray optimization and its parameter analysis. In *International Joint Conference on Computational Sciences and Optimization* (*CSO*) (pp. 918–922). Sanya, China. IEEE.

Shi, Y. (2011). Brain storm optimization algorithm. In Y. Tan, Y. Shi, & G. Wang, (Eds.). *ICSI 2011, Pat I, LNCS* (pp. 303–309). Berlin: Springer.

Simon, D. (2008). Biogeography-based optimization. *IEEE Transactions on Evolutionary Computation, 12*, 702–713.

Steinbuch, R. (2011). Bionic optimisation of the earthquake resistance of high buildings by tuned mass dampers. *Journal of Bionic Engineering, 8*, 335–344.

Su, M.-C., Su, S.-Y., & Zhao, Y.-X. (2009). A swarm-inspired projection algorithm. *Pattern Recognition, 42*, 2764–2786.

Sumathi, S., & Paneerselvam, S. (2010). *Computational intelligence paradigms: theory and applications using MATLAB*. Boca Raton: CRC Press, Taylor and Francis. ISBN 978-1-4398-0902-0.

Sun, J., & Lei, X. (2009). Geese-inspired hybrid particle swarm optimization algorithm. In *International Conference on Artificial Intelligence and Computational Intelligence* (pp. 134–138). IEEE.

Taherdangkoo, M., Shirzadi, M. H., & Bagheri, M. H. (2012). A novel meta-heuristic algorithm for numerical function optimization_blind, naked mole-rats (BNMR) algorithm. *Scientific Research and Essays, 7*, 3566–3583.

Taherdangkoo, M., Yazdi, M., & Bagheri, M. H. (2011). Stem cells optimization algorithm. *LNBI*, (pp. 394–403). Berlin: Springer.

Tan, Y., & Zhu, Y. (2010). Fireworks algorithm for optimization. In: Y. Tan, Y. Shi, & Tan, K. C. (Eds.). *ICSI 2010, Part I, LNCS,* (Vol. 6145, pp. 355–364). Berlin: Springer.

Tayarani, N. M. H., & Akbarzadeh, T. M. R. (2008). Magnetic optimization algorithms a new synthesis. In *IEEE Congress on Evolutionary Computation* (*CEC*) (pp. 2659–2664). IEEE.

Teodorović, D., & Dell'orco, M. (2005). Bee colony optimization: a cooperative learning approach to complex transportation problems. In *16th Mini-EURO Conference on Advanced OR and AI Methods in Transportation* (pp. 51–60).

Thammano, A., & Moolwong, J. (2010). A new computational intelligence technique based on human group formation. *Expert Systems with Applications, 37*, 1628–1634.

Theraulaz, G., Goss, S., Gervet, J., & Deneubourg, J. L. (1991). Task differentiation in polistes wasps colonies: a model for self-organizing groups of robots. In *First International Conference on Simulation of Adaptive Behavior* (pp. 346–355). Cambridge: MIT Press.

Wang, J., & Kusiak, A. (Eds.). (2001). *Computational intelligence in manufacturing handbook,* Boca Raton: CRC Press. ISBN 0-8493-0592-6.

Wedde, H. F., Farooq, M., & Zhang, Y. (2004). Beehive: an efficient fault-tolerant routing algorithm inspired by honey bee behavior. In: Dorigo, M. (Ed.). *ANTS 2004, LNCS*, (Vol. 3172, pp. 83–94). Berlin: Springer.

Wei, Z. H., Cui, Z. H., & Zeng, J. C. (2010, September 26–28). Social cognitive optimization algorithm with reactive power optimization of power system. In *2010 International Conference on Computational Aspects of Social Networks* (*CASoN*) (pp. 11–14). Taiyuan, China.

Whitten, K. W., Davis, R. E., Peck, M. L., & Stanley, G. G. (2014). *Chemistry,* Belmont: Brooks/Cole, Cengage Learning. ISBN-13: 978-1-133-61066-3.

Xie, L.-P. & Zeng, J.-C. (2009, June 12–14). A global optimization based on physicomimetics framework. *First ACM/SIGEVO Summit on Genetic and Evolutionary Computation* (*GEC*) (pp. 609–616). Shanghai, China. IEEE.

Xing, B., & Gao, W.-J. (2014). *Computational intelligence in remanufacturing,* Hershey: IGI Global. ISBN 978-1-4666-4908-8.

Yan, G.-W., & Hao, Z. (2012, July 7–9). A novel atmosphere clouds model optimization algorithm. In *International Conference on Computing, Measurement, Control and Sensor Network* (*CMCSN*) (pp. 217–220). Taiyuan, China. IEEE.

Yang, X. S. (2005). Engineering optimizations via nature-inspired virtual bee algorithms. In *IWINAC 2005, lNCS,* (Vol. 3562, pp. 317–323). Berlin: Springer.

Yang, C., Tu, X., & Chen, J. (2007). Algorithm of marriage in honey bees optimization based on the wolf pack search. *International Conference on Intelligent Pervasive Computing* (*IPC*) (pp. 462–467). IEEE.

Yang, F.-C., & Wang, Y.-P. (2007). Water flow-like algorithm for object grouping problems. *Journal of the Chinese Institute of Industrial Engineers, 24*, 475–488.

Yang, X.-S. (2008). *Nature-inspired metaheuristic algorithms*. Frome: Luniver Press. ISBN 978-1-905986-28-6.

Yang, X.-S. (2010a). *Engineering optimization: an introduction with metaheuristic applications.* Hoboken: Wiley. ISBN 978-0-470-58246-6.

Yang, X.-S. (2010b) A new metaheuristic bat-inspired clgorithm. Nature Inspired Cooperative Strategies for Optimization (NISCO 2010). *Studies in Computational Intelligence, SCI284* (pp. 65–74). Berlin: Springer.

Yang, X.-S. (2012). Flower pollination algorithm for global optimization. In *Unconventional Computation and Natural Computation, LNCS*, (pp. 240–249). Berlin: Springer.

Yang, X.-S., & Deb, S. (2009, December 9–11). Cuckoo search via Lévy flights. In *World Congress on Nature and Biologically Inspired Computing (NaBIC)* (pp. 210–214). India. IEEE.

Yang, X.-S., & Deb, S. (2010). Eagle strategy using Lévy walk and firefly algorithms for stochastic optimization. In: Gonzalez, J. R. (Ed.), *Nature Inspired Cooperative Strategies for Optimization (NISCO 2010), SCI 284,* (pp. 101–111). Berlin: Springer.

Yang, X.-S., et al. (2013). *Mathematical modeling with multidisciplinary applications*. Hoboken: Wiley. ISBN 978-1-118-29441-3.

Zaránd, G., Pázmándi, F., Pál, K. F., & Zimányi, G. T. (2002). Using hysteresis for optimization. *Physical Review Letters, 89,* 1501–1502.

Zheng, M., Liu, G.-X., Zhou, C.-G., Liang, Y.-C., & Wang, Y. (2010). Gravitation field algorithm and its application in gene cluster. *Algorithms for Molecular Biology, 5,* 1–11.

Zheng, Y., Chen, W., Dai, C., & Wang, W. (2009). Stochastic focusing search: a novel optimization algorithm for real-parameter optimization. *Journal of Systems Engineering and Electronics, 20,* 869–876.

Zhang, X., Chen, W., & Dai, C. (2008, April 6–9) Application of oriented search algorithm in reactive power optimization of power system. *DRPT2008* (pp. 2856–2861). Nanjing, China. DRPT.

Zhang, W., Luo, Q.m & Zhou, Y. (2009). A method for training RBF neural networks based on population migration algorithm. In *International Conference on Artificial Intelligence and Computational Intelligence (AICI)* (pp. 165–169). IEEE.

Zhang, X., Sun, B., Mei, T., & Wang, R. (2010, November 28–30) Post-disaster restoration based on fuzzy preference relation and bean optimization algorithm. *IEEE Youth Conference on Information Computing and Telecommunications (YC-ICT)* (pp. 271–274). IEEE.

Zhang, X., Huang, S., Hu, Y., Zhang, Y., Mahadevan, S., & Deng, Y. (2013). Solving 0–1 knapsack problems based on amoeboid organism algorithm. *Applied Mathematics and Computation, 219,* 9959–9970.

Zhu, C., & Ni, J. (2012, April 21–23). Cloud model-based differential evolution algorithm for optimization problems. In *Sixth International Conference on Internet Computing for Science and Engineering (ICICSE)* (pp. 55–59), Henan, China. IEEE.

Zungeru, A. M., Ang, L.-M., & Seng, K. P. (2012). Termite-hill: performance optimized swarm intelligence based routing algorithm for wireless sensor networks. *Journal of Network and Computer Applications, 35,* 1901–1917.

Part II
Biology-based CI Algorithms

Chapter 2
Bacteria Inspired Algorithms

Abstract In this chapter, we present a set of algorithms that are inspired by the different bacteria behavioural patterns, i.e., bacterial foraging algorithm (BFA), bacterial colony chemotaxis (BCC) algorithm, superbug algorithm (SuA), bacterial colony optimization (BCO) algorithm, and viral system (VS) algorithm. We first describe the general knowledge of bacteria foraging behaviour in Sect. 2.1. Then, the fundamentals and performances of BFA, BCC algorithm, SuA, BCO algorithm, and VS algorithm are introduced in Sects. 2.2 and 2.3, respectively. Finally, Sect. 2.4 summarises this chapter.

2.1 Introduction

It is fair to say that bacteria can be observed almost everywhere, from the most hospitable surroundings to the most hostile environment (Helden et al. 2012). Some of them are harmful but majority of them are beneficial to the nature. They comprise various attributes (e.g., shape, texture, and metabolism) and different behavioural patterns (e.g., foraging, reproduction, and movement) (Modrow et al. 2013; Giguère et al. 2013). In addition to purely scientific aspects related to the bacteria, those characteristics also initiate computer scientists to develop algorithms for the solution of optimization problems. The first attempts appeared in the late 1970s [e.g., bacterial chemotaxis algorithm that proposed by Bremermann (1974)]. Since then, similar ideas have attracted a steadily increasing amount of research. In this chapter, a set of bacteria inspired algorithms are collected and introduced as follows:

- Section 2.2: Bacterial Foraging Algorithm.
- Section 2.3.1: Bacterial Colony Chemotaxis.
- Section 2.3.2: Superbug Algorithm.
- Section 2.3.3: Bacterial Colony Optimization.
- Section 2.3.4: Viral System.

B. Xing and W.-J. Gao, *Innovative Computational Intelligence: A Rough Guide to 134 Clever Algorithms*, Intelligent Systems Reference Library 62,
DOI: 10.1007/978-3-319-03404-1_2,

The effectiveness of these newly developed algorithms are validated through the testing on a wide range of benchmark functions and engineering design problems, and also a detailed comparison with various traditional performance leading computational intelligence (CI) algorithms, such as particle swarm optimization (PSO), genetic algorithm (GA), differential evolution (DE), evolutionary algorithm (EA), fuzzy system (FS), ant colony optimization (ACO), and simulated annealing (SA).

2.1.1 Bacteria

Bacteria are micro-organisms that include a large domain. For example, there are roughly 5,000 operational taxonomic unit of bacteria for every gram of soil (Maheshwari 2012). In addition, they have a wide range of shapes, such as spheres, spirals, and rods. Also, they live in everywhere, such as in plants, animals, and human's body. Due to their diverse ecology, they exhibit multifarious functional characters beneficial in nature.

2.1.2 Bacterial Foraging Behaviour

The bacterial foraging behaviour (i.e., bacterial movement) has been studied by biologists for many years, but there are very few reports until 70s (Berg and Brown 1972; Adler 1975). During foraging, there are two types of bacterial motility: i.e., flagellum-dependent and flagellum-independent. In addition, biologists found that bacterial movement was not random and arbitrary. Instead, bacterial cells exhibited directed movement toward certain stimuli and away from others (Adler 1975). This behaviour is called bacterial chemotaxis in the literature. Since its discovery, chemotactic behaviour has stimulated the curiosity of numerous investigators. Nowadays, those research outcomes have been adopted by the models simulating bacterial foraging patterns (Paton et al. 2004).

2.2 Bacterial Foraging Algorithm

2.2.1 Fundamentals of Bacterial Foraging Algorithm

Bacterial foraging algorithm (BFA) was originally proposed by Passino (2002) in which the foraging strategy of *E. Coli* bacteria has been simulated. Typically, the BFA consists of four main mechanisms: chemotaxis, swarming, reproduction, and elimination-dispersal event. The main steps of BFA are outlined below (Passino 2002):

- Chemotaxis: The movement of *E. Coli* bacteria can be performed through two different ways: swimming which means the movement in the same direction, and tumbling refers the movement in a random direction. The movement of the ith bacterium after one step is given by Eqs. 2.1 and 2.2, respectively (Passino 2002):

$$\theta^i(j+1,\,k,l) = \theta^i(j,k,l) + C(i)\phi(j), \tag{2.1}$$

$$\begin{cases} \theta^i(j+1,k,l) > \theta^i(j,k,l), & \text{swimming in which } \phi(j) = \phi(j-1) \\ \theta^i(j+1,k,l) < \theta^i(j,k,l), & \text{tumbling in which } \phi(j) \in [0,2\pi] \end{cases}. \tag{2.2}$$

  Where $\theta^i(j,k,l)$ denotes the location of ith bacterium at jth chemotactic, kth reproductive and lth elimination and dispersal step, $C(i)$ is the length of unit walk, and $\phi(j)$ is the direction angle of the jth step.

- Swarming: Under the stresses circumstances, the bacteria release attractants to signal bacteria to swarm together, while they also release a repellant to signal others to be at a minimum distance from it. The cell to cell signalling can be represented by Eq. 2.3 (Passino 2002):

$$\begin{aligned} J_{cc}(\theta,P(j,k,l)) &= \sum_{i=1}^{S} J_{cc}^i(\theta,\theta^i(j,k,l)) \\ &= \sum_{i=1}^{S}\left[-d_{attract}\exp\left(-\omega_{attract}\sum_{m=1}^{P}(\theta_m-\theta_m^i)^2\right)\right] \\ &\quad + \sum_{i=1}^{S}\left[h_{repellant}\exp\left(-\omega_{repellant}\sum_{m=1}^{P}(\theta_m-\theta_m^i)^2\right)\right]. \end{aligned} \tag{2.3}$$

Where $J_{cc}(\theta,P(j,k,l))$ is the objective function value to be added to the actual objective function, S is the total number of bacteria, P is the number of variables to be optimized which are present in each bacterium, $\theta = [\theta_1,\theta_2,\ldots,\theta_P]^T$ denotes a point in the P-dimensional search domain, $d_{attract}$ is the depth of the attractant released by the cell, $\omega_{attract}$ is a measure of the width of the attractant signal, $h_{repellant}$ is the height of the repellant effect (i.e., $h_{repellant} = d_{attract}$), and $\omega_{repellant}$ is the measure of the width of the repellant.

- Reproduction: After N_c chemotaxis steps, the reproduction step should be performed. The fitness value of the bacteria is stored in an ascending order. The working principle is the least health bacteria eventually die and the remaining bacteria (i.e., healthiest bacteria) will be divided into two identical ones and placed at the same location.

- Dispersion and elimination: For the purpose to avoid local optima, dispersion and elimination process is performed after a certain number of reproduction steps. According to a present probability (p_{ed}), a bacterium is chosen to be dispersed and moved to another position within the environment.

Taking into account the key phases described above, the steps of implementing BFA can be summarized as follows (Passino 2002; Boussaïd et al. 2013; El-Abd 2012; Tang and Wu 2009):

- Step 1: Defining the optimization problem, and initializing the optimization parameters.
- Step 2: Iterative algorithm for optimization. In this step the bacterial population, chemotaxis loop ($j = j + 1$), reproduction loop ($k = k + 1$), and elimination and dispersal operations loop ($l = l + 1$) are performed.
- Step 3: If $j < N_c$, go to the chemotaxis process.
- Step 4: Reproduction.
- Step 5: If $k < N_{re}$, go to reproduction process.
- Step 6: Elimination-dispersal.
- Step 7: If $l < N_{ed}$, then go to elimination-dispersal process; otherwise end.

2.2.2 Performance of BFA

In order to show how the BFA performs, a set of experimental studies were adopted in the literature (Passino 2002). Computational results showed that BFA gives some promising results.

2.3 Emerging Bacterial Inspired Algorithms

In addition to the aforementioned BFA, the characteristics of this interesting biological organisms also motivate researchers to develop other bacterial inspired innovative CI algorithms.

2.3.1 Bacterial Colony Chemotaxis Algorithm

2.3.1.1 Fundamentals of Bacterial Colony Chemotaxis Algorithm

Bacterial colony chemotaxis (BCC) algorithm was originally proposed by Li et al. (2005) that is based on bacterial chemotaxis algorithm (Bremermann and Anderson 1991) and incorporated the communication mechanisms between the colony members. There are several variants and applications relative to BCC can be found in the literature (Li et al. 2009; Li and Li 2012; Sun et al. 2012; Lu et al.

2013; Irizarry 2011). The basic steps of BCC are as follows (Lu et al. 2013; Müller et al. 2002; Sun et al. 2012):

- Step 1: Initial Bacterial Positions.Generate N bacteria randomly in the search space, the velocity is assumed constant by Eq. 2.4 (Lu et al. 2013; Müller et al. 2002):

$$v = const. \tag{2.4}$$

- Step 2: Optimize by a single bacterium.

Compute the duration of the trajectory τ from the distribution of a random variable with an exponential probability density function via Eq. 2.5 (Lu et al. 2013; Müller et al. 2002):

$$P(X = \tau) = \frac{1}{T} e^{-\tau/T}, \tag{2.5}$$

where the expectation value $\mu = E(X) = T$, and the variance $\sigma^2 = Var(X) = T^2$. The time T is given by Eq. 2.6 (Müller et al. 2002):

$$T = \begin{cases} T_0 & \text{for } \frac{f_{pr}}{l_{pr}} \geq 0, \\ T_0\left(1 + b \cdot \left\{\frac{f_{pr}}{l_{pr}}\right\}\right) & \text{for } \frac{f_{pr}}{l_{pr}} < 0, \end{cases} \tag{2.6}$$

where T_0 is minimal mean time, f_{pr} is the difference between the actual and the previous function value, l_{pr} is the length of the previous step and $l_{pr} = |\vec{x}_{pr}|$, $\vec{x}_{pr}$ denotes vector connecting the previous and the actual position in the parameter space, and b is dimensionless parameter.

The position of a bacterium is defined by x with a radius and angles via Eq. 2.7 (Lu et al. 2013):

$$\begin{aligned} x_1 &= \gamma \prod_{s-1}^{n-1} \cos(\phi_s), \\ x_i &= \gamma \ \sin(\phi_{i-1})\gamma \prod_{s-i}^{n-1} \cos(\phi_s), \quad \text{for } i = 2, 3, \ldots, n, \\ x_n &= \gamma \ \sin(\cos(\phi_{n-1})), \end{aligned} \tag{2.7}$$

where γ is the radius, and $\Phi = \{\phi_1, \phi_2, \ldots, \phi_{n-1}\}$ denote angles. According to the Gaussian distribution, the angel (ϕ_i) between the previous and the new direction for turning left or right are defined by Eqs. 2.8 and 2.9, respectively (Lu et al. 2013):

$$P(X_i = \phi_i, v_i = \mu_i) = \frac{1}{\sigma_i\sqrt{2\pi}} \exp\left[-\frac{(\phi_i - v_i)^2}{2\sigma_i^2}\right], \tag{2.8}$$

$$P(X_i = \phi_i, v_i = -\mu_i) = \frac{1}{\sigma_i\sqrt{2\pi}}\exp\left[-\frac{(\phi_i - v_i)^2}{2\sigma_i^2}\right], \tag{2.9}$$

where $\phi_i \in [0°, 180°]$. The expectation value and the standard deviation are defined by Eqs. 2.10 and 2.11, respectively (Lu et al. 2013):

$$\begin{aligned} \mu &= E(X_i) = 62°(1 - \cos(\theta)), \\ \sigma_i &= \sqrt{Var(X_i)} = 26°(1 - \cos(\theta)), \end{aligned} \tag{2.10}$$

$$\cos(\theta) = \begin{cases} 0 & \text{for } \frac{f_{pr}}{l_{pr}} \geq 0, \\ \exp(-\tau_c/\tau_{pr}) & \text{for } \frac{f_{pr}}{l_{pr}} < 0, \end{cases} \tag{2.11}$$

where τ_{pr} is the duration of the previous step, and τ_c is the correlation time.

Compute the new position ($\vec{x}_{new}$) via Eqs. 2.12 and 2.13, respectively (Lu et al. 2013):

$$\vec{x}_{new} = \vec{x}_{old} + \vec{n}_u \cdot l, \tag{2.12}$$

$$l = v \cdot t, \tag{2.13}$$

where l is the length of the path, and $\vec{n}_u$ is the normalized new direction vector.

- Step 3: Optimize by the bacterial colony.

Initialize the number of bacteria colony and the parameters, such as the initial starting precision (ε_{begin}), the constant of updating precision (ε_{cons}), and the target precision (ε_{end}).

The best neighbour centre position of the ith bacterium is given via Eq. 2.14 (Sun et al. 2012):

$$center\_position(i) = rand() \cdot dis\left(\vec{x}_{i,t}, Center\left(\vec{x}_{i,t}\right)\right), \tag{2.14}$$

where $dis\left(\vec{x}_{i,t}, Center\left(\vec{x}_{i,t}\right)\right)$. is the distance between the bacterium i and the centre position, and $rand()$ is a random number meeting the uniform distribution in the interval of $(0, 1)$.

Based on the single bacterium movement process, the next position of ith bacterium can be defined by Eqs. 2.15 and 2.16, respectively (Sun et al. 2012):

$$individual\_position(i) = current\_position(i) + next(i), \tag{2.15}$$

$$next(i) = v \cdot d \cdot |\vec{n}_u|, \tag{2.16}$$

where v denotes the velocity, d is the duration, and $\vec{n}_u$ is the normalized new direction vector.

- Step 4: Bacterial colony move to the new location. Compare the objective function fitness value between *center_position*(i) and *individual_position*(i), then ith bacterium chooses the better value position to move.
- Step 5: If the target precision (ε_{end}) is gained end, otherwise turn to Step 2.

2.3.1.2 Performance of BCC

To illustrate the performance of BCC, a set of experimental studies were adopted in (Li and Li 2012; Sun et al. 2012). Computational results showed that the proposed algorithm performs well.

2.3.2 Superbug Algorithm

2.3.2.1 Fundamentals of Superbug Algorithm

Superbug algorithm (SuA) was originally proposed in (Anandaraman et al. 2012) that is based on the findings of epidemic and antiviral research (Lindsay and Holden 2006; Low et al. 1999; Bisen and Raghuvanshi 2013; Gong 2013; Desai and Meanwell 2013; Cook 2013; Morrow et al. 2012; Lahtinen et al. 2012; Helden et al. 2012). To implement SuA, the following steps need to be performed (Anandaraman et al. 2012):

- Step 1: Generating initial population.
- Step 2: Mutating the bacteria (two mechanisms are introduced in SuA, namely, inverse and pairwise interchange mutation).
- Step 3: Transferring gene between bacteria for the purpose of enhancing the fitness level.
- Step 4: Performing single point mutation on the modified bacteria.

2.3.2.2 Performance of SuA

The core concept underlying the SuA is the antibiotic resistance that a bacterium has developed during the procedure of mutation. To illustrate the performance of SuA, Anandaraman et al. (2012) applied it to the flexible manufacturing system scheduling problem. In comparison with other CI techniques (e.g., GA), the SuA showed a better performance in majority of the cases. At the end of their study, Anandaraman et al. (2012) suggested that it is possible to apply SuA to other multi-machine and dynamic job shop scheduling problems.

2.3.3 Bacterial Colony Optimization Algorithm

2.3.3.1 Fundamentals of Bacterial Colony Optimization Algorithm

Bacterial colony optimization (BCO) algorithm was originally proposed by Niu and Wang (2012) which is inspired by five basic behaviours of *E. Coli* bacteria in their whole lifecycle, i.e., chemotaxis, communication, elimination, reproduction, and migration. The main steps of BCO are outlined as follows (Niu and Wang 2012):

- Chemotaxis and communication model: In BCO, the movements of bacteria (i.e., runs and tumbles) are accompanied with communication mechanism in which three types of information (i.e., group information, personal previous information, and a random direction) are exchanged to guide them in direction and ways of movement. The bacterium runs or tumbles with communication process can be formulated via Eqs. 2.17 and 2.18, respectively (Niu and Wang 2012):

$$Position_i(T) = Position_i(T-1) + R_i \cdot \left(Run_{Info}\right) + R\Delta(i). \tag{2.17}$$

$$Position_i(T) = Position_i(T-1) + R_i \cdot \left(Tumb_{Info}\right) + R\Delta(i). \tag{2.18}$$

- Elimination and reproduction model: Based to the bacteria searching ability, each bacterium is marked with an energy degree. According to this level, the decisions of elimination and reproduction are defined via Eq. 2.19 (Niu and Wang 2012):

$$\begin{cases} \text{if } L_i > L_{given}, \text{ and } i \in \text{healthy}, & \text{then } i \in Candidate_{repr}, \\ \text{if } L_i < L_{given}, \text{ and } i \in \text{healthy}, & \text{then } i \in Candidate_{eli}, \\ \text{if } i \in \text{unhealthy}, & \text{then } i \in Candidate_{eli}, \end{cases} \tag{2.19}$$

  where L_i denotes the ith bacterium's level of energy.
- Migration model: To avoid local optimum, the BCO's migration model is defined via Eq. 2.20 (Niu and Wang 2012):

$$Position_i(T) = rand \cdot (ub - lb) + lb, \tag{2.20}$$

  where *ub* and *lb* are the upper and lower boundary, respectively, and $rand \in (0, 1)$.

  In addition, in the process of tumbling, a stochastic direction are incorporated into actually running process. Therefore, the position of each bacterium is updated via Eqs. 2.21–2.23, respectively (Niu and Wang 2012):

$$\begin{aligned} Position_i(T) = {} & Position_i(T-1) + C(i) \\ & * \left[f_i \cdot (G_{best} - Position_i(T-1)) + (1-f_i)\right. \\ & \left. * (P_{best_i} - Position_i(T-1)) + turbulent_i\right], \end{aligned} \quad (2.21)$$

$$\begin{aligned} Position_i(T) = {} & Position_i(T-1) + C(i) \\ & * \left[f_i * (G_{best} - Position_i(T-1)) + (1-f_i)\right. \\ & \left. * (P_{best_i} - Position_i(T-1))\right], \end{aligned} \quad (2.22)$$

$$C(i) = C_{\min} + \left(\frac{iter_{\max} - iter_j}{iter_{\max}}\right)^n \cdot (C_{\max} - C_{\min}), \quad (2.23)$$

where $turbulent_i$ is the turbulent direction variance of the ith bacterium, $f_i \in (0, 1)$, G_{best} and P_{best_i} are the globe best and personal best position of the ith bacterium, respectively, $C(i)$ is the chemotaxis step size of the ith bacterium, $iter_{\max}$ is the maximal number of iterations, and $iter_j$ is the current number of iterations.

2.3.3.2 Performance of BCO

To test the effectiveness of BCO, a set of well-known test functions were adopted in (Niu and Wang 2012). Compared with five other algorithms [i.e., PSO, GA, BFA, bacterial foraging optimization with linear decreasing chemotaxis step (BFO-LDC), and bacterial foraging optimization with nonlinear decreasing chemotaxis step (BFO-NDC)], computational results showed that BCO performs significantly better than four other algorithms (i.e., GA, BFA, BFO-LDC, and BFO-NDC) in all test functions, and BCO obtains better results than PSO in most of functions.

2.3.4 Viral System Algorithm

2.3.4.1 Fundamentals of Viral System Algorithm

Viral system (VS) algorithm was originally proposed by Cortés et al. (2008) which is based on viral infection processes. Two mechanisms called replication and infection are employed to the VS algorithm. The main steps of VS are as follows (Cortés et al. 2008, 2012; Ituarte-Villarreal and Espiritu 2011):

- Step 1: Initialisation the VS components. Each VS is defined by three components, i.e., a set of viruses, an organism, and an interaction between them. In addition, each virus includes four components, i.e., state, input, output, and process, and each organism includes two components, i.e., state and process. Overall, the VS can be described via Eqs. 2.24–2.27, respectively (Cortés et al. 2008):

$$VS = \langle Virus, Organism, Interaction \rangle, \tag{2.24}$$

$$Virus = \{Virus_1, Virus_2, \ldots, Virus_n\}, \tag{2.25}$$

$$Virus_i = \langle State_i, Input_i, Output_i, \Pr ocess_i \rangle, \tag{2.26}$$

$$Organism = \langle State_0, \Pr ocess_0 \rangle, \tag{2.27}$$

where $State_i$ denotes the characterises of the virus, $Input_i$ identifies the information that the virus can collect form the organism, $Output_i$denotes the actions that the virus can take, $Process_i$ represents autonomous behaviour of the virus that changing the $State_i$, $State_0$ characterises the organism state in each instant, consisting of clinical picture and the lowest healthy cell, and $Process_0$ represents the autonomous behaviour of the organism that tries to protect itself from the infection threat, consisting of antigen liberation.

- Step 2: Population construction. The set of feasible solutions in a specific space is given by Eq. 2.28 (Cortés et al. 2008):

$$K = \{x : g_i(x) \leq 0, \forall i = 1, 2, \ldots n\}, \tag{2.28}$$

where $x \in K$ denotes the feasible solutions and has been called a cell.

- Step 3: Define type of virus infection: selective or massive infection.

In case of massive infection: $Y - A$ cells of the neighbourhood are infected, and must be incorporated into the clinical picture. If there is not enough free space in the population, it will randomly erase the necessary cells from the $Y - A$ selected cells.

In case of selective infection: One only cell from the neighbourhood is selected according to the selective selection. The antigenic response of such cell is evaluated as a Bernoulli process (A). In case of antigenic response a lysogenic replication is initiated.

- Step 4: Define type of evolution of the virus: i.e., the lytic replication and the lysogenic replication.

In case of the lytic replication: This process starts only after a specific number of nucleus-capsids haven been replicated.

Calculate the limit number of nucleus-capsids replication (LNR) in a cell x by Eq. 2.29 (Cortés et al. 2008):

$$LNR_{cell-x} = LNR^0 \cdot \left(\frac{f(x) - f\left(\widehat{x}\right)}{f\left(\widehat{x}\right)} \right), \tag{2.29}$$

where $\widehat{x}$ is the cell that produces the best known result of the problem (in terms of $f(x)$), x is the infected cell being analysed, and LNR^0 is the initial value for LNR.

In each iteration, a number of virus replications (NR) takes place. The number of replications per iteration is calculated as function of a binomial variable (Z), adding its value to the total NR. The probability of replicating exactly z nucleus-capsids ($P(Z = z)$), as well as the average ($E(Z)$), and variance ($Var(Z)$) are given by Eqs. 2.30–2.32, respectively (Cortés et al. 2008):

$$P(Z = z) = \binom{LNR}{z} p_r^z (1 - p_r)^{LNR - z}, \tag{2.30}$$

$$E(Z) = p_r LNR, \tag{2.31}$$

$$Var(Z) = p_r(1 - p_r)LNR, \tag{2.32}$$

where p_r is the single probability of one replication, and Z is a binomial random variable, i.e., $Z = Bin(LNR, p_r)$.

Once the bacterium border is broken liberating the viruses, each one of the viruses has a probability (p_i) of infecting other new cells of the neighbourhood. The probability of infecting exactly y nucleus-capsids ($P(Y = y)$), as well as the average ($E(Y)$), and variance ($Var(Y)$) are given via Eqs. 2.33–2.35, respectively (Cortés et al. 2008):

$$P(Y = y) = \binom{|V(x)|}{y} p_i^y (1 - p_i)^{|V(x)| - y}, \tag{2.33}$$

$$E(Y) = p_i |V(x)|, \tag{2.34}$$

$$Var(Y) = p_i(1 - p_i)|V(x)|, \tag{2.35}$$

where $|V(x)|$ is the feasible solution of the neighbourhood, Y is a binomial random variable representing the cells infected by the virus in the neighbourhood, i.e., $Y = Bin(|V(x)|, p_i)$.

Each one of the infected cells in the clinical picture has a probability of developing antibodies against the infection based on a Bernoulli probability distribution ($p_{an} : A(x) = Ber(p_{an})$). So, the total population of infected cells generating antibodies is characterised by a binomial distribution ($p_{an} : A(population) = Bin(n, p_{an})$). The probability of finding exactly a immune cells ($P(A = a)$), as well as the average ($E(A)$), and variance ($Var(A)$) are given by Eqs. 2.36–2.38, respectively (Cortés et al. 2008):

$$P(A = a) = \binom{|V(x)|}{a} p_{an}^a (1 - p_{an})^{|V(x)| - a}, \tag{2.36}$$

$$E(A) = p_{an} |V(x)|, \tag{2.37}$$

$$Var(A) = p_{an}(1 - p_{an})|V(x)|, \tag{2.38}$$

where a denotes the immune cell, and x is an infected cell.

In the case of the lysogenic replication, calculate the limit number of interaction (LIT) in a cell x via Eq. 2.39 (Cortés et al. 2008):

$$LIT_{cell-x} = LIT^0 \cdot \left(\frac{f(x) - f(\hat{x})}{f(\hat{x})} \right), \tag{2.39}$$

where LIT^0 is the initial value for LIT.

- Step 5: Ending. The VS algorithm ended according to two criteria, i.e., the collapse and death of the organism, or the isolation of the virus.

2.3.4.2 Performance of VS

To test the proposed algorithm, a well-known NP-Compete problem, i.e., the Steiner problem, is adopted in (Cortés et al. 2008). Compared with TS and GA, VS clearly improves the results from GA and for several cases VS obtains better results than TS.

2.4 Conclusions

In this chapter, we introduced five bacteria inspired CI algorithms. They stem from two background: BFA, BCC algorithm an BCO algorithm are currently bred by the further understanding of bacterial foraging patterns, while SuA and VS algorithm are mainly motivated by the viral research. Although they are newly introduced CI method, we have witnessed the following rapid spreading of at least one of them, i.e., BFA:

First, several enhanced versions of BFA can be found in the literature as outlined below:

- Adaptive BFA (Majhi et al. 2009; Sanyal et al. 2011; Sathya and Kayalvizhi 2011d; Panigrahi and Pandi 2009).
- Amended BFA (Sathya and Kayalvizhi 2011a).
- BFA with varying population (Li et al. 2010).
- Fuzzy adaptive BFA (Venkaiah and Kumar 2011).
- Fuzzy dominance based BFA (Panigrahi et al. 2010).
- Hybrid BFA and differential evolution (Pandi et al. 2010).
- Hybrid BFA and genetic algorithm (Nayak et al. 2012; Kim et al. 2007).
- Hybrid BFA and particle swarm optimization (Hooshmand and Mohkami 2011; Saber 2012).
- Hybrid BFA, differential evolution, and particle swarm optimization (Vaisakh et al. 2012).

- Modified BFA (Verma et al. 2011; Sathya and Kayalvizhi 2011b; Hota et al. 2010; Deshpande et al. 2011; Biswas et al. 2010a, b).
- Multi-colony BFA (Chen et al. 2010).
- Multiobjective BFA (Panigrahi et al. 2011).
- Oppositional based BFA (Mai and Ling 2011).
- Other hybrid BFA (Lee and Lee 2012; Panda et al. 2011; Hooshmand et al. 2012; Rajni and Chana 2013).
- Quantum inspired bacterial swarming optimization (Cao and Gao 2012).
- Rule based BFA (Mishra et al. 2007).
- Self-adaptation BFA (Su et al. 2010).
- Simplified BFA (Muñoz et al. 2010).
- Synergetic bacterial swarming optimization (Chatzis and Koukas 2011).
- Velocity modulated BFA (Gollapudi et al. 2011).

Second, the BFA has also been successfully applied to a variety of optimization problems as listed below:

- Circuit design optimization (Chatterjee et al. 2010).
- Communication optimization (Su et al. 2010; Chen et al. 2010).
- Data mining (Lee and Lee 2012).
- Fuzzy system design optimization (Kamyab and Bahrololoum 2012).
- Image processing (Verma et al. 2011, 2013; Maitra and Chatterjee 2008; Sanyal et al. 2011; Panda et al. 2011; Sathya and Kayalvizhi 2011a, b, c, d).
- Inventory management (Deshpande et al. 2011).
- Manufacturing cell formation (Nouri et al. 2010; Nouri and Hong 2012).
- Motor control optimization (Bhushan and Singh 2011; Sakthivel and Subramanian 2012; Sakthivel et al. 2011).
- Nonlinear system identification (Majhi and Panda 2010).
- Power system optimization (Tang et al. 2006; Ulagammai et al. 2007; Mishra et al. 2007; Panigrahi and Pandi 2009; Pandi et al. 2010; Hota et al. 2010; Panigrahi et al. 2010; Ali and Abd-Elazim 2011; Tabatabaei and Vahidi 2011; Venkaiah and Kumar 2011; Panigrahi et al. 2011; Hooshmand and Mohkami 2011; Abd-Elazim and Ali 2012; Saber 2012; Hooshmand et al. 2012; Kumar and Jayabarathi 2012; Vaisakh et al. 2012).
- Robot control optimization (Turduev et al. 2010; Supriyono and Tokhi 2012).
- Scheduling optimization (Nayak et al. 2012; Vivekanandan and Ramyachitra 2012; Rajni and Chana 2013).
- Stock market prediction (Majhi et al. 2009).

Interested readers are referred to them, together with several excellent reviews (e.g., Tang and Wu 2009; Boussaïd et al. 2013; Agrawal et al. 2011; Niu et al. 2010a, b), as a starting point for a further exploration and exploitation of these bacteria inspired algorithms.

References

Abd-Elazim, S. M., & Ali, E. S. (2012). Coordinated design of PSSs and SVC via bacteria foraging optimization algorithm in a multimachine power system. *Electrical Power and Energy Systems, 41*, 44–53.

Adler, J. (1975). Chemotaxis in bacteria. *Annual Reviews of Biochemistry, 44*, 341–356.

Agrawal, V., Sharma, H. & Bansal, J. C. (2011, December 20–22). Bacterial foraging optimization: A survey. In *The International Conference on Soft Computing for Problem Solving (SOCPROS)* (pp. 227–242). Berlin: Springer.

Ali, E. S., & Abd-Elazim, S. M. (2011). Bacteria foraging optimization algorithm based load frequency controller for interconnected power system. *Electrical Power and Energy Systems, 33*, 633–638.

Anandaraman, C., Sankar, A. V. M., & Natarajan, R. (2012). A new evolutionary algorithm based on bacterial evolution and its applications for scheduling a flexible manufacturing system. *Jurnal Teknik Industri, 14*, 1–12.

Berg, H. C., & Brown, D. A. (1972). Chemotaxis in Escherichia coli analyzed by three-dimensional tracking. *Nature, 239*, 500–504.

Bhushan, B., & Singh, M. (2011). Adaptive control of DC motor using bacterial foraging algorithm. *Applied Soft Computing, 11*, 4913–4920.

Bisen, P. S. & Raghuvanshi, R. (2013). *Emerging epidemics: Management and control.* Hoboken: Wiley, ISBN 978-1-118-39323-9.

Biswas, A., Das, S., Abraham, A., & Dasgupta, S. (2010a). Analysis of the reproduction operator in an artificial bacterial foraging system. *Applied Mathematics and Computation, 215*, 3343–3355.

Biswas, A., Das, S., Abraham, A., & Dasgupta, S. (2010b). Stability analysis of the reproduction operator in bacterial foraging optimization. *Theoretical Computer Science, 411*, 2127–2139.

Boussaïd, I., Lepagnot, J., & Siarry, P. (2013). A survey on optimization metaheuristics. *Information Sciences, 237*, 82–117.

Bremermann, H. J. (1974). Chemotaxis and optimization. *Journal of Franklin Institute, 297*, 397–404.

Bremermann, H. J. & Anderson, R. W. (1991). How the brain adjusts synapses-maybe. In: Boyer, R. S. (Ed.) *Automated reasoning: Essays in honor of Woody Bledsoe*. Norwell: Kluwer.

Cao, J. & Gao, H. (2012). A quantum-inspired bacterial swarming optimization algorithm for discrete optimization problems. In: Y. Tan, Y. Shi, & Z. Ji (Eds.), *ICSI 2012, Part I, LNCS 7331* (pp. 29–36). Berlin: Springer.

Chatterjee, A., Fakhfakh, M., & Siarry, P. (2010). Design of second-generation current conveyors employing bacterial foraging optimization. *Microelectronics Journal, 41*, 616–626.

Chatzis, S. P., & Koukas, S. (2011). Numerical optimization using synergetic swarms of foraging bacterial populations. *Expert Systems with Applications, 38*, 15332–15343.

Chen, H., Zhu, Y., & Hu, K. (2010). Multi-colony bacteria foraging optimization with cell-to-cell communication for RFID network planning. *Applied Soft Computing, 10*, 539–547.

Cook, N. (Ed.). (2013). *Viruses in food and water: Risks, surveillance and control.* Cambridge: Woodhead Publishing Limited. ISBN 978-0-85709-430-8.

Cortés, P., García, J. M., Muñuzuri, J., & Onieva, L. (2008). Viral systems: A new bio-inspired optimisation approach. *Computers & Operations Research, 35*, 2840–2860.

Cortés, P., García, J. M., Muñuzuri, J. & Guadix, J. (2012). Viral system algorithm: foundations and comparison between selective and massive infections. *Transactions of the Institute of Measurement and Control*. doi:10.1177/0142331211402897.

Desai, M. C., & Meanwell, N. A. (Eds.). (2013). *Successful strategies for the discovery of antiviral drugs*. Cambridge: The Royal Society of Chemistry. ISBN 978-1-84973-657-2.

Deshpande, P., Shukla, D., & Tiwari, M. K. (2011). Fuzzy goal programming for inventory management: A bacterial foraging approach. *European Journal of Operational Research, 212*, 325–336.

El-Abd, M. (2012). Performance assessment of foraging algorithms vs. evolutionary algorithms. *Information Sciences, 182*, 243–263.

Giguère, S., Prescott, J. F., & Dowling, P. M. (Eds.). (2013). *Antimicrobial therapy in veterinary medicine*. Chichester: Wiley. ISBN 978-0-470-96302-9.

Gollapudi, S. V. R. S., Pattnaika, S. S., Bajpai, O. P., Devi, S., & Bakwad, K. M. (2011). Velocity modulated bacterial foraging optimization technique (VMBFO). *Applied Soft Computing, 11*, 154–165.

Gong, E. Y., et al. (2013). *Antiviral methods and protocols*. New York: Springer. ISBN 978-1-62703-483-8.

Helden, J. V., Toussaint, A., & Thieffry, D. (Eds.). (2012). *Bacterial molecular networks: Methods and protocols*. New York: Springer. ISBN 978-1-61779-360-8.

Hooshmand, R. A., & Mohkami, H. (2011). New optimal placement of capacitors and dispersed generators using bacterial foraging oriented by particle swarm optimization algorithm in distribution systems. *Electrical Engineering, 93*, 43–53.

Hooshmand, R.-A., Parastegari, M., & Morshed, M. J. (2012). Emission, reserve and economic load dispatch problem with non-smooth and non-convex cost functions using the hybrid bacterial foraging-Nelder–Mead algorithm. *Applied Energy, 89*, 443–453.

Hota, P. K., Barisal, A. K., & Chakrabarti, R. (2010). Economic emission load dispatch through fuzzy based bacterial foraging algorithm. *Electrical Power and Energy Systems, 32*, 794–803.

Irizarry, R. (2011). Global and dynamic optimization using the artificial chemical process paradigm and fast Monte Carlo methods for the solution of population balance models. In: Dritsas, I. (Ed.) *Stochastic optimization—Seeing the optimal for the uncertain*, (Chapter 16). Rijeka: InTech, ISBN 978-953-307-829-8.

Ituarte-Villarreal, C. M. & Espiritu, J. F. (2011). Wind turbine placement in a wind farm using a viral based optimization algorithm. In *41st International Conference on Computers & Industrial Engineering* (pp. 672–677). Los Angeles, CA, USA, 23–26 October 2011.

Kamyab, S., & Bahrololoum, A. (2012). Designing of rule base for a TSK-fuzzy system using bacterial foraging optimization algorithm (BFOA). *Procedia-Social and Behavioral Sciences, 32*, 176–183.

Kim, D. H., Abraham, A., & Cho, J. H. (2007). A hybrid genetic algorithm and bacterial foraging approach for global optimization. *Information Sciences, 177*, 3918–3937.

Kumar, K. S., & Jayabarathi, T. (2012). Power system reconfiguration and loss minimization for an distribution systems using bacterial foraging optimization algorithm. *Electrical Power and Energy Systems, 36*, 13–17.

Lahtinen, S., Salminen, S., Ouwehand, A., & Wright, A. V. (Eds.). (2012). *Lactic acid bacteria: Microbiological and functional aspects*. Boca Raton: Taylor & Francis Group LLC. ISBN 978-1-4398-3678-1.

Lee, C.-Y. & Lee, Z.-J. (2012). A novel algorithm applied to classify unbalanced data. *Applied Soft Computing*. doi:10.1016/j.asoc.2012.03.051.

Li, Y., & Li, G. (2012). A new mean shift algorithm based on bacterial colony chemotaxis. *International Journal of Fuzzy Systems, 14*, 257–263.

Li, W–. W., Wang, H., & Zou, Z. J. (2005). Function optimization method based on bacterial colony chemotaxis. *Journal of Circuits and Systems, 10*, 58–63.

Li, G.-Q., Liao, H.-L. & Chen, H.-H. (2009). Improved bacterial colony chemotaxis algorithm and its application in available transfer capability. In *Fifth International Conference on Natural Computation (ICNC)* (pp. 286–291), 14–16 August 2009, Tianjin, China, IEEE.

Li, M. S., Ji, T. Y., Tang, W. J., Wu, Q. H., & Saunders, J. R. (2010). Bacterial foraging algorithm with varying population. *BioSystems, 100*, 185–197.

Lindsay, J. A., & Holden, M. T. G. (2006). Understanding the rise of the superbug: Investigation of the evolution and genomic variation of *Staphylococcus aureus*. *Functional & Integrative Genomics, 6*, 186–201.

Low, D. E., Kellner, J. D., & Wright, G. D. (1999). Superbugs: How they evolve and minimize the cost of resistance. *Current Infectious Disease Reports, 1*, 464–469.

Lu, Z.-G., Feng, T., & Li, X.-P. (2013). Low-carbon emission/economic power dispatch using the multi-objective bacterial colony chemotaxis optimization algorithm considering carbon capture power plant. *Electrical Power and Energy Systems, 53*, 106–112.

Maheshwari, D. K. (Ed.). (2012). *Bacteria in agrobiology: Plant probiotics*. Berlin: Springer. ISBN 978-3-642-27514-2.

Mai, X.-F., & Ling, L. (2011). Bacterial foraging optimization algorithm based on opposition-based learning. *Energy Procedia, 13*, 5726–5732.

Maitra, M., & Chatterjee, A. (2008). A novel technique for multilevel optimal magnetic resonance brain image thresholding using bacterial foraging. *Measurement, 41*, 1124–1134.

Majhi, B., & Panda, G. (2010). Development of efficient identification scheme for nonlinear dynamic systems using swarm intelligence techniques. *Expert Systems with Applications, 37*, 556–566.

Majhi, R., Panda, G., Majhi, B., & Sahoo, G. (2009). Efficient prediction of stock market indices using adaptive bacterial foraging optimization (ABFO) and BFO based techniques. *Expert Systems with Applications, 36*, 10097–10104.

Mishra, S., Tripathy, M., & Nanda, J. (2007). Multi-machine power system stabilizer design by rule based bacteria foraging. *Electric Power Systems Research, 77*, 1595–1607.

Modrow, S., Falke, D., Truyen, U., & Schätzl, H. (2013). *Molecular virology*. Berlin: Springer. ISBN 978-3-642-20717-4.

Morrow, W. J. W., Sheikh, N. A., Schmidt, C. S., & Davies, D. H. (Eds.). (2012). *Vaccinology: Principles and practice*. Chichester: Blackwell Publishing Ltd. ISBN 978-1-4051-8574-5.

Müller, S. D., Marchetto, J., Airaghi, S., & Koumoutsakos, P. (2002). Optimization based on bacterial chemotaxis. *IEEE Transactions on Evolutionary Computation, 6*, 16–29.

Muñoz, M. A., Halgamuge, S. K., Alfonso, W. & Caicedo, E. F. (2010). Simplifying the bacteria foraging optimization algorithm. In *IEEE World Congress on Computational Intelligence (WCCI)* (pp. 4095–4101), 18–23 July 2010, CCIB, Barcelona, Spain, IEEE.

Nayak, S. K., Padhy, S. K., & Panigrahi, S. P. (2012). A novel algorithm for dynamic task scheduling. *Future Generation Computer Systems, 28*, 709–717.

Niu, B., & Wang, H. (2012). Bacterial colony optimization. *Discrete Dynamics in Nature and Society, 2012*, 1–28.

Niu, B., Fan, Y., Tan, L., Rao, J. & Li, L. (2010a). A review of bacterial foraging optimization Part I: background and development. *Advanced intelligent computing theories and applications, communications in computer and information science* (Vol. 93, Part 26, pp. 535–543). Berlin: Springer.

Niu, B., Fan, Y., Tan, L., Rao, J. & Li, L. (2010b). A review of bacterial foraging optimization Part II: Applications and challenges. *Advanced intelligent computing theories and applications, communications in computer and information science* (Vol. 93, Part 26, pp. 544–550). Berlin: Springer.

Nouri, H. & Hong, T. S. (2012). A bacteria foraging algorithm based cell formation considering operation time. *Journal of Manufacturing Systems*, http://dx.doi.org/10.1016/j.jmsy.2012.03.001.

Nouri, H., Tang, S. H., Tuah, B. T. H., & Anuar, M. K. (2010). BASE: A bacteria foraging algorithm for cell formation with sequence data. *Journal of Manufacturing Systems, 29*, 102–110.

Panda, R., Naik, M. K., & Panigrahi, B. K. (2011). Face recognition using bacterial foraging strategy. *Swarm and Evolutionary Computation, 1*, 138–146.

Pandi, V. R., Biswas, A., Dasgupta, S., & Panigrahi, B. K. (2010). A hybrid bacterial foraging and differential evolution algorithm for congestion management. *European Transactions on Electrical Power, 20*, 862–871.

Panigrahi, B. K., & Pandi, V. R. (2009). Congestion management using adaptive bacterial foraging algorithm. *Energy Conversion and Management, 50*, 1202–1209.

Panigrahi, B. K., Pandi, V. R., Das, S., & Das, S. (2010). Multiobjective fuzzy dominance based bacterial foraging algorithm to solve economic emission dispatch problem. *Energy, 35*, 4761–4770.

Panigrahi, B. K., Pandi, V. R., Sharma, R., Das, S., & Das, S. (2011). Multiobjective bacteria foraging algorithm for electrical load dispatch problem. *Energy Conversion and Management, 52*, 1334–1342.

Passino, K. M. (2002). Biomimicry of bacterial foraging for distributed optimization and control. *IEEE Control System Management, 22*, 52–67.

Paton, R., Gregory, R., Vlachos, C., Palmer, J. W., Suanders, J., & Wu, Q. H. (2004). Evolvable social agents for bacterial systems modelling. *IEEE Transactions on Nanobioscience, 3*, 208–216.

RAJNI, & CHANA, I. (2013). Bacterial foraging based hyper-heuristic for resource scheduling in grid computing. *Future Generation Computer Systems, 29*, 751–762.

Saber, A. Y. (2012). Economic dispatch using particle swarm optimization with bacterial foraging effect. *Electrical Power and Energy Systems, 34*, 38–46.

Sakthivel, V. P., & Subramanian, S. (2012). Bio-inspired optimization algorithms for parameter determination of three-phase induction motor. *COMPEL: The International Journal for Computation and Mathematics in Electrical and Electronic Engineering, 31*, 528–551.

Sakthivel, V. P., Bhuvaneswari, R., & Subramanian, S. (2011). An accurate and economical approach for induction motor field efficiency estimation using bacterial foraging algorithm. *Measurement, 44*, 674–684.

Sanyal, N., Chatterjee, A., & Munshi, S. (2011). An adaptive bacterial foraging algorithm for fuzzy entropy based image segmentation. *Expert Systems with Applications, 38*, 15489–15498.

Sathya, P. D., & Kayalvizhi, R. (2011a). Amended bacterial foraging algorithm for multilevel thresholding of magnetic resonance brain images. *Measurement, 44*, 1828–1848.

Sathya, P. D., & Kayalvizhi, R. (2011b). Modified bacterial foraging algorithm based multilevel thresholding for image segmentation. *Engineering Applications of Artificial Intelligence, 24*, 595–615.

Sathya, P. D., & Kayalvizhi, R. (2011c). Optimal multilevel thresholding using bacterial foraging algorithm. *Expert Systems with Applications, 38*, 15549–15564.

Sathya, P. D., & Kayalvizhi, R. (2011d). Optimal segmentation of brain MRI based on adaptive bacterial foraging algorithm. *Neurocomputing, 74*, 2299–2313.

Su, T.-J., Cheng, J.-C., & Yu, C.-J. (2010). An adaptive channel equalizer using self-adaptation bacterial foraging optimization. *Optics Communications, 283*, 3911–3916.

Sun, J.-Z., Geng, G.-H., Wang, S.-Y., & Zhou, M.-Q. (2012). Chaotic hybrid bacterial colony chemotaxis algorithm based on tent map. *Journal of Software, 7*, 1030–1037.

Supriyono, H. & Tokhi, M. O. (2012). Parametric modelling approach using bacterial foraging algorithms for modelling of flexible manipulator systems. *Engineering Applications of Artificial Intelligence*, http://dx.doi.org/10.1016/j.engappai.2012.03.004.

Tabatabaei, S. M., & Vahidi, B. (2011). Bacterial foraging solution based fuzzy logic decision for optimal capacitor allocation in radial distribution system. *Electric Power Systems Research, 81*, 1045–1050.

Tang, W. J., & Wu, Q. H. (2009). Biologically inspired optimization: A review. *Transactions of the Institute of Measurement and Control, 31*, 495–515.

Tang, W. J., Li, M. S., He, S., Wu, Q. H. & Saunders, J. R. (2006). Optimal power flow with dynamic loads using bacterial foraging algorithm. In *International Conference on Power System Technology (PowerCon)* (pp. 1–5), 2006. IEEE.

Turduev, M., Kirtay, M., Sousa, P., Gazi, V. & Marques, L. Chemical concentration map building through bacterial foraging optimization based search algorithm by mobile robots. In *IEEE International Conference on Systems, Man, and Cybernetics (IEEE SMC)* (pp. 3242–3249), 10–13 October 2010, Istanbul, Turkey, IEEE.

Ulagammai, M., Venkatesh, P., Kannan, P. S., & Padhy, N. P. (2007). Application of bacterial foraging technique trained artificial and wavelet neural networks in load forecasting. *Neurocomputing, 70*, 2659–2667.

Vaisakh, K., Praveena, P., Rao, S. R. M., & Meah, K. (2012). Solving dynamic economic dispatch problem with security constraints using bacterial foraging PSO-DE algorithm. *Electrical Power and Energy Systems, 39*, 56–67.

Venkaiah, C., & Kumar, D. M. V. (2011). Fuzzy adaptive bacterial foraging congestion management using sensitivity based optimal active power re-scheduling of generators. *Applied Soft Computing, 11*, 4921–4930.

Verma, O. P., Hanmandlu, M., Kumar, P., Chhabra, S., & Jindal, A. (2011). A novel bacterial foraging technique for edge detection. *Pattern Recognition Letters, 32*, 1187–1196.

Verma, O. P., Hanmandlu, M., Sultania, A. K. & Parihar, A. S. (2013). A novel fuzzy system for edge detection in noisy image using bacterial foraging. *Multidimensional Systems and Signal Processing*. doi: 10.1007/s11045-011-0164-1.

Vivekanandan, K., & Ramyachitra, D. (2012). Bacteria foraging optimization for protein sequence analysis on the grid. *Future Generation Computer Systems, 28*, 647–656.

Chapter 3
Bat Inspired Algorithms

Abstract In this chapter, we present two algorithms that are inspired by the behaviours of bats, i.e., bat algorithm (BaA) and bat intelligence (BI) algorithm. We first describe the general knowledge of the foraging behaviour of bats in Sect. 3.1. Then, the fundamentals and performances of the BaA and BI algorithm are introduced in Sects. 3.2 and 3.3, respectively. Finally, Sect. 3.4 summarises in this chapter.

3.1 Introduction

Bats exhibit mysterious behaviours that have long since attracted the attention of human beings. Probably one of the most noticeable behaviours visible to us is they are capable to orientate their environments and food acquisition without relying on their eyesight (Merritt 2010). This is accomplished by bats continuously emit echolocation signals (Yovel et al. 2008). Through analyzing the returning echoes in the auditory system, bats can easily identify their surroundings and locate preys. By continuously observing and studying the abilities of bats, recently, computer scientists proposed several bat inspired algorithms (i.e., bat algorithm (BaA) and bat intelligence (BI) algorithm) to solve well-known optimization problems such as image processing, scheduling, clustering, and data mining.

3.1.1 Foraging Behaviour of Bats

Traditionally, families of bats are divided into two major groups, i.e., megabats and microbats. Typically, bats have three habitats as they feed through the whole year, i.e., roost, foraging habitats, and commuting habitats. The foraging habitats are used to find food, while the commuting habitats are used to travel between roosts and foraging habitats (Mills et al. 2010). In general, the foraging behaviour of bats can be divided into three phase: search phase, pursuit phase, and capture phase

B. Xing and W.-J. Gao, *Innovative Computational Intelligence: A Rough Guide to 134 Clever Algorithms*, Intelligent Systems Reference Library 62, DOI: 10.1007/978-3-319-03404-1_3,

(Kalko 1995). Also, they are very flexible in their foraging behaviour, such as feeding by aerial hawking (catching insects on the wing), gleaning (picking insets from vegetation), and pouncing on prey close to the ground. At each phase, they show different behaviours based on a perceptual system, i.e., bats echolocation.

3.1.2 Characteristics of Echolocation

One of the unique characteristics of bats from other animals is their echolocation capacity, i.e., emitting ultrasonic pulses and listening the echoes (Mills et al. 2010). According to the different pulses, they can detect prey, avoid obstacles, and even locate their roosting crevices in the dark (Yang 2010). For example, as bats fly through the night, their echolocation calls bounce off the environment, helping the bats find their way to and from their habitats. Typically, each bat emits about 10–20 pulse every second. Though each pulse lasts a few thousands of a second, it has a constant frequencies which normally range from 25 to 150 kHz (Yang 2010). Amazingly, when bats hunt for prey, the rate of pulse emission can reach roughly 200 pulses per second when they fly near their prey (Yang 2010). In contrast to the microbats use vocalization to emit sonar signals, the megabats use their tongue to emit very brief wideband echolocation (Yovel et al. 2011). Furthermore, although echolocation was first identified in bats, it is also used by some other species, such as dolphin and toothed whales.

3.2 Bat Algorithm

3.2.1 Fundamentals of Bat Algorithm

Bat algorithm (BaA) was originally proposed by Yang (2010) which is based on the echolocation behaviour of bats. To implement the proposed algorithm, a set of approximations or idealized rules are given as follows (Yang 2010):

- Based on the echolocation behaviour, the bats can detect the distance and the differentiate between food and background barriers as well, even in the darkness.
- Bats usually fly randomly to search for prey. According to that, some numerical parameters associated with the bats' foraging behaviour are defined first, such as velocity (v_i) at position (x_i) with a fixed frequency ($f_{\min}$), varying wavelength (λ) and loudness (A_o).
- The loudness can varies from a large positive (A_o) when searching for prey to a minimum constant value ($A_{\min}$) when homing towards the prey.

The basic steps of implementing BaA are described as follows.

3.2.1.1 Movement of Virtual Bats

Just like the real bats, the virtual bats fly randomly by adjusting its frequency first and updating its velocity and position next. The new locations/solutions (x_i^t) and velocities (v_i^t) at time step t are given via Eqs. (3.1)–(3.3) respectively (Yang 2010):

$$f_i = f_{\min} + (f_{\max} - f_{\min}) \cdot \beta, \tag{3.1}$$

$$v_i^t = v_i^{t-1} + \left(x_i^t - x_*\right) \cdot f_i, \tag{3.2}$$

$$x_i^t = x_i^{t-1} + v_i^t, \tag{3.3}$$

where $\beta \in [0, 1]$ is a random vector drawn from a uniform distribution, f_i is the velocity increment, $f_{\min}$ and $f_{\max}$ are the lower and upper bounds imposed for the frequency range of virtual bats, respectively, and x_* is the current global best location.

For the local search part, the random walks are used to generate a new solution for each bat from the current population via Eq. 3.4 (Yang 2010):

$$x_{new} = x_{old} + \varepsilon A^t, \tag{3.4}$$

where $\varepsilon \in [-1, 1]$ is a random number, while $A^t = \langle A_i^t \rangle$ is the average loudness of all the bats at this time step.

3.2.1.2 Loudness and Pulse Emission

The general rule is that the loudness decreases while the rate of pulse emission increases, when a bat has found its prey. The loudness and the rate of pulse emission are updated via Eqs. (3.5) and (3.6), respectively (Yang 2010):

$$A_i^{t+1} = \alpha A_i^t, \tag{3.5}$$

$$r_i^{t+1} = r_i^0[1 - \exp(-\gamma t)], \tag{3.6}$$

where A_i^t and A_i^{t+1} are the previous and updated values of the loudness for the ith bat, α and γ are constants, and r_i^{t+1} is the pulse rate of the ith bat at iteration $t + 1$. For any $0 < \alpha < 1$ and $\gamma > 0$, we have Eq. (3.7) (Yang 2010):

$$A_i^t \to 0,\ r_i^t \to r_i^0,\ \text{as } t \to \infty. \tag{3.7}$$

Summarizing the steps in BaA yields to (Yang 2010; Carbas and Hasancebi 2013):

- Step 1: Initializing bat population [such as position (x_i) and velocities (v_i)].
- Step 2: Initializing parameters, such as pulse frequency (f_i), pulse rates (r_i), and loudness parameters (A_i).

- Step 3: Evaluating bats in the initial population according to the fitness function.
- Step 4: Generating candidate bats through random flying and local search.
- Step 5: Evaluating candidate bats according to the fitness function.
- Step 6: Echolocation parameters update.
- Step 7: Rank the bats and find the current best x_*.
- Step 8: Termination. The steps 4–7 are employed in the same way until a termination criterion is met.

3.2.2 Performance of BaA

To text the effectiveness of BaA, a set of well-known test functions were adopted in (Yang 2010), namely, Rosenbrock's function, De Jong's standard sphere function, Ackley's function, Michalewica's test function, Schewfer's function, Rastrigin's function, Easom's function, Griewangk's function, and Shubert's test function. Compared with other algorithms [i.e., particle swarm optimization (PSO) and genetic algorithm (GA)], computational results showed that BaA performs significantly better than other algorithms in terms of accuracy and efficiency.

3.3 Emerging Bat Inspired Algorithms

In addition to the aforementioned BaA, the characteristics of this interesting animal also motivate researchers to develop other bat inspired innovative computational intelligence (CI) algorithms.

3.3.1 Bat Intelligence Algorithm

3.3.1.1 Fundamentals of Bat Intelligence (BI) Algorithm

Bat intelligence (BI) algorithm was originally proposed in (Malakooti et al. 2013b) that mimics the prey hunting behaviours of bats. For BI, a set of signals are generated not only based on the echolocation characteristic, but also the constant absolute target direction (CATD) technique, where the bat maintains same pursuit angle to the prey, has been incorporated. Also, during the prey hunting process, after seizing and consuming a prey, bats can proceed to capture other preys in the foraging area. The main steps of BI are described as follows (Malakooti et al. 2013b):

- Step 1: Initializing the parameters and population.
- Step 2: The fitness function value (i.e., signal strength) corresponding to each candidate solution is calculated.

- Step 3: Generating a set of solutions containing common elements.
- Step 4: Select the best solution.
- Step 5: Choosing a common element and include it in the common element list.
- Step 6: Checking the terminating condition. If it is reached, go to step 3 for new beginning. If a specified termination criteria is satisfied, stop and return the best solution.

3.3.1.2 Performance of BI

To illustrate the performance of BI, a multiple objective energy aware multiprocessor scheduling problem with three objectives (i.e., makespan, tardiness, and energy consumption) are presented in (Malakooti et al. 2013a). Compared with GA, the BI algorithm shows better performance to move towards the optimal solution.

3.4 Conclusions

In this chapter, we introduced two bat inspired CI algorithms. Although they are newly introduced CI method, we have witnessed the following rapid spreading of at least one of them, i.e., BaA:

First, several enhanced versions of BaA can be found in literature as outlined below:

- Differential operator and Lévy flight BaA (Xie et al. 2013).
- Modified BaA (Damodaram and Valarmathi 2012).

Second, the BaA has also been successfully applied to a variety of optimization problems as listed below:

- Data mining (Damodaram and Valarmathi 2012).
- Gas turbine generator exergy analysis (Lemma and Hashim 2011).
- Motor control (Bora et al. 2012).
- Scheduling optimization (Musikapun and Pongcharoen 2012).
- Steel space frames design optimization (Carbas and Hasancebi 2013).

Interested readers are referred to them, together with several excellent reviews [e.g., (Yang 2013; Schnitzer 2002; Sureja 2012)], as a starting point for a further exploration and exploitation of bat inspired algorithms.

References

Bora, T. C., Coelho, L. D. S., & Lebensztajn, L. (2012). Bat-inspired optimization approach for the brushless DC wheel motor problem. *IEEE Transactions on Magnetics, 48*, 947–950.

Carbas, S., & Hasancebi, O. (2013, May 19–24). Optimum design of steel space frames via bat inspired algorithm, *10th World Congress on Structural and Multidisciplinary Optimization* (pp. 1–10). Orlando, Florida, USA.

Damodaram, R., & Valarmathi, M. L. (2012). Phishing website detection and optimization using modified bat algorithm. *International Journal of Engineering Research and Applications, 2*, 870–876.

Kalko, E. K. V. (1995). Insect pursuit, prey capture and echolocation in pipistrelle bats. *Animal Behaviour, 50*, 861–880.

Lemma, T.A. & Hashim, F.B.M. (2011). Use of fuzzy systems and bat algorithm for exergy modeling in a gas turbine generator. IEEE Colloquium on Humanities, Science and Engineering Research (CHUSER) (pp. 305–310). IEEE, Penang, 5–6 December 2013.

Malakooti, B., Kim, H. & Sheikh, S. (2013a). Bat intelligence search with application to multi-objective multiprocessor scheduling optimization. *International Journal of Advanced Manufacturing Technology,* (Vol. 60, pp. 1071–1086). DOI 10.1007/s00170-011-3649-z.

Malakooti, B., Sheikh, S., Al-Najjar, C. & Kim, H. (2013b). Multi-objective energy aware multiprocessor scheduling using bat intelligence. *Journal of Intelligent Manufacturing, 24*, 805–819. DOI 10.1007/s10845-012-0629-6.

Merritt, J. F. (2010). *The biology of small mammals*. Baltimore: John Hopkins University Press.

Mills, D. S., Marchant-Forde, J. N., McGreevy, P. D., Morton, D. B., Nicol, C. J., Phillips, C. J. C., et al. (Eds.). (2010). *The encyclopedia of applied animal behaviour and welfare*. Wallingford: CAB International. ISBN 978-0-85199-724-7.

Musikapun, P., & Pongcharoen, P. (2012). Solving multi-stage multi-machine multi-product scheduling problem using bat algorithm. *2nd International Conference on Management and Artificial Intelligence* (pp. 98–102). IACSIT.

Schnitzer, M. J. (2002). Amazing algorithms. *Nature, 416*, 683.

Sureja, N. (2012). New inspirations in nature: a survey. *International Journal of Computer Applications and Information Technology, I*, 21–24.

Xie, J., Zhou, Y., & Chen, H. (2013). A novel bat algorithm based on differential operator and Lévy flights trajetory. *Computational Intelligence and Neuroscience*, 1–13.

Yang, X.-S. (2010). A new metaheuristic bat-inspired algorithm. *Nature inspired cooperative strategies for optimization* (*NISCO 2010*), Studies in Computational Intelligence, SCI 284 (pp. 65–74). Berlin: Springer.

Yang, X.-S. (2013). Bat algroithm: literature review and applications. *International Journal of Bio-Inspired Computation, 5*, 141–149.

Yovel, Y., Franz, M. O., Stilz, P., & Schnitzler, H.-U. (2008). Plant classification from bat-like echolocation signals. *PLoS Computational Biology, 4*, 1–13.

Yovel, Y., Geva-Sagiv, M., & Ulanovsky, N. (2011). Click-based echolocation in bats: not so primitive after all. *Journal of Comparative Physiology A, 197*, 515–530.

Chapter 4
Bee Inspired Algorithms

Abstract In this chapter, we present a set of algorithms that are inspired by different honeybees behavioural patterns, i.e., artificial bee colony (ABC) algorithm, honeybees mating optimization (HBMO) algorithm, artificial beehive algorithm (ABHA), bee colony optimization (BCO) algorithm, bee colony inspired algorithm (BCiA), bee swarm optimization (BSO) algorithm, bee system (BS) algorithm, BeeHive algorithm, bees algorithm (BeA), bees life algorithm (BLA), bumblebees algorithm, honeybee social foraging (HBSF) algorithm, OptBees algorithm, simulated bee colony (SBC) algorithm, virtual bees algorithm (VBA), and wasp swarm optimization (WSO) algorithm. We first describe the general knowledge of honeybees in Sect. 4.1. Then, the fundamentals and performances of these algorithms are introduced in Sects. 4.2, 4.3, and 4.4, respectively. Finally, Sect. 4.5 summarises in this chapter.

4.1 Introduction

Honeybee is a typical social insect that works together in a highly structured social order to finish different kinds of jobs such as bee dance (communication), bee foraging, queen bee, task selection, collective decision making, nest site selection, mating, floral/pheromone laying, and navigation that have long since attracted the attention of human beings (Janson et al. 2005; Latty et al. 2009; Slaa and Hughes 2009; Landa and Tullock 2003). Based on those features, many models have been developed for intelligent systems and applied to solve combinatorial optimization problems. In this chapter, a set of bee inspired algorithms are collected and introduced as follows:

- Section 4.2: Artificial Bee Colony.
- Section 4.3: Honeybees Mating Optimization.
- Section 4.4.1: Artificial Beehive Algorithm.
- Section 4.4.2: Bee Colony Optimization.

B. Xing and W.-J. Gao, *Innovative Computational Intelligence: A Rough Guide to 134 Clever Algorithms*, Intelligent Systems Reference Library 62,
DOI: 10.1007/978-3-319-03404-1_4, © Springer International Publishing Switzerland 2014

- Section 4.4.3: Bee Colony-inspired Algorithm.
- Section 4.4.4: Bee Swarm Optimization.
- Section 4.4.5: Bee System.
- Section 4.4.6: BeeHive.
- Section 4.4.7: Bees Algorithm.
- Section 4.4.8: Bees Life Algorithm.
- Section 4.4.9: Bumblebees Algorithm.
- Section 4.4.10: Honeybee Social Foraging.
- Section 4.4.11: OptBees.
- Section 4.4.12: Simulated Bee Colony.
- Section 4.4.13: Virtual Bees Algorithm.
- Section 4.4.14: Wasp Swarm Optimization.

The effectiveness of these newly developed algorithms are validated through the testing on a wide range of benchmark functions and engineering design problems, and also a detailed comparison with various traditional performance leading computational intelligence (CI) algorithms, such as particle swarm optimization (PSO), genetic algorithm (GA), differential evolution (DE), evolutionary algorithm (EA), fuzzy system (FS), ant colony optimization (ACO), and simulated annealing (SA).

4.1.1 Foraging Behaviour of Bees

Among other algorithms inspired by honeybees, probably one of the most noticeable behaviours visible to us is the foraging of each individual bee. Foraging process includes two main modes of behaviour: recruitment of nectar source and abandonment of a source (Tereshko and Lee 2002). It starts with some scout bees leaving the hive in order to search food source to gather nectar. After finding food (i.e., flowers), scout bees return to the hive and inform their hive-mates about the richness of the flower (i.e., quantity and quality) and the distance of the flower to the hive (i.e., location) through a special movements called "dance", such as round dance, waggle dance, and tremble dance depending on the distance information of the source. Typically, she dances on different areas in an attempt to "advertise" food locations (by touching her antennae) and encourage more remaining bees to collect nectar from her source. After the dancing show, more foraging bees will leave the hive to collect nectar follow one of the dancing scout bees. Upon arrive, the foraging bee stores the nectar in her honey stomach and returns to the hive unloading the nectar to empty honeycomb cells. The described process continues repeatedly until the scout bees explore new areas with potential food sources.

4.1.2 Marriage Behaviour of Bees

Another famous behaviour of bees is the marriage (i.e., mating) behaviour. In the kingdom of honeybees, queens are specialized in egg-laying. A colony may contain one queen or more during its life cycle. Typically, mating occur in flight and 10–40 m above ground. It begins when the queen flights far away from the nest performing the mating flight during which the drones follow the queen and mate with her in the air. Normally, the queen mates with 12 ± 7 drones, and after the mating process, the drones die.

4.1.3 Dancing and Communication Behaviour of Bees

In general, honeybees (i.e., scout bees) perform a series of movements (i.e., dancing) to exchange information (such as the location, quantity and quality of food sources) and persuade their nestmates to follow them. There are two types of dances, i.e., round dance when food is very close and waggle dance. In addition, according to the speed of the dances, honeybees transmit the distance information, i.e., if dance is faster, then the food distance is smaller (Bitam et al. 2010).

4.2 Artificial Bee Colony Algorithm

4.2.1 Fundamentals of Artificial Bee Colony Algorithm

Artificial bee colony (ABC) algorithm was recently proposed in Karaboga and Basturk (2007), Karaboga (2005). The basic idea of designing ABC is to mimic the foraging behaviour (such as exploration, exploitation, recruitment and abandonment) of honeybees. Typically, ABC algorithm consists of two groups of bees: employed artificial bees (i.e., current exploiting foragers) and unemployed artificial bees (i.e., looking for a food source to exploit). The latter will be classified further in two groups: scouts who are searching the environment surrounding the nest for new food sources, and onlookers that are waiting in the nest and finding a food source through the information shared by employed artificial bees (Karaboga and Basturk 2008). In ABC, it is assumed that each food source position corresponds to a possible solution and the nectar amount of a food source corresponds to the quality (fitness) of the associated solution. The main steps of the ABC algorithm are listed as below (Karaboga and Basturk 2008; Karaboga and Akay 2009b; Senthilnath et al. 2011):

- Initialization: The initial population can be defined as $P(G = 0)$ of SN solutions (food source positions), where SN denotes the size of employed bees or onlooker bees. Moreover, each solution $x_{ij}(i = 1, 2, \ldots, SN;\ j = 1, 2, \ldots, D)$ is a D-dimensional vector. Here, D is the number of optimization parameters.

- Then, placing the employed bees on the food sources in the memory and updating feasible food source. In order to produce a candidate food position from the old one (x_{ij}) in memory, the memory by employed bees is updated via Eq. 4.1 (Karaboga and Basturk 2008; Karaboga and Akay 2009b; Senthilnath et al. 2011):

$$v_{ij} = x_{ij} + \phi_{ij}\left(x_{ij} - x_{kj}\right), \quad j \in \{1, 2, \ldots, D\}, k \in \{1, 2, \ldots, SN\} \wedge k \neq i, \tag{4.1}$$

where v_{ij} is a new feasible dimension value of the food sources that is modified from its previous food sources value (x_{ij}) based on a comparison with the randomly selected neighbouring food source value (x_{kj}), and ϕ_{ij} is a random number between $[-1, 1]$ to adjust the production of neighbour food sources around x_{ij} and represents the comparison of two food positions visually.

- Next, placing the onlooker bees on the food sources in the memory. The onlooker bee chooses a probability value associated with that food source (p_i) via Eq. 4.2 (Karaboga and Basturk 2008; Karaboga and Akay 2009b; Senthilnath et al. 2011):

$$p_i = \frac{fit_i}{\sum_{n=1}^{SN} fit_n}, \tag{4.2}$$

where fit_i is the fitness value of the solution i which is proportional to the nectar amount of the food source in the position; SN denotes the size of employed bees or onlooker bees. Clearly, the higher fit_i is, the greater the probability is of selecting x_{ij}.

- Updating feasible food source, by onlooker bees using Eq. 4.1.
- Adjusting by sending the scout bees in order to discovering new food sources. The operation of scout bees explore a new food source can be defined by Eq. 4.3 (Karaboga and Basturk 2008; Karaboga and Akay 2009b; Senthilnath et al. 2011):

$$x_i^j = x_{\min}^j + rand[0, 1]\left(x_{\max}^j - x_{\min}^j\right), \tag{4.3}$$

where $x_{\min}$ and $x_{\max}$ is the lower and upper limit respectively of the search scope on each dimension. Here the value of each component in every x_i vector should be clamped to the range $[x_{\min}, x_{\max}]$ to reduce the likelihood of scout bees leaving the search space.

- Memorizing the best food source found so far.
- Finally, checking whether the stopping criterion is met. If yes, terminating the algorithm; otherwise, restarting the main procedure of the ABC.

Summarizing the steps in standard ABC yields to (Karaboga and Akay 2009a; Karaboga and Akay 2009b; Karaboga and Basturk 2008; Senthilnath et al. 2011):

- Step 1: The employed bees will be randomly sent to the food sources and evaluating their nectar amounts. If an employed bee finds a better solution, she will update her memory; otherwise, she counts the number of the searches around the source in her memory.

- Step 2: If all employed bees complete the search process, the nectar and position information of the food sources will be shared with the onlooker bees.
- Step 3: An onlooker bee does not have any source in her memory and thus she will evaluate all the information from employed bees and choose a probably profitable food source (recruitment).
- Step 4: After arriving at the selected area, the onlooker bee searches the neighbourhood of the source and if she finds a better solution, she will update the food source position just as an employed bee does. The criterion for determination of a new food source is based on the comparison process of food source positions visually.
- Step 5: Stopping the exploitation process of the sources abandoned by the employed/onlooker bees if the new solution cannot be further improved through a predetermined number of trials limit. At this moment, the employed/onlooker bees become scout bees.
- Step 6: Sending the scouts into the search area for discovering new food sources (exploration), randomly.
- Step 7: Memorizing the best food source found so far.

These seven steps are repeated until a termination criterion (e.g., maximum cycle number) is satisfied.

4.2.2 Performance of ABC

In order to evaluate the ABC algorithm, Karaboga and Basturk (2007) employed five high dimensional benchmark testing functions. In comparison with other CI algorithms (e.g., EA, PSO, GA), the simulation results demonstrated that the ABC algorithm has the capability of getting out of a local minimum trap which make it a promising candidate in dealing with multivariable, multimodal function optimization tasks.

4.3 Honeybee Mating Optimization Algorithm

4.3.1 Fundamentals of Honeybee Mating Optimization Algorithm

In general, a honeybee community consists of three types of members: the queen, male honeybees (or drones), and neuter/undeveloped female honeybees (or workers). The honeybees mating optimization (HBMO) algorithm was proposed in Abbass (2001a, b) to simulate the social behaviour found among honeybees. In the original HBMO, a drone mates with a queen probabilistically through the use of an annealing function as stated via Eq. 4.4 (Niknam et al. 2011):

$$\Pr(D) = \exp\left(\frac{-\Delta(F)}{V_{queen}(t)}\right), \tag{4.4}$$

where $\Pr(D)$ denotes the probability for a drone D adding the sperm to the queen's spermatheca, $\Delta(F)$ represents the absolute difference between the fitness of the drone and the queen, and $V_{queen}(t)$ stands for the velocity of the queen at time t. After each transition in space, the velocity and the energy of the queen decreases based on Eq. 4.5 (Niknam et al. 2011; Marinakis et al. 2008):

$$\begin{array}{l} V_{queen}(t+1) = \alpha \times V_{queen}(t) \\ E_{queen}(t+1) = \alpha \times E_{queen}(t) \end{array}, \tag{4.5}$$

where α denotes the decreasing factor which is a positive real number within the interval of $[0, 1]$. The amount of speed and energy reduction after each transition and each step is controlled through this parameter.

In HBMO algorithm, the broods that are generated through the mating process between the queen and a drone are calculated through Eq. 4.6 (Niknam et al. 2011):

$$\begin{array}{l} X_{Brood,j} = X_{queen} + \beta \times \left(X_{queen} - D_i\right) \\ X_{Brood,j} = \left[x_{broodj,1}, x_{broodj,1}, \ldots, x_{broodj,Ng}\right]_{1\times Ng} \end{array}, \tag{4.6}$$

where D_i represents the ith drone stored in the spermatheca of the queen, and $\beta \in [0, 1]$ refers to the mating factor.

Summarizing the steps in HBMO yields to (Horng 2010; Haddad et al. 2006; Afshar et al. 2007; Boussaïd et al. 2013):

- Step 1: Initializing parameters and population.
- Step 2: Mating flight process, where a queen (best solution) selects drones probabilistically to form the spermatheca (list of drones). A drone then selected from the list randomly for the creation of broods.
- Step 3: Creation of new broods by crossover the drone's genotypes with the queens.
- Step 4: Use of workers to conduct local search on broods (trial solutions).
- Step 5: Adaptation of worker's fitness, based on the amount of improvement achieved on broods.
- Step 6: Replacement of weaker queen by fitter broods.
- Step 7: Termination.

4.3.2 *Performance of HBMO*

In Marinakis et al. (2008), by utilizing the HBMO algorithm, the authors made an attempt to solve vehicle routing problem. Overall fourteen benchmark problems from the literature were chosen to test the performance of HBMO. Each instance of the problem set consists of between 51 and 200 nodes including the depot. The location of the nodes is defined by their Cartesian coordinates and the travel cost from node i to j is assumed to be associated with the corresponding Euclidean distance. Each selected problem includes the capacity constraints while the problems 6–10, 13, and 14 also have the restrictions of the maximum length of travel route and non zero service times. For the first ten problems, nodes are randomly placed over a square, while for the remaining four problems, nodes are distributed in clusters and the depot is not located in the centre. The experimental results obtained through the HBMO were compared with other twenty methods, among them, half is the most efficient metaheuristic approaches, and the other half is the most efficient nature inspired algorithms. Through the comparison, it can be observed that the HBMO ranks the 2nd and the 1st place among the ten metaheuristic methods and ten nature inspired algorithms, respectively.

4.4 Emerging Bee Inspired Algorithms

In addition to the aforementioned two bee inspired algorithms, i.e., ABC and HBMO, the characteristics of this interesting insect also motivate researchers to develop several other bee inspired innovative CI algorithms.

4.4.1 *Artificial Beehive Algorithm*

4.4.1.1 Fundamentals of Artificial Beehive Algorithm

Artificial beehive algorithm (ABHA) was originally proposed by Muñoz et al. (2009). Based on a set of behavioural rules of each individual bee (e.g., the individual oriented model), the ABHA is used to solve continuous optimization problem. In order to implement ABHA, the following procedures need to be followed (Muñoz et al. 2009):

- Initializing parameters, such as the current position of the individual ($\theta(t)$), the current cost value ($J(\theta(t))$), the past cost value ($J(\theta(t-1))$), the abandon tendency ($0 < p_{ab} < 1$), and the homing motivation ($0 < p_h < 1$).
- Defining the bees' states. Typically, there are four states for each individual, i.e., novice state, experimented state, search state, and food source state. The details of each state are as follows (Muñoz et al. 2009):

State 1: Novice state. In this state, the bee is in the "nest" (i.e., an abstract position represented only by the state of the bee, where the information is exchanged) and does not have information about a source. It performs a random search or follow a dance if it is available. The current position of each bee can be defined via Eq. 4.7 (Muñoz et al. 2009):

$$\theta_i(t) = NaN(t), \tag{4.7}$$

where *NaN* represents not a number.

State 2: Experimented state. In this state, the bee is still in the "nest" but it has some information about a food source. If the information means high quality (i.e., a good food source), it can be transmitted to other individuals through a dancing that denoted as a selection probability (p_{si}) via Eq. 4.8 (Muñoz et al. 2009):

$$\begin{cases} p_{si} = -\left(\frac{1}{\max(J_j - \min(J_j))}\right)(J_i - \max(J_i)) & \text{good information} \\ \text{random search or follow a dance} & \text{bad information} \end{cases}, \tag{4.8}$$

where p_{si} is a selection probability that i indicates one of the jth individuals with available dances.

State 3: Search state. In this state, the bee leaves its nest and looks for a better foraging source than the current. The bee's position is updated via Eq. 4.9 (Muñoz et al. 2009):

$$\theta_i(t+1) = \theta_i + SS(i)\psi(t), \tag{4.9}$$

where $SS(i)$ is the step size at the direction $\psi(t)$.

State 4: Food source state. Finally, the bee decides which source is the best.

- Calculating the probabilities that are used to balance the exploitation and exploration characteristics of the algorithm.

4.4.1.2 Performance of ABHA

To test the effectiveness of ABHA, a set of well-known test functions were adopted in Muñoz et al. (2009), namely, Grienwank function, Rastrigin function, Ackley function, De Jong F2 function, Schewefel function, and Schaffer F6 function. Compared with other CI algorithms, such as PSO and its variants, computational results showed that ABHA achieved good results. Also, in a few cases it is capable with a better performance than the compared algorithms.

4.4.2 Bee Colony Optimization

4.4.2.1 Fundamentals of Bee Colony Optimization Algorithm

Bee colony optimization (BCO) algorithm was originally proposed in Teodorović and Dell'Orco (2005). In BCO, when the foragers return to the hive, a waggle dance is performed by each forager, then the other bees based on a probability follow the foragers. In order to implement BCO, the following procedures need to be followed (Bonyadi et al. 2008; Teodorović 2009b; Teodorović and Dell'Orco 2005; Teodorović et al. 2011):

- Step 1: Initialization: Assigning an empty solution to each bee within the colony.
- Step 2: For each bee, performing the forward pass mechanism, i.e., fulfilling the following subtasks. (a) Setting $k = 1$; (b) Evaluating all possible constructive movements; (c) Selecting one movement using the roulette wheel strategy, (d) Letting $k = k + 1$. If $k \leq NC$, redoing the subtask (b).
- Step 3: All bees return to the hive, i.e., starting the backward pass mechanism.
- Step 4: Evaluating (partial) objective function value carried by each bee.
- Step 5: For each bee, determining whether to carry on with its own exploration and becoming a recruiter, or turning to a follower. In BCO, a loyalty decision strategy is introduced at this step, i.e., at the beginning of each new round of forward pass, the probability of the bth bee is loyal to its previously obtained partial solution. This probability can be expressed via Eq. 4.10 (Teodorović et al. 2011):

$$p_b^{u+1} = e^{-\frac{O_{\max} - O_b}{u}}, \quad b = 1, 2, \ldots, B, \tag{4.10}$$

  where O_b denotes the normalized value of the objective function carried by the bth bee, O_b represents the maximum over all normalized values of partial solution to be evaluated, and the ordinary number of the forward pass is indicated by u.
- Step 6: For each follower bee, selecting a new solution from its recruiters using the roulette wheel strategy. In BCO, a recruiting mechanism is performed at this step. For each unemployed bee, it will decided to follow which recruiter based on a certain probability. The probability that b's partial solution would be selected by any unallocated bee is defined by Eq. 4.11 (Teodorović et al. 2011):

$$p_b = \frac{O_b}{\sum_{k=1}^{R} O_k}, \quad b = 1, 2, \ldots, R, \tag{4.11}$$

  where O_k denotes the normalized value for the objective function of the kth announced partial solution, and the number of recruiters is represented by R.
- Step 7: If the solutions are not finished, returning to Step 2.
- Step 8: Evaluating all solutions and looking for the best one.

- Step 9: Checking whether the termination criterion is met. If not, going back to Step 2; otherwise, outputting the best available solution.

In addition to the standard BCO detailed in this section, several enhanced versions of BCO can also be found in the literature as outlined below:

- Autonomous BCO (Zeng et al. 2010).
- Multiobjective BCO (Low et al. 2009).

4.4.2.2 Performance of BCO

In order to see how the BCO algorithm performs, Teodorović and Dell'Orco (2005) tested it on a ride-matching problem in which the ridesharing is one of the popular travel demand management methodologies. The preliminary experimental results showed that the performance of BCO is very promising.

Apart from the original case study described above, the BCO algorithm has also been successfully applied to a variety of optimization problems as listed below:

- *p*-Center problem (Davidović et al. 2011).
- Scheduling optimization (Chong et al. 2006; Wong et al. 2008).
- Software maintenance (Kaur and Goyal 2011).
- Transportation (Teodorović 2008; Teodorović and Dell'Orco 2005).
- Travelling salesman problem (Bonyadi et al. 2008).

Interested readers are referred to these selected variants and representative applications, together with several excellent reviews [e.g., (Teodorović 2008; Bitam et al. 2010; Teodorović et al. 2011; Goyal 2012; Figueira and Talbi 2013; Teodorović 2009a)], for a further exploration and exploitation of the BCO algorithm.

4.4.3 *Bee Colony-inspired Algorithm*

4.4.3.1 Fundamentals of Bee Colony-Inspired Algorithm

Bee colony inspired algorithm (BCiA) was originally proposed in Häckel and Dippold (2009). In order to implement BCiA, the following control procedures need to be followed (Häckel and Dippold 2009):

- First of all, initializing both populations P_1 and P_2 by scout bees.
- Stage 1: Once the bees in P_1 finish the solution construction, they will communicate with their counterparts in P_2. At this stage, if the quality of a solution φ^{eb} in P_1 is better than the worst one $\varphi^{worst'}$ found in P_2 and also if this solution is not yet contained in P_2, then the solution of $\varphi^{worst'}$ is replaced by φ^{eb}.
- Stage 2: If the algorithm reaches the second stage, in a similar way, the bees in P_2 will conduct the solution construction task with a subsequent feedback to

their peers in P_1. At this stage, if the quality of a solution $\varphi^{eb'}$ in P_2 is better than the worst one φ^{worst} on found in P_1 and also in not yet included in P_1, then the solution of φ^{worst} is replaced by $\varphi^{eb'}$.
- Afterwards, checking the solutions' age. If, after the maximum iteration numbers, the quality of a solution does not improve, this solution is removed. An old solution in P_1 is replaced by the solution carried by a scout bee, while a solution of P_2 is replaced by the best solution found in P_1, but not yet included in P_2.
- Finally, checking whether the stopping criterion is met. If not, the above mentioned procedure will be repeated; otherwise, the algorithm terminates.

4.4.3.2 Performance of BCiA

In order to see how the BCiA performs, Häckel and Dippold (2009) tested it on a classic vehicle routing problem with time windows. The preliminary experimental results showed that the performance of BCiA is very promising, in particular for the smaller test entities of the benchmarks.

4.4.4 Bee Swarm Optimization

4.4.4.1 Fundamentals of Bee Swarm Optimization Algorithm

Bee swarm optimization (BSO) algorithm was originally proposed by Akbari et al. (2009). Typically, the proposed algorithm includes three types of bees, i.e., experienced forager, onlooker, and scout bees. Each type of bees had a distinct moving pattern which are used by the bees to adjust their flying trajectories. In order to implement BSO, the following procedures need to be followed (Akbari et al. 2009; Niknam and Golestaneh 2013; Sotelo-Figueroa et al. 2010):

- Initializing the population. The initial size of population (β) is determined manually that involves three types of bees via Eq. 4.12 (Akbari et al. 2009):

$$\beta = \xi \cup \kappa \cup \vartheta, \tag{4.12}$$

where ξ represents the sets of experienced forager bees, κ denotes the sets of onlooker bees, and ϑ is the sets of scout bees.

In addition, each bee (i) is associated with a position vector via Eq. 4.13 (Akbari et al. 2009):

$$\vec{x}(\beta, i) = (x(\beta, i_1), x(\beta, i_2), \ldots, x(\beta, i_D)), \tag{4.13}$$

where $\vec{x}(\beta, i)$ denotes a feasible solution in an d-dimensional search space ($S \subset R^D$) that need to be optimized.

- Initializing parameters. A set of parameters will be initialized at this phase, such as the maximum number of iterations ($Iter_{\max}$), the number of the bees ($n(\beta)$), and the initialization function via Eq. 4.14 (Akbari et al. 2009):

$$\vec{x}_0(\beta, i) = Init(i, S), \forall i \in \beta, \tag{4.14}$$

where $Init(i, S)$ represents the initialization function which associates a random position to the ith bee in the search space S.

- Creating the fitness function. The fitness function is given via Eq. 4.15 (Akbari et al. 2009):

$$\vec{x}(\beta, i) = fit(\vec{x}(\beta, i)). \tag{4.15}$$

- Updating phase.

1. The position of an experienced forager bee (ξ) is updated via Eq. 4.16 (Akbari et al. 2009):

$$\begin{aligned}\vec{x}_{new}(\xi, i) = \vec{x}_{old}(\xi, i) + \omega_b r_b\left(\vec{b}(\xi, i) - \vec{x}_{old}(\xi, i)\right) \\ + \omega_e r_e\left(\vec{e}(\xi, \cdot) - \vec{x}_{old}(\xi, i)\right),\end{aligned} \tag{4.16}$$

   where r_b and r_e are random variables of uniform distribution in range of $[0, 1]$, ω_b and ω_e denote the best food source found by the ith bee and the elite bee, respectively, and $\vec{x}_{new}(\xi, i)$ represents the position vector of the new food source found by the experienced forager.
   Overall, the whole equation can be divided into three parts, the first part in the right side represents the position vectors of the old food sources found by the experienced forager, the second parts in the right side represents the cognitive knowledge that attract the experienced forager towards the best position ever found by the bee, and the third parts in the right side represents the social knowledge that attract the experienced forage towards the best position ($\vec{e}(\xi, \cdot)$) which is found by the interesting elite bee.
2. The onlooker bees (κ) use only the social knowledge provided by experienced forager bees to adjust their moving trajectory in the next iteration. Their positions are updated via Eq. 4.17 (Akbari et al. 2009):

$$\vec{x}_{new}(\kappa, i) = \vec{x}_{old}(\kappa, i) + \omega_e r_e\left(\vec{e}(\xi, i) - \vec{x}_{old}(\kappa, i)\right), \tag{4.17}$$

   where $\vec{x}_{new}(\kappa, i)$ represents the position of the new food source which is selected by the onlooker bee (i), $\omega_e r_e$ is the parameter that probabilistically controls the attraction of the onlooker bees towards their interesting food source area, and $\vec{e}(\xi, i)$ is the position vector of the interesting elite bee for onlooker bees (p_{jd}) that is determined through Eq. 4.18 (Akbari et al. 2009):

$$p_j = \frac{fit\left(\vec{x}(\xi, j)\right)}{\sum_{c=1}^{n(\xi)} fit\left(\vec{x}(\xi, c)\right)}, \tag{4.18}$$

where $fit\left(\vec{x}(\xi, i)\right)$ is the fitness value of the food source which is found by the experienced forager bee (j), and $n(\xi)$ is the number of experienced forager bees.

3. The positions of the scout bees (ϑ) are updated via Eq. 4.19 (Akbari et al. 2009):

$$\vec{x}_{new}(\vartheta, i) = \vec{x}_{old}(\vartheta, i) + Rw(\tau, x_{old}(\vartheta, i)), \tag{4.19}$$

where $\vec{x}_{old}(\vartheta, i)$ represents the position of the abandoned food source, and Rw is a random walk function that depends on the current position of the scout bee and the radius search τ. Typically, the initial value of radius $\tau \in (\tau_{\min} < \tau < \tau_{\max})$ is defined as a percentage of $|X_{\max} - X_{\min}|$, where $X_{\max}$ and $X_{\min}$ are the maximum and minimum value of the search space along a dimension.

- Information selecting. It will be determined via Eq. 4.20 (Akbari et al. 2009):

$$\begin{cases} \text{if} \quad fit\left(\vec{x}(\xi, i)\right) > fit\left(\vec{b}(\xi, i)\right) & \text{then} \quad \vec{b}(\xi, i) = \vec{x}(\xi, i) \\ \text{if} \quad fit\left(\vec{b}(\xi, i)\right) > fit\left(\vec{e}(\xi, \cdot)\right) & \text{then} \quad \vec{e}(\xi, \cdot) = \vec{b}(\xi, i) \end{cases}, \tag{4.20}$$

where $\vec{b}(\xi, i)$ denotes the position of the best food source that an experienced forager (i) can remember, and $\vec{e}(\xi, \cdot)$ represents the position of the best food source that the elite bee can find.

4.4.4.2 Performance of BSO

To evaluate the effectiveness of BSO, six analytical benchmark functions were employed in Akbari et al. (2009). Compared with other bee inspired algorithms (such as ABC), computational results showed that the proposed algorithm outperforms the others investigated in this study.

4.4.5 Bee System

4.4.5.1 Fundamentals of Bee System Algorithm

Bee system (BS) algorithm was originally proposed in Sato and Hagiwara (1997). In order to implement BS, the following procedures need to be followed (Lučić 2002; Sato and Hagiwara 1997; Lučić and Teodorović 2003):

- In BS, the first global search is performed by *pop_G*. The purpose of this function is to find as broad as possible to escape from a local optimum. If, for successive G_{SC} generations, one chromosome is found to be the best, it will be regarded as a very good solution point around which there may exist the global best. In BS, this solution is call superior chromosome which will be kept for local search.
- Concentrated crossover: All chromosomes in pop_L_k are made couple with SC_k at the beginning of local search and the crossover mechanism is performed. This concentrated crossover transfers information about the kth superior chromosome to all other chromosomes in the kth population denoted by pop_L_k.
- Migration: In BS, an individual bee is randomly selected per predetermined generation G_{mig} for emigrating to its neighbourhood population. Through this strategy, each population manages to search independently and cooperatively.
- Pseudo-simplex approach: In BS, for a more efficient and effective search, a pseudo-simplex mechanism is employed.

First, picking up three best so far chromosomes and name them C_1, C_2, and C_3 according to their corresponding fitness value.

Then, translating them into three vectors, i.e., $\mathbf{X}_1$, $\mathbf{X}_2$, and $\mathbf{X}_3$. Next, calculating the middle point $\mathbf{X}_0$ of $\mathbf{X}_1$ and $\mathbf{X}_2$ according to Eq. 4.21 (Sato and Hagiwara 1997):

$$\mathbf{X}_0 = \frac{\mathbf{X}_1 + \mathbf{X}_2}{2}. \tag{4.21}$$

Right after this, computing $\mathbf{X}_{ref}$ and $\mathbf{X}_{cont}$, respectively, based on Eqs. 4.22 and 4.23, respectively (Sato and Hagiwara 1997):

$$\mathbf{X}_{ref} = (1+\alpha)\mathbf{X}_0 - \alpha\mathbf{X}_3, \tag{4.22}$$

$$\mathbf{X}_{cont} = (1-\beta)\mathbf{X}_0 + \beta\mathbf{X}_3, \tag{4.23}$$

where the reflection ration and contraction ration are denoted by α and β, respectively.

Next, exchanging $\mathbf{X}_{ref}$ and $\mathbf{X}_{cont}$ into chromosomes and set them as C_{ref} and C_{cont}, respectively.

Finally introducing C_1, C_{ref}, and C_{cont} to the initial population where the crossover and mutation have already applied.

- Turning back to global search: The local search will be terminated once the predetermined number of generations is reached. If the best solution found so far does not meet the stopping criterion, the algorithm will repeat.

4.4.5.2 Performance of BS

In order to see how the BS algorithm performs, Sato and Hagiwara (1997) tested it on nine benchmark test functions selected from the literature. Compared with other CI approaches (e.g., GA), the preliminary experimental results showed that BS outperforms GA in all cases which make it a very promising optimization algorithms in dealing with highly complex multivariate functions.

4.4.6 BeeHive

4.4.6.1 Fundamentals of BeeHive Algorithm

BeeHive algorithm was originally proposed in Wedde et al. (2004) and Farooq (2006) which is a novel network routing algorithm. It is inspired from the dance language and foraging behaviour of honeybees. The main characteristics of BeeHive algorithm can be concluded as follows: (Farooq 2009; Wedde and Farooq 2006; Farooq and Caro 2008):

- Step 1: During the start up stage, all nodes in the networks begin with a process named foraging region formation. The first generation of short distance bee agents are launched at this stage for propagating their nomination in their neighbourhood.
- Step 2: By comparing the information received from a short distance bee agent, a node will determined whether to resign as a representative node and join the foraging region.
- Step 3: Once the former representative node quits, the other nodes will activate an election mechanism
- Step 4: The nodes continues to announce the next generations of short distance bee agents by following pre-described steps until the network is split into disjoint foraging regions, and overlapping with the foraging zones.
- Step 5: After the Step 4 is executed, the BeeHive algorithm gets into a normal phase.
- Step 6: When a replica of a specific bee agent reaches a site, it will update the local stored routing information, and then, except of being sent back to the node where the replica comes from, it will be continuously flooded.
- Step 7: Representative nodes only generate long distance bee agents that could be received by the neighbours.
- Step 8: In BeeHive algorithm, bee agents employ priority queues mechanism for the purpose of a quick routing information dissemination.
- Step 9: Each node carries the current routing information not only for reaching nodes within its foraging zone, but also for reaching the representative nodes of foraging regions.
- Step 10: Choosing the next hop for a data packet in a stochastic manner.

- Step 11: The goodness of a neighbour j of node i for arriving at a destination d, denoted by g_{jd}, is defined via Eq. 4.24 (Wedde and Farooq 2006):

$$g_{jd} = \frac{\frac{1}{p_{jd}+q_{jd}}}{\sum_{k=1}^{N}\left(\frac{1}{p_{kd}+q_{kd}}\right)}, \tag{4.24}$$

where the propagation and queuing delays are denoted by p_{jd} and q_{jd}, respectively.
- Step 12: In BeeHive algorithm, three kinds of routing tables, namely, intra foraging zone, inter foraging region, and foraging region membership are allocated to each node i.

4.4.6.2 Performance of BeeHive Algorithm

In order to evaluate the BeeHive algorithm, the Japanese Internet backbone scenario was employed in Wedde et al. (2004). Through an extensive comparison with two other state of the art routing algorithms, BeeHive showed a quite attractive overall performance. Interested readers are referred to Farooq (2009) for a more detailed explanation about the working principles and applications of the BeeHive algorithm.

4.4.7 Bees Algorithm

4.4.7.1 Fundamentals of Bees Algorithm

Bees algorithm (BeA) was originally proposed in Pham et al. (2006). The basic working procedures of BeA are listed as follows (Pham et al. 2006; Karaboga and Akay 2009a; Pham and Castellani 2009; El-Abd 2012b):

- Step 1: Initializing population with random solutions. At this stage, the BA requires several parameters to be set such as number of scout bees (n), number of sites selected out of n visited sites (m), number of the best sites out of m selected sites (e), number of bees recruited for the best e sites (nep), number of bees recruited for the other ($m-e$) selected sites (nsp), and initial size of patches (ngh) which includes the site, its neighbourhood, and the stopping criterion.
- Step 2: Assessing the fitness of the population.
- Step 3: Forming new population while stopping criterion is not met.
- Step 4: Selecting sites for neighborhood search. The bees with the highest fitness values are chosen as "selected bees" at Step 4, and accordingly, the sites visited by them are chosen for neighborhood search.
- Step 5: Recruiting bees for selected sites (more bees for best s sites), and evaluating the fitness value. In Step 5, the BeA conducts search in the

neighborhood of the selected sites. More bees will be assigned to search around the best *e* sites. The bees can either be selected directed based on the value of fitness associated with the sites they are visiting, or the fitness values will be used to determine the probability of the bees being selected. Exploration of the surroundings of the best *e* sites represents more suitable solutions can be made available through recruiting more bees to follow them than the other selected bees. This differential recruitment mechanism, along with scouting strategy, is the core operation of the BeA algorithm.
- Step 6: Choosing the fittest bee from each patch. Pham et al. (2006) introduced a constraint at this stage that is for each patch, only the bees with the highest fitness value can be selected to from the next bee population. The purpose of introducing such restriction is to reduce the number of points that are going to be explored.
- Step 7: Assigning the remaining bees to do random search, and evaluating their fitness. In the bee population, the remaining bees are assigned randomly around the search space looking for new possible solution candidates.
- Step 8: Terminating the loop when stopping criterion is met. All aforementioned steps will be executed repeatedly until a stooping criterion is met. At the end of each iteration, the population of bee colony consists of two parts: representatives from each selected patch, and other scout bees performing random searches.

In addition to the standard BeA detailed in this section, several enhanced versions of BeA can also be found in the literature as outlined below:

- Binary BeA (Xu et al. 2010b).
- Distributed BeA (Jevtić et al. 2012).
- Hybrid BeA (Shafia et al. 2011; Lien and Cheng 2012).
- Multi-objective BeA (Pham and Ghanbarzadeh 2007).
- Neighbourhood enhanced BeA (Ahmad 2012).

4.4.7.2 Performance of BeA

In order to see how BeA performs, Pham et al. (2006) first employed two standard benchmark functions, namely, Shekel's Foxholes function and inverted Schwefel's function with six dimensions, for testing purpose. Furthermore, eight other benchmark functions selected from the literature were introduced for validating BeA. Compared with other CI methods (e.g., GA and ACO), the overall performance of BeA is quite competitive.

Recently, Xu et al. (2010b) utilized BeA to study a group of reconfigurable mobile robots which are designed to provide daily service in hospital environments for different kinds of tasks such as guidance, cleaning, delivery, and monitoring. The fulfilment of each job requires an associated functional module that can be installed onto various robot platforms via a standard connection interface.

Since the classic BeA focuses mainly on single-objective functional optimization problems, a variant called binary BeA was proposed in Xu et al. (2010b) to deal with the multi-objective multi-constraint combinatorial optimization task. In binary BeA, a bee is describe as two binary matrixes **MR** and **RH**, standing for how to assign the M tasks to the R robots and the R robots to the H homes, respectively. The size of **MR** is $M \times R$ in which its R columns represent the R robots, while the M missions is represented by the M rows. Xu et al. (2010b) evaluated the proposed algorithm with an example problem (20 missions, 8 robots, and 4 homes) with a size of $8^{20} \times 4''^{8} = 2^{76}$ combinations. At first 12 stochastic solutions are obtained by scout bees through global search in which six elite bees survive after the non-dominated selection. The final experiments demonstrated that the proposed algorithm is a suitable candidate tool in treating workload balancing issue among a team of swarm robots.

Apart from the case study described above, the BeA has also been successfully applied to a variety of optimization problems as listed below:

- Controller design optimization (Jones and Bouffet 2008).
- Construction site layout optimization (Lien and Cheng 2012).
- Data clustering (Shafia et al. 2011).
- File search optimization (Dhurandher et al. 2011).
- Filter design optimization (Pham and Koç 2010).
- Manufacturing system optimization (Pham et al. 2007a, Ramírez et al. 2010).
- Mechanical design optimization (Pham et al. 2009).
- Robot control optimization (Xu et al. 2010b; Jevtić et al. 2012).
- Scheduling optimization (Pham et al. 2007b).

4.4.8 Bees Life Algorithm

4.4.8.1 Fundamentals of Bees Life Algorithm

Bees life algorithm (BLA) was recently proposed by Bitam and Mellouk (2013) to solve a vehicular ad hoc network (VANET) problem, in particular, the quality of service multicast routing problem (QoS-MRP). Two bees' behaviours are employed in the proposed algorithm, i.e., reproduction and foraging behaviours. In addition, the former incorporated the crossover and mutation operators, while the latter used a neighbourhood search approach. In order to implement BLA, the following procedures need to be followed (Bitam and Mellouk 2013):

- Initializing bees' population.
- The fitness function value corresponding to each candidate solution is calculated.
- Reproduction process. In this process, based on the crossover and mutation operators, two new individuals (i.e., queen bee and drone bee) are selected. The queen starts breeding broods, then those broods will be evaluated according to

the fitness function, after that, the best fittest brood will be considered as the new queen for next iteration. Also, the drone bees and worker bees are updated.
- Foraging process. In this process, the worker bees are looking for food sources. Each worker will be mimicked as one region. During the recruitment process, the best food source will be founded.
- Ranking and selecting. Only the highest fitness will be selected to form the next bee population.
- Evaluating the fitness of population (i.e., queen, drone, and worker bees).
- Termination.

4.4.8.2 Performance of BLA

To evaluate the effectiveness of BLA, a series of tests based on VANET simulation scenario are conducted in Bitam and Mellouk (2013). Compared with other CI algorithms (such as GA, BeA, and HBMO), computational results showed that the proposed algorithm outperforms the others in terms of the solution quality and complexity.

4.4.9 Bumblebees Algorithm

4.4.9.1 Fundamentals of Bumblebees Algorithm

Bumblebees algorithm was originally proposed in Comellas and Martínez-Navarro (2009). It is a more simple and efficient version than the one introduced in Comellas and Gallegos (2005) as angels and mortals. In bumblebees algorithm, the bumblebees are employed to play the role of the mortals, while the static food cells are used to replace the angels. The food cells together with the fixed position of nest creates a simulating environment which can influence the bumblebees' behaviour and in turn helps to move the optimization process forward. In bumblebees algorithm, the following operators are defined (Comellas and Martínez-Navarro 2009):

- *The world*: In the proposed algorithm, the habitat of the colony of bumblebees is an artificial world consisting of a toroidal square grid with $n \times n$ cells. There are four possible states assigned to each individual cell, namely, empty, with food, with a bumblebee, or containing a nest.
- *Movement*: Initially, all bumblebees are located in the next. At each generation, each single bumblebee will fly out of the nest one by one and randomly move to any other positions around its current position.
- *Bumblebees' birth*: During the algorithm initialization, all bumblebees are positioned inside the nest and each bumblebee is associated with a randomly generated solution. The record of some of the best solutions found so far is kept

by the queen who will pass one of such solution to a newly born bumblebee every few generation.

- *Mutation*: Similarly to GA, the mutation mechanism is also introduced in bumblebees algorithm with a slightly modification of applying mutation on all individual bumblebees at each generation.
- *The reaper*: In order to avoid the over- or de-population, a reaper mechanism is employed in bumblebees algorithm, i.e., one unit of life is subtracted from a bumblebee's life at each generation. When the age of a bumblebee arrives at 0, this individual will be removed from the current artificial world.

4.4.9.2 Performance of Bumblebees Algorithm

In order to test the performance of the bumblebees algorithm, Comellas and Martínez-Navarro (2009) tested it on the classic graph colouring problem. In comparison with other CI methods (e.g., GA), the bumblebees algorithm offered a better solution quality.

4.4.10 Honeybee Social Foraging Algorithm

4.4.10.1 Fundamentals of Honeybee Social Foraging Algorithm

Honeybee social foraging (HBSF) algorithm was recently proposed in Quijano and Passino (2010). In order to implement the HBSF algorithm, the following procedures need to be followed (Quijano and Passino 2010; Nakrani and Tovey 2003; Nakrani and Tovey 2004; Scholz-Reiter et al. 2008):

- Foraging profitability landscape: For $i = 1, 2, \ldots, B$, bee i is denoted by $\theta^i \in \Re^2$, a position expression within 2-dimensional space. The foraging profitability landscape is represented by $J_f(\theta)$ which has a value falling within the range of $[0, 1]$. It is proportional to the profitability of nectar at a location indicated by $\theta \in \Re^2$.
- Bee roles and expeditions: Setting $x_j(k)$ as the number of bees at site j at k. A normal option for expressing this is via Eq. 4.25 (Quijano and Passino 2010):

$$s_j(k) = \frac{a_j}{x_j(k)}, \tag{4.25}$$

where the amount of nutrients per second at the jth site is denoted by a_j. Suppose that there are $B_f(k)$ deployed forager bees. Initially letting $B_f(0) = 0$ for the reason of no foraging sites are being discovered. Then the foraging profitability assessment used by an individual deployed forager bee can be expressed via Eq. 4.26 (Quijano and Passino 2010):

$$F^i(k) = \begin{cases} 1 & \text{if } J_f(\theta^i(k)) + w_f^i(k) \geq 1 \\ J_f(\theta^i(k)) + w_f^i(k) & \text{if } 1 > J_f(\theta^i(k)) + w_f^i(k) > \varepsilon_n \, , \\ 0 & \text{if } J_f(\theta^i(k)) + w_f^i(k) \leq \varepsilon_n \end{cases} \tag{4.26}$$

where $w_f^i(k)$ stands for the profitability assessment noise.

- Dance strength determination: In HBSF, the dance strength is denoted by $L_f^i(k)$, i.e., the number of waggle runs of bee i at step k. Setting $F_q^i(k)$ as the amount of nectar gathered under the profitability assessment $F^i(k)$. Then the total quantity of nectar influx to the hive at step k, denoted by $F_{tq}(k)$ can be defined via Eq. 4.27 (Quijano and Passino 2010):

$$F_{tq}(k) = \sum_{i=1}^{B} F_q^i(k) = \alpha \sum_{i=1}^{B} F^i(k) = \alpha F_t(k). \tag{4.27}$$

- Explorer allocation and forager recruitment: In HBSF, the explorer allocation process is designed to simultaneously happen with the recruitment of observer bees to forage sites. Then the probability that the dance of bee i will be followed by an observer bee is defined via Eq. 4.28 (Quijano and Passino 2010):

$$p_i(k) = \frac{L_f^i(k)}{\sum_{i=1}^{B_f(k)} L_f^i(k)}. \tag{4.28}$$

4.4.10.2 Performance of HBSF

To evaluate the effectiveness of the HBSF algorithm, an engineering application, namely, dynamic resource allocation for multizone temperature control problem was adopted by Quijano and Passino (2010). Computational results showed that the proposed algorithm is able to achieve an ideal free distribution situation which could maximize the uniform temperature allocation.

4.4.11 OptBees

4.4.11.1 Fundamentals of OptBees Algorithm

OptBees algorithm was originally proposed in Maia et al. (2012) that is based on the processes of collective decision-making by bee colonies to solve multimodal continuous optimization problem. In order to implement the OptBees algorithm, the following procedures need to be followed (Maia et al. 2012):

- Determination of the recruiter bees: Typically, the probability of being a recruiter bee is associated with each bee in the swarm is given via Eq. 4.29 (Maia et al. 2012):

$$p_i = \left(\frac{p_{\max} - p_{\min}}{Q_{\max} - Q_{\min}} \right) \cdot (Q_i - Q_{\min}) + p_{\min}, \tag{4.29}$$

where $p_{\min}$ and $p_{\max}$ are the minimum and maximum probabilities of a bee be a recruiter, respectively, Q_i represents the quality of the site explored by bee i, and $Q_{\min}$ and $Q_{\max}$ represent the minimum and maximum site qualities at the current iteration. After these procedures, the number non-recruiter bees in the swarm is defined via Eq. 4.30 (Maia et al. 2012):

$$M = N - r, \tag{4.30}$$

where N is the total number of bees, r denotes the number of recruiter bees, and M represents the number of non-recruiter bees.

- Determination of the recruited and scout bees: The number of recruited bees (n) is calculated by Eq. 4.31 (Maia et al. 2012):

$$n = [p_{rec} \cdot M], \tag{4.31}$$

where p_{rec} is the percentage of non-recruiter bees that will be actually recruited, M is the number of non-recruiter bees, and $[\cdot]$ denotes the nearest integer function. So, the number of scout bees (S) is given by Eq. 4.32 (Maia et al. 2012):

$$S = M - n, \tag{4.32}$$

- Recruitment process: Due to two-dimensional search spaces, two equal probabilities' level are employed in this process via Eq. 4.33 (Maia et al. 2012):

$$\begin{cases} x_i = x_i + \alpha \cdot \mathbf{U} \otimes (y - x_i) \\ x_i = x_i + u \cdot \alpha \cdot (y - x_i) \end{cases}, \tag{4.33}$$

where α is the recruitment rate, is the recruited bee, x_i is the recruiter bee, y is a random number with uniform distribution in the interval $[0, 1]$, $\mathbf{U}$ is a vector whose elements are random numbers with uniform distribution in the interval $[0, 1]$, and $\otimes$ denotes the element-wise product.

- Exploration process: In this process, the scout bees are moved to a random point (i.e., a new region in the search space).

4.4.11.2 Performance of OptBees

To evaluate the effectiveness of the OptBees algorithm, five minimization problems were adopted by Maia et al. (2012). Computational results showed that the proposed algorithm is capable of generating and maintaining the diversity and consequently obtaining multiple local optima solutions without losing the ability of global optimization.

4.4.12 *Simulated Bee Colony Algorithm*

4.4.12.1 Fundamentals of Simulated Bee Colony Algorithm

Simulated bee colony (SBC) algorithm was originally proposed in McCaffrey and Dierking (2009) to extract rule sets from clustered categorical data. In SBC, each bee is viewed as an object with a memory matrix and it is modelled as an array. In order to implement the SBC algorithm, the following procedures need to be followed (McCaffrey and Dierking 2009):

- Step 1: Initialization.
- Step 2: Repeating. First, placing the active bees on the food sources in the memory. Then, putting the inactive bees on the food source in the memory. Next, sending the scout bees to the search area for discovering new food sources. Finally, memorizing the best food source found so far.
- Step 3: Termination.

4.4.12.2 Performance of SBC

To evaluate the effectiveness of SBC, the proposed algorithm was tested with six benchmark data sets in McCaffrey and Dierking (2009). Computational results showed that SBC can successfully discover the underlying rule set for all six test data sets.

4.4.13 *Virtual Bees Algorithm*

4.4.13.1 Fundamentals of Virtual Bees Algorithm

Virtual bees algorithm (VBA) was originally proposed by Yang (2005) and recently is used in Khan et al. (2010) for solving controller design problem. For VBA, the population of bees is associated with a memory bank, a food source. Also, all the memories communicate between bees with a waggle dance procedure. In order to implement VBA, the following procedures need to be followed (Khan et al. 2010; Yang 2005):

- Initializing the population.
- The fitness function value corresponding to each candidate solution is calculated.

- Defining a criterion for communicating the direction and distance.
- Updating the new position of each individual. The new position is updated via Eq. 4.34 (Khan et al. 2010; Yang 2005):

$$\begin{cases} x_k^{i+1} = x_k^i \cdot (1+\beta) + x_{best} \cdot \beta + \alpha \cdot (rand(i) - 0.5) \\ y_k^{i+1} = y_k^i \cdot (1+\beta) + y_{best} \cdot \beta + \alpha \cdot (rand(i) - 0.5) \end{cases}, \tag{4.34}$$

  where α and β are two positive constants called randomness amplitude and speed of convergence, respectively, x_{best} and y_{best} are best parameters in the ith iteration, and $rand(i)$ is a random number in the interval $[0, 1]$.
- Ranking the candidate solutions.
- Checking termination criterion.

4.4.13.2 Performance of VBA

To evaluate the effectiveness of VBA, De Jong's test function and keane's multi-peaked bumpy function were adopted in Yang (2005). Compared with other CI algorithms (such as GA), computational results showed that the proposed algorithm is capable of solving multilevel optimization problems in which many local minimums are involved.

4.4.14 Wasp Swarm Optimization

4.4.14.1 Fundamentals of Wasp Swarm Optimization Algorithm

Wasp swarm optimization (WSO) algorithm was originally proposed in Theraulaz et al. (1991) that is based on some behaviours found in wasp colony (Karsai and Wenzel 2000; Lucchetta et al. 2008).

The basic idea of WSO was to mimic a wasp colony behaviour, in particular according to the importance of individual wasp to the whole colony, assigning the resources to different wasp (Fan and Zhong 2012; Theraulaz et al. 1991). Therefore in WSO algorithm, resources will be allocated to individual candidate solutions and such allocation is completed in a randomly manner where the strength of each option controls its chosen probability. In Cicirello and Smith (2004), a tournament process was utilized to implement this stochastic selection process: the weakest option (for example a) challenges the second weakest option (for instance b) and the winning probability of a over b is determined through $p_{ab} = s_a^2 / \left(s_a^2 + s_b^2\right)$. The winner of this challenge (say a) will carry on to challenges the third weakest option (denoted by c), and wins with a probability of $p_{ac} = s_a^2 / \left(s_a^2 + s_c^2\right)$. The challenge will continue until the final winner is selected. In some situations, it is more convenient to allocate costs instead of strengths to the individual wasps, i.e., the lower the cost, the higher the strength of a wasp. In this case, the winning probability of wasp i over j can be defined via Eq. 4.35 (Cicirello and Smith 2004):

$$p_{ij} = \frac{s_j^2}{s_i^2 + s_j^2},\ i, j = 1, \ldots, c. \tag{4.35}$$

4.4.14.2 Performance of WSO

In Song et al. (2005), the authors utilized WSO to find a trade-off between the distribution cost and the service level in the context of dynamic vehicle routing problem with time windows (DVRPTW for short). In real-world environment, a real-time request impacts more on one vehicle route while less on the other vehicle routes. In order to take the geographical location information and time window information of the customers into account, a wasp-like agent strategy is employed in their study to determine when to re-optimize the vehicle routes. The vehicle considered in Song et al. (2005) is associated with a wasp named vehicle wasp which is used to control the status of the vehicle and it has a set of response thresholds as stated in Eq. 4.36 (Song et al. 2005):

$$\theta_v = \{\theta_{v,1}, \theta_{v,2}, \ldots, \theta_{v,n}\}, \tag{4.36}$$

where $\theta_{v,i}$ denotes the response threshold value of the vehicle wasp to a new demand i.

Based on the simulation data for static VRPTW found in the literature, the authors re-constructed them for DVRPTW through producing request randomly. The experimental results showed that when the dynamic property is less than 80 %, the WSO algorithm can generates a better solution. Nevertheless, when the dynamic property is greater than 80 %, the WSO algorithm failed to provide a suitable solution. At the end of the study, the authors claimed that WSO algorithm could fit the practical distribution environment well as long as the route that needs to be re-optimized does not deviate far from the original planned route.

Apart from the case study described above, the WSO algorithm has also been successfully applied to a variety of optimization problems as listed below:

- Data clustering (Runkler 2008).
- Decentralized control optimization (Theraulaz and Bonabeau 1995; Baker 1998; Karsai 1999; Anderson and Bartholdi 2000; Cicirello and Smith 2004; Wang 2009).
- Image processing (Fan and Zhong 2012).
- Logistics system optimization (Pinto et al. 2005).
- Scheduling optimization (Wang et al. 2006).
- Vehicle routing problem (Song et al. 2005).
- Maximum satisfiability problem (Pinto et al. 2006; Cao et al. 2009; Anonymous 2010).

4.5 Conclusions

In this chapter, we introduced a set of CI algorithms which are based on the behaviour of the honeybees. These algorithms are mainly divided into two categories, i.e., the foraging behaviour and the mating behaviour. Although they are newly introduced CI methods, we have witnessed the following rapid spreading of at least two of them, i.e., ABC and HBMO.

First, several enhanced versions of ABC can be found in the literature as outlined below:

- ABC* (Forghany et al. 2012).
- Cauchy distribution-based ABC (Rajasekhar et al. 2011).
- Chaotic ABC (Xu et al. 2010a; Alatas 2010).
- Cooperative ABC (El-Abd 2010).
- DisABC (Kashan et al. 2012).
- Discrete ABC (Tasgetiren et al. 2011; Tasgetiren et al. 2010; Karabulut and Tasgetiren 2012; Koc et al. 2012; Tasgetiren et al. 2013).
- Elitist ABC (Mezura-Montes and Velez-Koeppel 2012).
- Enhanced ABC (Tsai et al. 2009).
- Global best ABC (Gao et al. 2012; Li et al. 2012; Jadhav and Roy 2013).
- Hybrid ABC (Tien and Li 2012; Shi et al. 2010; Vitorino et al. 2012; Kang et al. 2013; Zhang et al. 2013).
- Improved ABC (Karaboğa and Çetinkaya 2011; Gao et al. 2011; Gao and Liu 2011; Cheng and Jiang 2012).
- Mean mutation operator based ABC (Sharma et al. 2012a).
- Memetic ABC (Fister et al. 2012).
- Micro ABC (Rajasekhar et al. 2012).
- Modified ABC (Anandhakumar et al. 2011; Mezura-Montes et al. 2010; Gao and Liu 2012; Sharma et al. 2012a; Gao et al. 2013).
- Multi-hive ABC (Zhang et al. 2012).
- Multiobjective ABC (Akbari et al. 2012; Chaves-González et al. 2013).
- Mutable smart ABC (Gorji-Bandpy and Mozaffari 2012).
- Opposition-based ABC (El-Abd 2012a).
- Parallel ABC (Narasimhan 2009; Subotic et al. 2010).
- Penalty guided ABC (Hsieh and Yeh 2012).
- Rosenbrock ABC (Kang et al. 2011).
- Vector evaluated ABC (Omkar et al. 2011).
- Chaotic improved HBMO (Niknam et al. 2011).
- Hybrid HBMO (Niknam et al. 2008; Niknam 2009).
- Maximum entropy based HBMO (Horng 2010).
- Modified HBMO (Olamaei et al. 2012; Niknam et al. 2012).
- Multiobjective HBMO (Niknam 2011).

Second, the ABC algorithm has also been successfully applied to a variety of optimization problems as listed below:

- Artificial neural network training (Dhahri et al. 2012).
- Automated software refactoring (Koc et al. 2012).
- Circuit design optimization (Zhang and Ye 2012).
- Composite structures optimization (Omkar et al. 2011).
- Controller design optimization (Rajasekhar et al. 2012).
- Disassembly line balancing (Kalayci and Gupta 2013).
- Expert system design (Babu et al. 2011).
- Filter design optimization (Karaboğa and Çetinkaya 2011).
- Fuel cell research (Zhang et al. 2013).
- Gene research (Forghany et al. 2012; Chaves-González et al. 2013; lez-Álvarez et al. 2013).
- Image processing (Ma et al. 2011; Akay 2013; Hancer et al. 2012; Akay and Kirmizi 2012).
- Manufacturing optimization (Samanta and Chakraborty 2011; Yildiz 2013; Ajorlou and Shams 2012).
- Portfolio optimization (Chen et al. 2012).
- Power system optimization (Hemamalini and Simon 2010; Hong 2011; Subramanian et al. 2011; Anandhakumar et al. 2011; Ayan and Kılıç 2012; Bommirani and Thenmalar 2013; Jadhav and Roy 2013).
- Redundancy allocation problem (Hsieh and Yeh 2012).
- Robot control (Xu et al. 2010a).
- Scheduling optimization (Tasgetiren et al. 2011; Sundar and Singh 2012; Tasgetiren et al. 2010; Tasgetiren et al. 2013).
- Supply chain optimization (Kumar et al. 2010).
- Thermal engine optimization (Gorji-Bandpy and Mozaffari 2012).
- Travelling salesman problem (Karabulut and Tasgetiren 2012).
- Unstable periodic orbits detection (Gao et al. 2012).

Third, the relative HBMO algorithm applications can be found below:

- Image processing (Horng 2010).
- Power system optimization (Niknam et al. 2008; Niknam 2009; Niknam et al. 2011; Niknam 2011; Olamaei et al. 2012; Niknam et al. 2012).
- Travelling salesman problem (Marinakis et al. 2011).
- Vehicle routing problem (Marinakis et al. 2008; Marinakis et al. 2010).
- Water resource management (Haddad et al. 2006; Afshar et al. 2007; Haddad et al. 2009; Haddad et al. 2010).

Interested readers are referred to them together with several excellent reviews [e.g., (Karaboga and Akay 2009c; Karaboga et al. 2012; Teodorović 2009a; Teodorović et al. 2011; Goyal 2012)] as a starting point for a further exploration and exploitation of the honeybees inspired algorithms.

References

Abbass, H. A. (2001a, May 27–30). MBO: Marriage in honey bees optimization. A haplometrosis polygynous swarming approach. In *IEEE Proceedings of the Congress on Evolutionary Computation* (pp. 207–214). Seoul, South Korea.

Abbass, H. A. (2001b). A monogenous MBO approach to satisfiability. In *Proceeding of the International Conference on Computational Intelligence for Modelling, Control and Automation (CIMCA)*. Las Vegas, NV, USA.

Afshar, A., Haddad, O. B., Mariño, M. A., & Adams, B. J. (2007). Honey-bee mating optimization (HBMO) algorithm for optimal reservoir operation. *Journal of the Franklin Institute, 344*, 452–462.

Ahmad, S. A. (2012). *A study of search neighbourhood in the bees algorithm.* Unpublished doctoral thesis, Cardiff University.

Ajorlou, S., & Shams, I. (2013). Artificial bee colony algorithm for CONWIP production control system in a multi-product multi-machine manufacturing environment. *Journal of Intelligent Manufacturing, 24*, 1145–1156. doi:10.1007/s10845-012-0646-5.

Akay, B., (2013). A study on particle swarm optimization and artificial bee colony algorithms for multilevel thresholding. *Applied Soft Computing, 13*, 3066–3091. (http://dx.doi.org/10.1016/j.asoc.2012.03.072).

Akay, B., & Kirmizi, I. (2012, June 10–15). Structural optimization of wavelet packets using swarm algorithms. In *IEEE World Congress on Computational Intelligence (WCCI)* (pp. 1–5). Brisbane, Australia.

Akbari, R., Mohammadi, A., & Ziarati, K. (2009). A powerful bee swarm optimization algorithm. In *IEEE 13th International Multitopic Conference (INMIC)* (pp. 1–6).

Akbari, R., Hedayatzadeh, R., Ziarati, K., & Hassanizadeh, B. (2012). A multi-objective artificial bee colony algorithm. *Swarm and Evolutionary Computation, 2*, 39–52.

Alatas, B. (2010). Chaotic bee colony algorithms for global numerical optimization. *Expert Systems with Applications, 37*, 5682–5687.

Anandhakumar, R., Subramanian, S., & Ganesan, S. (2011). Modified ABC algorithm for generator maintenance scheduling. *International Journal of Computer and Electrical Engineering, 3*, 812–819.

Anderson, C., & Bartholdi, J. J. (2000, September 19–20). Centralized versus decentralized control in manufacturing: Lessons from social insects. In *International Conference on Complexity and Complex Systems in Industry* (pp. 92–105). UK: University of Warwick.

Anonymous. (2010). WAIST: Wasp inspired scheduling for real-time strategy games. *Especial Videojogos, 11*, 1–7.

Ayan, K., & KıLıÇ, U. (2012). Artificial bee colony algorithm solution for optimal reactive power flow. *Applied Soft Computing, 12*, 1477–1482.

Babu, M. S. P., Ramjee, M., Narayana, S. S. V. N. L., & Murty, S. N. V. R. (2011). Sheep and goat expert system using artificial bee colony (ABC) algorithm and particle swarm optimization (PSO) algorithm. In *IEEE 2nd International Conference on Software Engineering and Service Science (ICSESS)* (pp. 51–54).

Baker, A. D. (1998). A survey of factory control algorithms that can be implemented in a multi-agent heterarchy: Dispatching, scheduling, and pull. *Journal of Manufacturing Systems, 17*, 297–320.

Bitam, S., & Mellouk, A. (2013). Bee life-based multi constraints multicast routing optimization for vehicular ad hoc networks. *Journal of Network and Computer Applications, 36*, 981–991.

Bitam, S., Batouche, M., & Talbi, E.-G. (2010, April 19–23). A survey on bee colony algorithms. In *IEEE International Symposium on Parallel and Distributed Processing, Workshops and Phd Forum (IPDPSW)* (pp. 1–8). Atlanta, GA.

Bommirani, B., & Thenmalar, K. (2013). Optimization technique for the economic dispatch in power system operation. *International Journal of Computer and Information Technology, 2*, 158–163.

Bonyadi, M. R., Azghadi, M. R., & Shah-Hosseini, H. (2008). Population-based optimization algorithms for solving the travelling salesman problem. In F. Greco (Ed.), *Travelling salesman problem, Chapter 1* (pp. 1–34). Vienna: In-Tech.

Boussaïd, I., Lepagnot, J., & Siarry, P. (2013). A survey on optimization metaheuristics. *Information Sciences, 237*, 82–117.

Cao, Y., Yang, Y., & Wang, H. (2009). Intelligent job shop scheduling based on MAS and integrated routing wasp algorithm and scheduling wasp algorithm. *Journal of Software, 4*, 487–494.

Chaves-González, J. M., Vega-Rodríguez, M. A., & Granado-Criado, J. M. (2013). A multiobjective swarm intelligence approach based on artificial bee colony for reliable DNA sequence design. *Engineering Applications of Artificial Intelligence, 26*, 2045–2057. (http://dx.doi.org/10.1016/j.engappai.2013.04.011).

Chen, A. H. L., Liang, Y.-C., & Liu, C.-C. (2012, June 10–15). An artificial bee colony algorithm for the cardinality-constrained portfolio optimization problems. In *IEEE World Congress on Computational Intelligence (WCCI)* (pp. 1–8). Brisbane, Australia.

Cheng, X., & Jiang, M. (2012). An improved artificial bee colony algorithm based on Gussian mutation and chaos disturbance. In Y. Tan, Y. Shi, & Z. Ji (Eds.), *ICSI 2012, Part I* (Vol. 7331, pp. 326–333)., LNCS Berlin, Heidelberg: Springer.

Chong, C. S., Low, M. Y. H., Sivakumar, A. I., & Gay, K. L. (2006). A bee colony optimization algorithm to job shop scheduling. In *IEEE Proceedings of 2006 Winter Simulation Conference* (pp. 1954–1961).

Cicirello, V. A., & Smith, S. F. (2004). Wasp-like agents for distributed factory coordination. *Autonomous Agents and Multi-Agent Systems, 8*, 237–266.

Comellas, F., & Gallegos, R. (2005). Angels & mortals: A new combinatorial optimization algorithm. *Study on Fuzziness Soft Computing, 166*, 397–405.

Comellas, F., & Martínez-Navarro, J. (2009). Bumblebees: A multiagent combinatorial optimization algorithm inspired by social insect behaviour. In *Proceedings of First ACM/SIGEVO Summit on Genetic and Evolutionary Computation (GEC)* (pp. 811–814).

Davidović, T., Ramljak, D., Šelmić, M., & Teodorović, D. (2011). Bee colony optimization for the *p*-center problem. *Computers and Operations Research, 38*, 1367–1376.

Dhahri, H., Alimi, A. M., & Abraham, A. (2012, June 10–15). Designing beta basis function neural network for optimization using artificial bee colony (ABC). In *IEEE World Congress on Computational Intelligence (WCCI)* (pp. 1–7). Brisbane, Australia.

Dhurandher, S. K., Misra, S., Pruthi, P., Singhal, S., Aggarwal, S., & Woungang, I. (2011). Using bee algorithm for peer-to-peer file searching in mobile ad hoc networks. *Journal of Network and Computer Applications, 34*, 1498–1508.

El-Abd, M. (2010, July 18–23). A cooperative approach to the artificial bee colony algorithm. In *IEEE World Congress on Computational Intelligence (WCCI)* (pp. 124–128). Barcelona, Spain: CCIB.

El-Abd, M. (2012a, June 10–15). Generalized opposition-based artificial bee colony algorithm. In *IEEE World Congress on Computational Intelligence (WCCI)* (pp. 1–4). Brisbane, Australia.

El-Abd, M. (2012b). Performance assessment of foraging algorithms vs. evolutionary algorithms. *Information Sciences, 182*, 243–263.

Fan, H., & Zhong, Y. (2012). A rough set approach to feature selection based on wasp swarm optimization. *Journal of Computational Information Systems, 8*, 1037–1045.

Farooq, M. (2006). *From the wisdom of the hive to intelligent routing in telecommunication networks: A step towards intelligent network management through natural engineering*. Unpublished doctoral thesis, Universität Dortmund.

Farooq, M. (2009). *Bee-inspired protocol engineering: From nature to networks*. Berlin, Heidelberg: Springer. ISBN 978-3-540-85953-6.

Farooq, M., & Caro, G. A. D. (2008). Routing protocols for next-generation networks inspired by collective behaviors of insect societies: An overview. In C. Blum & D. Merkle (Eds.), *Swarm intelligence: Introduction and application* (pp. 101–160). Berlin, Heidelberg: Springer.

Figueira, J. R., & Talbi, E.-G. (2013). Emergent nature inspired algorithms for multi-objective optimization. *Computers and Operations Research, 40*, 1521–1523.

Fister, I., Fister, I. J., Brest, J., & Žumer, V. (2012, June 10–15). Memetic artificial bee colony algorithm for large-scale global optimization. In *IEEE World Congress on Computational Intelligence (WCCI)* (pp. 1–8). Brisbane, Australia.

Forghany, Z., Davarynejad, M., & Snaar-Jagalska, B. E. (2012, June 10–15). Gene regulatory network model identification using artificial bee colony and swarm intelligence. In *IEEE World Congress on Computational Intelligence (WCCI)* (pp. 1–6). Brisbane, Australia.

Gao, W., & Liu, S. (2011). Improved artificial bee colony algorithm for global optimization. *Information Processing Letters, 111*, 871–882.

Gao, W.-F., & Liu, S.-Y. (2012). A modified artificial bee colony algorithm. *Computers and Operations Research, 39*, 687–697.

Gao, W.-F., Liu, S.-Y., & Jiang, F. (2011). An improved artificial bee colony algorithm for directing orbits of chaotic systems. *Applied Mathematics and Computation, 218*, 3868–3879.

Gao, F., Fei, F.-X., Deng, Y.-F., Qi, Y.-B., & Balasingham, I. (2012). A novel non-Lyapunov approach through artificial bee colony algorithm for detecting unstable periodic orbits with high orders. *Expert Systems with Applications, 39*, 12389–12397. (http://dx.doi.org/10.1016/j.eswa.2012.04.083).

Gao, W., Liu, S., & Huang, L. (2012). A global best artificial bee colony algorithm for global optimization. *Journal of Computational and Applied Mathematics,* 236, 2741–2753.

Gao, W.-F., Liu, S.-Y., & Huang, L.-L. (2013). A novel artificial bee colony algorithm with Powell's method. *Applied Soft Computing, 13,* 3763–3775. (http://dx.doi.org/10.1016/j.asoc.2013.05.012).

Gorji-Bandpy, M., & Mozaffari, A. (2012). Multiobjective optimization of irreversible thermal engine using mutable smart bee algorithm. *Applied Computational Intelligence and Soft Computing, 12*, 1–13.

Goyal, S. (2012). The applications survey: Bee colony. *Engineering Science and Technology: An International Journal, 2*, 293–297.

Häckel, S., & Dippold, P. (2009, July 8–12). The bee colony-inspired algorithm (BCiA)–a two-stage approach for solving the vehicle routing problem with time windows. In *Proceedings of GECCO'09* (pp. 25–32). Montréal, Québec, Canada.

Haddad, O. B., Afshar, A., & Mariño, M. A. (2006). Honey-bees mating optimization (HBMO) algorithm: A new heuristic approach for water resources optimization. *Water Resources Management, 20*, 661–680.

Haddad, O. B., Afshar, A., & Mariño, M. A. (2009). Optimization of non-convex water resource problems by honey-bee mating optimization (HBMO) algorithm. *Engineering Computations: International Journal for Computer-Aided Engineering and Software: Practice and Experience, 26*, 267–280.

Haddad, O. B., Mirmomeni, M., & Mariño, M. A. (2010). Optimal design of stepped spillways using the HBMO algorithm. *Civil Engineering and Environmental Systems, 27*, 81–94.

Hancer, E., Ozturk, C., & Karaboga, D. (2012, June 10–15) Artificial bee colony based image clustering method. In *IEEE World Congress on Computational Intelligence (WCCI)* (pp. 1–5). Brisbane, Australia.

Hemamalini, S., & Simon, S. P. (2010). Artificial bee colony algorithm for economic load dispatch problem with non-smooth cost functions. *Electric Power Components and Systems, 38*, 786–803.

Hong, W.-C. (2011). Electric load forecasting by seasonal recurrent SVR (support vector regression) with chaotic artificial bee colony algorithm. *Energy, 36*, 5568–5578.

Horng, M.-H. (2010). A multilevel image thresholding using the honey bee mating optimization. *Applied Mathematics and Computation, 215*, 3302–3310.

Hsieh, T.-J., & Yeh, W.-C. (2012). Penalty guided bees search for redundancy allocation problems with a mix of components in series–parallel systems. *Computers and Operations Research, 39*, 2688–2704.

Jadhav, H. T., & Roy, R. (2013). Gbest guided artificial bee colony algorithm for environmental/economic dispatch considering wind power. *Expert Systems with Applications, 40*, 6385–6399. (http://dx.doi.org/10.1016/j.eswa.2013.05.048).

Janson, S., Middendorf, M., & Beekman, M. (2005). Honeybee swarms: How do scouts guide a swarm of uninformed bees. *Animal Behaviour, 70*, 349–358.

Jevtić, A., Gutiérrez, Á., Andina, D., & Jamshidi, M. (2012). Distributed bees algorithm for task allocation in swarm of robots. *IEEE Systems Journal, 6*, 296–304. (http://dx.doi.org/10.1109/JSYST.2011.2167820"10.1109/JSYST.2011.2167820).

Jones, K. O., & Bouffet, A. (2008, June 12–13). Comparison of bees algorithm, ant colony optimisation and particle swarm optimisation for PID controller tuning. In *Proceedings of International Conference on Computer Systems and Technologies (CompSysTech)* (pp. IIIA. 9-1–IIIA.9-6). Gabrovo, Bulgaria.

Kalayci, C. B., & Gupta, S. M. (2013). Artificial bee colony algorithm for solving sequence-dependent disassembly line balancing problem. *Expert Systems with Applications, 40*, 7231–7241. (http://dx.doi.org/10.1016/j.eswa.2013.06.067).

Kang, F., Li, J., & Ma, Z. (2011). Rosenbrock artificial bee colony algorithm for accurate global optimization of numerical functions. *Information Sciences, 181*, 3508–3531.

Kang, F., Li, J., & Li, H. (2013). Artificial bee colony algorithm and pattern search hybridized for global optimization. *Applied Soft Computing, 13*, 1781–1791.

Karaboga, D. (2005). *An idea based on honey bee swarm for numerical optimization*. Technical Report TR06, Computer Engineering Department, Engineering Faculty, Erciyes University.

Karaboga, D., & Akay, B. (2009a) Artificial bee colony (ABC), harmony search and bees algorithms on numerical optimization. In *Proceedings of IPROMS Conference* (pp. 1–6).

Karaboga, D., & Akay, B. (2009b). A comparative study of artificial bee colony algorithm. *Applied Mathematics and Computation, 214*, 108–132.

Karaboga, D., & Akay, B. (2009c). A survey: Algorithms simulating bee swarm intelligence. *Artificial Intelligence Review, 31*, 61–85.

Karaboga, D., & Basturk, B. (2007). A powerful and efficient algorithm for numerical function optimization: Artificial bee colony (ABC) algorithm. *Journal of Global Optimization, 39*, 459–471.

Karaboga, D., & Basturk, B. (2008). On the performance of artificial bee colony (ABC) algorithm. *Applied Soft Computing, 8*, 687–697.

Karaboğa, N., & Çetinkaya, M. B. (2011). A novel and efficient algorithm for adaptive filtering: Artificial bee colony algorithm. *Turkey Journal of Electrical Engineering and Computer Science, 19*, 175–190.

Karaboga, D., Gorkemli, B., Ozturk, C., & Karaboga, N. (2012). A comprehensive survey: Artificial bee colony (ABC) algorithm and applications. *Artificial Intelligence Review*. doi: 10.1007/s10462-012-9328-0.

Karabulut, K., & Tasgetiren, M. F. (2012, June 10–15). A discrete artificial bee colony algorithm for the traveling salesman problem with time windows. In *IEEE World Congress on Computational Intelligence (WCCI)* (pp. 1–7). Brisbane, Australia.

Karsai, I. (1999). Decentralized control of construction behavior in paper wasps: An overview of the stigmergy approach. *Artificial Life, 5*, 117–136.

Karsai, I., & Wenzel, J. W. (2000). Organization and regulation of nest construction behavior in *Metapolybia* wasps. *Journal of Insect Behavior, 13*, 111–140.

Kashan, M. H., Nahavandi, N., & Kashan, A. H. (2012). DisABC: A new artificial bee colony algorithm for binary optimization. *Applied Soft Computing, 12*, 342–352.

Kaur, A., & Goyal, S. (2011). A bee colony optimization algorithm for fault coverage based regression test suite prioritization. *International Journal of Advanced Science and Technology, 29*, 17–29.

Khan, L., Ullah, I., Saeed, T., & Lo, K. L. (2010). Virtual bees algorithm based design of damping control system for TCSC. *Australian Journal of Basic and Applied Sciences, 4*, 1–18.

Koc, E., Ersoy, N., Camlidere, Z. S., & Kilic, H. (2012). A Web-service for automated software refactoring using artificial bee colony optimization. In Y. Tan, Y. Shi, & Z. Ji (Eds.), *ICSI 2012, Part I* (Vol. 7331, pp. 318–328)., LNCS Berlin, Heidelberg: Springer.

Kumar, S. K., Tiwari, M. K., & Babiceanu, R. F. (2010). Minimisation of supply chain cost with embedded risk using computational intelligence approaches. *International Journal of Production Research, 48*, 3717–3739.

Landa, J. T., & Tullock, G. (2003). Why ants do but honeybees do not construct satellite nests. *Journal of Bioeconomics, 5*, 151–164.

Latty, T., Duncan, M., & Beekman, M. (2009). High bee traffic disrupts transfer of directional information in flying honeybee. *Animal Behaviour, 78*, 117–121.

Lez-Álvarez, D. L. G., Vega-Rodríguez, M. A., Gómez-Pulido, J. A., & Sánchez-Pérez, J. M. (2013). Comparing multiobjective swarm intelligence metaheuristics for DNA motif discovery. *Engineering Applications of Artificial Intelligence, 26*, 314–326.

Li, G., Niu, P., & Xiao, X. (2012). Development and investigation of efficient artificial bee colony algorithm for numerical function optimization. *Applied Soft Computing, 12*, 320–332.

Lien, L.-C., & Cheng, M.-Y. (2012). A hybrid swarm intelligence based particle-bee algorithm for construction site layout optimization. *Expert Systems with Applications, 39*, 9642–9650.

Low, M. Y. H., Chandramohan, M., & Choo, C. S. (2009). Application of multi-objective bee colony optimization algorithm to automated red teaming. In *Proceedings of IEEE 2009 Winter Simulation Conference* (pp. 1798–1808).

Lucchetta, P., Bernstein, C., Théry, M., Lazzari, C., & Desouhant, E. (2008). Foraging and associative learning of visual signals in a parasitic wasp. *Animal Cognition, 11*, 525–533.

Lučić, P. (2002). *Modeling transportation problems using concepts of swarm intelligence and soft computing*. Unpublished doctoral thesis, Virginia Polytechnic Instituute and State University.

Lučić, P., & Teodorović, D. (2003). Computing with bees: Attacking complex transportation engineering problems. *International Journal on Artificial Intelligence Tools, 12*, 375–394.

Ma, M., Liang, J., Guo, M., Fan, Y., & Yin, Y. (2011). SAR image segmentation based on artificial bee colony algorithm. *Applied Soft Computing, 11*, 5205–5214.

Maia, R. D., Castro, L. N. D., & Caminhas, W. M. (2012, June 10–15). Bee colonies as model for multimodal continuous optimization: The OptBees algorithm. In *IEEE World Congress on Computational Intelligence (WCCI)* (pp. 1–8). Brisbane, Australia.

Marinakis, Y., Marinaki, M., & Dounias, G. (2008). Honey bees mating optimization algorithm for the vehicle routing problem. *Studies in Computational Intelligence (SCI), 129*, 139–148. (Berlin, Heidelberg: Springer).

Marinakis, Y., Marinaki, M., & Dounias, G. (2010). Honey bees mating optimization algorithm for large scale vehicle routing problems. *Natural Computing, 9*, 5–27.

Marinakis, Y., Marinaki, M., & Dounias, G. (2011). Honey bees mating optimization algorithm for the Euclidean traveling salesman problem. *Information Sciences, 181*, 4684–4698.

Mccaffrey, J. D., & Dierking, H. (2009). An empirical study of unsupervised rule set extraction of clustered categorical data using a simulated bee colony algorithm. In G. Governatori, J. Hall, & A. Paschke (Eds.), *RuleML 2009* (Vol. 5858, pp. 182–193)., LNCS Berlin, Heidelberg: Springer.

Mezura-Montes, E., & Velez-Koeppel, R. E. (2012, July 18–23). Elitist artificial bee colony for constrained real-parameter optimization. In *IEEE World Congress on Computational Intelligence (WCCI)* (pp. 2068–2075). Barcelona, Spain: CCIB.

Mezura-Montes, E., Damián-Araoz, M., & Cetina-Domíngez, O. (2010, July 18–23). Smart flight and dynamic tolerances in the artificial bee colony for constrained optimization. In *IEEE World Congress on Computational Intelligence (WCCI)* (pp. 4118–4125). Barcelona, Spain: CCIB.

Muñoz, M. A., López, J. A., & Caicedo, E. (2009). An artificial beehive algorithm for continuous optimization. *International Journal of Intelligent Systems, 24*, 1080–1093.

Nakrani, S., & Tovey, C. (2003, December 15–17). On honey bees and dynamic allocation in an Internet server colony. In *Proceedings of 2nd International Workshop on the Mathematics and Algorithms of Social Insects* (pp. 1–8). Atlanta, Georgia.

Nakrani, S., & Tovey, C. (2004). On honey bees and dynamic server allocation in internet hosting centers. *Adaptive Behavior, 12*, 223–240.

Narasimhan, H. (2009). Parallel artificial bee colony (PABC) algorithm. In *IEEE World Congress on Nature and Biologically Inspired Computing (NaBIC)* (pp. 306-311).

Niknam, T. (2009). An efficient hybrid evolutionary algorithm based on PSO and HBMO algorithms for multi-objective distribution feeder reconfiguration. *Energy Conversion and Management, 50*, 2074–2082.

Niknam, T. (2011). An efficient multi-objective HBMO algorithm for distribution feeder reconfiguration. *Expert Systems with Applications, 38*, 2878–2887.

Niknam, T., & Golestaneh, F. (2013). Enhanced bee swarm optimization algorithm for dynamic economic dispatch. *IEEE Systems Journal, 7*, 754–762. doi:10.1109/JSYST.2012.2191831.

Niknam, T., Olamaie, J., & Khorshidi, R. (2008). A hybrid algorithm based on HBMO and fuzzy set for multi-objective distribution feeder reconfiguration. *World Applied Sciences Journal, 4*, 308–315.

Niknam, T., Mojarrad, H. D., Meymand, H. Z., & Firouzi, B. B. (2011). A new honey bee mating optimization algorithm for non-smooth economic dispatch. *Energy, 36*, 896–908.

Niknam, T., Fard, A. K., & Seifi, A. (2012). Distribution feeder reconfiguration considering fuel cell-wind-photovoltaic power plants. *Renewable Energy, 37*, 213–225.

Olamaei, J., Niknam, T., Badali, S., & Arefi, (2012). Distribution feeder reconfiguration for loss minimization based on modified honey bee mating optimization algorithm. *Energy Procedia, 14*, 304–311.

Omkar, S. N., Senthilnath, J., Khandelwal, R., Naik, G. N., & Gopalakrishnan, S. (2011). Artificial bee colony (ABC) for multi-objective design optimization of composite structures. *Applied Soft Computing, 11*, 489–499.

Pham, D. T., & Castellani, M. (2009). The bees algorithm: Modelling foraging behaviour to solve continuous optimization problems. *Proceedings of the Institution of Mechanical Engineers, Part C: Journal of Mechanical Engineering Science, 223*, 2919–2938.

Pham, D. T., & Ghanbarzadeh, A. (2007). Multi-objective optimisation using the bees algorithm. In *Third International Virtual Conference on Intelligent Production Machines and Systems (IPROMS)* (pp. 1–5). Dunbeath, Scotland: Whittles.

Pham, D. T., & Koç, E. (2010). Design of a two-dimensional recursive filter using the bees algorithm. *International Journal of Automation and Computing, 7*, 399–402.

Pham, D. T., Ghanbarzadeh, A., Koç, E., Otri, S., Rahim, S., & Zaidi, M. (2006). The bees algorithm—a novel tool for complex optimisation problems. In *Proceedings of Second International Virtual Conference on Intelligent production machines and systems (IPROMS)* (pp. 454–459). Oxford: Elsevier.

Pham, D. T., Afify, A. A., & Koç, E. (2007a). Manufacturing cell formation using the bees algorithm. In *Third International Virtual Conference on Intelligent Production Machines and Systems (IPROMS)* (pp. 1–6). Dunbeath, Scotland: Whittles.

Pham, D. T., Ghanbarzadeh, A., Koç, E., Otri, S., Rahim, S., & Zaidi, M. (2007b). Using the bees algorithm to schedule jobs for a machine. In *Proceedings of Eighth International Conference on Laser metrology, CMM and machine tool performance (LAMDAMAP)* (pp. 430–439). Euspen, UK: Cardiff.

Pham, D. T., Ghanbarzadeh, A., Otri, S., & Koç, E. (2009). Optimal design of mechanical components using the bees algorithm. *Proceedings of the Institution of Mechanical Engineers, Part C: Journal of Mechanical Engineering Science, 223*, 1051–1056.

Pinto, P., Runkler, T. A., & Sousa, J. M. (2005). Wasp swarm optimization of logistic systems. In B. Ribeiro, R. F. Albrecht, A. Dobnikar, D. W. Pearson, & N. C. Steele (Eds.), *International Conference on Adaptive and Natural Computing Algorithms, Coimbra, Portugal* (pp. 264–267). Wien: Springer.

Pinto, P. C., Runkler, T. A., & Sousa, J. M. C. (2006, September 11–13). Agent based optimization of the MAX-SAT problem using wasp swarms. In *7th Portuguese Conference on Automatic Control (CONTROLO)* (pp. 1–6). Lisboa, Portugal: Instituto Superior Técnico.

Quijano, N., & Passino, K. M. (2010). Honey bee social foraging algorithms for resource allocation: Theory and application. *Engineering Applications of Artificial Intelligence, 23*, 845–861.

Rajasekhar, A., Pant, M., & Abraham, A. (2011). Cauchy movements for artificial bees for finding better food sources. In *Third World Congress on Nature and Biologically Inspired Computing (NaBIC)* (pp. 279–284).

Rajasekhar, A., Das, S., & Suganthan, P. N. (2012, June 10–15). Design of fractional order controller for a servohydraulic positioning system with micro artificial bee colony algorithm. In *IEEE World Congress on Computational Intelligence (WCCI)* (pp. 1–8). Brisbane, Australia.

Ramírez, F. J., Lee, J. Y., Packianather, M. S., & Pham, D. T. (2010, November 15–26). Enhancing multi-stage deep-drawing processes through the novel use of the bees-algorithm. In *Proceedings of 6th IPROMS Virtual Conference* (pp. 1–6).

Runkler, T. A. (2008). Wasp swarm optimization of the c-means clustering model. *International Journal of Intelligent Systems, 23*, 269–285.

Samanta, S., & Chakraborty, S. (2011). Parametric optimization of some non-traditional machining processes using artificial bee colony algorithm. *Engineering Applications of Artificial Intelligence, 24*, 946–957.

Sato, T., & Hagiwara, M. (1997). Bee system: Finding solution by a concentrated search. In *IEEE International Conference on Systems, Man, and Cybernetics (SMC)* (pp. 3954–3959).

Scholz-Reiter, B., Jagalski, T., & Bendul, J. C. (2008). Autonomous control of a shop floor based on bee's foraging behaviour. In H.-D. Haasis (Ed.), *Dynamics in logistics* (pp. 415–423). Berlin: Springer.

Senthilnath, J., Omkar, S. N., Mani, V., Tejovanth, N., Diwakar, P. G., & Archana, S. B. (2011). Multi-spectral satellite image classification using glowworm swarm optimization. In *IEEE International Geoscience and Remote Sensing Symposium (IGARSS)* (pp. 47–50).

Shafia, M. A., Moghaddam, M. R., & Tavakolian, R. (2011). A hybrid algorithm for data clustering using honey bee algorithm, genetic algorithm and *k*-means method. *Journal of Advanced Computer Science and Technology Research, 1*, 110–125.

Sharma, T. K., Pant, M., & Bansal, J. C. (2012a, June 10–15). Artificial bee colony with mean mutation operator for better exploitation. In *IEEE World Congress on Computational Intelligence (WCCI)* (pp. 1–7). Brisbane, Australia.

Sharma, T. K., Pant, M., & Bansal, J. C. (2012b, June 10–15). Some modifications to enhance the performance of artificial bee colony. In *IEEE World Congress on Computational Intelligence (WCCI)* (pp. 1–8). Brisbane, Australia.

Shi, X., Li, Y., Li, H., Guan, R., Wang, L., & Liang, Y. (2010). An integrated algorithm based on artificial bee colony and particle swarm optimization. In *Sixth International Conference on Natural Computation (ICNC)* (pp. 2586–2590).

Slaa, E. J., & Hughes, W. O. H. (2009). Local enhancement, local inhibition, ravesdropping, and the parasitism of social insect communication. In S. Jarau & M. Hrncir (Eds.), *Ecological, behavioral, and theoretical approaches, Chapter 8* (pp. 147–164). Boca Raton, FL: CRC Press, Taylor & Francis Group, LLC. ISBN ISBN 978-1-4200-7560-1.

Song, J., Hu, J., Tian, Y., & Xu, Y. (2005, September 13–16). Re-optimization in dynamic vehicle routing problem based on wasp-like agent strategy. In *Proceedings of 8th International Conference on Intelligent Transportation Systems* (pp. 688–693). Vienna, Austria.

Sotelo-Figueroa, M. A., Baltazar-Flores, M. D. R., Carpio, J. M., & Zamudio, V. (2010). A comparation between bee swarm optimization and greedy algorithm for the knapsack problem with bee reallocation. In *Ninth Mexican International Conference on Artificial Intelligence* (pp. 22–27).

Subotic, M., Tuba, M., & Stanarevic, N. (2010). Parallelization of the artificial bee colony (ABC) algorithm. In *Recent Advances in Neural Networks, Fuzzy Systems and Evolutionary Computing* (191–196).

Subramanian, S., Anandhakumar, R., & Ganesan, S. (2011). Generator maintenance management using bio-inspired search algorithm. *International Journal of Energy Sector Management, 5*, 522–544.

Sundar, S., & Singh, A. (2012). A swarm intelligence approach to the early/tardy scheduling problem. *Swarm and Evolutionary Computation, 4*, 25–32.

Tasgetiren, M. F., Pan, Q.-K., Suganthan, P. N., & Chen, A. H.-L. (2010, July 18–23) A discrete artificial bee colony algorithm for the permutation flow shop scheduling problem with total flowtime criterion. In *IEEE World Congress on Computational Intelligence* (pp. 137–144). Barcelona, Spain: CCIB.

Tasgetiren, M. F., Pan, Q.-K., Suganthan, P. N., & Chen, A. H.-L. (2011). A discrete artificial bee colony algorithm for the total flowtime minimization in permutation flow shops. *Information Sciences, 181*, 3459–3475.

Tasgetiren, M. F., Pan, Q.-K., Suganthan, P. N., & Oner, A. (2013). A discrete artificial bee colony algorithm for the no-idle permutation flowshop scheduling problem with the total tardiness criterion. *Applied Mathematical Modelling, 37*, 6758–6779.

Teodorović, D. (2008). Swarm intelligence systems for transportation engineering: Principles and applications. *Transportation Research Part C, 16*, 651–667.

Teodorović, D. (2009a). Bee colony optimization (BCO). In C. P. Lim, L. C. Jain, & S. Dehuri (Eds.), *Innovations in swarm intelligence* (Vol. 248, pp. 39–60)., SCI Berlin, Heidelberg: Springer.

Teodorović, D. (2009b). Bee colony optimization (BCO). In C. P. Lim, L. C. Jain, & S. Dehuri (Eds.), *Innovations in swarm intelligence*. Berlin, Heidelberg: Springer.

Teodorović, D., & Dell'Orco, M. (2005). Bee colony optimization: A cooperative learning approach to complex transportation problems. In *16th Mini-EURO Conference on Advanced OR and AI Methods in Transportation* (pp. 51–60).

Teodorovic, D. U. Š. A. N., Davidovic, T., & Selmic, M. (2011). Bee colony optimization: The applications survey. *ACM Transactions on Computational Logic, 1529*, 3785.

Tereshko, V., & Lee, T. (2002). How information mapping patterns determine foraging behaviour of a honeybee colony. *Open Systems and Information Dynamics, 9*, 181–193.

Theraulaz, G., & Bonabeau, E. (1995). Coordination in distributed building. *Science, 269*, 686–688.

Theraulaz, G., Goss, S., Gervet, J., & Deneubourg, J. L. (1991). Task differentiation in polistes wasps colonies: A model for self-organizing groups of robots. In *First International Conference on Simulation of Adaptive Behavior* (pp. 346–355). Cambridge, MA: MIT Press.

Tien, J. P., & Li, T. H. S. (2012). Hybrid Taguchi-chaos of multilevel immune and the artificial bee colony algorithm for parameter identification of chaotic systems. *Computers and Mathematics with Applications, 64*(5), 1108–1119. doi:10.1016/j.camwa.2012.03.029.

Tsai, P.-W., Pan, J.-S., Liao, B.-Y., & Chu, S.-C. (2009). Enhanced artificial bee colony optimization. *International Journal of Innovative Computing, Information and Control, 5*, 1–12.

Vitorino, L. N., Ribeiro, S. F., & Bastos-Filho, C. J. A. (2012, June 10–15). A hybrid swarm intelligence optimizer based on particles and artificial bees for high-dimensional search spaces. In *IEEE World Congress on Computational Intelligence (WCCI)* (pp. 1–6). Brisbane, Australia.

Wang, Z. (2009). *Cooperative construction*. Unpublished master thesis, The Ohio State University.

Wang, D.-Z., Zhang, J.-S., Wan, F., & Zhu, L. (2006, May 27–29). A dynamic task scheduling algorithm in grid environment. In *5th WSEAS International Conference on Telecommunications and Informatics* (pp. 273–275). Istanbul, Turkey.

Wedde, H. F., & Farooq, M. (2006). A comprehensive review of nature inspired routing algorithms for fixed telecommunication networks. *Journal of Systems Architecture, 52*, 461–484.

Wedde, H. F., Farooq, M., & Zhang, Y. (2004). Beehive: An efficient fault-tolerant routing algorithm inspired by honey bee behavior. In M. Dorigo (Ed.), *ANTS 2004* (Vol. 3172, pp. 83–94)., LNCS Berlin, Heidelberg: Springer.

Wong, L.-P., Puan, C. Y., Low, M. Y. H., & Chong, C. S. (2008) Bee colony optimization algorithm with big valley landscape exploitation for job shop scheduling problems. In *Proceedings of IEEE 2008 Winter Simulation Conference* (pp. 2050–2058).

Xu, C., Duan, H., & Liu, F. (2010a). Chaotic artificial bee colony approach to uninhabited combat air vehicle (UCAV) path planning. *Aerospace Science and Technology, 14*, 535–541.

Xu, S., Ji, Z., Pham, D. T., & Yu, F. (2010b). Bio-inspired binary bees algorithm for a two-level distribution optimisation problem. *Journal of Bionic Engineering, 7*, 161–167.

Yang, X. S. (2005). Engineering optimizations via nature-inspired virtual bee algorithms. In José Mira José & R. Álvarez (Eds.), *Artificial intelligence and knowledge engineering applications: A bioinspired approach*. Berlin Heidelberg: Springer.

Yildiz, A. R. (2013). A new hybrid artificial bee colony algorithm for robust optimal design and manufacturing. *Applied Soft Computing, 13*, 2906–2912. (http://dx.doi.org/10.1016/j.asoc.2012.04.013).

Zeng, F., Decraene, J., Low, M. Y. H., Hingston, P., Cai, W., Zhou, S., & Chandramohan, M. (2010, July 18–23). Autonomous bee colony optimization for multi-objective function. In *IEEE World Congress on Computational Intelligence (WCCI)* (pp. 1279–1286). Barcelona, Spain: CCIB.

Zhang, H., & Ye, D. (2012). An artificial bee colony algorithm approach for routing in VLSI. In Y. Tan, Y. Shi, & Z. Ji (Eds.), *ICSI 2012, Part I* (Vol. 7331, pp. 334–341)., LNCS Berlin, Heidelberg: Springer.

Zhang, W., Wang, N., & Yang, S. (2013). Hybrid artificial bee colony algorithm for parameter estimation of proton exchange membrane fuel cell. *International Journal of Hydrogen Energy, 38*, 5796–5806.

Zhang, H., Zhu, Y., & Yan, X. (2102, June 10–15). Multi-hive artificial bee colony algorithm for constrained multi-objective optimization. In *IEEE World Congress on Computational Intelligence (WCCI)* (pp. 1–8). Brisbane, Australia.

Chapter 5
Biogeography-based Optimization Algorithm

Abstract In this chapter, we introduce a novel optimization algorithm called biogeography-based optimization (BBO) which is inspired by the science of biogeography. We first describe the general knowledge of the science of biogeography in Sect. 5.1. Then, the fundamentals and performance of BBO are introduced in Sect. 5.2. Finally, Sect. 5.3 summarises this chapter.

5.1 Introduction

Biogeography is a branch of geography in which the past and present distribution of the world's species are studied. In the area of biogeography, two prominent biologist, i.e., MacArthur and Wilson, in their book MacArthur and Wilson (1967) showed that the patterns of species richness of and area can be explained through a combination of historical factors, such as habitat area, immigration rate, and extinction rate. Inspired by that, recently, Simon (2008) proposed a new computational intelligence (CI) algorithm, called biogeography-based optimization (BBO) algorithm.

5.1.1 Science of Biogeography

The biogeography can be defined as a study of distribution of life forms (such as, plant, human, and animal species) in nature over time and space, such as the immigration and emigration of species between habitats. In general, it can be divided into two areas: i.e., ecological biogeography and historical biogeography. The former is used to deal with the current distribution patterns, while the latter is used to concern with long-term and large-scale distributions (Hobbs et al. 2013). In addition, some experimental studies showed that biogeography are closely linked to the evolution and the ecology. In other words, based on the science of biogeography, we can understand why the species are in their present locations and in developing protecting the world's natural habitats.

B. Xing and W.-J. Gao, *Innovative Computational Intelligence: A Rough Guide to 134 Clever Algorithms*, Intelligent Systems Reference Library 62,
DOI: 10.1007/978-3-319-03404-1_5,

5.2 Biogeography-based Optimization Algorithm

5.2.1 *Fundamentals of Biogeography-based Optimization Algorithm*

Biogeography-based optimization (BBO) algorithm was originally proposed in (Simon 2008) which is based on the mathematical models of biogeography that is studied by Robert MacArthur and Edward Wilson (MacArthur and Wilson 1967). In principle, there are two main operators in BBO, i.e., migration and mutation. Before optimizing, each individual of population is evaluated and then follows migration and mutation step to reach global minima. In order to implement BBO, the following components need to be taken into account (Simon 2008):

- In migration, the information is shared between habitats that depend on emigration rates μ and immigration rates λ of each solution. The immigration rate λ and the emigration rate μ are functions of the number of species in the habitat. The equilibrium number of species is S_0 at which point the immigration and emigration rates are equal (Simon 2008). They can be calculated through Eqs. 5.1 and 5.2, respectively (Simon 2008):

$$\lambda_s = I\left(1 - \frac{S}{S_{\max}}\right), \quad \text{for } 0 \le S \le S_{\max}, \tag{5.1}$$

$$\mu_s = E\frac{S}{S_{\max}}, \quad \text{for } 0 \le S \le S_{\max}, \tag{5.2}$$

where $S_{\max}$ denotes the largest possible number of species, S is the number of species, I is the maximum immigration rate, and E is the maximum emigration rate.

In addition, each solution is modified depending on probability that is a user defined parameter. Each individual has its own λ and μ and are functions of the number of species S in the habitat. Poor solutions accept more useful information from good solution, which improve the exploitation ability of algorithm.

- Meanwhile, BBO takes the advantage of mutation. The mutation is used to increase the diversity of the population to get the good solutions. The aim of this scheme is to make an island with low habitat suitability index (HSI) more likely to mutate its suitability index variables (SIVs). It can be defined by Eq. 5.3 (Simon 2008):

$$m(S) = m_{\max}\left(\frac{1 - P_s}{P_{\max}}\right), \tag{5.3}$$

where m_{max} is a user defined parameter, $m(S)$ is the mutation rate for a habitat that contains S species, and P_{max} is the maximum probability.

Taking into account two key operators described above, the steps of implementing standard BBO algorithm can be summarized as follows (Simon 2008):

- Step1: Initializing the BBO parameters.
- Step 2: Initializing a stochastic set of habitats where each potential solution to the target problem is linked to a corresponding habitat.
- Step 3: For each habitat, mapping the HSI to the number of species (denoted by S), the immigration rate (indicated by λ), and the emigration rate (represented by μ).
- Step 4: Using immigration and emigration to adjust each non-elite habitat in a probabilistic manner.
- Step 5: For each habitat, first updating the probability of its species amount through Eq. 5.4 (Simon 2008), next mutating each non-elite habitat according to its probability (see Eq. 5.3), and then recalculating each HSI.

$$P_s = \begin{cases} -(\lambda_s + \mu_s)P_s + \mu_{s+1}P_{s+1} & S = 0 \\ -(\lambda_s + \mu_s)P_s + \lambda_{s-1}P_{s-1} + \mu_{s+1}P_{s+1} & 1 \le S \le S_{max} - 1, \\ -(\lambda_s + \mu_s)P_s + \lambda_{s-1}P_{s-1} & S = S_{max} \end{cases} \tag{5.4}$$

  where P_s is the probability of a habitat that contains exactly S species, P_{s+1} is the probability of a habitat that contains $S + 1$ species, P_{s-1} is the probability of a habitat that contains $S - 1$ species, and λ_s and μ_s are the immigration and emigration rates, respectively, for a habitat that contains S species.
- Step 6: Returning to Step 3 for the next iteration. This loop can be stopped based on a predetermined termination criterion.

5.2.2 Performance of BBO

In order to show how the BBO algorithm performs, Simon (2008) first tested it on a set of benchmark functions, such as Ackley function, Griewank function, Rastrigin function and Rosenbrock function. In comparison with other CI techniques (e.g., ant colony optimization (ACO), genetic algorithm (GA), particle swarm optimization (PSO), etc.), the BBO was able to offer better results in most cases. Then a real world aircraft engine health estimation problem was further employed by Simon (2008). Again, BBO offered better results than other competing algorithms in this case study.

5.3 Conclusions

In this chapter, we introduced a newly developed CI algorithm, i.e., BBO. In BBO, each candidate solution is represented by a vector variable of the optimization problem. It is considered as a "habitat" or an "island" in biogeography and their features (include vegetation, rainfall, topographic diversity, temperate, etc.) that characterize habitability are called SIV. The fitness of each solution is called its HSI and depends on many features of the habitat. A good solution indicates an island with a high HSI, which are well suited as habitats for biological species. Habitats with a high HSI tend to have a large number of species and more likely to share their features (SIVs), while those with a low HSI have a small number of species and tend to accept features of other solutions. Furthermore, habitats with a high HSI have a low species immigration rate and have a high emigration rate, because on one side they are already nearly saturated with species, but on the other side they have many opportunities to emigrate to neighbouring habitats, as animals ride flotsam, fly or swim to neighbouring islands. Through this kind of probabilistic evolution, BBO searches for a good solution to an optimization problem. Although it is a newly introduced CI method, we have witnessed the following rapid spreading of BBO:

First, several enhanced versions of BBO can be found in the literature as outlined below:

- Accelerated BBO (Lohokare et al. 2010, 2012).
- BBO with elitism (Simon et al. 2009).
- BBO with ensemble of migration models (Ma et al. 2012b).
- Binary BBO (Zhao et al. 2012).
- Biogeography migration algorithm (Mo and Xu 2010a).
- Blended BBO (Ma and Simon 2011b).
- Constrained BBO (Boussaïd et al. 2012).
- Dynamic system model of BBO (Simon 2011a).
- Enhanced BBO (Pattnaik et al. 2010).
- Hybrid ant colony optimization and BBO (Goel et al. 2012).
- Hybrid BBO with bacterial foraging algorithm (Lohokare et al. 2009c).
- Hybrid BBO with differential evolution (Boussaïd et al. 2011a; Bhattacharya and Chattopadhyay 2010d, 2011b; Mahdad and Srairi 2011; Wang and Xu 2011; Gong et al. 2011; Boussaïd et al. 2011b).
- Hybrid BBO with differential mutation (Wang and Cai 2011).
- Hybrid BBO with evolutionary strategy (Rathi et al. 2011; Du et al. 2009).
- Hybrid particle swarm optimization and BBO (Kundra and Sood 2010; Goel et al. 2011b).
- Improved BBO (Chatterjee et al. 2012).
- Markov chains of BBO (Ma and Simon 2011a; Simon et al. 2011a, b).
- Modified BBO (Lohokare et al. 2009b; Kanoongo and Jain 2012; Ma et al. 2009; Ma 2010).

- Modified BBO based on predator-prey concepts (Silva et al. 2010, 2012; Silva and Coelho 2010).
- Multiobjective BBO (Jamuna and Swarup 2012; Ma et al. 2012a).
- Multi-operator BBO (Li and Yin 2012).
- Oppositional BBO (Bhattacharya and Chattopadhyay 2010e, f; Ergezer et al. 2009; Yang et al. 2011; Ergezer and Simon 2011; Ergezer and Sikder 2011).
- Perturb BBO (Li et al. 2011).
- Real-coded BBO (Gong et al. 2010).
- Simplified BBO (Simon 2011b).

Second, the BBO algorithm has also been successfully applied to a variety of optimization problems as listed below:

- Antenna design optimization (Lohokare et al. 2009a, b; Singh et al. 2010; Sharaqa and Dib 2011; Goudos et al. 2012; Panduro et al. 2006).
- Communication network optimization (Ashrafinia et al. 2011a, b, c; Boussaïd et al. 2011a).
- Data mining (Nikumbh et al. 2012).
- Image processing (Panchal et al. 2009; Gupta and Panchal 2011; Goel et al. 2011a; Panchal et al. 2011; Sinha et al. 2012; Goel et al. 2011b, 2012).
- Knapsack problem (Zhao et al. 2012).
- Machining process optimization (Mukherjee and Chakraborty 2013; Mukherjee et al. 2012).
- Motor design optimization (Silva et al. 2012).
- Parameter estimation (Wang and Xu 2011).
- Path planning (Silva et al. 2010; Huang et al. 2012; Kundra and Sood 2010).
- Power system optimization (Jamuna and Swarup 2011b; Roy et al. 2009a, b, 2010a, b, c, d, 2011; Bhattacharya and Chattopadhyay 2009a, b, c, 2010a, b, c, d, e, f, g, 2011a, b, 2012; Rarick et al. 2009; Silva and Coelho 2010; Lohokare et al. 2010; Rathi et al. 2011; Jamuna and Swarup 2011a, 2012; Pandit 2012; Rabiee et al. 2012; Kankanala et al. 2012; Mohammed and Talaq 2012; Gupta et al. 2012; Mahdad and Srairi 2011; Kanoongo and Jain 2012; Mandal et al. 2011).
- Remanufacturing (Gao et al. 2013).
- Scheduling (Rahmati and Zandieh 2012).
- Travelling salesman problem (Mo and Xu 2010a, b; Song et al. 2010).
- Virtual simulation optimization (Gardner and Simon 2009).

Interested readers are referred to them as a starting point for a further exploration and exploitation of the BBO algorithm.

Overall, it is still too early to claim that BBO is one of the best CI algorithms, but the preliminary studies proved that it is indeed very competitive tool for solving optimization problems.

References

Ashrafinia, S., Naeem, M., & Lee, D. (2011a). A low complexity evolutionary algorithm for multi-user MIMO detection. In *IEEE Symposium on Computational Intelligence in Multicriteria Decision-Making (MDCM)* (pp. 8–13). IEEE.

Ashrafinia, S., Pareek, U., Naeem, M., & LEE, D. (2011b). Biogeography-based optimization for joint relay assignment and power allocation in cognitive radio systems. In *IEEE Symposium on Swarm Intelligence (SIS)* (pp. 1–8). IEEE.

Ashrafinia, S., Pareek, U., Naeem, M., & LEE, D. (2011c). Source and relay power selection using biogeography-based optimization for cognitive radio systems. In *IEEE Vehicular Technology Conference (VTC Fall)* (pp. 1–5). IEEE.

Bhattacharya, A. & Chattopadhyay, P. K. (2009a). Biogeography-based optimization and its application to nonconvex economic emission load dispatch problems. In *8th International Conference on Advances in Power System Control, Operation and Management (APSCOM)* (pp. 1–6). IEEE.

Bhattacharya, A. & Chattopadhyay, P. K. (2009b). Economic dispatch solution using biogeography-based optimization. In *Annual IEEE India Conference (INDICON)* (pp. 1–4). IEEE.

Bhattacharya, A. & Chattopadhyay, P. K. (2009c). Non convex economic load dispatch problem solution using biogeography-based optimization. In *8th International Conference on Advances in Power System Control, Operation and Management (APSCOM)* (pp. 1–6). IEEE.

Bhattacharya, A., & Chattopadhyay, P. K. (2010a). Application of biogeography-based optimization for solving multi-objective economic emission load dispatch problems. *Electric Power Components and Systems, 38*, 340–365.

Bhattacharya, A., & Chattopadhyay, P. K. (2010b). Biogeography-based optimization for different economic load dispatch problems. *IEEE Transactions on Power Systems, 25*, 1064–1077.

Bhattacharya, A. & Chattopadhyay, P. K. (2010c). Biogeography-based optimization for solution of optimal power flow problem. In *International Conference on Electrical Engineering/ Electronics Computer Telecommunications and Information Technology (ECTI-CON)* (pp. 435–439). IEEE.

Bhattacharya, A., & Chattopadhyay, P. K. (2010d). Hybrid differential evolution with biogeography-based optimization for solution of economic load dispatch. *IEEE Transactions on Power Systems, 25*, 1955–1964.

Bhattacharya, A. & Chattopadhyay, P. K. (2010e). Oppositional biogeography-based optimization for multi-objective economic emission load dispatch. In *Annual IEEE India Conference (INDICON)* (pp. 1–6). IEEE.

Bhattacharya, A., & Chattopadhyay, P. K. (2010f). Solution of economic power dispatch problems using oppositional biogeography-based optimization. *Electric Power Components and Systems, 38*, 1139–1160.

Bhattacharya, A., & Chattopadhyay, P. K. (2010g). Solving complex economic load dispatch problems using biogeography-based optimization. *Expert Systems with Applications, 37*, 3605–3615.

Bhattacharya, A., & Chattopadhyay, P. K. (2011a). Application of biogeography-based optimisation to solve different optimal power flow problems. *IET Generation, Transmission and Distribution, 5*, 70–80.

Bhattacharya, A., & Chattopadhyay, P. K. (2011b). Hybrid differential evolution with biogeography-based optimization algorithm for solution of economic emission load dispatch problems. *Expert Systems with Applications, 38*, 14001–14010.

Bhattacharya, A., & Chattopadhyay, P. K. (2012). Closure to discussion of "Hybrid differential evolution with biogeography-based optimization for solution of economic load dispatch". *IEEE Transactions on Power Systems, 27*, 575.

Boussaïd, I., Chatterjee, A., Siarry, P., & Ahmed-Nacer, M. (2011a). Hybridizing biogeography-based optimization with differential evolution for optimal power allocation in wireless sensor networks. *IEEE Transactions on Vehicular Technology, 60*, 2347–2353.

Boussaïd, I., Chatterjee, A., Siarry, P., & Hmed-Nacer, M. (2011b). Two-stage update biogeography-based optimization using differential evolution algorithm (DBBO). *Computers and Operations Research, 38*, 1188–1198.

Boussaïd, I., Chatterjee, A., Siarry, P., & Ahmed-Nacer, M. (2012). Biogeography-based optimization for constrained optimization problems. *Computers and Operations Research*, http://dx.doi.org/10.1016/j.cor.2012.04.012.

Chatterjee, A., Siarry, P., Nakib, A., & Blanc, R. (2012). An improved biogeography based optimization approach for segmentation of human head CT-scan images employing fuzzy entropy. *Engineering Applications of Artificial Intelligence*, doi:10.1016/j.engappai.2012.02.007.

Du, D., Simon, D., & Ergezer, M. (2009, October). Biogeography-based optimization combined with evolutionary strategy and immigration refusal. In *IEEE International Conference on Systems, Man, and Cybernetics (SMC)* (pp. 997–1002). San Antonio, TX, USA: IEEE.

Ergezer, M. & Sikder, I. (2011, Dec 22–24). Survey of oppositional algorithms. In *14th International Conference on Computer and Information Technology (ICCIT 2011)*, Dhaka, Bangladesh (pp. 623–628). IEEE.

Ergezer, M. & Simon, D. (2011). Oppositional biogeography-based optimization for combinatorial problems. In *IEEE Congress on Evolutionary Computation (CEC)* (pp. 1496–1503). IEEE.

Ergezer, M., Simon, D., & Du, D. (2009). Oppositional biogeography-based optimization. In *IEEE International Conference on Systems, Man, and Cybernetics (SMC)* (pp. 1009–1014). San Antonio, TX, USA: IEEE.

Gao, W.-J., Xing, B., & Marwala, T. (2013). Computational intelligence in used products retrieval and reproduction. *International Journal of Swarm Intelligence Research, 4*, 78–125.

Gardner, B. G. & Simon, D. (2009). Evolutionary algorithm Sandbox: A web-based graphical user interface for evolutionary algorithms. In *IEEE International Conference on Systems, Man, and Cybernetics (SMC)* (pp. 577–582). San Antonio, TX, USA: IEEE.

Goel, L., Gupta, D., & Panchal, V. K. (2011a). Performance governing factors of biogeography based land cover feature extraction: An analytical study. In *World Congress on Information and Communication Technologies (WICT)* (pp. 165–170). IEEE.

Goel, S., Sharma, A., & Goel, A. (2011b). Development of swarm based hybrid algorithm for identification of natural terrain features. In *Proceedings of the 2011 International Computational Intelligence and Communication Networks (CICN)* (pp. 293–296). IEEE.

Goel, L., Gupta, D., & Panchal, V. K. (2012). Hybrid bio-inspired techniques for land cover feature extraction: A remote sensing perspective. *Applied Soft Computing, 12*, 832–849.

Gong, W., Cai, Z., Ling, C. X., & Li, H. (2010). A real-coded biogeography-based optimization with mutation. *Applied Mathematics and Computation, 216*, 2749–2758.

Gong, W., Cai, Z., & Ling, C. X. (2011). DE/BBO: A hybrid differential evolution with biogeography-based optimization for global numerical optimization. *Soft Computing, 15*, 645–665.

Goudos, S. K., Baltzis, K. B., Siakavara, K., Samaras, T., Vafiadis, E., & Sahalos, J. N. (2012). Reducing the number of elements in linear arrays using biogeography-based optimization. In *6th European Conference on Antennas and Propagation (EUCAP)* (pp. 1615–1618). IEEE.

Gupta, N. & Panchal, V. K. (2011). Artificial intelligence for mixed pixel resolution. In *IEEE International Geoscience and Remote Sensing Symposium (IGARSS)* (pp. 2801–2804). IEEE.

Gupta, M., Gupta, N., Swarnkar, A., & Niazi, K. R. (2012). Network constrained economic load dispatch using biogeography based optimization. In *Students Conference on Engineering and Systems (SCES)* (pp. 1–4). IEEE.

Hobbs, R. J., Higgs, E. S., & Hall, C. M. (Eds.). (2013). *Novel ecosystems: Intervening in the new ecological world order,* River Street, Hoboken, NJ 07030-5774, USA: Wiley, ISBN 978-1-118-35422-3.

Huang, N., Liu, G., & He, B. (2012). Path planning based on Voronoi diagram and biogeography-based optimization. In Y. Tan., Y. Shi. & Z. Ji (Eds.), *ICSI 2012, Part I, LNCS 7331* (pp. 225–232). Berlin Heidelberg: Springer.

Jamuna, K., & Swarup, K. S. (2011a). Biogeography based optimization for optimal meter placement for security constrained state estimation. *Swarm and Evolutionary Computation, 1*, 89–96.

Jamuna, K. & Swarup, K. S. (2011b, July 20–22). Power system observability using biogeography based optimization. In *Second International Conference on Sustainable Energy and Intelligent System (SEISCON)*, Dr. M.G.R. University, Maduravoyal, Chennai, Tamil Nadu, India (pp. 384–389). IEEE.

Jamuna, K., & Swarup, K. S. (2012). Multi-objective biogeography based optimization for optimal PMU placement. *Applied Soft Computing, 12*, 1503–1510.

Kankanala, P., Srivastava, S. C., Srivastava, A. K., & Schulz, N. N. (2012). Optimal control of voltage and power in a multi-zonal MVDC shipboard power system. *IEEE Transactions on Power Systems, 27*, 642–650.

Kanoongo, S. & Jain, P. (2012). Biogeography based optimization for different economic load dispatch problem with different migration models. In *IEEE Students' Conference on Electrical, Electronics and Computer Science (SCEECS)* (pp. 1–4). IEEE.

Kundra, H., & Sood, M. (2010). Cross-country path finding using hybrid approach of PSO and BBO. *International Journal of Computer Applications, 7*, 15–19.

Li, X. & Yin, M. (2012). Multi-operator based biogeography based optimization with mutation for global numerical optimization. *Computers and Mathematics with Applications*. doi:10.1016/j.camwa.2012.04.015.

Li, X., Wang, J., Zhou, J., & Yin, M. (2011). A perturb biogeography based optimization with mutation for global numerical optimization. *Applied Mathematics and Computation, 218*, 598–609.

Lohokare, M. R., Pattnaik, S. S., Devi, S., Bakwad, K. M., & Joshi, J. G. (2009a). Parameter calculation of rectangular microstrip antenna using biogeography-based optimization. In *Applied Electromagnetics Conference (AEMC)* (pp. 1–4). IEEE.

Lohokare, M. R., Pattnaik, S. S., Devi, S., Panigrahi, B. K., Bakwad, K. M., & Joshi, J. G. (2009b). Modified BBO and calculation of resonant frequency of circular microstrip antenna. In *World Congress on Nature and Biologically Inspired Computing (NaBIC)* (pp. 487–492). IEEE.

Lohokare, M. R., Pattnaik, S. S., Devi, S., Panigrahi, B. K., Das, S., & Bakwad, K. M. (2009c). Intelligent biogeography-based optimization for discrete variables. In *World Congress on Nature and Biologically Inspired Computing (NaBIC)* (pp. 1088–1093). IEEE.

Lohokare, M., Panigrahi, B. K., Pattanaik, S. S., Devi, S., & Mohapatra, A. (2010). Optimal load dispatch using accelerated biogeography-based optimization. In *2010 Joint International Conference on Power Electronics, Drives and Energy Systems (PEDES) and 2010 Power India* (pp. 1–5). IEEE.

Lohokare, M. R., Panigrahi, B. K., Pattnaik, S. S., Devi, S., & Mohapatra, A. (2012). Neighborhood search-driven accelerated biogeography-based optimization for optimal load dispatch. *IEEE Transactions on Systems, Man, and Cybernetics—Part C: Applications and Reviews*, doi:10.1109/TSMCC.2012.2190401.

Ma, H. (2010). An analysis of the equilibrium of migration models for biogeography-based optimization. *Information Sciences, 180*, 3444–3464.

Ma, H., & Simon, D. (2011a). Analysis of migration models of biogeography-based optimization using Markov theory. *Engineering Applications of Artificial Intelligence, 24*, 1052–1060.

Ma, H., & Simon, D. (2011b). Blended biogeography-based optimization for constrained optimization. *Engineering Applications of Artificial Intelligence, 24*, 517–525.

Ma, H., Ni, S., & Sun, M. (2009, Dec 16–18). Equilibrium species counts and migration model tradeoffs for biogeography-based optimization. In *48th IEEE Conference on Decision and Control and the 28th Chinese Control Conference*, Shanghai, P.R.China, (pp. 3306–3310). IEEE.

Ma, H.-P., Ruan, X.-Y., & Pan, Z.-X. (2012a). Handling multiple objectives with biogeography-based optimization. *International Journal of Automation and Computing, 9*, 30–36.

Ma, H., Fei, M., Ding, Z., & Jin, J. (2012b). Biogeography-based optimization with ensemble of migration models for global numerical optimization. In *IEEE World Congress on Computational Intelligence (WCCI)*, 10–15 June, Brisbane, Australia (pp. 1–8). IEEE.

Macarthur, R. & Wilson, E. O. (1967). *The theory of island biogeography*. Princeton, NJ: Princeton University Press.

Mahdad, B. & Srairi, K. (2011). Differential evolution based dynamic decomposed strategy for solution of large practical economic dispatch. In *10th International Conference on Environment and Electrical Engineering (EEEIC)* (pp. 1–5). IEEE.

Mandal, K. K., Bhattacharya, B., Tudu, B., & Chakraborty, N. (2011). A novel population-based optimization algorithm for optimal distribution capacitor planning. In *International Conference on Energy, Automation, and Signal (ICEAS)* (pp. 1–6). IEEE.

Mo, H. & Xu, L. (2010a). Biogeography based optimization for traveling salesman problem. In *Sixth International Conference on Natural Computation (ICNC)* (pp. 3143–3147). IEEE.

Mo, H. & Xu, L. (2010b). Biogeography migration algorithm for traveling salesman problem. In Y. Tan., Y. Shi., & K. C. Tan (Eds.), *ICSI 2010, Part I, LNCS 6145* (pp. 405–414). Berlin Heidelberg: Springer.

Mohammed, Z. & Talaq, J. (2012). Unit commitment by biogeography based optimization method. In *16th IEEE Mediterranean Electrotechnical Conference (MELECON)* (pp. 551–554). IEEE.

Mukherjee, R. & Chakraborty, S. (2013). Selection of the optimal electrochemical machining process parameters using biogeography-based optimization algorithm. *International Journal of Advanced Manufacturing Technology*, doi:10.1007/s00170-012-4060-0.

Mukherjee, R., Chakraborty, S., & Samanta, S. (2012). Selection of wire electrical discharge machining process parameters using non-traditional optimization algorithms. *Applied Soft Computing, 12*, 2506–2516.

Nikumbh, S., Ghosh, S., & Jayaraman, V. K. (2012). Biogeography-based informative gene selection and cancer classification using SVM and random forests. In *IEEE World Congress on Computational Intelligence (WCCI)*, 10–15 June, Brisbane, Australia (pp. 187–192). IEEE.

Panchal, V. K., Goel, S., & Bhatnagar, M. (2009). Biogeography based land cover feature extraction. In *World Congress on Nature and Biologically Inspired Computing (NaBIC)* (pp. 1588–1591). IEEE.

Panchal, V. K., Bhugra, D., Goel, S., & Singhania, V. (2011). Study on the behaviour of BBO over natural terrain features. In *3rd International Conference on Electronics Computer Technology (ICECT)* (pp. 28–32). IEEE.

Pandit, M. (2012). Discussion of "Hybrid differential evolution with biogeography-based optimization for solution of economic load dispatch". *IEEE Transactions on Power Systems, 27*, 574–575.

Panduro, M., Mendez, A., Dominguez, R., & Romero, G. (2006). Design of non-uniform circular antenna arrays for side lobe reduction using the method of genetic algorithms. *International Journal of Electronic Communication, 60*, 713–717.

Pattnaik, S. S., Lohokare, M. R., & Devi, S. (2010). Enhanced biogeography-based optimization using modified clear duplicate operator. In *Second World Congress on Nature and Biologically Inspired Computing*, 15–17 Dec, Kitakyushu, Fukuoka, Japan (pp. 715–720). IEEE.

Rabiee, A., Mohammadi-Ivatloo, B., & Ehsan, M. (2012). Discussion of "Hybrid differential evolution with biogeography-based optimization for solution of economic load dispatch". *IEEE Transactions on Power Systems, 27*, 574.

Rahmati, S. H. A., & Zandieh, M. (2012). A new biogeography-based optimization (BBO) algorithm for the flexible job shop scheduling problem. *International Journal of Advanced Manufacturing Technology, 58*, 1115–1129.

Rarick, R., Simon, D., Villaseca, F. E., & Vyakaranam, B. (2009, October). Biogeography-based optimization and the solution of the power flow problem. In *IEEE International Conference on Systems, Man, and Cybernetics (SMC)*, San Antonio, TX, USA, (pp. 1003–1008). IEEE.

Rathi, A., Agarwal, A., Sharma, A., & Jain, P. (2011). A new hybrid technique for solution of economic load dispatch problems based on biogeography based optimization. In *IEEE Region 10 Conference TENCON* (pp. 19–24). IEEE.

Roy, P. K., Ghoshal, S. P., & Thakur, S. S. (2009a). Biogeography based optimization technique applied to multi-constraints economic load dispatch problems. In *Transmission and Distribution Conference and Exposition: Asia and Pacific* (pp. 1–4). IEEE.

Roy, P. K., Ghoshal, S. P., & Thakur, S. S. (2009b). Biogeography based optimization to solve economic load dispatch considering valve point effects. In *World Congress on Nature and Biologically Inspired Computing (NaBIC)* (pp. 1213–1218). IEEE.

Roy, P. K., Ghoshal, S. P., & Thakur, S. S. (2010a). Biogeography-based optimization for economic load dispatch problems. *Electric Power Components and Systems, 38*, 166–181.

Roy, P. K., Ghoshal, S. P., & Thakur, S. S. (2010b). Biogeography based optimization for multi-constraint optimal power flow with emission and non-smooth cost function. *Expert Systems with Applications, 37*, 8221–8228.

Roy, P. K., Ghoshal, S. P., & Thakur, S. S. (2010c). Combined economic and emission dispatch problems using biogeography-based optimization. *Electrical Engineering, 92*, 173–184.

Roy, P. K., Ghoshal, S. P., & Thakur, S. S. (2010d). Multi-objective optimal power flow using biogeography-based optimization. *Electric Power Components and Systems, 38*, 1406–1426.

Roy, P. K., Ghoshal, S. P., & Thakur, S. S. (2011). Optimal reactive power dispatch considering flexible AC transmission system devices using biogeography-based optimization. *Electric Power Components and Systems, 39*, 733–750.

Sharaqa, A. & Dib, N. (2011). Design of linear and circular antenna arrays using biogeography based optimization. In *IEEE Jordan Conference on Applied Electrical Engineering and Computing Technologies (AEECT)* (pp. 1–6). IEEE.

Silva, M. D. A. C. E. & Coelho, L. D. S. (2010). Biogeography-based optimization combined with predator-prey approach applied to economic load dispatch. In *Eleventh Brazilian Symposium on Neural Networks (SBRN)* (pp. 164–169). IEEE.

Silva, M. A. C., Coelho, L. D. S., & Freire, R. Z. (2010). Biogeography-based optimization approach based on predator-prey concepts applied to path planning of 3-DOF robot manipulator. In *IEEE Conference on Emerging Technologies and Factory Automation (ETFA)* (pp. 1–8). IEEE.

Silva, M. D. A. C. E., Coelho, L. D. S., & Lebensztajn, L. (2012). Multiobjective biogeography-based optimization based on predator-prey approach. *IEEE Transactions on Magnetics, 48*, 951–954.

Simon, D. (2008). Biogeography-based optimization. *IEEE Transactions on Evolutionary Computation, 12*, 702–713.

Simon, D. (2011a). A dynamic system model of biogeography-based optimization. *Applied Soft Computing, 11*, 5652–5661.

Simon, D. (2011b). A probabilistic analysis of a simplified biogeography-based optimization algorithm. *Evolutionary Computation, 19*, 167–188.

Simon, D., Ergezer, M., & Du, D. (2009). Population distributions in biogeography-based optimization algorithms with elitism. In *IEEE International Conference on Systems, Man, and Cybernetics (SMC)*, San Antonio, TX, USA, (pp. 991–996). IEEE.

Simon, D., Ergezer, M., Du, D., & Rarick, R. (2011a). Markov models for biogeography-based optimization. *IEEE Transactions on Systems, Man, and Cybernetics—Part B: Cybernetics, 41*, 299–306.

Simon, D., Rarick, R., Ergezer, M., & Du, D. (2011b). Analytical and numerical comparisons of biogeography-based optimization and genetic algorithms. *Information Sciences, 181*, 1224–1248.

Singh, U., Kumar, H., & Kamal, T. S. (2010). Design of Yagi-Uda antenna using biogeography based optimization. *IEEE Transactions on Antennas and Propagation, 58*, 3375–3379.

Sinha, S., Bhola, A., Panchal, V. K., Singhal, S., & Abraham, A. (2012, June 10–15). Resolving mixed pixels by hybridization of biogeography based optimization and ant colony optimization. In *IEEE World Congress on Computational Intelligence (WCCI)*, Brisbane, Australia (pp. 1–6). IEEE.

Song, Y., Liu, M., & Wang, Z. (2010). Biogeography-based optimization for the traveling salesman problems. In *Third International Joint Conference on Computational Science and Optimization (CSO)* (pp. 295–299). IEEE.

Wang, Y. & Cai, Z. (2011, Aug 12–14). A novel hybrid biogeography-based optimization with differential mutation. In *International Conference on Electronic and Mechanical Engineering and Information Technology (EMEIT)* (pp. 2710–2714). IEEE.

Wang, L., & Xu, Y. (2011). An effective hybrid biogeography-based optimization algorithm for parameter estimation of chaotic systems. *Expert Systems with Applications, 38*, 15103–15109.

Yang, X., Cao, J., Li, K., & Li, P. (2011, Oct 19–21). Improved opposition-based biogeography optimization. In *Fourth International Workshop on Advanced Computational Intelligence (IWACI)*, Wuhan, Hubei, China (pp. 642–647). IEEE.

Zhao, B., Deng, C., Yang, Y., & Peng, H. (2012). Novel binary biogeography-based optimization algorithm for the knapsack problem. In Y. Tan., Y. Shi., & Z. Ji (Eds.), *ICSI 2012, Part I, LNCS 7331* (pp. 217–224). Berlin Heidelberg: Springer.

Chapter 6
Cat Swarm Optimization Algorithm

Abstract In this chapter, we present a new population-based method, called cat swarm optimization (CSO) algorithm, which imitates the natural behaviour of cats. We first describe the general knowledge of the behaviour of cats in Sect. 6.1. Then, the fundamentals and performance of CSO are introduced in Sect. 6.2. Next, some selected variations of CSO are explained in Sect. 6.3. Right after this, Sect. 6.4 presents a representative CSO application. Finally, Sect. 6.5 summarises this chapter.

6.1 Introduction

Cats exhibit fascinated social behaviours that have long since attracted the attention of human beings. When we were young, we may have observed that cats have a strong curiosity towards moving objects. We may have also discovered that even though cats spend most of their time in resting, they always remain alert and can possess a good hunting skill. Inspired by these behavioural pattern, Chu and Tsai (2007) proposed a new optimization algorithm called cat swarm optimization (CSO) that involves two modes (i.e., seeking and tracing) of operations for solving complex optimization problems.

6.1.1 Behaviour of Cats

Nowadays, one of the most popular companion animals can be found in homes throughout the world is the domestic cats. However, cat social behaviour is complex and incompletely understood. Recently, researchers have focused their attention on two categories, i.e., resting behaviour and chasing behaviour.

B. Xing and W.-J. Gao, *Innovative Computational Intelligence: A Rough Guide to 134 Clever Algorithms*, Intelligent Systems Reference Library 62,
DOI: 10.1007/978-3-319-03404-1_6,

6.1.1.1 Cat Sleeping/Resting Behaviour

You may interesting about why the cats sleep so much? Several issues are involved (Mills et al. 2010): first, this is because they have different daily sleep-wake cycle that we do. That means, they may sleep too much during the day when we are awake, and spend so much time awake at night when we need to sleep. Second, cats rely strongly on "resting" behaviour to hunt (i.e., sleep time can help cats increase the amount of energy required). Third, cat usually keep "one eye open" condition which means when the cat "sleeping", he still has a very high level of alertness so that he can spring up and into action at a moment's notice.

6.1.1.2 Cat Hunting/Chasing Behaviour

As we know, hunting is one of cats' instinct characteristics and is always a big part of their life. Once a cat is ready for hunting, he will slowly move his body into a crouch and wait, eyes glues on the prey. At this moment, we can say that he is a perfect example in exercising the art of patience. Just before the pounce, he sizes up the distance between himself and the prey by shaking her head so this can better judge his leap. He moves incredibly fast and finally catches his prey. Furthermore, in the wild, the cat usually need to hunt a lot of smaller prey throughout the day to keep himself nourished. To do that, he must be able to rest between each hunt in order to pounce quickly when his target approaches.

6.2 Fundamentals of Cat Swarm Optimization Algorithm

Cat swarm optimization (CSO) algorithm is based on a swarm of N individuals, each evolving in M dimensional space with its coordinates representing a potential solution to a problem with M attributes. In general, there are two sub-modes (i.e., seeking mode and tracing mode) are combined to define the flow of CSO. To achieve an appropriate balance between the exploitation of the search position gathered so far (i.e., seeking mode) and the exploration of untraced or relatively unexplored search space regions (i.e., tracing mode), a mixed ration (MR) is employed. Normally, due to the cats are resting most of the time, the value of MR is very small. Furthermore, according to the memory structures which described in the seeking mode and the velocities behaviour which involved in the tracing mode, the positions of each cat is updated.

6.2.1 Rest and Alert-Seeking Mode

Just like the resting behaviour of the real cats, the objective of the artificial cats in the seeking mode is to observe their environment for the best place to move to given its current position. It presents a higher exploitative capacity. The working principle is very simple that involves two steps: i.e., improvement and update. In general, the seeking behaviour is based on four essential factors, which are described as follows (Chu and Tsai 2007):

- Seeking memory pool (SMP): It gives the size of memory of each cat in which the cats should improve it.
- Seeking range of the selected dimension (SRD): By taking into account the mutative range of cats' positions which composed of M dimensions, SRD guarantees that it not be out of range when a dimension is selected for mutation.
- Counts of dimension to change (CDC): It controls how many of the dimensions will be varied.
- Self position consideration (SPC): It is a Boolean valued variable to decide whether the present position is a candidate solution. Note that no matter the value of SPC is true of false, the value of SMP will not be influenced.

The working procedure of seeking mode is as follows (Chu and Tsai 2007):

- Make j copies of the present position of cat_k, where $j = SMP$. If the value of SPC is true, let $j = (SMP - 1)$, then retain the present position as one of the candidates.
- Whenever a candidate point enters, the position of each copy will be updated according to CDC by randomly adding or subtracting SRD percents the present position value.
- Calculate the fitness value (FS) of all candidate points.
- If all fitness values are equal, set all the selecting probability of each candidate point be 1; otherwise, the selecting probability of each candidate point will be calculated via Eq. 6.1 (Chu and Tsai 2007):

$$P_i = \frac{|FS_i - FS_b|}{FS_{\max} - FS_{\min}}, \quad \text{where } 0 < i < j, \tag{6.1}$$

where FS_b is the best solution so far, $FS_{\max}$is the largest FS in the candidates, and $FS_{\min}$ is the smallest one.

If $FS_b = FS_{\max}$, that means the goal of the fitness function is to find the minimum solution, otherwise, it will be set for $FS_b = FS_{\min}$.

- Randomly pick a newly generated solution to move to and replaces the position of cat_k.

6.2.2 Movement-Tracing Mode

The tracing mode can be seen as an exploration mechanism that avoids quick convergence. (Chu and Tsai 2007) assumed that each cat chase a prey or any moving object according to its velocity. The action of tracing mode can be described as follows (Chu and Tsai 2007):

- Update the velocities for every dimension ($v_{k,\,d}$) via Eq. 6.2 (Chu and Tsai 2007):

$$v_{k,\,d} = v_{k,\,d} + r_1 \cdot c_1 \cdot \left(x_{best,d} - x_{k,\,d}\right), \quad d = 1,\, 2, \ldots, M, \tag{6.2}$$

  where $x_{best,\,d}$ is the position of the best solution found by the cat, $x_{k,\,d}$ is the position of cat_k, c_1 is a constant and r_1 is a random value in the range of $[0, 1]$.
- Check if the velocities are in the range of maximum velocity. In case the new velocity is over-range, it is set equal to the limit.
- Update the position of cat_k via Eq. 6.3 (Chu and Tsai 2007):

$$x_{k,\,d} = x_{k,d} + v_{k,\,d}. \tag{6.3}$$

Overall, the CSO algorithm can be summarized as follows (Chu and Tsai 2007):

- Step 1: Create N cats. For each cat_k, define its positions and velocities via Eqs. 6.4 and 6.5, respectively (Chu and Tsai 2007):

$$x_{k,\,d} = \left(x_{k,\,1}, x_{k,\,2}, \ldots, x_{k,\,d}\right), \tag{6.4}$$

$$v_{k,\,d} = \left(v_{k,\,1}, v_{k,\,2}, \ldots, v_{k,\,d}\right), \tag{6.5}$$

  where $d(1 \le d \le D)$ represents the dimensions.
- Step 2: Initialize the flag of each cat. Randomly pick number of cats and set them into tracing mode according to MR, and the others set into seeking mode.
- Step 3: Evaluate the fitness value of each cat and keep the best position ($x_{best,\,d}$) into the memory.
- Step 4: Move the cats according to their flags, if cat_k is in seeking mode, apply the cat to the seeking mode process, otherwise apply it to the tracing mode process. Both process steps are presented above.
- Step 5: Reinitialize the flag of each cat.
- Step 6: Check the termination condition, if satisfied, terminate the program, otherwise, repeat Step 3 to Step 5.

6.2.3 Performance of CSO

To evaluate the performance of the CSO algorithm, Chu and Tsai (2007) proposed six test functions. Compared with particle swarm optimization (PSO) and PSO with weighting factor (PSO-WF), the CSO algorithm presented a better results of finding the global best solution.

6.3 Selected CSO Variants

Although CSO algorithm is a new member of computational intelligence (CI) family, a number of CSO variations have been proposed in the literature for the purpose of further improving the performance of CSO. This section gives an overview to some of these CSO variants which have been demonstrated to be very efficient and robust.

6.3.1 Parallel CSO Algorithm

Tsai et al. (2008) and (2012) pointed out that one of the active research directions in CSO would be to develop the effective parallelization of CSO. The reason for that is due to the CSO algorithm includes many individual and local procedures.

In 2008, Tsai et al. (2008) proposed a parallel structure of the CSO algorithm, called Parallel CSO (PCSO). In general, PCSO has the same operation as the CSO method for the process of the seeking mode. The main difference between both algorithms is reflected in tracing mode. In CSO, each cat move forward to the global best solution directly, while in PCSO, at each construction step all the cats first move forward to the local best solution of its own group. Furthermore, the individuals of the PCSO algorithm is initially divided into several subpopulations that occasionally exchange solutions through an information exchange process. The main benefit of this procedure is that performs a cooperation mechanism between subpopulations. By comparing the local best solutions collected from the parallel groups, a global best solution can be discovered.

The parallel tracing mode process can be described as follows (Tsai et al. 2008):

- Update the velocities for every dimension ($v_{k,d}(t)$) for the cat_k at the current iteration via Eq. 6.6 (Tsai et al. 2008):

$$v_{k,d}(t) = v_{k,d}(t-1) + r_1 \cdot c_1 \cdot \left[x_{lbest,d}(t-1) - x_{k,d}(t-1)\right], \quad d = 1, 2, \ldots, M, \tag{6.6}$$

where t denotes the iteration number, $x_{lbest,\,d}(t-1)$ denotes the position of the cat who has the best fitness value at the previous iteration in the group that cat_k belongs to, and M denotes the dimension of the solution space.

- Check if the velocities are in the range of maximum velocity. The new velocity is bounded to the maximum velocity in case the new velocity is over-range.
- Update the position of cat_k via Eq. 6.7 (Tsai et al. 2008):

$$x_{k,\,d}(t) = x_{k,\,d}(t-1) + v_{k,\,d}(t). \tag{6.7}$$

Furthermore, in terms of the information exchanging process, Tsai et al. (2008) employed a parameter called ECH to control the whole procedure. The process can be performed through the following steps:

- Pick up a group of subpopulations sequentially and sort the individuals in this group according to their fitness values.
- Randomly select a local best solution from an unrepeatable group.
- The individual whose fitness value is the worst in the group is replaced by the selected local best solution.
- Repeatedly perform Step 1 to Step 3 G times (i.e., the number of subgroups) to let every group receives a local best solution from the others.

Taking into account some basic rules described above, the procedures of the PCSO algorithm can be summarized as follows (Tsai et al. 2008):

- Step 1: Create N cats, randomly sprinkle the cats into the M-dimensional solution space within the constrain ranges of the initial value and randomly collect them into G groups. Meanwhile, generate the velocities for each dimension.
- Step 2: Initialize the flag of each cat. Randomly pick number of cats and set them into tracing mode according to MR, where $MR \in [0, 1]$, and the others set into seeking mode.
- Step 3: Evaluate the fitness value of each cat by taking the coordinates into the fitness function which represents the benchmark and the characteristics of the problem to be solved. After calculating, record the coordinate $x_{best,\,d}$ and the fitness value of the cat which has the best fitness value found so far.
- Step 4: Move the cats according to their flags, if cat_k is in seeking mode, apply the cat to the seeking mode process, otherwise apply it to the parallel tracing mode process. Both process steps are presented above.
- Step 5: Reinitialize the flag of each cat and separate them into statuses that indicating the seeking or the tracing by re-pick $[N \times (1 - MR)]$ cats to move in the seeking mode and $(N \times MR)$ cats to move in the parallel tracing mode.
- Step 6: Check whether the number of iterations reaches a predefined iteration number, if satisfied, apply the information exchanging process.

- Step 7: Check the termination condition. If satisfied, output the coordinate which represents the found best solution and stop; otherwise, go to Step 3.

6.3.1.1 Performance of PCSO

To test the performance of PCSO, three test functions (i.e., Rosenbrock function, Rastrigrin function, and Griewank function) were adopted in (Tsai et al. 2008) and compared with standard CSO, PSO, and PSO-WF. Computational results showed that PCSO performs better then CSO and much better than PSO when the population size is small and the iteration is less. Furthermore, the PCSO algorithm revealed a good convergence and searching ability.

6.3.2 Multiobjective CSO Algorithm

Recently, in order to deal with multi-criteria optimization problems, Pradhan and Panda (2012) developed a new multiobjective cat swarm optimization (called MOCSO), in which the concept of external archive and Pareto dominance is incorporated. The basic idea of the Multiobjective CSO (MOCSO) algorithm utilized the major structure of the CSO method. That means, two modes of operations (i.e., seeking mode and tracing mode) are mathematically modelled as well.

6.3.2.1 Seeking Mode of MOCSO

A term used in this mode is SMP that representing the number of copies of a cat produced in seeking mode. The main steps involved in this mode are as follows (Pradhan and Panda 2012):

- Create $T(=SMP)$ copies of jth cat, i.e., Y_{kd} where $(1 \leq k \leq T)$ and $(1 \leq d \leq D)$. D is the total number of dimensions.
- Apply a mutation operator to Y_k.
- Evaluate the fitness of all mutated copies.
- Update the contents of the archive with the position of those mutated copies which represent non-dominated solutions.
- Pick a candidate randomly from T copies and place it at the position of jth cat.

6.3.2.2 Tracing Mode of MOCSO

In this mode, the rapid chase of the cat is mathematically modelled as a large change in its position. The global best position of the cat is represented as

$X_g = (X_{g1}, X_{g2}, \ldots, X_{gd})$. The main steps involved in tracing mode are outlined as follows (Pradhan and Panda 2012):

- Compute the new velocity of ith cat using Eq. 6.8 (Pradhan and Panda 2012):

$$V_{id} = w \times V_{id} + c \times r \times (X_{gd} - X_{id}), \tag{6.8}$$

  where w is the inertia weight, c is the acceleration constant, and r is a random number uniformly distributed in the range $[0, 1]$. The global best (X_g) is selected randomly from the external archive.
- Compute the new position of ith cat using Eq. 6.9 (Pradhan and Panda 2012):

$$X_{id} = X_{id} + V_{id}. \tag{6.9}$$

- If the new position of ith cat corresponding to any dimension goes beyond the search space, then the corresponding boundary value is assigned to that dimension and the velocity corresponding to that dimension is multiplied by -1 to continue the search in the opposite direction.
- Evaluate the fitness of the cats.
- Update the contents of the archive with the position of those cats which represent non-dominated vectors.
- Overall, the procedures of the MOCSO algorithm are summarized as follows (Pradhan and Panda 2012):
- Step 1: Randomly initialize the position of cats in D-dimensional space, i.e., X_{id} representing position of ith cat in dth dimension.
- Step 2: Randomly initialize the velocity of cats, i.e., V_{id}.
- Step 3: According to *MR*, cats are randomly picked from the population and their flag is set to seeking mode, and for others the flag is set to tracing mode.
- Step 4: Evaluate the fitness of each cat.
- Step 5: Store the position of the cats representing non-dominated solutions in the archive.
- Step 6: If the ith cat is in seeking mode, apply the cat to the seeking mode process, otherwise apply it to the tracing mode process.
- Step 7: Check the termination condition, if satisfied, terminate the program. Otherwise, repeat Steps 3 to 5.

6.3.2.3 Performance of MOCSO

To quantify the efficient of MOCSO, some performance metrics, namely, set coverage metric, generational distance, maximum Pareto-optimal front error, Spacing, and Spread are tested in (Pradhan and Panda 2012). In addition, two multiobjective tested function are proposed. The computational results are compared with the multiobjective particle swarm optimization (MOPSO) and an improved version of non-dominated sorting genetic algorithm (NSGA-II). After several experiments, the performance showed that the proposed algorithm has a superior quality of solution that can cover the full Pareto front.

6.4 Representative CSO Application

In this section, we will introduce how the CSO algorithm can be adapted to solve scheduling optimization problem.

6.4.1 Aircraft Schedule Recovery Problem

The studies relation to the airline problems can be classified into three groups: fleeting-related problems, routing-related problems, and aircraft recovery problems (Sarac et al. 2006). The aircraft schedule recovery problem fell into the last category and it is a typical NP-hard optimization problem. It is happened when the established aircraft schedule has to be changed due to some inevitable reasons, such as the weather or mechanical problems. The flight schedule can be described via Eq. 6.10 (Tsai et al. 2012):

$$\begin{gathered} S = \left\{ s_{ij} \middle| s_{ij} = \left\langle d_{ij}, a_{ij}, \hat{p}_{ij}, \bar{p}_{ij}, \hat{t}_{ij}, \bar{t}_{ij} \right\rangle \right\} \\ \text{subject to} \qquad , \\ 1 \le i \le M,\ 1 \le j \le N \end{gathered} \tag{6.10}$$

where S represents the flight schedule, s_{ij} denotes the assignment of the jth flight to the ith airplane, d_{ij} stands for the duty name of s_{ij}, a_{ij} is the airplane code, $\hat{p}_{ij}$ indicates the origin of s_{ij}, $\bar{p}_{ij}$ denotes the destination of s_{ij}, $\hat{t}_{ij}$ is the departure time from $\hat{p}_{ij}$, $\bar{t}_{ij}$ represents the arrival time at airport $\bar{p}_{ij}$, M stands for the total number of the airplanes, and the number of flights assigned to the airplane is denoted by N.

The recovered flight schedule is denoted by S' with the same definition for all the elements. The constrains and the objective of the flight schedule recovery are defined by five formulas, shown as follows:

- The total flight delay time is defined by Eq. 6.11 (Liu et al. 2009):

$$\phi_1(S, S') = \sum_{i=1}^{M} \sum_{j=1}^{N} x_{ij}, \quad \text{where } x_{ij} = \begin{cases} 0 & \text{if } \hat{t}'_{ij} \le \hat{t}_{ij} \\ \hat{t}'_{ij} - \hat{t}_{ij} & \text{otherwise} \end{cases}. \tag{6.11}$$

- The flight duty swap is expressed by Eq. 6.12 (Liu et al. 2009):

$$\phi_2(S, S') = \sum_{i=1}^{M} \sum_{j=1}^{N} x_{ij}, \tag{6.12}$$

where $x_{ij} = \begin{cases} 1 & \text{if } \left[(j=1) \wedge \left(a'_{ij} \neq i\right)\right] \vee \left[(j \neq 1) \wedge \left(a'_{ij-1} \neq a'_{ij}\right)\right] \\ 0 & \text{otherwise} \end{cases}$.

- The delayed time variance is defined by Eq. 6.13 (Liu et al. 2009):

$$\phi_3(S, S') = \frac{1}{MN} \sum_{i=1}^{M} \sum_{j=1}^{N} \left(x_{ij} - \omega\right)^2, \quad \text{where} \begin{cases} \omega = \frac{1}{MN} \sum_{i=1}^{M} \sum_{j=1}^{N} x_{ij} \\ x_{ij} = \begin{cases} 0 & \text{if } \hat{t}'_{ij} \leq \hat{t}_{ij} \\ \hat{t}'_{ij} - \hat{t}_{ij} & \text{otherwise} \end{cases} \end{cases} \quad (6.13)$$

- The delayed flight number objective is express in Eq. 6.14 (Liu et al. 2009):

$$\phi_4(S, S') = \sum_{i=1}^{M} \sum_{j=1}^{N} x_{ij}, \quad \text{where } x_{ij} = \begin{cases} 0 & \text{if } \hat{t}'_{ij} \leq \hat{t}_{ij} \\ 1 & \text{otherwise} \end{cases}. \quad (6.14)$$

- The objective of the number of long-delayed flights over 30 min is defined by Eq. 6.15 (Liu et al. 2009):

$$\phi_5(S, S') = \sum_{i=1}^{M} \sum_{j=1}^{N} x_{ij}, \quad \text{where } x_{ij} = \begin{cases} 1 & \text{if } \left(\hat{t}'_{ij} - \hat{t}_{ij}\right) \geq 30 \\ 0 & \text{otherwise} \end{cases}. \quad (6.15)$$

The integrated description of those five constraints and the objectives is shown in Eq. 6.16 (Tsai et al. 2012):

$$\begin{aligned} &\text{Minimize: } \lambda(S, S') = \sum_{x=1}^{5} \phi_x(S, S') \\ &\text{Subject to: } s_{ij} \in S, s'_{ij} \in S', 1 \leq i \leq M, 1 \leq j \leq N \end{aligned}, \quad (6.16)$$

where S represents the flight schedule, S' denotes the recovered flight schedule, s_{ij} stands for the assignment of the jth flight to the ith airplane, s'_{ij} is the recovered assignment of the jth flight to the ith airplane, M stands for the total number of the airplanes, and the number of flights assigned to the airplane is denoted by N.

To solve the problem, Tsai et al. (2012) proposed an enhanced version of PCSO, called EPCSO, in which the orthogonal array of the Taguchi method is integrated into the tracing mode process of the PCSO method. The computational results showed that the EPCSO method gets higher accuracies with less running time.

6.5 Conclusions

The roots of the CSO algorithm lay in ethological metaphors of computing models. In general, it can be generalized into two modes, i.e., resting behaviour (seeking mode) and moving behaviour (tracing mode). In essence, the CSO algorithm maintains a population of cats (the swarm), where each cat is defined by its own position in a M dimensional search space and represents a potential solution to the optimization problem (i.e., either minimum or maximum) at hand. The cats start at random positions and move about the search space to look for the prey, i.e., the objective function. The movements of the cats depend on their flags, i.e., if the cats in the seeking mode, their movement according to the memory of the good solutions in the past obtained; on the other hand, if the cats in the tracing mode, their movement according to their velocities for each dimension. The final solution would be the best position of one of the cats.

Although it is a newly introduced CI method, we have witnessed the CSO algorithm being successfully applied to a variety of optimization problems as listed below:

- Adaptive plant modelling (Panda et al. 2011a).
- Cclustering problem (Santosa and Ningrum 2009).
- Economic dispatch problem (Chen et al. 2011).
- Graph colouring problem (Bacarisas and Yusiong 2011).
- IIR system identification (Panda et al. 2011b).
- Image quality problem (Wang et al. 2012; Kalaiselvan et al. 2011; Cui et al. 2013).

Interested readers are referred to them as a starting point for a further exploration and exploitation of cats inspired algorithm.

References

Bacarisas, N. D., & Yusiong, J. P. T. (2011). The effects of varying the fitness function on the efficiency of the cat swarm optimization algorithm in solving the graph coloring problem. *Annals of Computer Science Series, 9*, 17–38.

Chen, J. C., Hwang, J. C., & Pan, J. S. (2011). CSO algorithm for economic dispatch decision of hybrid generation system. *Journal of Energy and Power Engineering, 5*, 73–749.

Chu, S.-C., & Tsai, P.-W. (2007). Computational intelligence based on the behavior of cats. *International Journal of Innovative Computing, Information and Control, 3*, 163–173.

Cui, S.-Y., Wang, Z.-H., Tsai, P.-W., Chang, C.-C., & Yue, S. (2013). Single bitmap block truncation coding of color images using cat swarm optimization. In J.-S. Pan, H.-C. Huang, L. C. Jain, & Y. Zhao (Eds.), *Recent advances in information hiding and applications*. Berlin: Springer.

Kalaiselvan, G., Lavanya, A., & Natrajan, V. (2011). *Enhancing the performance of watermarking based on cat swarm optimization method.* International Conference on Recent Trends in Information Technology (ICRTIT), MIT, Anna University, Chennai, 3–5 June, pp. 1081–1086. IEEE.

Liu, T. K., Chen, C. H., & Chou, J. H. (2009). Optimization of short-haul aircraft schedule recovery problems using a hybrid multiobjective genetic algorithm. *Expert Systems with Applications, 37*, 2307–2315.

Mills, D. S., Marchant-Forde, J. N., Mcgreevy, P. D., Morton, D. B., Nicol, C. J., Phillips, C. J. C., Sandøe, P. & Swaisgood, R. R. (Eds.). (2010). *The encyclopedia of applied animal behaviour and welfare*. Nosworthy Way, Wallingford, Oxfordshire OX10 8DE, UK: CAB International, ISBN 978-0-85199-724-7.

Panda, G., Pradhan, P. M., & Majhi, B. (2011a). Direct and inverse modeling of plants using cat swarm optimization. In B. K. Panigrahi, Y. Shi, & M.-H. Lim (Eds.), *Handbook of swarm intelligence*. Berlin: Springer.

Panda, G., Pradhan, P. M., & Majhi, B. (2011b). IIR system identification using cat swarm optimization. *Expert Systems with Applications, 38*, 12671–12683.

Pradhan, P. M., & Panda, G. (2012). Solving multiobjective problems using cat swarm optimization. *Expert Systems with Applications, 39*, 2956–2964.

Santosa, B. & Ningrum, M. K. (2009). *Cat swarm optimization for clustering*. IEEE International Conference of Soft Computing and Pattern Recognition (SOCPAR), pp. 54–59.

Sarac, A., Batta, R., & Rump, C. M. (2006). A branch-and-price approach for operational aircraft maintenance routing. *European Journal of Operational Research, 175*, 1850–1869.

Tsai, P.-W., Pan, J.-S., Chen, S.-M., & Liao, B.-Y. (2012). Enhanced parallel cat swarm optimization based on the Taguchi method. *Expert Systems with Applications, 39*, 6309–6319.

Tsai, P.-W., Pan, J.-S., Chen, S.-M., Liao, B.-Y. & Hao, S.-P. (2008). Parallel cat swarm optimization. Seventh International Conference on Machine Learning and Cybernetics, 12–15 July, Kunming, China, pp. 3328–3333. IEEE.

Wang, Z.-H., Chang, C.-C., & Li, M.-C. (2012). Optimizing least-significant-bit substitution using cat swarm optimization strategy. *Information Sciences, 192*, 98–108.

Chapter 7
Cuckoo Inspired Algorithms

Abstract In this chapter, a set of cuckoo inspired (CS) optimization algorithms are introduced. We first, in Sect. 7.1, describe the general knowledge of cuckoos. Then, the fundamentals and performance of CS are introduced in Sect. 7.2. Next, the selected variants of CS are outlined in Sect. 7.3 which is followed by a presentation of representative CS application in Sect. 7.4. Right after this, Sect. 7.5 introduces an emerging algorithm, i.e., cuckoo optimization algorithm (COA), which also falls within this category. Finally, Sect. 7.6 draws the conclusions of this chapter.

7.1 Introduction

Edward Jenner (1749–1823) is best remembered, by most of us, for experimentally inoculating his patients with cowpox as an approach of stopping the smallpox and for his subsequent promulgation of antigenic vaccination. Nevertheless, this country doctor and natural historian's first academic achievement, and the work that earned him the fellowship of the Royal Society of London, was a research regarding the common cuckoo (Winfree 1999). In this section, we will provide a quick overview of some interesting lifestyle of cuckoo particularly focusing on its brood parasitism.

7.1.1 Cuckoo: A Brood Parasite

One of the most surprising behavioural patterns exhibited by cuckoos is their host-parasite evolution which normally called brood parasite, i.e., laying their eggs in the nests of other species and depending on them to raise its offspring (Servedio and Lande 2003; Planqué et al. 2002; Langmore et al. 2009). Ornithologists Edward Jenner (1749–1823) discovered that brood parasitism in birds is

B. Xing and W.-J. Gao, *Innovative Computational Intelligence: A Rough Guide to 134 Clever Algorithms*, Intelligent Systems Reference Library 62,
DOI: 10.1007/978-3-319-03404-1_7,

widespread and phylogenetically diverse (Winfree 1999). In addition, there is an especially interesting pattern that most female cuckoos (such as *Tapera*) can mimic their eggs in colour and pattern of their host species (Payne et al. 2005). This evidence has been proved by (Davies and Brooke 1998). They have shown experimentally that host species parasitized with mimetic eggs reject nonmimetic eggs from their nests. Of course, some host birds can engage direct conflict with the intruding cuckoos. For example, if a host bird discovers the eggs are not their own, it will wither throw these alien eggs away or simply abandon its nest and build a new nest elsewhere. However, biologists convinced that due to the inherent characteristics (i.e., evolutionary lag), hosts might need to take longer to evolve complete rejection (Davies and Brooke 1998). As a result, the mimic characteristic by cuckoos reduces the probability of their eggs being abandoned and thus increases their reproductively.

7.2 Fundamentals of the Cuckoo Search Algorithm

Cuckoo search (CS) is an optimization algorithm developed by Yang and Deb (2009). It was inspired by cuckoos' breeding behaviour of parasitic cuckoo species, and in the meantime combining the Lévy flight behaviour discovered in some birds and fruit flies. In addition, this algorithm uses a balanced combination of a local random walk with permutation and the global explorative random walk, controlled by a switching parameter which is related to similarity of the egg/solution to the existing egg/solution (Gandomi et al. 2012). As a result, more similar eggs will be more likely to be survive and be part of the next generation.

7.2.1 Characteristics of Lévy Flight

Broadly speaking, Lévy flights, named after the French mathematician Paul Lévy, are a class of random walks whose step length occurs with a power-law frequency in contrast to a conventional random walk for which larger steps are exponentially rare (Viswanathan et al. 2002; Brown et al. 2007). Often, the step lengths are chosen from a generalized Lévy probability density distribution with a power-law tail by Eq. 7.1 (Yang and Deb 2011):

$$L(s) = |s|^{-1-\beta}, \tag{7.1}$$

where $0 < \beta \leq 2$ is an index.

Mathematically speaking, a simple version of Lévy distribution can be defined via Eq. 7.2 (Yang 2010)

$$L(s,\gamma,\mu) = \begin{cases} \sqrt{\frac{\gamma}{2\pi}}\exp\left[-\frac{\gamma}{2(s-\mu)}\right]\frac{1}{(s-\mu)^{3/2}}, & \text{if} \quad 0<\mu<s<\infty \\ 0 & \text{if} \quad s\leq 0 \end{cases}, \tag{7.2}$$

where $\mu > 0$ is a minimum step, γ is a scale parameter.

Clearly, as $s \to \infty$, we have a special case of the generalized Lévy distribution as Eq. 7.3 (Yang and Deb 2011)

$$L(s,\gamma,\mu) \approx \sqrt{\frac{\gamma}{2\pi}}\frac{1}{s^{3/2}}. \tag{7.3}$$

For general case, the Lévy distribution function should be defined in terms of Fourier transform as Eq. 7.4 (Yang and Deb 2011):

$$F(k) = \exp\left[-\alpha|k|^{\beta}\right], \quad 0<\beta\leq 2, \tag{7.4}$$

where α is a scale parameter.

The inverse of this integral is given via Eq. 7.5 (Yang and Deb 2011):

$$L(s) = \frac{1}{\pi}\int_0^{\infty}\cos(ks)\exp\left[-\alpha|k|^{\beta}\right]dk. \tag{7.5}$$

Obviously, the Eq. 7.5 can be estimated only when $s \to \infty$. We have Eq. 7.6 (Yang and Deb 2011):

$$L(s) \to \frac{\alpha\beta\Gamma(\beta)\sin(\pi\beta/2)}{\pi|s|^{1+\beta}}, \quad s \to \infty. \tag{7.6}$$

Here $\Gamma(z)$ is the Gamma function as Eq. 7.7 (Yang and Deb 2011):

$$\Gamma(z) = \int_0^{\infty} t^{z-1}e^{-t}dt. \tag{7.7}$$

In the case when $z = n$ is an integer, we have $\Gamma(n) = (n-1)!$.

The Lévy distribution has been observed in many systems, such as financial, physical (Shlesinger et al. 1995; Figueiredo et al. 2004; Nakao 2000), and biological (Reynolds 2006; Viswanathan et al. 2002; Hanert 2012). In the context of biological, Lévy flights concept has been broad studied due to there is evidence that most biological organisms perform random walks by searching for the target objects, such as predator–prey, mating partner, and pollinator-flower (Viswanathan et al. 2002).

Inspired by that, a number of algorithmic approaches were developed. The algorithmic examples which demonstrated the characteristics of Lévy flights please refer to (Pavlyukevich 2007; Yang and Deb 2009, 2010a; Wang et al. 2013; Xie et al. 2013). But, why this concept is so efficiency? One of the main reasons is due to the fact that the variance of Lévy flights ($\sigma^2(t) \sim t^{3-\beta}, \quad 1 \le \beta \le 2$) increase much faster than the linear relationship ($\sigma^2(t) \sim t$) of Brownian random walks (Yang and Deb 2011). In other words, it ensures diversification of solution space.

7.2.2 Standard CS Algorithm

To simplicity, in practice three idealized rules are utilized (Yang and Deb 2009):

- Each cuckoo lays one egg at a time, and dumps it in a randomly chosen nest.
- The best nests with high quality of eggs (solutions) will be part of the next generations.
- The number of available host nests remain fixed during the whole solution process, and a host can discover an alien egg with a probability $P_a \in [0, 1]$. This implies that the fraction P_a of n nests is replaced by new nests (with new random solutions).

To implement a CS algorithm, the following steps have to be specified.

- First, an initial population of n host nests $x_i(i = 1, 2, \ldots, n)$ is generated randomly.
- Second, to generate new solutions ($x_i^{(t+1)}$) for the ith cuckoo, a Lévy flight is performed using Eq. 7.8 (Yang and Deb 2009):

$$x_i^{(t+1)} = x_i^{(t)} + \alpha \oplus \text{Lévy } (\lambda), \tag{7.8}$$

where $\alpha(\alpha > 0)$ is the step size which should be related to the scales of the problem of interest. In most case, we can use $\alpha = 1$. The product $\oplus$ means entry-wise multiplications. The parameter Lévy (λ) is the length of random walk and can be drawn from a Lévy distribution for large steps via Eq. 7.9 (Yang and Deb 2009):

$$\text{Lévy} \sim \mu = \mathrm{t}^{-\lambda}, (1 < \lambda \le 3), \tag{7.9}$$

which has an infinite variance with an infinite mean. Here the consecutive jumps/steps of a cuckoo essentially form a random walk process which obeys a power-law step-length distribution with a heavy tail.

In addition, the fitness of the cuckoo is calculated using Eq. 7.10 (Yang and Deb 2009):

$$f(x), x = (x_1, x_2, \ldots x_u)^T. \tag{7.10}$$

Taking into account three basic rules described above, the steps of implementing CS algorithm can be summarized as follows (Yang and Deb 2009):

- Step 1: Randomly generating the initial population of n host nests $x_i(i = 1, 2, \cdots, n)$.
- Step 2: Repeat till stopping criteria met. First, randomly select a cuckoo via Lévy flight. Second, calculate its fitness function (F_i). Third, randomly select a nest among n available nests (for example j). Fourth, if $(F_i > F_j)$, then replace the j by the new solution. Fifth, a fraction (P_a) of worse nests are abandoned and new ones are built. Sixth, calculate fitness and keep the best solutions (or nests with quality solutions). Seventh, rank the solutions and store the current best nest as optimal fitness value.
- Step 3: Post process and visualize results.

7.2.3 Performance of CS

In order to show how the CS algorithm performs, Yang and Deb (2009) used a series of test functions (i.e., both standard and stochastic), namely, De Jong's first function, Rosenbrock's function, Schwefel's function, Ackley's function, Rastrigin's function, Easom's function, Griewangk's function, and two stochastic test functions which designed by Yang (2010). Compared with conventional computational intelligence (CI) techniques [such as genetic algorithm (GA) and particle swarm optimization (PSO)], the CS algorithm is much more efficient in finding the global optima with higher success rates.

7.3 Selected CS Variants

Although CS algorithm is a new member of CI family, a number of CS variations have been proposed in the literature for the purpose of further improving the performance of CS. This section gives an overview to some of these CS variants which have been demonstrated to be very efficient and robust.

7.3.1 Modified CS (MCS) Algorithm

Yang and Deb (2009) pointed out as the cuckoo's egg is very similar to a host's eggs, it is worth to do a random walk in a biased way with some random step sizes. In the light of this statement, a modification of the standard CS (called MCS) was made by Walton et al. (2011) with the aim to speed up convergence. The modification involves two parts (Walton et al. 2011):

- First, the size of the Lévy flight step size has been reset. In CS, α is constant and the value $\alpha = 1$ is used (Yang and Deb 2009). In the MCS, the value of α is decreases as the number of generations increases. The main reason for this choice is to encourage more localised searching as the individuals (i.e., eggs) get closer to the solution. It will be constructed by applying the following simple procedure: (1) an initial value of the Lévy flight step size $A = 1$ is chosen and, (2) at each generation, a new Lévy flight step is calculated using Eq. 7.11 (Walton et al. 2011):

$$\alpha = A/\sqrt{G}, \tag{7.11}$$

where G is the generation number. This exploratory search is only performed on the fraction of nests to be abandoned.

- Second, an additional step of information exchange between the top eggs has been added. Instead of the random walk with permutation, for the MCS, (Walton et al. 2011) represented a sorted method, i.e., a fraction of the eggs with the best fitness are put into a group of top eggs. At each of the top eggs, a second egg in this group is picked at random and a new egg is then generated on the line connecting these two top eggs. The new egg position is calculated using the inverse of the golden ration (see Eq. 7.12), such that it is closer to the egg with the best fitness (Walton et al. 2011):

$$\varphi = \left(1 + \sqrt{5}\right)/2, \tag{7.12}$$

where φ is the inverse of the golden ration.

In the case that both eggs have the same fitness, the new egg is generated at the midpoint. Furthermore, if the same egg is picked twice, a local Lévy flight step size is performed by using Eq. 7.13 (Walton et al. 2011):

$$\alpha = A/G^2. \tag{7.13}$$

7.3.1.1 Performance of MCS

To judge the performance of the MCS algorithm, (Walton et al. 2011) employed seven functions (i.e., Rosenbrock's function, De Jong's function, Rastrigin's function, Schwefel's function, Ackley's function, Griewank's function, and Easom's 2D function) as objective functions which have been used by (Yang and Deb 2009) to test the CS algorithm. Experimental results showed that the MCS algorithm outperform the CS algorithm for all of the standard test examples, especially as the number of dimensions is increased.

7.3.2 Multiobjective CS (MOCS) Algorithm

To deal with multi-criteria optimization problems, in 2011 the CS algorithm's finder Yang and Deb (2011) proposed a new CS algorithm called MOCS for multiobjective optimization. This approach used some weighted sum method to combine multiple objectives to a single objective. In addition, Yang and Deb (2011) employed many different solution points to increase the computational efficient. The goal is to find a solution that gives the best compromise between the various objectives.

For multiobjective optimization problems with k different objectives, Yang and Deb (2011) modify the first and last rules to incorporate multiobjective needs:

- Each cuckoo lays k eggs at a time, and dumps them in a randomly chosen nest. Egg k corresponds to the solution to the kth objective.
- The best nests with high quality of eggs (solutions) will be part of the next generations.
- Each nest will be abandoned with a probability p_a and a new nest with k eggs will be built, according to the similarities/differences of the host eggs. Some random mixing can be used to generate diversity. This implies that the fraction P_a of n nests is replaced by new nests (with new random solutions).

Yang and Deb (2011) pointed out that based on the first rule, the MOCS algorithm built new solutions by performing randomized walks or Lévy flight. At the same time, a crossover operator is carried out over solutions through selective random permutation and generates new solutions. For each nest, there can be k solutions which are generated in the same way as Eq. 7.8. In addition, the second rule can be seen as an elitist strategy allows the MOCS algorithm to find best solutions and ensure the algorithm converge properly. Finally, as an additional means for diversifying the search, the third rule introduced the mutation concept which borrowed from evolutionary computation, so that the worst solutions are discarded with a probability and new solutions are generated. For the MOCS, the mutation is a vectorized operator that combines Lévy flight or differential quality of the solutions.

Based on these three rules, the following steps have to be specified.

- First, an initial population of n host nests $x_i (i = 1, 2, \ldots, n)$ and each with k eggs are generated randomly.
- Second, to generate new solutions ($x_i^{(t+1)}$) for the ith cuckoo, a Lévy flight is performed using Eq. 7.14 (Yang and Deb 2011):

$$x_i^{(t+1)} = x_i^{(t)} + \alpha \oplus \text{Lévy } (\beta), \tag{7.14}$$

where $\alpha (\alpha > 0)$ is the step size which should be related to the scales of the problem of interest. In most case, we can use $\alpha = 1$. In order to accommodate the difference between solution quality, we can also use Eq. 7.15 (Yang and Deb 2011):

$$\alpha = \alpha_0 \left(x_j^{(t)} - x_i^{(t)} \right), \tag{7.15}$$

where α_0 is a constant, while $\left(x_j^{(t)} - x_i^{(t)} \right)$ represents the difference of two randomly solutions. The product $\oplus$ means entry-wise multiplications. The parameter Lévy (β) is the length of random walk and can be drawn from a Lévy distribution for large steps via Eq. 7.16 (Yang and Deb 2011):

$$\text{Lévy} \sim \mu = t^{-1-\beta}, \quad (0 < \beta \leq 2), \tag{7.16}$$

which has an infinite variance with an infinite mean. Here the consecutive jumps/steps of a cuckoo essentially form a random walk process which obeys a power-law step-length distribution with a heavy tail.

From the implementation point of view, obviously, the generation of steps (s) is very important which obey the chosen Lévy distribution. In fact, there are a few ways of achieving this. One of the most simple scheme is discussed in detail by (Yang 2010; Yang and Deb 2010b) and can be summarized via Eq. 7.17 (Yang and Deb 2011):

$$s = \alpha_0 \left(x_j^{(t)} - x_i^{(t)} \right) \oplus \text{Lévy}(\beta) \sim 0.01 \frac{\mu}{|v|^{1/\beta}} \left(x_j^{(t)} - x_i^{(t)} \right), \tag{7.17}$$

where μ and v are drawn from normal distribution via Eqs. 7.18 and 7.19, respectively (Yang and Deb 2011):

$$\mu \sim N\left(0, \sigma_\mu^2\right), \tag{7.18}$$

$$v \sim N\left(0, \sigma_v^2\right), \tag{7.19}$$

where the parameters (i.e., σ_μ^2 and σ_v^2) can be defined by Eqs. 7.20 and 7.21, respectively (Yang and Deb 2011):

$$\sigma_\mu^2 = \left\{ \frac{\Gamma(1+\beta)\sin(\pi\beta/2)}{\Gamma[(1+\beta)/2]\beta 2^{(\beta-1)/2}} \right\}^{1/\beta}, \tag{7.20}$$

$$\sigma_v^2 = 1, \tag{7.21}$$

where Γ is the standard Gamma function.

Overall, it is worth mentioning that there are three unique features of the MOCS algorithm (Yang and Deb 2011): exploration by Lévy flight, mutation by a combination of Lévy flight and vectorized solution difference, crossover by selective random permutation, and elitism.

7.3.2.1 Performance of MOCS

To validate the MOCS algorithm, Yang and Deb (2011) used a set of multi-objective test functions, namely, Schaffer's MIn-Min (SCH) function, ZDT1 function, ZDT2 function, ZDT3 function, and LZ function. Compared with convention CI techniques (such as vector evaluated genetic algorithm (VEGA), NSGA-II, multiobjective differential evolution (MODE), and differential evolution for multiobjective optimization (DEMO)), the MOCS algorithm performed well for almost all these test problems.

7.4 Representative CS Application

As we know, in real life most of the problems belong to a class of combinatorial (discrete) or numerical optimization which can be further divided into the following categories: routing, assignment, scheduling, and subset problems. In this section, we introduced how the CS algorithm can be adapted to solve those problems, in particular, the scheduling optimization problem.

7.4.1 Scheduling Optimization Problem

Generally speaking, scheduling is concerned with the allocation of scarce resources to tasks over time. It is an important decision-making task in the manufacturing industries. Burnwal and Deb (2013) studied the scheduling problems arising in five flexible manufacturing cells (FMCs). Each cell is serviced by one to three robots for material handling inside the cell and thus also called a robotic cell. The objective is to allocate the resource to the jobs so as to minimize the total penalty cost and machine idle time.

To validate the performance of the CS algorithm, a comparison between the CS algorithm and other conventional methods (i.e., GA, PSO) has been proposed by Burnwal and Deb (2013). Computational results showed that the CS algorithm is better than the aforementioned methods.

7.5 Emerging Cuckoo Inspired Algorithms

In addition to the aforementioned CS algorithm, the lifestyle of cuckoo also motivates researchers to develop another cuckoo inspired innovative CI algorithm.

7.5.1 Fundamentals of the Cuckoo Optimization Algorithm

Cuckoo optimization algorithm (COA) was proposed by Rajabioun (2011). Similar to conventional evolutionary CI algorithms, the proposed COA also starts with an initial population of cuckoos which have some eggs to be laid in some host birds' nest. The fate of these eggs can be roughly classified as follows (Rajabioun 2011):

- The similarity degree between the cuckoo eggs and the host bird's own eggs is high: This class of cuckoo eggs enjoys a higher chance of being hatched by the host bird and thus growing up as an adult cuckoo;

- The similarity degree between the cuckoo eggs and the host bird's own eggs is low: This class of cuckoo eggs suffers a higher chance of being disposed by the host bird.

Based on this rule, the number of hatched eggs reveals the suitability of the nests in the chosen area. In other words, the more cuckoo eggs being hatched in an area, the more gain (or profit) is linked to that area. Basically, COA works as follows (Rajabioun 2011).

7.5.1.1 Initial Cuckoo Habitat Generation

When dealing with an optimization problem, the values of problem variables are normally formed as an array. Unlike the terminologies such as "chromosome" and "particle position" commonly found in GA and PSO, in COA, "habitat" is used to describe this array. In a N_{var}-dimension optimization problem, a habitat can be regarded as an array of $1 \times N_{\text{var}}$, representing the present living location of a cuckoo. The general form of this array can be expressed via Eq. 7.22 (Rajabioun 2011):

$$\text{Habitat} = [x_1, x_2, \ldots, x_{N_{\text{var}}}], \tag{7.22}$$

where a floating point number is used to represent the value of each variable $(x_1, x_2, \ldots, x_{N_{\text{var}}})$.

Through the evaluation of a profit function f_p at a habitat of $(x_1, x_2, \ldots, x_{N_{\text{var}}})$, the profit of a habitat can be acquired via Eq. 7.23 (Rajabioun 2011):

$$\text{Profit} = f_p(\text{habitat}) = f_p(x_1, x_2, \ldots, x_{N_{\text{var}}}). \tag{7.23}$$

As we can see that COA is an algorithm that can be used for profit maximization problems. In order to minimize a cost function via COA, we can simply maximize the profit function by Eq. 7.24 (Rajabioun 2011):

$$\text{Profit} = -\text{Cost}(\text{habitat}) = -f_c(x_1, x_2, \ldots, x_{N_{\text{var}}}). \tag{7.24}$$

In COA, a matrix of candidate habitat with the size of $N_{\text{pop}} \times N_{\text{var}}$ is first generated. Right after this, some amount of eggs (randomly selected numbers) are supposed for each of these initial cuckoo habitats. The upper- and lower-boundary of the egg numbers are set to 20 and 5, respectively. This setting is based on the real world cuckoo breeding-style. Meanwhile Rajabioun (2011) also named a variable which is called "Egg Laying Radius (ELR)". The ELR is introduced to mimic the other habit of real cuckoo, i.e., laying eggs within a maximum distance from its habitat. Mathematically, ELR is defined as Eq. 7.25 (Rajabioun 2011):

$$\text{ELR} = \alpha \times \frac{\text{Number of current cuckoo's eggs}}{\text{Total number of eggs}} \times (\text{var}_{hi} - \text{var}_{low}), \tag{7.25}$$

where var_{hi} and var_{low} are upper- and lower-limit for variables; α is an integer (designed to address the maximum value of ELR).

7.5.1.2 Cuckoo Eggs' Placement

Within the range of ELR, each cuckoo begins laying eggs in some randomly chosen host birds' nest. Once this procedure is done, certain amount of cuckoo eggs could not survive which means they will be spotted by host birds and thus being discarded. Therefore, in COA, a variable of $p\%$ is designed to address this issue, i.e., $p\%$ of all cuckoo eggs which have less profit values will be killed.

7.5.1.3 Adult Cuckoos' Emigration

When young cuckoos become mature enough to have their own eggs, they often emigrate to another habitat. The newly selected habitat which normally has more desirable host birds and also sufficient food supply. In COA, when different cuckoo groups are formed in various areas, the society with the best profit value is often selected as the goal point where other cuckoos can move to. However, once the adult cuckoos scatter in a broad area, it is often not easy to pinpoint the place where a cuckoo comes from. To deal with this problem, Rajabioun (2011) employed the K-means clustering method for the purpose of grouping cuckoos. Once the cuckoo groups are classified, their corresponding mean value of profit can be calculated. By comparing these results, the one with the maximum mean value will be regarded as the goal group and thus the group's habitat is selected as the new target habitat for other ready-to-emigrate cuckoos.

In addition to abovementioned grouping problem, the inventor of COA also took the following scenario into account: When flying toward goal habitat, the cuckoos may stop at halfway or deviate from their original direction. Two parameters, λ and φ are engineered to assist a cuckoo in looking for more candidate positions across the area. According to (Rajabioun 2011), their expression can be found via Eq. 7.26:

$$\begin{aligned} \lambda &\sim U(0,1) \\ \varphi &\sim U(-\omega,\omega) \end{aligned}, \tag{7.26}$$

where $\lambda \sim U(0,1)$ represents an uniformly distributed random number between 0 and 1; ω is a parameter which control the deviation degree from the target habitat. Rajabioun (2011) suggested that an ω of $\pi/6$ would be sufficient for leading to a good convergence.

7.5.1.4 Cuckoo Population's Control

It is normally unreasonable to assume that the population of cuckoo will grow forever. Since there are many influencing factors such as being killed by predators and the limitation of food which will limit the total number of cuckoos in a certain territory. Being aware this fact, Rajabioun (2011) also introduced N_{max} to express the maximum amount of cuckoos that can live in an area.

7.5.1.5 Convergence

After certain round of iterations, a habitat with the maximum egg similarity to the host birds' and also the maximum food supply will be found by the majority of cuckoo population. Under such situation, this habitat will generate the maximum profit ever. In COA, a convergence criterion that is more than 95 % of all cuckoos move to the same habitat will be used for terminating the algorithm (Rajabioun 2011).

Overall, taking into account some basic rules described above, the procedures of COA can be summarized as follows (Rajabioun 2011):

- Step 1: Initializing cuckoo habitats with some random points on the profit function.
- Step 2: Dedicating some eggs to each cuckoo.
- Step 3: Defining ELR for each cuckoo.
- Step 4: Letting cuckoos lay eggs within their corresponding ELR.
- Step 5: Killing those eggs that have been recognized by the host birds.
- Step 6: Letting the survived eggs hatch and chicks grow.
- Step 7: Evaluating the habitat of each newly grown cuckoo.
- Step 8: Limiting the total number of cuckoo population in the area and eliminating those who live in worst habitats.
- Step 9: Clustering cuckoos, looking for best group, and choosing the goal habitat.
- Step 10: Letting new cuckoo population emigrate toward goal habitat.
- Step 11: Evaluating if the stopping criterion is met, if not, go to Step 2.

7.5.1.6 Performance of COA

In order to test the performance of the COA algorithm, Rajabioun (2011) employed the following problem set:

- Four Benchmark functions which are selected from the literature;
- One 10-dimensional Rastrigin function; and
- A real world case study, namely, multi-input multi-output (MIMO) distillation column process.

For the first two problem sets, the comparison was made among COA, standard PSO, and GA (equipped with roulette wheel selection and uniform cross-over). The results indicated that COA outperform its competitive algorithms for all five benchmark cost functions.

In terms of the MIMO process case study, the obtained results via COA are compared with GA and a non-CI approach called decentralized relay feedback (DRF). Considering all the parameter value found by these methods, the COA algorithm is clear the best of all three.

7.6 Conclusions

In this chapter, two CI algorithms (i.e., CS and COA) inspired by the breeding behaviour of real cuckoos were introduced. In terms of CS, like other population-based algorithms, the algorithm use reproduction operators to explore the search space. The working principle is that each individual (i.e., egg) in the algorithm represents directly a solution to the problem under consideration. If the cuckoo egg is very similar to the host's, then this egg will survive and be part of the next generation. The objective is to find the new and potentially better solutions (Yang and Deb 2009). In addition, the unique characteristics of the CS over other conventional methods lie in that it dependents only on a relatively small number of parameters to define and determine the algorithm's performance. Therefore, it is very easy to implement. Although they are newly introduced CI methods, we have witnessed the following rapid spreading of at least one of them, i.e., CS:

First, in addition to the selected variant, several enhanced versions of CS can also be found in the literature as outlined below:

- Hybrid CS (Chandrasekaran and Simon 2012; Dhivya et al. 2011; Ghodrati and Lotfi 2012a, b; Perumal et al. 2011).
- Enhanced CS (Li et al. 2013).
- Discrete CS (Ouaarab et al. 2013).
- Modified CS (Salimi et al. 2012; Tuba et al. 2011).
- Improved CS (Valian et al. 2011).

Second, apart from the representative application, the CS algorithm has also been successfully applied to a variety of optimization problems as listed below:

- Chemical engineering problem (Bhargava et al. 2013).
- Clustering problem (Senthilnath et al. 2012).
- Data gathering problem (Dhivya and Sundarambal 2011).
- Image segmentation problem (Agrawal et al. 2013).
- Distributed generation allocation problem (Moravej and Akhlaghi 2013)
- Energy efficient computation of data fusion problem (Dhivya et al. 2011).
- Milling operation optimization problem (Yildiz 2013).

- Feedforward Neural network training problem (Valian et al. 2011).
- Scheduling problem (Chandrasekaran and Simon 2012; Rabiee and Sajedi 2013).
- Semantic web service composition problem (Chifu et al. 2012).
- Software implementation problem (Bacanin 2011, 2012; Kalpana and Jeyakumar 2011; Perumal et al. 2011).
- Structural optimization problems (Yang and Deb 2010a, b; Durgun and Yildiz 2012; Gandomi et al. 2013; Yang and Deb 2011; Kaveh and Bakhshpoori 2011; Gandomi et al. 2012; Kaveh et al. 2012; Bulatović et al. 2013).
- System reliability optimization problem (Valian et al. 2013).
- Travelling salesman problem (Ouaarab et al. 2013).

Regarding the COA, the applications is still very limited at this stage. Interested readers are referred to the studies mentioned in this chapter together with several excellent reviews [e.g. (Yang and Deb 2013; Xing et al. 2013; Civicioglu and Besdok 2013; Walton et al. 2013)] as a starting point for further exploration and exploitation of the cuckoo inspired algorithms.

References

Agrawal, S., Panda, R., Bhuyan, S., & Panigrahi, B. K. (2013). Tsallis entropy based optimal multilevel thresholding using cuckoo search algorithm. *Swarm and Evolutionary Computation, 11*, 16–30.

Bacanin, N. (2011, April 28–30). *An object-oriented software implementation of a novel cuckoo search algorithm*. European Computing Conference (ECC '11), *Paris* (pp. 245–250). Paris: WSEAS Press.

Bacanin, N. (2012). Implementation and performance of an object-oriented software system for cuckoo search algorithm. *International Journal of Mathematics and Computers in Simulation, 6*, 185–193.

Bhargava, V., Fateen, S. E. K., & Bonilla-Petriciolet, A. (2013). Cuckoo search: A new nature-inspired optimization method for phase equilibrium calculations. *Fluid Phase Equilibria, 337*, 191–200.

Brown, C. T., Liebovitch, L. S., & Glendon, R. (2007). Lévy flights in Dobe Ju/'hoansi foraging patterns. *Human Ecology, 35*, 129–138.

Bulatović, R. R., Đorđević, S. R., & Đorđević, V. S. (2013). Cuckoo search algorithm: a metaheuristic approach to solving the problem of optimum synthesis of a six-bar double dwell linkage. *Mechanism and Machine Theory, 61*, 1–13.

Burnwal, S., & Deb, S. (2013). Scheduling optimization of flexible manufacturing system using cuckoo search-based approach. *International Journal of Advanced Manufacturing Technology, 64*, 951–959.

Chandrasekaran, K., & Simon, S. P. (2012). Multi-objective scheduling problem: Hybrid approach using fuzzy assisted cuckoo search algorithm. *Swarm and Evolutionary Computation*. doi:10.1016/j.swevo.2012.01.001.

Chifu, V. R., Pop, C. B., Salomie, I., Suia, D. S., & Niculici, A. N. (2012). Optimizing the semantic web service composition process using cuckoo search. In F. M. T. Brazier (Ed.), *Intelligent distributed computing V, SCI 382, Berlin* (pp. 93–102). Berlin: Springer.

Civicioglu, P., & Besdok, E. (2013). A conceptual comparison of the cuckoo-search, particle swarm optimization, differential evolution and artificial bee colony algorithms. *Artificial Intelligence Review, 39*, 315–346.

Davies, N. B., & Brooke, M. D. L. (1998). Cuckoos versus hosts: Experimental evidence for coevolution. In S. I. Rothstein & S. K. Robinson (Eds.), *Parasitic birds and their hosts: Studies in coevolution*. Oxford: Oxford University Press.

Dhivya, M., & Sundarambal, M. (2011). Cuckoo search for data gathering in wireless sensor networks. *International Journal of Mobile Communications, 9*, 642–656.

Dhivya, M., Sundarambal, M., & Anand, L. N. (2011). Energy efficient computation of data fusion in wireless sensor networks using cuckoo based particle approach (CBPA). *International Journal of Communications, Network and System Sciences, 4*, 249–255.

Durgun, İ., & Yildiz, A. R. (2012). Structural design optimization of vehicle components using cuckoo search algorithm. *Materials Testing, 54*, 185–188.

Figueiredo, A., Gleria, I., Matsushita, R., & Silva, S. D. (2004). Lévy flights, autocorrelation, and slow convergence. *Physica A, 337*, 369–383.

Gandomi, A. H., Talatahari, S., Yang, X.-S., & Deb, S. (2012). Design optimization of truss structures using cuckoo search algorithm. *The Structural Design of Tall and Special Buildings*. doi:10.1002/tal.1033.

Gandomi, A. H., Yang, X.-S., & Alavi, A. H. (2013). Cuckoo search algorithm: a metaheuristic approach to solve structural optimization problems. *Engineering with Computers*. doi:10.1007/s00366-011-0241-y.

Ghodrati, A., & Lotfi, S. (2012a). A hybrid CS/GA algorithm for global optimization. In: K. Deep (Ed.), *Proceedings of the International Conference on SocProS 2011, AISC 130, India* (pp. 397–404). India: Springer.

Ghodrati, A., & Lotfi, S. (2012b). A hybrid CS/PSO algorithm for global optimization. In: J.-S. Pan, S.-M. Chen & N. T. Nguyen (Eds.), *ACIIDS 2012, Part III, LNAI 7198, Berlin* (pp. 89–98). Berlin: Springer.

Hanert, E. (2012). Front dynamics in a two-species competition model driven by Lévy flights. *Journal of Theoretical Biology, 300*, 134–142.

Kalpana, A. M., & Jeyakumar, A. E. (2011). An questionnaire based assessment method for process improvement in Indian small scale software organizations. *European Journal of Scientific Research, 60*, 379–395.

Kaveh, A., & Bakhshpoori, T. (2011). Optimum design of steel frames using cuckoo search algorithm with lévy flights. *The Structural Design of Tall and Special Buildings*. doi:10.1002/tal.754.

Kaveh, A., Bakhshpoori, T., & Ashoory, M. (2012). An efficient optimization procedure based on cuckoo search algorithm for practical design of steel structures. *International Journal of Optimization in Civil Engineering, 2*, 1–14.

Langmore, N. E., Stevens, M., Maurer, G., & Kilner, R. M. (2009). Are dark cuckoo eggs cryptic in host nests? *Animal Behaviour, 78*, 461–468.

Li, X., Wang, J., & Yin, M. (2013). Enhancing the performance of cuckoo search algorithm using orthogonal learning method. *Neural Computing and Applications*. doi:10.1007/s00521-013-1354-6.

Moravej, Z., & Akhlaghi, A. (2013). A novel approach based on cuckoo search for DG allocation in distribution network. *Electrical Power and Energy Systems, 44*, 672–679.

Nakao, H. (2000). Multi-scaling properties of truncated Lévy flights. *Physics Letters A, 266*, 282–289.

Ouaarab, A., Ahiod, B., & Yang, X.-S. (2013). Discrete cuckoo search algorithm for the travelling salesman problem. *Neural Computing and Applications*. doi:10.1007/s00521-013-1402-2.

Pavlyukevich, I. (2007). Lévy flights, non-local search and simulated annealing. *Journal of Computational Physics, 226*, 1830–1844.

Payne, R. B., Sorenson, M. D., & Klitz, K. (2005). *The cuckoos*. Oxford: Oxford University Press.

Perumal, K., Ungati, J. M., Kumar, G., Jain, N., Gaurav, R., & Srivastava, P. R. (2011). Test data generation: a hybrid approach using cuckoo and tabu Search. In: B. K. Panigrahi (Ed.), *Swarm, Evolutionary, and Memetic Computing (SEMCCO), Part II, LNCS 7077, Berlin* (pp. 46–54). Berlin: Springer.

Planqué, R., Britton, N. F., Franks, N. R., & Peletier, M. A. (2002). The adaptiveness of defence strategies against cuckoo parasitism. *Bulletin of Mathematical Biology, 64*, 1045–1068.

Rabiee, M., & Sajedi, H. (2013). Job scheduling in grid computing with cuckoo optimization algorithm. *International Journal of Computer Applications, 62*, 38–43.

Rajabioun, R. (2011). Cuckoo optimization algorithm. *Applied Soft Computing, 11*, 5508–5518.

Reynolds, A. M. (2006). Cooperative random Lévy flight searches and the flight patterns of honeybees. *Physics Letters A, 354*, 384–388.

Salimi, H., Giveki, D., Soltanshahi, M. A., & Hatami, J. (2012). Extended mixture of MLP experts by hybrid of conjugate gradient method and modified cuckoo search. *International Journal of Artificial Intelligence and Applications, 3*, 1–13.

Senthilnath, J., Das, V., Omkar, S. N., & Mani, V. (2012). Clustering using levy flight cuckoo search. In: J. C. Bansal (Ed.), *Proceedings of Seventh International Conference on Bio-Inspired Computing: Theories and Applications (BIC-TA), Advances in Intelligent Systems and Computing, India* (Vol. 202, pp. 65–75). India: Springer.

Servedio, M. R., & Lande, R. (2003). Coevolution of an avian host and its parasitic cuckoo. *Evolution, 57*, 1164–1175.

Shlesinger, M. F., Zaslavsky, G. M., & Frisch, U. (Eds.). (1995). *Lévy flights and related topics in physics*. Berlin: Springer.

Tuba, M., Subotic, M., & Stanarevic, N. (2011, April 28–30). Modified cuckoo search algorithm for unconstrained optimization problems. *European Computing Conference (ECC '11), Paris* (pp. 263–268). Paris: WSEAS Press.

Valian, E., Mohanna, S., & Tavakoli, S. (2011). Improved cuckoo search algorithm for feedforward neural network training. *International Journal of Artificial Intelligence and Applications, 2*, 36–43.

Valian, E., Tavakoli, S., Mohanna, S., & Haghi, A. (2013). Improved cuckoo search for reliability optimization problems. *Computers and Industrial Engineering, 64*, 459–468.

Viswanathan, G. M., Bartumeus, F., Buldyrev, S. V., Catalan, J., Fulco, U. L., Havlin, S., et al. (2002). Lévy flight random searches in biological phenomena. *Physica A, 314*, 208–213.

Walton, S., Hassan, O., & Morgan, K. (2013). Selected engineering applications of gradient free optimisation using cuckoo search and proper orthogonal decomposition. *Archives of Computational Methods in Engineering, 20*, 123–154.

Walton, S., Hassan, O., Morgan, K., & Brown, M. R. (2011). Modified cuckoo search: a new gradient free optimisation algorithm. *Chaos, Solitons and Fractals, 44*, 710–718.

Wang, G., Guo, L., Gandomi, A. H., Cao, L., Alavi, A. H., Duan, H., et al. (2013). Lévy-flight krill herd algorithm. *Mathematical Problems in Engineering, 2013*, 1–14.

Winfree, R. (1999). Cuckoos, cowbirds and the persistence of brood parasitism. *Trends in Ecology and Evolution, 14*, 338–343.

Xie, J., Zhou, Y., & Chen, H. (2013). A novel bat algorithm based on differential operator and Lévy flights trajectory. *Computational Intelligence and Neuroscience, 2013*, 1–13.

Xing, B., Gao, W.-J., & Marwala, T. (2013, April 15–19). An overview of cuckoo-inspired intelligent algorithms and their applications. *IEEE Symposium Series on Computational Intelligence (IEEE SSCI), Singapore* (pp. to appear). Singapore: IEEE.

Yang, X.-S. (2010). *Engineering optimization: an introduction with metaheuristic applications*, Hoboken: Wiley, Inc. ISBN 978-0-470-58246-6.

Yang, X.-S., & Deb, S. (2009, December 9–11). Cuckoo search via Lévy flights. *World Congress on Nature and Biologically Inspired Computing (NaBIC), India* (pp. 210–214). India: IEEE.

Yang, X.-S., & Deb, S. (2010a). Eagle strategy using Lévy walk and firefly algorithms for stochastic optimization. In: J. R. Gonzalez (Ed.), *Nature Inspired Cooperative Strategies for Optimization (NISCO 2010), SCI 284, Berlin* (pp. 101–111). Berlin: Springer.

Yang, X.-S., & Deb, S. (2010b). Engineering optimisation by cuckoo search. *International Journal of Mathematical Modelling and Numerical Optimisation, 1*, 330–343.
Yang, X.-S., & Deb, S. (2011). Multiobjective cuckoo search for design optimization. *Computers and Operations Research.* doi:10.1016/j.cor.2011.09.026.
Yang, X.-S., & Deb, S. (2013). Cuckoo search: recent advances and applications. *Neural Computing and Applications*. doi:10.1007/s00521-013-1367-1.
Yildiz, A. R. (2013). Cuckoo search algorithm for the selection of optimal machining parameters in milling operations. *International Journal of Advanced Manufacturing Technology*. doi:10.1007/s00170-012-4013-7.

Chapter 8
Luminous Insect Inspired Algorithms

Abstract In this chapter, we present three algorithms that are inspired by the flashing behaviour of luminous insects, i.e., firefly algorithm (FA), glowworm swarm optimization (GlSO) algorithm, and bioluminescent swarm optimization (BiSO) algorithm. We first describe the general knowledge of the luminous insects in Sect. 8.1. Then, the fundamentals, performances and selected applications of FA, GlSO algorithm and BiSO algorithm are introduced in Sects. 8.2, 8.3 and 8.4, respectively. Finally, Sect. 8.5 summarises this chapter.

8.1 Introduction

The flashing light of luminous insects is an amazing sight in the summer sky. More information about firefly flash code evolution please refer to (Buck and Case 2002). In this chapter, we presented three algorithms that are inspired by the flashing behaviour of luminous insects, i.e., firefly algorithm (FA), glowworm swarm optimization (GlSO) algorithm, and bioluminescent swarm optimization (BiSO) algorithm.

8.2 Firefly Algorithm

8.2.1 Fundamentals of Firefly Algorithm

Firefly algorithm (FA) is a nature-inspired, optimization algorithm which is based on the social (flashing) behaviour of fireflies, or lighting bugs, in the summer sky in the tropical temperature regions (Yang 2008, 2009, 2010b). In the FA, physical entities (fireflies) are randomly distributed in the search space. They carry a bio-luminescence quality, called luciferin, as a signal to communicate with other fireflies, especially to prey attractions (Babu and Kannan 2002). In detail, each

B. Xing and W.-J. Gao, *Innovative Computational Intelligence: A Rough Guide to 134 Clever Algorithms*, Intelligent Systems Reference Library 62, DOI: 10.1007/978-3-319-03404-1_8,

firefly is attracted by the brighter glow of other neighbouring fireflies. The attractiveness decreases as their distance increases. If there is no brighter one than a particular firefly, it will move randomly. Its main merit is the fact that the FA uses mainly real random numbers and is based on the global communication among the swarming particles (i.e., the fireflies), and as a result, it seems more effective in multi-objective optimization.

Normally, FA uses the following three idealized rules to simplify its search process to achieve an optimal solution (Yang 2010b):

- Fireflies are unisex so that one firefly will be attracted to other fireflies regardless of their sex, that means no mutation operation will be done to alter the attractiveness fireflies have for each other;
- The sharing of information or food between the fireflies is proportional to the attractiveness that increases with a decreasing Cartesian or Euclidean distance between them due to the fact that the air absorbs light. Thus for any two flashing fireflies, the less bright one will move towards the brighter one. If there is no brighter one than a particular firefly, it will move randomly; and
- The brightness of a firefly is determined by the landscape of the objective function. For the maximization problems, the light intensity is proportional to the value of the objective function.

Furthermore, there are two important issues in the FA that are the variation of light intensity or brightness and formulation of attractiveness. Yang (2008) simplifies a firefly's attractiveness β (determined by its brightness *I*) which in turn is associated with the encoded objective function. As light intensity and thus attractiveness decreases as their distance from the source increases, the variations of light intensity and attractiveness should be monotonically decreasing functions.

- Variation of light intensity: Suppose that there exists a swarm of n fireflies, and x_i, $i = 1, 2, \ldots, n$ represents a solution for a firefly i initially positioned randomly in the space, whereas $f(x_i)$ denotes its fitness value. In the simplest form, the light intensity $I(r)$ varies with the distance r monotonically and exponentially. That is determined by Eq. 8.1 (Yang 2008, 2009, 2010b):

$$I = I_0 e^{-\gamma r_{ij}}, \tag{8.1}$$

where I_0 is the original light intensity, γ is the light absorption coefficient, and$r$ is the distance between firefly i and firefly j at x_i and x_j as Cartesian distance $r_{ij} = \|x_i - x_j\| = \sqrt{\sum_{k=1}^{d} (x_{i,k} - x_{j,k})^2}$ or the ℓ_2-norm, where $x_{i,k}$ is the kth

component of the spatial coordinate x_i of the ith firefly and d is the number of dimensions we have, for $d = 2$, we have $r_{ij} = \sqrt{(x_i - x_j)^2 + (y_i - y_j)^2}$.

- Movement toward attractive firefly: A firefly attractiveness is proportional to the light intensity seen by adjacent fireflies (Yang 2008). Each firefly has its distinctive attractiveness β which implies how strong it attracts other members of the swarm. However, the attractiveness is relative; it will vary with the distance between two fireflies. The attractiveness function $\beta(r)$ of the firefly is determined via Eq. 8.2 (Yang 2008, 2009, 2010b):

$$\beta = \beta_0 e^{-\gamma r_{ij}^2}, \tag{8.2}$$

where β_0 is the attractiveness at $r = 0$, and γ is the light absorption coefficient which controls the decrease of the light intensity.

The movement of a firefly i at location x_i attracted to another more attractive (brighter) firefly j at location x_j is determined by Eq. 8.3 (Yang 2008, 2009, 2010a):

$$x_i(t+1) = x_i(t) + \beta_0 e^{-\gamma r_{ij}^2}\left(x_j - x_i\right) + \alpha\varepsilon_i, \tag{8.3}$$

where the first term is the current position of a firefly, the second term is used for considering a firefly's attractiveness to light intensity seen by adjacent fireflies, and the third term is randomization with the vector of random variables ε_i being drawn from a Gaussian distribution, in case there are not any brighter ones. The coefficient α is a randomization parameter determined by the problem of interest.

- Special cases: From Eq. 8.3, it is easy to see that there exit two limit cases when γ is small or large, respectively (Yang 2008, 2009, 2010b). When γ tends to zero, the attractiveness and brightness are constant $\beta = \beta_0$ which means the light intensity does not decrease as the distance r between two fireflies increases. Therefore, a firefly can be seen by all other fireflies, a single local or global optimum can be easily reached. This limiting case corresponds to the standard particle swarm optimization algorithm.

On the other hand, when γ is very large, then the attractiveness (and thus brightness) decreases dramatically, and all fireflies are short-sighted or equivalently fly in a deep foggy sky. This means that all fireflies move almost randomly, which corresponds to a random search technique.

In general, the FA corresponds to the situation between these two limit cases, and it is thus possible to fine-tune these parameters, so that FA can find the global optima as well as all the local optima simultaneously in a very effective manner. A further advantage of FA is that different fireflies will work almost independently, it is thus particular suitable for parallel implementation. It is even better than genetic algorithm and particle swarm optimization because fireflies aggregate more closely around each optimum. It can be anticipated that the interactions between different sub-regions are minimal in parallel implementation.

Overall, taking into account the basic information described above, the steps of implementing FA can be summarized as follows (Yang 2009; Jones and Boizanté 2011):

- Step 1: Generate initial the population of fireflies placed at random positions within the n-dimensional search space.
- Step 2: Initialize the parameters, such as the light absorption coefficient (γ).

- Step 3: Define the light intensity (I_i) of each firefly (x_i) as the value of the cost function ($f(x_i)$).
- Step 4: For each firefly (x_i), compare its light intensity with the light intensity of every other firefly (i.e., x_j).
- Step 5: If ($I_j > I_i$), then move firefly x_i towards x_j in n-dimensions.
- Step 6: Calculate the new values of the cost function for each firefly and update the light intensity.
- Step 7: Rank the fireflies and determine the current best.
- Step 8: Repeat Steps 3–7 until the termination criteria is satisfied.

8.2.2 *Performance of FA*

To test the performance of FA, a set of benchmark functions are adopted in (Yang 2009), namely, Michalewicz function, Rosenbrock function, De Jong function, Schwefel function, Ackley function, Rastrigin function, Easom function, Griewank function, Shubert function, and Yang function. Compared with other computational intelligence (CI) algorithms [such as particle swarm optimization (PSO) and genetic algorithm (GA)], computational results showed that FA is much more efficient in finding the global optima with higher success rates.

8.3 Glowworm Swarm Optimization Algorithm

8.3.1 *Fundamentals of Glowworm Swarm Optimization Algorithm*

Also inspired by luminous insect, the glowworm swarm optimization (GlSO) algorithm was originally proposed by Krishnanand and Ghose (2005) to deal with multimodal problems. Just like ants, elephants, mice, and snakes, glowworms also use some chemical substances, called luciferin, as signals for indirect communication. By sensing luciferin, glowworms can be attracted by strongest luciferin concentrations. In this way, the final optimization results can be found.

Typically, each iteration of the GlSO algorithm consists of two phases, namely, a luciferin-update phase and a movement phase. In addition, for GlSO, there is a dynamic decision range update rule that is used to adjust the glowworms' adaptive neighbourhoods. The details are listed as below (Krishnanand and Ghose 2009):

- Luciferin-update phase: It is the process by which the luciferin quantities are modified. The quantities value can either increase, as glowworms deposit luciferin on the current position, or decrease, due to luciferin decay. The luciferin update rule is given via Eq. 8.4 (Krishnanand and Ghose 2009):

$$l_i(t+1) = (1-\rho) \cdot l_i(t) + \gamma \cdot J \cdot [x_i \cdot (t+1)], \tag{8.4}$$

where $l_i(t)$ denotes the luciferin level associated with the glowworm i at time t, ρ is the luciferin decay constant $(0<\rho<1)$, γ is the luciferin enhancement constant, and $J(x_i(t))$ stands for the value of the objective function at glowworm i's location at time t.

- Movement phase: During this phase, glowworm i chooses the next position j to move to using a bias (i.e., probabilistic decision rule) toward good-quality individual which has higher luciferin value than its own. In addition, based on their relative luciferin levels and availability of local information, the swarm of glowworms can be partitioned into subgroups that converge on multiple optima of a given multimodal function. The probability of moving toward a neighbour is given by Eq. 8.5 (Krishnanand and Ghose 2009):

$$p_{ij}(t) = \frac{l_j(t) - l_i(t)}{\sum_{k \in N_i(t)} [l_k(t) - l_i(t)]}, \tag{8.5}$$

where $j \in N_i(t)$, $N_i(t) = \{j : d_{ij}(t) < r_d^i(t);\ l_i(t) < l_{js}(t)\}$ is the set of neighbours of glowworm i at time t, $d_{ij}(t)$ denotes the Euclidean distance between glow worms i and j at time t, and $r_d^i(t)$ stands for the variable neighbourhood range associated with glowworm i at time t.

Based on Eq. 8.5, the discrete-time model of the glowworm movements can be stated via Eq. 8.6 (Krishnanand and Ghose 2009):

$$x_i(t+1) = x_i(t) + s\left[\frac{x_j(t) - x_i(t)}{\left\|x_j(t) - x_i(t)\right\|}\right], \tag{8.6}$$

where $x_i(t) \in \mathbf{R}^m$ is the location of glowworm i at time t in the m-dimensional real space, $\|\cdot\|$ denotes the Euclidean norm operator, and s (>0) is the step size.

- Neighbourhood range update rule: In addition to the luciferin value update rule that is illustrated in the movement phase, in GlSO the glowworms use a radial range [i.e., $(0 < r_d^i \le r_s)$] update rule to explore an adaptive neighbourhood (i.e., to detect the presence of multiple peaks in a multimodal function landscape). Let r_0 be the initial neighbourhood range of each glowworm [i.e., $r_d^i(0) = r_0\ \forall i$], then the updating rule is given via Eq. 8.7 (Krishnanand and Ghose 2009):

$$r_d^i(t+1) = \min\{r_s, \max\{0, r_d^i(t) + \beta(n_t - |N_i(t)|)\}\}, \tag{8.7}$$

where β is a constant parameter, and $n_t \in N$ is a parameter used to control the number of neighbours.

Furthermore, in order to escape the dead-lock situation (i.e., all the glowworms converge to suboptimal solutions), Krishnanand and Ghose (2009) employed a local search mechanism.

The working principle is described as follows: during the movement phase, each glowworm moves a distance of step size (s) toward a neighbour. Hence, when

$d_{ij}(t) < s$, glowworm i leapfrogs over the position of a neighbour j and becomes a leader to j. In the next iteration, glowworm i remains stationary and j overtakes the position of glowworm i, thus regaining its leadership. In this way, the GlSO algorithm converges to a state in which all the glowworms construct the optimal solution over and over again.

Typically, by taking into account the basic rules described above, the steps of implementing the GlSO algorithm can be summarized as follows (Krishnanand and Ghose 2009):

- Step 1: Initialize the parameters.
- Step 2: Initiation population of N candidate solution is randomly generated all over the search space.
- Step 3: The fitness function value corresponding to each candidate solution is calculated.
- Step 4: Perform the iteration procedures that include luciferin update phase, movement phase, and decision range update phase.
- Step 5: Check if maximum iteration is reached, go to step 3 for new beginning. If a specified termination criteria is satisfied, stop and return the best solution.

8.3.2 *Performance of GlSO*

To evaluate the performance of the GlSO algorithm, a set of multimodal test functions haven been proposed in Krishnanand and Ghose (2009), such as Peaks function, Rastrigin's function, Circles function, Plateaus function, Equal-peaks-A function, Random-peaks function, Himmelblau's function, Equal-peaks-B function, and Staircase function. Compared with niche particle swarm optimization (NichePSO), the GlSO algorithm presented a better results in terms of the number of peaks captured.

8.3.3 *Selected GlSO Variants*

Although GlSO is a new member of CI family, a number of GlSO variations have been proposed in the literature for the purpose of further improving the performance of GlSO. This section gives an overview to some of these GlSO variants which have been demonstrated to be very efficient and robust.

8.3.3.1 Niching GlSO with Mating Behaviour (MNGSO)

As we know, GlSO is developed to solve multimodal function optimization problem which is characterized by the existence of more than one global optimal solution. To increase the search robustness, speed up the convergence, and get more precise solutions, Huang et al. (2011) proposed a new variant of GlSO, called MNGSO, in which a niching strategy and mating behaviour are incorporated.

Generally speaking, niching is a concept developed in the genetic algorithm (GA) community (Angus 2008). Some of the better known niching methods include crowding (Mahfoud 1995), fitness sharing (Goldberg and Richardson 1987), and clearing (Petrowski 1996). Nowadays, Niching strategy has been used extensively in the filed of CI to find multiple solutions at the same time, such as niching for ant colony optimization (ACO) (Angus 2008, 2009), and NichePSO (Engelbrecht 2007).

The basic operating principle of MNGSO is using restricted competition selection (RCS) dynamic niching strategy (Lee et al. 1999), which is a variation of crowding to search several local optimal synchronously. The detail procedures of RCS are as follows (Huang et al. 2011):

- Initialize N subpopulations and mark the best individuals of every subpopulation with p_{nbest}.
- When the distance (d_{ij}) between p_{ibest} and p_{jbest} (where p_{ibest} and p_{jbest} are best individuals of two different subpopulations) is shorter than R_{niche} (where R_{niche} is the radius of niche), then compare their fitness, set 0 to the lower one and keep the value of the other. The R_{niche} can be updated via Eq. 8.8 (Huang et al. 2011):

$$R_{niche}^{(t+1)} = R_{niche}^{t} - R_{niche}^{t} \times c, \tag{8.8}$$

where c is a constant used for adjusting the decay rate.
- Randomly initialize the best individuals who are set to 0, and reset its local-decision range r_d to r_s. In addition, reselect the best one in its niche, then return to Step 1 until the distance (d_{ij}) of any two best individuals respectively belongs to two different niches is lesser than the radius of niche.

In addition, Huang et al. (2011) added a mating behaviour to the MNGSO algorithm in order to get more precise solutions. The formula of updating mate-decision range ($mate\_rs$) is via Eq. 8.9 (Huang et al. 2011):

$$mate\_rs = (1 - constrap) mate\_rs, \tag{8.9}$$

where $constrap$ denotes the contractibility rate.

The steps of implementing the MNGSO algorithm can be summarized as follows (Huang et al. 2011):

- Step 1: Initialize the parameters.
- Step 2: Update luciferin of all the glowworm.
- Step 3: Calculate the neighbours of each glowworm.
- Step 4: Select $j (j \in N_i(t))$ as the movement direction of glowworm i by roulette, and update the position of i.
- Step 5: Implement the RCS niching strategy, determine the best individuals of every niching subgroups.

- Step 6: Implement mating behaviour to the best individual of each niche.
- Step 7: When the predetermined iterations for eliminating reached, the worst niching subgroup is eliminated and updated.
- Step 8: Check if maximum iteration is reached, go to step 2 for new beginning. If a specified termination criteria is satisfied stop and return the best solution.

8.3.3.2 Performance of MNGSO

To verify the availability and feasibility of MNGSO, a set of standard functions are tested in Huang et al. (2011). Compared with PSO, PSO with chaos (CPSO), artificial fish swarm algorithm (AFSA), and AFSA with chaos (CAFSA), the experimental results showed that MNGSO is an effective global algorithm for finding optimal results.

8.3.4 Representative GlSO Applications

The applications of GlSO can be found in many areas, in this section, wireless sensor networks (WSNs) is selected as an example and summarized in the following section. Recently, WSNs are becoming a rapidly developing area in both research community and civilian applications, such as target acquisition, forest fire prevention, structural health measurement, and surveillance. In general, a WSN includes a large number of small wireless devices (i.e., sensor nodes) in which each one has high precision to acquire some physical data (Benini et al. 2006). Among others, one of the key features in a WSN is the coverage issue including energy saving (Anastasi et al. 2009), connectivity (Raghavan and Kumara 2007), and deployment of wireless sensor nodes (Pradhan and Panda 2012).

8.3.4.1 Sensor Deployment Approach Using GlSO

To ensure that the area of targets of interest can be covered, an optimized sensor deployment scheme is an essential guide for anyone interested in wireless communications. Recently, Liao et al. (2011) proposed a GlSO-based deployment approach to enhance the coverage after an initial random placement of sensors. In details, each sensor node is mimicked as a glowworm and emitted by luciferin. The intensity of luciferin is based on the distance between a sensor and its neighbours. By using the probabilistic mechanism, each sensor node selects its neighbours which has lower intensity of lucifein and decides to move towards one of them. In this way, the coverage of sparsely covered areas can be minimized.

To validate the performance of the GlSO algorithm, a comparison with the virtual force algorithm (VFA) has been illustrated. Computational results showed that the GlSO algorithm can improve the coverage rate with limited senor movement.

8.4 Emerging Luminous Insect Inspired Algorithms

In addition to the aforementioned FA and GlSO algorithms, the characteristics of this interesting insect also motivate researchers to develop another luminous insect inspired innovative CI algorithm.

8.4.1 *Fundamentals of Bioluminescent Swarm Optimization Algorithm*

Bioluminescent swarm optimization (BiSO) algorithm was proposed by Oliveira et al. (2011). Although BiSO can be loosely regarded as a hybridization of PSO and GlSO, several characteristics have made it unique. For example, apart from the basic characteristics of GlSO (such as luciferin update rule and stochastic neighbour movement rule), Oliveira et al. (2011) proposed a set of new features, namely stochastic adaptive step sizing, global optimum attraction, leader movement, and mass extinction. In addition, the BiSO algorithm is incorporated with two local search techniques, i.e., local unimodal sampling (LUS) and single-dimension perturbation search (SDPS). The following subsections give us a detailed description about some of these unique features.

8.4.1.1 Luciferin-Update Phase

Instead of using fitness-based function $(J(x_i(t)))$ to evaluate the luciferin value between the glowworms as proposed by the GlSO, BiSO uses luciferin-based attraction which is controlled by luciferin decay constant (ρ) and the luciferin enhancement constant (γ), respectively. The luciferin update rule is given by Eq. 8.10 (Oliveira et al. 2011):

$$l_i(t+1) = (1-\rho) \cdot l_i(t) + \gamma \cdot f(x_i(t)), \tag{8.10}$$

where $l_i(t)$ denotes the luciferin level associated with the glowworm i at time t, ρ is the luciferin decay constant $(0<\rho<1)$, γ is the luciferin enhancement constant, and $f(x_i(t))$ stands for the value of the objective function at glowworm i's location at time t.

8.4.1.2 Stochastic Adaptive Step Sizing

- In BiSO, the following equation is employed to calculated the next location of a given artificial luminous insect via Eq. 8.11 (Oliveira et al. 2011).

$$x_i(t+1) = x_i(t) + rand \cdot s \cdot \left[\frac{x_j(t) - x_i(t)}{\|x_j(t) - x_i(t)\|}\right] + c_g \cdot rand \cdot s \cdot \left[\frac{g(t) - x_i(t)}{\|g(t) - x_i(t)\|}\right], \tag{8.11}$$

where the artificial luminous insect's current position is denoted by $x_i(t)$, *rand* represents a random number which falls within $[0, 1]$, the artificial luminous insect's current step size is indicated by s, c_g is a constant which is used to express the global best attraction, and $g(t)$ stands for the global best location.

- In GlSO, a fixed step size is normally used, whereas BiSO alters the step size in a random manner which is similar to PSO. Apart from this, the maximum step in BiSO is adaptive governed by Eq. 8.12 (Oliveira et al. 2011):

$$s = s_0 \cdot \frac{1}{1 + c_s \cdot l_i(t)}, \tag{8.12}$$

where s_0 stands for the maximum step, $l_i(t)$ denotes the amount of luciferin of an artificial luminous insect, and c_s represents a slowing constant.

8.4.1.3 Global Optimum Attraction

Like PSO, BiSO employed a global optimum factor (c_g) to enhance the neighbour selection. In other words, the selecting of next location is governed by two factors: the current step size and an attractive force. By using a combination of these two factors, every node tries to maximize its value while maintaining the required number of neighbours.

8.4.1.4 Mass Extinction

To prevent early stagnation, Oliveira et al. (2011) proposed a mechanism called mass extinction to counteract this effect. It works by reinitializing all or part of the particles, but keeping the best-so-far value (i.e., global optima). That means, in BiSO, the Luciferin value is reinitialized each time when the system approaches stagnation or no improved solution has been generated for a certain number of iterations, except the global best location ($g(t)$). The parameter eT is used to control this procedure.

8.4.1.5 Local Search Procedures

Local search is usually used to find high-quality solutions to combinatorial optimization problems in reasonable time. In BiSO, Oliveira et al. (2011) applied two local search method, i.e., LUS and SDPS. The former one is embedded at each iteration meaning the default movement for the best particle, called weak one, while the latter one is embedded at each *IR* iterations for searching an improved solution within the neighbourhood of the current solution, called strong one.

The steps of implementing the BiSO algorithm can be summarized as follows (Oliveira et al. 2011):

- Step 1: Initialize the parameters.
- Step 2: Randomly generate the bioluminescent particle population.
- Step 3: Perform the iteration procedures that include luciferin update phase, movement phase, step size update phase, and local search phase.
- Step 4: Check if maximum iteration is reached, go to Step 3 for new beginning. If a specified termination criteria is satisfied stop and return the best solution.

8.4.2 Performance of BiSO

The BiSO algorithm has been tested by four well-known benchmark functions, namely, Rastrigin function, Griewank function, Schaffer function, and Rosenbrock function in (Oliveira et al. 2011). Compared with PSO, the BiSO algorithm presented a better results of finding the global best solution.

8.5 Conclusions

In this chapter, three CI methods are introduced, namely, FA, GlSO algorithm, and BiSO algorithm. The general idea behind those algorithms is similar, such as all algorithms are inspired by the luminous insects, and the updating rule is proportional to the higher value of objective function. However, the actual procedures is very different. For example, FA is proposed as a general optimization algorithm, GlSO algorithm is designed to capture multiple peaks in mulitmodal functions (i.e., without the aim of finding the global best), and BiSO can be loosely regarded as a hybridization of PSO algorithm and GlSO algorithm. The main difference between GlSO algorithm and BiSO algorithm lies in the finding of global optimum. Although FA, GlSO algorithm, and BiSO algorithm are newly introduced CI methods, we have witnessed the following rapid spreading of these luminous insect inspired algorithms:

First, in addition to the selected variants detailed in this chapter, several enhanced version of FA and GlSO algorithm can also be found in the literature are outlined below:

- Chaos enhanced FA (Yang 2011).
- Discrete FA (Sayadi et al. 2010).
- Enhanced FA (Niknam et al. 2012).
- Lévy-flight FA (Yang 2010a).
- Multiobjective FA (Yang 2013).

- Definite updating search domains based GlSO (Liu et al. 2011).
- Hierarchical multi-subgroups based GlSO (He et al. 2013).
- Hybrid GlSO (Zhou et al. 2013; Gong et al. 2011).
- Improved GlSO (Wu et al. 2012; He and Zhu 2011).
- Local search based GlSO (Zhao et al. 2012b).
- MapReduce based GlSO (Aljarah and Ludwig 2013a).
- Metropolis criterion based GlSO (Zhao et al. 2012a).
- Modified GlSO (Oramus 2010; Zhang et al. 2011).

Second, the FA has also been successfully applied to a variety of optimization problems as listed below:

- Artificial neural network training (Horng et al. 2012).
- Continuous constrained optimization (Łukasik and Żak 2009).
- Data clustering (Senthilnath et al. 2011a).
- Image processing (Horng and Liou 2011; Horng 2012).
- Linear array antenna design optimizaiton (Basu and Mahanti 2011).
- Multimodal optimization (Yang 2009).
- Multivariable proportional-integral-derivative control (Coelho and Mariani 2012).
- Power system (Apostolopoulos and Vlachos 2011; Niknam et al. 2012; Yang et al. 2012).
- Scheuling optimization (Sayadi et al. 2010).
- Sematic Web service composition optimization (Pop et al. 2011a, 2011b).
- Stock market price forecasting (Kazem et al. 2013).
- Structure design optimization (Gomes 2011; Gandomi et al. 2011; Talatahari et al. 2012; Miguel and Miguel 2012).
- Structure design optimization (Talatahari et al. 2012).

Third, apart from the representative GlSO applications, it has also been successfully applied to a variety of optimization problems as arrayed below:

- Data clustering (Aljarah and Ludwig 2013b; Huang and Zhou 2011; Tseng 2008).
- Image processing (Senthilnath et al. 2011b).
- Injection mould water channel location optimization (Chiang 2012).
- Multi-dimensional knapsack problem (Gong et al. 2011).
- Robotics control (Krishnanand and Ghose 2005; Krishnanand et al. 2006).
- Wireless sensor networks (Krishnanand and Ghose 2005).

Interested readers please refer to them together with several excellent reviews [e.g., (Fister et al. 2013)] as a starting point for a further exploration and exploitation of luminous insect inspired algorithms.

References

Aljarah, I., Ludwig, S. A. (2013a, April 15–19). A MapReduce based glowworm swarm optimization approach for multimodal functions. In *IEEE Symposium Series on Computational Intelligence (SSCI 2013)*, Singapore (pp. 22–31). *IEEE*.

Aljarah, I., Ludwig, S. A. (2013b, June 20–23) A new clustering approach based on glowworm swarm optimization. In *IEEE Congress on Evolutionary Computation*, Cancún, México (pp. 2642–2649). IEEE.

Anastasi, G., Conti, M., Francesco, M. D., & Passarella, A. (2009). Energy conservation in wireless sensor networks: a survey. *Ad Hoc Networks, 7*, 537–568.

Angus, D. (2009). Niching for ant colony optimisation. In A. Lewis (Ed.), *Biologically-inspired optimisation methods, SCI 210*. Berlin Heidelberg: Springer.

Angus, D. J. (2008). *Niching ant colony optimisation*. Doctor of Philosophy, Swinburne University of Technology.

Apostolopoulos, T., & Vlachos, A. (2011). Application of the firefly algorithm for solving the economic emissions load dispatch problem. *International Journal of Combinatorics, 523806*, 1–23.

Babu, B. G., & Kannan, M. (2002). Lightning bugs. *Resonance, 7*, 49–55.

Basu, B., & Mahanti, G. K. (2011). Fire fly and artificial bees colony algorithm for synthesis of scanned and broad-side linear array antenna. *Progress In Electromagnetics Research B, 32*, 169–190.

Benini, L., Farella, E., & Guiducci, C. (2006). Wireless sensor networks: enabling technology for ambient intelligence. *Microelectronics Journal, 37*, 1639–1649.

Buck, J., & Case, J. (2002). Physiological links in firefly flash code evolution. *Journal of Insect Behavior, 15*, 51–68.

Chiang, Y.-S. (2012). *Water channel location optimization of injection molding using glowworm swarm algorithm with variable step*. Unpublished Master Thesis (in Chinese), Tatung University.

Coelho, L. D. S., & Mariani, V. C. (2012). Firefly algorithm approach based on chaotic Tinkerbell map applied to multivariable PID controller tuning. *Computers and Mathematics with Applications, 64*, 2371–2382.

Engelbrecht, A. P. (2007). *Computational intelligence: An introduction*, West Sussex, England: Wiley, ISBN 978-0-470-03561-0.

Fister, I., Jr Fister, I., Yang, X.-S. & Brest, J. (2013). A comprehensive review of firefly algorithm. *Swarm and Evolutionary Computation*. http://dx.doi.org/10.1016/j.swevo.2013.06.001i

Gandomi, A. H., Yang, X.-S., & Alavi, A. H. (2011). Mixed variable structural optimization using firefly algorithm. *Computers and Structures, 89*, 2325–2336.

Goldberg, D. E. & Richardson, J. (1987). Genetic algorithms with sharing for multimodal function optimization. In *2nd International Conference on Genetic Algorithm* (pp. 41–49).

Gomes, H. M. (2011). A firely metaheuristic algorithm for structural size and shape optimization with dynamic constraints. *Mecánica Computacional*, 30, 2059–2074.

Gong, Q., Zhou, Y., & Luo, Q. (2011). Hybrid artificial glowworm swarm optimization algorithm for solving multi-dimensional knapsack problem. *Procedia Engineering, 15*, 2880–2884.

He, D.-X., & Zhu, H.-Z. (2011). An improved glowworm swarm optimization algorithm for high-dimensional function optimization. *Energy Procedia, 13*, 5657–5664.

He, L., Tong, X., & Huang, S. (2013). Glowworm swarm optimization algorithm based on hierarchical multi-subgroups. *Journal of Information and Computational Science, 10*, 1245–1251.

Horng, M.-H. (2012). Vector quantization using the firefly algorithm for image compression. *Expert Systems with Applications, 39*, 1078–1091.

Horng, M.-H., Lee, Y.-X., Lee, M.-C. & Liou, R.-J. (2012). Firefly meta-heuristic algorithm for training the radia basis function network for data classification and disease diagnosis. In: R.

Parpinelli (Ed.), *Theory and new applications of swarm intelligence,* Chap. 7 (pp. 115–132). Rijeka, Croatia: In-Tech. ISBN 978-953-51-0364-6.

Horng, M.-H., & Liou, R.-J. (2011). Multilevel minimum cross entropy threshold selection based on the firefly algorithm. *Expert Systems with Applications, 38*, 14805–14811.

Huang, K., Zhou, Y., & Wang, Y. (2011). Niching glowworm swarm optimization algorithm with mating behavior. *Journal of Information and Computational Science, 8*, 4175–4184.

Huang, Z., & Zhou, Y. (2011). Using glowworm swarm optimization algorithm for clustering analysis. *Journal of Convergence Information Technology, 6*, 78–85.

Jones, K. O., & Boizanté, G. (2011, June 16–17). Comparison of firefly algorithm optimisation, particle swarm optimisation and differential evolution. *International Conference on Computer Systems and Technologies (CompSysTech)*, (pp. 191–197). Vienna, Austria.

Kazem, A., Sharifi, E., Hussain, F. K., Saberi, M. & Hussain, O. K. (2013). Support vector regression with chaos-based firefly algorithm for stock market price forecasting. *Applied Soft Computing, 13*, 947–958. http://dx.doi.org/10.1016/j.asoc.2012.09.024.

Krishnanand, K. N., Amruth, P., Guruprasad, M. H., Bidargaddi, S. V. & Ghose, D. (2006, May). Glowworm-inspired robot swarm for simultaneous taxis towards multiple radiation sources. In *IEEE International Conference on Robotics and Automation (ICRA)*, Orlando, Florida, USA, (pp. 958–963). IEEE.

Krishnanand, K. N. & Ghose, D. (2005). Detection of multiple source locations using a glowworm metaphor with applications to collective robotics. In *IEEE Swarm Intelligence Symposium (SIS)* (pp. 84–91). *IEEE.*

Krishnanand, K. N., & Ghose, D. (2009). Glowworm swarm optimization for simultaneous capture of multiple local optima of multimodal functions. *Swarm Intelligence, 3*, 87–124.

Lee, C. G., Cho, D. H., & Jung, H. K. (1999). Niche genetic algorithm with restricted competition selection for multimodal function optimization. *IEEE transaction on Magnetics, 35*, 1122–1125.

Liao, W.-H., Kao, Y., & Li, Y.-S. (2011). A sensor deployment approach using glowworm swarm optimization algorithm in wireless sensor networks. *Expert Systems with Applications, 38*, 12180–12188.

Liu, J., Zhou, Y., Huang, K., Ouyang, Z., & Wang, Y. (2011). A glowworm swarm optimization algorithm based on definite updating search domains. *Journal of Computational Information Systems, 7*, 3698–3705.

Łukasik, S., & Żak, S. (2009). Firefly algorithm for continuous constrained optimization tasks. *Computational collective intelligence. semantic web, social networks and multiagent systems LNCS 5796,* (pp. 97–106). Berlin: Spinger.

Mahfoud, S. W. (1995). *Niching methods for genetic algorithms.* Doctor of Philosophy, University of Illinois.

Miguel, L. F. F., & Miguel, L. F. F. (2012). Shape and size optimization of truss structures considering dynamic constraints through modern metaheuristic algorithms. *Expert Systems with Applications, 39*, 9458–9467.

Niknam, T., Azizipanah-Abarghooee, R., Roosta, A., & Amiri, B. (2012). A new multi-objective reserve constrained combined heat and power dynamic economic emission dispatch. *Energy, 42*, 530–545.

Oliveira, D. R. D., Parpinelli, R. S. & Lopes, H. S. (2011). Bioluminescent swarm optimization algorithm. *Evolutionary Algorithms,* Chap. 5 (pp. 71–84). Eisuke Kita: InTech.

Oramus, P. (2010). Improvements to glowworm swarm optimization algorithm. *Computer Science, 11*, 7–20.

Petrowski, A. (1996). A clearing procedure as a niching method for genetic algorithms. In *IEEE International Conference on Evolutionary Computation*, (pp. 798–803).

Pop, C. B., Chifu, V. R., Salomie, I., Baico, R. B., Dinsoreanu, M., & Copil, G. (2011a). A hybrid firefly-inspired approach for optimal semantic Web service composition. *Scalable Computing Practice and Experience, 12*, 363–369.

Pop, C. B., Chifu, V. R., Salomie, I., Baico, R. B., Dinsoreanu, M. & Copil, G. (2011b, 19–21 September). A hybrid firefly-inspired approach for optimal semantic Web service composition. *3rd Workshop on Software Services: Semantic-based software services*, Szczecin, Poland, (pp. 1–6).

Pradhan, P. M., & Panda, G. (2012). Connectivity constrained wireless sensor deployment using multi objective evolutionary algorithms and fuzzy decision making. *Ad Hoc Networks, 10*, 1134–1145.

Raghavan, U. N., & Kumara, S. R. T. (2007). Decentralised topology control algorithms for connectivity of distributed wireless sensor networks. *International Journal of Sensor Networks, 2*, 201–210.

Sayadi, M. K., Ramezanian, R., & Ghaffari-Nasab, N. (2010). A discrete firefly meta-heuristic with local search for makespan minimization in permutation flow shop scheduling problems. *International Journal of Industrial Engineering Computations, 1*, 1–10.

Senthilnath, J., Omkar, S. N., & Mani, V. (2011a). Clustering using firefly algorithm: Performance study. *Swarm and Evolutionary Computation, 1*, 164–171.

Senthilnath, J., Omkar, S. N., Mani, V., Tejovanth, N., Diwakar, P. G., & Archana, S. B. (2011b). Multi-spectral satellite image classification using glowworm swarm optimization. In *IEEE International Geoscience and Remote Sensing Symposium (IGARSS)* (pp. 47–50). IEEE.

Talatahari, S., Gandomi, A. H. & Yun, G. J. (2012). Optimum design of tower structures using firefly algorithm. *The Structural Design of Tall and Special Buildings.* (DOI:10.1002/tal.1043).

Tseng, K.-T. (2008). *A glowworm algorithm for solving data clustering problems (in Chinese).* Unpublished Master Thesis, Tatung University.

Wu, B., Qian, C., Ni, W., & Fan, S. (2012). The improvement of glowworm swarm optimization for continuous optimization problems. *Expert Systems with Applications, 39*, 6335–6342.

Yang, X.-S. (2008). *Nature-inspired metaheuristic algorithms*. UK: Luniver Press. ISBN 978-1-905986-28-6.

Yang, X.-S. (2009). Firefly algorithms for multimodal optimization. In O. Watanabe, & T. Zeugmann, (Eds.), *SAGA 2009, LNCS 5792,* (pp. 169–178). Berlin Heidelberg: Springer.

Yang, X.-S. (2010a). Firefly algorithm, Lévy flights and global optimization. In M. Bramer, (Ed.) *Research and development in intelligent systems.* 26, 209–218. London, UK: Springer-Verlag.

Yang, X.-S. (2010b). Firefly algorithm, stochastic test functions and design optimisation. *International Journal of Bio-Inspired Computation, 2*, 78–84.

Yang, X.-S. (2011). Chaos-enhanced firefly algorithm with automatic parameter tuning. *International Journal of Swarm Intelligence Research, 2*, 1–11.

Yang, X.-S. (2013). Multiobjective firefly algorithm for continuous optimization. *Engineering with Computers, 29*, 175–184. (DOI 10.1007/s00366-012-0254-1).

Yang, X.-S., Hosseini, S. S. S., & Gandomi, A. H. (2012). Firefly algorithm for solving non-convex economic dispatch problems with valve loading effect. *Applied Soft Computing, 12*, 1180–1186.

Zhang, Y.-L., Ma, X.-P., Gu, Y., & Miao, Y.-Z. (2011) A modified glowworm swarm optimization for multimodal functions. In *Chinese Control and Decision Conference (CCDC),* (pp. 2070–2075). IEEE.

Zhao, G., Zhou, Y., Luo, Q., & Wang, Y. (2012a). A glowworm swarm optimization algorithm based on metropolis criterion. *International Journal of Advancements in Computing Technology, 4*, 149–155.

Zhao, G., Zhou, Y., & Wang, Y. (2012b). The glowworm swarm optimization algorithm with local search operator. *Journal of Information & Computational Science, 9*, 1299–1308.

Zhou, Y., Zhou, G., & Zhang, J. (2013). A hybrid glowworm swarm optimization algorithm for constrained engineering design problems. *Applied Mathematics and Information Sciences, 7*, 379–388.

Chapter 9
Fish Inspired Algorithms

Abstract In this chapter, we present several fish algorithms that are inspired by some key features of the fish school/swarm, namely, artificial fish school algorithm (AFSA), fish school search (FSS), group escaping algorithm (GEA), and shark-search algorithm (SSA). We first provide a short introduction in Sect. 9.1. Then, the detailed descriptions regarding AFSA and FSS can be found in Sects. 9.2 and 9.3, respectively. Next, Sect. 9.4 briefs two emerging fish inspired algorithms, i.e., GEA and SSA. Finally, Sect. 9.5 summarises in this chapter.

9.1 Introduction

In a water area, fishes are most likely distributed around the region where foods are most abundant and desire to stay close to the swarm. Biological research shows that the position of individual fish in swarm will adjust at any time with the external environment and its own state (Ban et al. 2009; Braithwaite 2006). In addition, the fish try to complete their food foraging process and keep the balance in such factors as momentum, hunger degree and fear degree, etc. According to those instinctive behaviours, in this chapter, several fish behaviour inspired algorithms are collected and introduced as follows:

- Section 9.2: Artificial Fish School Algorithm.
- Section 9.3: Fish School Search Algorithm.
- Section 9.4.1: Group Escaping Algorithm.
- Section 9.4.2: Shark-Search Algorithm.

The effectiveness of these newly developed algorithms are validated through the testing on a wide range of benchmark functions and engineering design problems, and also a detailed comparison with various traditional performance leading computational intelligence (CI) algorithms, such as particle swarm optimization (PSO), genetic algorithm (GA), differential evolution (DE), evolutionary algorithm (EA), fuzzy system (FS), ant colony optimization (ACO), and simulated annealing (SA).

B. Xing and W.-J. Gao, *Innovative Computational Intelligence: A Rough Guide to 134 Clever Algorithms*, Intelligent Systems Reference Library 62,
DOI: 10.1007/978-3-319-03404-1_9,

9.2 Artificial Fish School Algorithm

9.2.1 Fundamentals of Artificial Fish School Algorithm

Artificial fish swarm algorithm (AFSA), which was proposed in Li (2003), is a stochastic search optimization algorithm inspired by the natural social behaviour of fish schooling. In principle, AFSA is started first in a set of random generated potential solutions, and then performs the search for the optimum one interactively (Zhang et al. 2006a). The main steps of AFSA are outlined as follows (Li 2003; Wang et al. 2005; Luo et al. 2007):

Assuming in an n-dimensional searching space, there is a group composed of K articles of artificial fish (AF).

- Situation of each individual AF can be expressed as vector $X = (x_1, x_2, \ldots, x_k)$ is denoted the current state of AF, where $x_k (k = 1, 2, \ldots, k)$ is control variable.
- $Y = f(X)$ is the fitness or objective function of X, which can represent food concentration (FC) of AF in the current position.
- $d_{ij} = \|X_i - X_j\|$ is denoted the Euclidean distance between fishes.
- *Visual* and *Step* are denoted respectively the visual distance of AF and the distance that AF can move for each step.
- $X_v = (x_1^v, x_2^v, \ldots, x_k^v)$ is the visual position at some moment. If the state at the visual position is better than the current state, it goes forward ad step in this direction, and arrives the X_{next} state, otherwise, continues an inspecting tour in the vision.
- *try-number* is attempt times in the behaviour of prey.
- δ is the condition of jamming $(0 < \delta < 1)$.

The basic behaviours of AF inside water are defined as follows (Li 2003; Wang et al. 2005; Luo et al. 2007):

- Chasing trail behaviour (AF_Follow): When a fish finds the food dangling quickly after a fish, or a group of fishes, in the swarm that discovered food. If $Y_j > Y_i$ and $n_f/n < \delta$, then the AF_Follow behaviour is defined by Eq. 9.1 (Li 2003; Wang et al. 2005; Luo et al. 2007; Neshat et al. 2012a, b):

$$X_i^{(t+1)} = X_i^{(t)} + \frac{X_j - X_i^{(t)}}{\left\|X_j - X_i^{(t)}\right\|} \cdot Step \cdot rand(\,). \tag{9.1}$$

- Gathering behaviour (AF_Swarm): In order to survive and avoid hazards, the fish will naturally assemble in groups. There are three rules while fish gathering: firstly, a fish will try to keep a certain distance with each other to avoid crowding (i.e., Compartmentation Rule); secondly, a fish will try to move in a similar direction with its surrounding partners (i.e., Unification Rule); finally, a fish will try to move to the centre of its surrounding partners (i.e., Cohesion Rule). If $Y_c > Y_i$ and $n_f/n < \delta$, then the AF_Swarm behaviour is defined by Eq. 9.2 (Li 2003; Wang et al. 2005; Luo et al. 2007; Neshat et al. 2012a, b):

$$X_i^{(t+1)} = X_i^{(t)} + \frac{X_c - X_i^{(t)}}{\left\| X_c - X_i^{(t)} \right\|} \cdot Step \cdot rand(\ ), \tag{9.2}$$

where X_c denotes the centre position of AF, X_i be the AF current state, n_f be the number of its companions in the current neighbourhood ($d_{ij} < Visual$), and n is the total fish number.

- Random searching behaviour (AF_Random): This is a basic biological behaviour that tendts to the food. Generally the fish perceives the concentration of food in water to determine the movement by vision or sense and then chooses the tendency. The effect of this behaviour is similar to that of mutation operator in genetic algorithm (GA). It is defined by Eq. 9.3 (Neshat et al. 2012a, b):

$$X_i^{(t+1)} = X_i^{(t)} + Visual \cdot rand(\ ). \tag{9.3}$$

- Leaping behaviour (AF_Leap): When a fish 'stagnates' in a region, it looks for food in other regions defining the leaping behaviour. It can be defined by Eq. 9.4 (Neshat et al. 2012a, b):

$$\begin{cases} \text{if } (FC_{best}(m) - FC_{best}(n)) < eps \\ \text{then } X_{some}^{(t+1)} = X_{some}^{(t)} + \beta \cdot Visual \cdot rand(\,) \end{cases}, \tag{9.4}$$

where β is a parameter or a function that can makes some fish have other abnormal actions (values), *eps* is a smaller constant, and *FC* represents the food concentration.

- Foraging behaviour (AF_Prey): As feeding the fish, they will gradually move to the place where food is increasing. It is defined by Eqs. 9.5 and 9.6, respectively (Li 2003; Wang et al. 2005; Luo et al. 2007; Neshat et al. 2012a, b):

$$X_j = X_i + Visual \cdot rand(\ ), \tag{9.5}$$

where X_i be the AF current state and select a state X_j randomly in its visual distance, *rand*( ) is a random function in the range, and *Visual* represents the visual distance.

$$\begin{cases} \text{if } Y_i < Y_j, & \text{then } X_i^{(t+1)} = X_i^{(t)} + \frac{X_j - X_i^{(t)}}{\|X_j - X_i^{(t)}\|} \cdot Step \cdot rand(\,) \\ \text{if } Y_i > Y_j, & \text{then } \begin{cases} X_j = X_i + Visual \cdot rand(\,) \\ X_i^{(t+1)} = X_i^{(t)} + Visual \cdot rand(\,) \end{cases} \end{cases}, \tag{9.6}$$

where Y is the food concentration (objective function value), $X_i^{(t+1)}$ represents the AF's next state, and *Step* denotes the distance that AF can move for each step.

Taking into account the above mentioned behaviours, the steps of implementing AFSA can be summarized as follows (Neshat et al. 2012a, b):

- Step 1: Generate the initial fish swarm randomly in the search space.
- Step 2: Initialize the parameters.
- Step 3: Evaluate the fitness value of each AF.
- Step 4: Selecting behaviour. Each AF simulate the swarming and following behaviour, respectively, and select the best behaviour to perform by comparing the function values, the default is searching food behaviour.
- Step 5: Update the function value of the AF again.
- Step 6: Check the termination condition.

9.2.2 Performance of AFSA

In Li et al. (2012), the authors presented a real-world case study, i.e., transporting dangerous goods for Zhengzhou Coal Material Supply and Marketing Company, to test the performance of AFSA in solving vehicle routing problem. The comparison was made between AFSA and other selected methods, e.g., GA. The experimental results showed that the proposed algorithm performs well.

9.3 Fish School Search Algorithm

9.3.1 Fundamentals of Fish School Search Algorithm

Fish school search (FSS) algorithm was originally proposed in Bastos-Filho et al. (2008, 2009a) based on the simulation of social behaviour of biologic fish. In FSS, the search space is bounded and each possible position in the search space represents a possible solution for the problem. The success of a fish during the search process is indicated by its weight, so promising areas can be inferred from regions where bigger ensembles of fish are located (Janecek and Tan 2011a), and the amount of food that a fish eats depends on the improvement in its fitness and the largest improvement in the fitness of the entire school. Moreover, as any other intelligent technique based on population, FSS greatly benefits from the collective emerging behaviour that increases mutual survivability and achieve synergy (e.g., finding locations with lots of food) (Bastos-Filho et al. 2009a). Briefly, FSS consists of the following three operators, namely, feeding, swimming, and breeding.

- Feeding operator: The feeding operator determines the variation of the fish weight. All fish are born with the same weight and start with weight equal to 1. That means fish can increase or decrease their weight depending, respectively, on the success or failure of the individual movement. The fish weight variation is proportional to the normalized difference between the evaluation of the fitness function at current and new position which can be described by Eqs. 9.7 and 9.8, respectively (Bastos-Filho et al. 2009a):

$$w_i(t+1) = w_i(t) + \Delta f(i)/\max(\Delta f), \tag{9.7}$$

$$\Delta f(i) = f[\mathbf{x}_i(t+1) - f[\mathbf{x}_i(t)]], \tag{9.8}$$

  where $w_i(t)$ is the weight of the fish, $\mathbf{x}_i(t)$ is the position of the fish, $f[\mathbf{x}_i(t)]$ evaluates the fitness function (i.e., amount of food) in $\mathbf{x}_i(t)$, and $\Delta f(i)$ is the fitness difference.
- Swimming operator: For fish, swimming is directly related to all important individual and collective behaviours. In FSS, the swimming behaviour is considered to be an elaborate form of reaction regarding survivability. It aims at mimicking the coordinated and only apparent collective contained movement produced by all the fishes in the school. Normally, the swimming patterns of the fish school are the result of a combination of three different causes:

  (1) Individual movements occur for each fish at every cycle of the FSS algorithm. In each iteration, each fish randomly chooses a new position in its neighbourhood which is determined by the assessment of the food density (Notice that food here is a metaphor for the evaluation of candidate solutions in the search process). The next candidate position is determined by adding to each dimension of the current position a random number generated by a uniform distribution in the interval $[-1, 1]$ multiplied by a predetermined step as shown via Eqs. 9.9 and 9.10, respectively (Bastos-Filho et al. 2009a):

$$n_i(t) = \mathbf{x}_i(t) + rand(-1, 1) \cdot step_{ind}, \tag{9.9}$$

$$step_{ind}(t+1) = step_{ind}(t) - \frac{\left(step_{ind\_initial} - step_{ind\_final}\right)}{iterations}, \tag{9.10}$$

  where $step_{ind}$ denotes the predetermined step, $n_i(t)$ is the neighbour position of the fish i, and $step_{ind\_initial}$ and $step_{ind\_final}$ are the initial and the final individual movement step, respectively.

(2) After all fishes have moved individually, a weighted average of individual movements based on the instantaneous success of all fishes of the school is computed. This means that fishes that had successful individual movements (i.e., to regions of the space search in which it was discovered the large amounts of food) influence the resulting direction of movement more than other ones. When the overall direction is computed, each fish is repositioned according to Eqs. 9.11 and 9.12, respectively (Bastos-Filho et al. 2009a):

$$\mathbf{m}(t) = \frac{\sum_{i=1}^{N} \Delta\mathbf{x}_i \Delta f_i}{\sum_{i=1}^{N} \Delta f_i}, \tag{9.11}$$

$$\mathbf{x}_i(t+1) = \mathbf{x}_i(t) + \mathbf{m}(t), \tag{9.12}$$

where Δx_i is the displacement of the fish i due do the individual movement in the FSS cycle, and $\mathbf{m}(t)$ is the resulting direction.

(3) The collective-volatile movement controls the granularity of the search executed by the fish school based on the incremental weight variation of the fish school. When the whole school is achieving better results, the movement approximates the fish aiming to accelerate the convergence toward a good region. On the contrary, the movement spreads the fish away from the barycentre (i.e., inward drift) of the school and the school has more chances to escape from a local optimum which can be expressed via Eqs. 9.13, 9.14 and 9.15, respectively (Bastos-Filho et al. 2009a):

$$\mathbf{x}(t+1) = \mathbf{x}(t) + step_{vol} \cdot rand(0,1) \frac{(\mathbf{x}(t) - \mathbf{b}(t))}{\text{distance}(\mathbf{x}(\mathrm{t}), \mathbf{b}(t))}, \tag{9.13}$$

$$\mathbf{x}(t+1) = \mathbf{x}(t) - step_{vol} \cdot rand(0,1) \frac{(\mathbf{x}(t) - \mathbf{b}(t))}{\text{distance}(\mathbf{x}(\mathrm{t}), \mathbf{b}(t))}, \tag{9.14}$$

$$\mathbf{b}(t) = \frac{\sum_{i=1}^{N} \mathbf{x}_i w_i(t)}{\sum_{i=1}^{N} w_i(t)}, \tag{9.15}$$

where $step_{vol}$ is a parameter called volitive step, and distance( ) is a function which returns the Euclidean distance between the barycentre and the fish current position.

- Breeding operator: The breeding operator is responsible for refining the search performed. The selection of candidates for breeding considers all fishes that have reached a predefined threshold. The winner is the fish that presents the maximum ration of weight over distance in relation to the breeding candidate. Furthermore, it was also conceived to allow automatic transitioning from exploration to exploitation abilities.

Taking into account the above mentioned behaviours, the steps of implementing FSS can be summarized as follows (Bastos-Filho et al. 2008, 2009a):

- Step 1: Initializing locations (x_i) randomly for all fish, setting all weights (w_i) to one.
- Step 2: Starting the repeat loop.
- Step 3: Performing swimming 1. Calculating random individual movement for each individual fish.
- Step 4: Executing feeding operation. Updating weights for all fish based on new locations.
- Step 5: Performing swimming 2. Collectively instinctive moving towards overall direction.
- Step 6: Performing swimming 3. Collectively volitive moving dilation or contraction.
- Step 7: Checking the termination condition.

9.3.2 Performance of FSS

In order to verify FSS, a set of benchmark testing functions, such as Rosenbrock function, Rastrigin function, Griewank function, and Ackley function were employed in Bastos-Filho et al. (2008) for testing purpose. The experimental results demonstrated that FSS is very competitive in solving unstructured high dimensional spaces.

9.4 Emerging Fish Inspired Algorithms

In addition to the aforementioned AFSA and FSS, the different behaviour patterns of fish also motivates researchers to develop another fish inspired innovative CI algorithm.

9.4.1 Group Escaping Algorithm

9.4.1.1 Fundamentals of Group Escaping Algorithm

Group escaping algorithm (GEA) was originally proposed in Min and Wang (2010). It simulates the phenomenon of a school of fish changing their moving directions without any explicit centralized communications. The algorithm was proposed for solving robot swarm decentralized control issue. In order to implement GEA, the following components need to be taken into account (Min and Wang 2010):

- Component 1: Escape mode. In this mode, the moment applied to change a robot's direction is defined by Eq. 9.16 (Min and Wang 2010):

$$M_{re\_i} = SGN(\psi_i)F_m \cos\left(\frac{\psi_i}{2}\right). \tag{9.16}$$

Accordingly, by adding a constant extra propulsion force (denoted by F_e) to the system, dynamics equation of the ith robot in escape mode is shown in Eq. 9.17 (Min and Wang 2010):

$$\begin{cases} m_i \frac{d\vec{v}_i}{dt} = (F_p + F_e)\vec{H}_i + \sum\limits_{j,j\neq i} e_{ij}\vec{F}_{KH_{ij}} - \gamma\vec{v} \\ I_i \frac{d^2\alpha_i}{dt^2} = \sum\limits_{j,j\neq i} \left(e_{c\_\text{ij}} M_{c\_\text{ij}} + e_{ij} M_{d\_\text{ij}}\right) + M_{re\_\text{i}} - D_m \frac{d\alpha_i}{dt} \end{cases}. \tag{9.17}$$

- Component 2: Local interaction for group escaping control. In case of emergency, interaction moment M_{c_ij} (normally used at the time of normal schooling) is not suitable to perform escape mode transition, and thus a strong interaction moment M_{esp_ij} is employed according to Eq. 9.18 (Min and Wang 2010):

$$M_{ecp\_i} = K_e\alpha_{ij} + C_e\frac{d\alpha_{ij}}{dt}. \tag{9.18}$$

- Component 3: Group escaping states. In GEA, two particular states are designed for group escaping control. Mathematically, these two states can be defined by Eqs. 9.19 and 9.20, respectively (Min and Wang 2010):

$$\text{Mode 1:} \begin{cases} M_{c\_ij} = 0.0 \\ M_{ecp\_i} = K_e\alpha_{ij} + C_e\frac{d\alpha_{ij}}{dt} \\ M_{d\_ij} = 0.0 \end{cases}. \tag{9.19}$$

$$\text{Mode 2:} \begin{cases} M_{c\_ij} = 0.0 \\ M_{ecp\_i} = 0.0 \\ M_{d\_ij} = 0.0 \end{cases}. \tag{9.20}$$

9.4.1.2 Performance of GEA

To evaluate the proposed GEA, a simulation of 50 robots in virtual environment was first conducted by Min and Wang (2010), later on, the authors set up a real world testing environment consisting of four prototype robots. The overall results demonstrated the effectiveness of the proposed GEA.

9.4.2 Shark-Search Algorithm

9.4.2.1 Fundamentals of Shark-Search Algorithm

Shark-search algorithm (SSA) was originally proposed in Hersovici et al. (1998) for enhancing the Web browsing and search engine performance (Hillis et al. 2013; Cho et al. 1998; Jarvis 2009). There are several variants and applications of SSA can be found in the literature Luo et al. (2005), Chen et al. (2007), Sun et al. (2009). The SSA is built on its predecessor called "fish-search" algorithm and the key principles underlying them are the following: The algorithm first takes an input as a seed URL (standing for uniform resource locator) and search query. Then it dynamically sets up a priority sequence for the next URLs to be explored. At each step, the first node is popped from the list and attended. As each file's text becomes available, it will be analyzed by a scoring component and then evaluated for its relevance or irrelevance to the search query. Putting it simply, in SSA, when relevant information (standing for food) is discovered, searching agents (i.e., fish) reproduce and keep looking for food. They will die when the food is in absent condition or encountering polluted water (poor bandwidth situation). The original fish-search algorithm suffers from the following limitations (Hersovici et al. 1998):

- First, the relevance score is assigned in a discrete way.
- Second, the differentiation degree of the priority of the pages in the list is very low.
- Third, the number of addressed children is reduced by arbitrary using the width parameter.

Bearing this in mind, there are several improvements provided in SSA (Hersovici et al. 1998):

- Improvement 1: A similarity engine is introduced to rank the document relevance degree.
- Improvement 2: Refining the computation of the potential score of the children by taking the following two factors into account. First, propagating ancestral relevance scores deeper down the hierarchical structure. Second, making use of the meta-information carried by the links to the files.

9.4.2.2 Performance of SSA

To evaluate the efficacy of SSA, a measure called "getting as many relevant documents with the shorted delays" was proposed in Hersovici et al. (1998). By testing SSA on four case studies, the significant improvements were experimentally verified which offer SSA an advantage to replace original fish-search algorithm in dealing with dynamic and precise searches within the limited time range.

9.5 Conclusions

In this chapter, several fish inspired CI algorithms are introduced. Although they are newly joined members of the CI community, we have witnessed the following rapid spreading of at least two of them, i.e., AFSA and FSS:

First, several enhanced versions of AFSA and FSS can be found in the literature as outlined below:

- Adaptive meta-cognitive ASFA (Xu et al. 2009).
- Augmented Lagrangian ASFA (Rocha et al. 2011b).
- Chaotic ASFA (Nie et al. 2010; Huang and Lin 2008; Ma and Wang 2009; Zhu and Jiang 2009; Chen and Tian 2010).
- Extended ASFA (Li and Qian 2003).
- Fuzzy adaptive ASFA (Yazdani et al. 2010b).
- Hybrid ASFA (Neshat et al. 2011; Zheng and Lin 2012; Wei et al. 2010; Huang et al. 2009; Zhang et al. 2011; Wang and Ma 2011; Zhou and Huang 2009; Gao et al. 2009; Yazdani et al. 2010a; Jiang and Cheng 2010; Wu et al. 2011b).
- Improved ASFA (Luo et al. 2010; He et al. 2009b; Cheng et al. 2009; Yu et al. 2012; Tian and Tian 2009; Li et al. 2008; Yuan et al. 2010; Peng 2011; Tian and Liu 2009; Ma and Lei 2010; Bing and Wen 2010; Jiang et al. 2009).
- Knowledge-based ASFA (Gao et al. 2010; Wu et al. 2011a).
- Modified ASFA (Farzi 2009; Fernandes et al. 2009; Tsai and Lin 2011; Rocha et al. 2011a; Yazdani et al. 2012).
- Multiobjective ASFA (Jiang and Zhu 2011).
- Mutation-based ASFA (Rocha and Fernandes 2011a, b).
- Niche ASFA (Zhu et al. 2010).
- Parallel ASFA (Hu et al. 2011).
- Quantum ASFA (Zhu and Jiang 2010).
- Tabu-based ASFA (Xu and Liu 2010; Yu and He 2011)
- Hybrid FSS (Cavalcanti-Júnior et al. 2011, 2012; Turabieh and Abdullah 2011).
- Improved FSS (Madeiro et al. 2011; Bastos-Filho et al. 2009b; Janecek and Tan 2011a; Lacerda and Neto 2013).

Second, the AFSA has also been successfully applied to a variety of optimization problems as listed below:

- Artificial neural network training (Wang et al. 2005; Zhang et al., 2006a).
- Control optimization (Luo et al. 2007; Chen et al. 2008; Tian et al. 2009a; Luo et al. 2010; Cheng and Hong 2012).

- Data mining (Zhang et al. 2006b; Wang et al. 2009; He et al. 2009b; Cheng et al. 2009; Zhou and Liu 2009; Xiao 2010; Huang and Wang 2010; Xu and Liu 2010; Neshat et al. 2011; Sun et al. 2011; Yang et al. 2011; Zhu et al. 2012).
- Fault diagnosis (Wang and Xia 2010; Yu et al. 2012).
- Finance (Niu et al. 2010; Shen et al. 2011; Zheng and Lin 2012).
- Function optimization (He et al. 2009a; Hu et al. 2010; Wei et al. 2010; Shi and Shang 2010; Rocha et al. 2011b; Jiang and Zhu 2011).
- Image processing (Song et al. 2008; Tian et al. 2009b).
- Network optimization (Yu et al. 2005; Jiang et al. 2007a; Tian and Tian 2009; Liu et al. 2009a, b; Nie et al. 2010; Huang et al. 2009; Yu and He 2011; Song et al. 2010; Zhang et al. 2011; Yu et al. 2011; Wang et al. 2012).
- Power system optimization (Li et al. 2008; Yuan et al. 2010; Peng 2011).
- Robot control optimization (Tian and Liu 2009; Ma and Lei 2010; Zheng and Li 2010; Feng et al. 2010).
- Scheduling optimization (Xue et al. 2004; Huang and Lin 2008; Wang et al. 2008; Farzi 2009; Cai 2010; Bing and Wen 2010; Gao and Chen 2010; Wang and Ma 2011).
- Signal processing (Jiang and Yuan 2005; Jiang et al. 2007b; Qi et al. 2009).
- Steiner tree problem (Ma and Liu 2009).
- Vehicle routing problem (Li et al. 2012)

Third, the relative FSS algorithm applications can be found below:

- Graphic processing units testing (Lins et al. 2012).
- Non-negative matrix factorization (Janecek and Tan 2011b).
- Timetabling (Turabieh and Abdullah 2011).
- Turnround time improving (Banerjee and Caballé 2011)

Interested readers are referred to them together with several excellent reviews [e.g., (Neshat et al. 2012a, b)], as a starting point for a further exploration and exploitation of fish inspired algorithms.

References

Ban, X., Yang, Y., Ning, S., Lv, X. & Qin, J. (2009, August 20–24). A self-adaptive control algorithm of the artificial fish formation. IEEE International Conference on Fuzzy Systems (FUZZ-IEEE) (pp. 1903–1908). Korea.

Banerjee, S. & Caballé, S. (2011) Exploring fish school algorithm for improving turnaround time: an experience of content retrieval. Third International Conference on Intelligent Networking and Collaborative Systems (INCoS), pp. 842–847.

Bastos-Filho, C.J.A., Lima-Neto, F.B.D., Lins, A.J.C.C., Nascimento, A.I.S. & Lima, M.P. (2008). A novel search algorithm based on fish school behavior. IEEE International Conference on Systems, Man and Cybernetics (SMC), pp. 2646–2651.

Bastos-Filho, C.J.A., Lima-Neto, F.B.D., Lins, A.J.C.C., Nascimento, A.I.S. & Lima, M.P. (2009a). Fish school search. In Chiong, R. (ed.) *Nature-Inspired Algorithms for Optimisation, SCI 193,* (pp. 261–277). Berlin: Springer.

Bastos-Filho, C.J.A., Lima-Neto, F.B.D., Sousa, M.F.C., Pontes, M.R. & Madeiro, S.S. (2009b). On the influence of the swimming operators in the fish school search algorithm. IEEE International Conference on Systems, Man, and Cybernetics (SMC), October, San Antonio, TX, USA, pp. 5012–5017.

Bing, D. & Wen, D. (2010). Scheduling arrival aircrafts on multi-runway based on an improved artificial fish swarm algorithm. International Conference on Computational and Information Sciences (ICCIS), pp. 499–502.

Braithwaite, V. A. (2006). Cognitive ability in fish. *Behaviour and Physiology of Fish, 24,* 1–37.

Cai, Y. (2010). Artificial fish school algorithm applied in a combinatorial optimization problem. *International Journal of Intelligent Systems and Applications, 1,* 37–43.

Cavalcanti-Júnior, G.M., Bastos-Filho, C.J.A., Lima-Neto, F.B. & Castro, R.M.C.S. (2011). A hybrid algorithm based on fish school search and particle swarm optimization for dynamic problems. In Tan, Y. (ed.) *ICSI 2011, Part II, LNCS 6729,* (pp. 543–552). Berlin: Springer.

Cavalcanti-Júnior, G.M., Bastos-Filho, C.J.A. & Lima-Neto, F.B.D. (2012). Volitive Clan PSO—an approach for dynamic optimization combining particle swarm optimization and fish school search. In Parpinelli, R. (ed.) *Theory and New Applications of Swarm Intelligence, Chap. 5,* (pp. 69–86). 51000 Rijeka, Croatia: In-Tech, ISBN 978-953-51-0364-6.

Chen, Z., & Tian, X. (2010). Artificial fish-swarm algorithm with chaos and its application. *Second International Workshop on Education Technology and Computer Science (ETCS), 1,* 226–229.

Chen, Z., Ma, J., Lei, J., Yuan, B. & Lian, L. (2007). An improved shark-search algorithm based on multi-information. Fourth International Conference on Fuzzy Systems and Knowledge Discovery (FSKD), pp. 1–5.

Chen, X., Sun, D., Wang, J., & Liang, J. (2008). Time series forecasting based on novel support vector machine using artificial fish swarm algorithm. *Fourth International Conference on Natural Computation (ICNC), 2,* 206–211.

Cheng, Z. & Hong, X. (2012). PID controller parameters optimization based on artificial fish swarm algorithm. Fifth International Conference on Intelligent Computation Technology and Automation (ICICTA), pp. 265–268.

Cheng, Y., Jiang, M. & Yuan, D. (2009). Novel clustering algorithms based on improved artificial fish swarm algorithm. Sixth International Conference on Fuzzy Systems and Knowledge Discovery (FSKD), pp. 141–145.

Cho, J., Garcia-Molina, H., & Page, L. (1998). Efficient crawling through URL ordering. *Computer Networks and ISDN Systems, 30,* 161–172.

Farzi, S. (2009). Efficient job scheduling in grid computing with modified artificial fish swarm algorithm. *International Journal of Computer Theory and Engineering, 1,* 13–18.

Feng, X., Yin, J., Xu, M., Zhao, X. & Wu, B. (2010). The algorithm optimization on artificial fish-swarm for the target area on simulation robots. 2nd International Conference on Signal Processing Systems (ICSPS), pp. V3-87–V3-89.

Fernandes, E.M.G.P., Martins, T.F.M.C., Maria, A. & Rocha, A.C. (2009, 30 June–3 July). Fish swarm intelligent algorithm for bound constrained global optimization. International Conference on Computational and Mathematical Methods in Science and Engineering (CMMSE) (pp. 461–472).

Gao, Y. & Chen, Y. (2010). The optimization of water utilization based on artificial fish-swarm algorithm. IEEE Sixth International Conference on Natural Computation (ICNC), pp. 4415–4419.

Gao, W., Zhao, H., Song, C. & Xu, J. (2009). Mixed using artificial fish-particle swarm optimization algorithm for hyperspace basing on local searching. IEEE 3rd International Conference on Bioinformatics and Biomedical Engineering (ICBBE), pp. 1–4.

Gao, X. Z., Wu, Y., Zenger, K. & Huang, X. (2010). A knowledge-based artificial fish-swarm algorithm. IEEE 13th International Conference on Computational Science and Engineering (CSE), pp. 327–332.

He, D., Qu, L. & Guo, X. (2009a). Artificial fish-school algorithm for integer programming. International Conference on Information Engineering and Computer Science (ICIECS), pp. 1–4.

He, S., Belacel, N., Hamam, H. & Bouslimani, Y. (2009b). Fuzzy clustering with improved artificial fish swarm algorithm. IEEE International Joint Conference on Computational Sciences and Optimization (CSO), pp. 317–321.

Hersovici, M., Jacovi, M., Maarek, Y. S., Pelleg, D., Shtalhaim, M., & Ur, S. (1998). The shark-search algorithm. An application: tailored Web site mapping. *Computer Networks and ISDN Systems, 30*, 317–326.

Hillis, K., Petit, M. & Jarrett, K. (2013). *Google and the culture of search.* Routledge: Taylor & Francis. ISBN 978-0-415-88300-9.

Hu, J., Zeng, X. & Xiao, J. (2010). Artificial fish school algorithm for function optimization. IEEE 2nd International Conference on Information Engineering and Computer Science (ICIECS), pp. 1–4.

Hu, Y., Yu, B., Ma, J. & Chen, T. (2011). Parallel fish swarm algorithm based on GPU-acceleration. IEEE 3rd International Workshop on Intelligent Systems and Applications (ISA), pp. 1–4.

Huang, Y., & Lin, Y. (2008). Freight prediction based on BP neural network improved by chaos artificial fish-swarm algorithm. *International Conference on Computer Science and Software Engineering, 5*, 1287–1290.

Huang, Z.-J. & Wang, B.-Q. (2010). A novel swarm clustering algorithm and its application for CBR retrieval. IEEE 2nd International Conference on Information Engineering and Computer Science (ICIECS), pp. 1–5.

Huang, R., Tawfik, H., Nagar, A. & Abbas, G. (2009). A novel hybrid QoS multicast routing based on clonal selection and artificial fish swarm algorithm. Second International Conference on Developments in eSystems Engineering (DESE), pp. 47–52.

Janecek, A. & Tan, Y. (2011a). Feeding the fish—weight update strategies for the fish school search algorithm. In Tan, Y. (Ed.) *ICSI 2011, Part II, LNCS 6729,* (pp. 553–562). Berlin: Springer.

Janecek, A., & Tan, Y. (2011b). Swarm intelligence for non-negative matrix factorization. *International Journal of Swarm Intelligence Research, 2*, 12–34.

Jarvis, J. (2009). *What whould Google do?,* 55 Avenue Road, Suite 2900, Toronto, ON, M5R, 3L2. Canada: HarperCollins Publishers Ltd., ISBN 978-0-06-176472-1.

Jiang, M. & Cheng, Y. (2010, July 6–9). Simulated annealing artificial fish swarm algorithm. IEEE 8th World Congress on Intelligent Control and Automation (WCICA) (pp. 1590–1593). Jinan, China.

Jiang, M. & Yuan, D. (2005). Wavelet threshold optimization with artificial fish swarm algorithm. International Conference on Neural Networks and Brain (ICNN&B), vol. 1, pp. 569–572.

Jiang, M. & Zhu, K. (2011). Multiobjective optimization by artificial fish swarm algorithm. IEEE International Conference on Computer Science and Automation Engineering (CSAE), pp. 506–511.

Jiang, M., Wang, Y., Pfletschinger, S., Lagunas, M.A. & Yuan, D. (2007a). Optimal multiuser detection with artificial fish swarm algorithm. In Huang, D.-S., Heutte, L. & Loog, M. (Eds.) *ICIC 2007, CCIS 2,* (pp. 1084–1093). Berlin: Springer.

Jiang, M., Wang, Y., Rubio, F. & Yuan, D. (2007b). Spread spectrum code estimation by artificial fish swarm algorithm. IEEE International Symposium on Intelligent Signal Processing (WISP), pp. 1–6.

Jiang, M., Yuan, D. & Cheng, Y. (2009). Improved artificial fish swarm algorithm. Fifth International Conference on Natural Computation, pp. 281–285.

Lacerda, M.G.P.D. & Neto, F.B.D.L. (2013). A new heuristic of fish school segregation for multi-solution optimization of multimodal problems. Second International Conference on Intelligent Systems and Applications (INTELLI 2013), pp. 115–121. IARIA.

Li, X.-L. (2003). *A new intelligent optimization method—artificial fish school algorithm (in Chinese with English abstract).* Unpublished Doctoral Thesis, Zhejiang University.

Li, X.-L., & Qian, J.-X. (2003). Studies on artificial fish swarm optimization algorithm based on decomposition and coordination techniques. *Journal of Circuits and Systems, 8*, 1–6.

Li, G., Sun, H. & Lv, Z. (2008, April 6–9). Study of available transfer capability based on improved artificial fish swarm algorithm. Third International Conference on Electric Utility Deregulation and Restructuring and Power Technologies (DRPT) (pp. 999–1003), Nanjing, China.

Li, Z., Guo, H., Liu, L., Yang, J. & Yuan, P. (2012). Resolving single depot vehicle routing problem with artificial fish swarm algorithm. In Tan, Y., Shi, Y. & Ji, Z. (Eds.) *ICSI 2012, Part I, LNCS 7332* (pp. 422–430). Berlin: Springer.

Lins, A.J.C.C., Bastos-Filho, C.J.A., Nascimento, D.N.O., Junior, M.A.C.O. & Lima-Neto, F.B.D. (2012). Analysis of the performance of the fish school search algorithm running in graphic processing units. In Parpinelli, R. (Ed.) *Theory and New Applications of Swarm Intelligence, Chap. 2*, (pp. 17–32). Janeza Trdine 9, 51000 Rijeka, Croatia: InTech, ISBN 978-953-51-0364-6.

Liu, C.-B., Wang, H.-J., Luo, Z.-P., Yu, X.-Q. & Liu, L.-H. (2009a). QoS multicast routing problem based on artificial fish-swarm algorithm. IEEE First International Workshop on Education Technology and Computer Science (ETCS), pp. 814–817.

Liu, T., Hou, Y.-B., Qi, A.-L. & Chang, X.-T. (2009b). Feature optimization based on artificial fish-swarm algorithm in intrusion detections. IEEE International Conference on Networks Security, Wireless Communications and Trusted Computing (NSWCTC), pp. 542–545.

Luo, F.-F., Chen, G.-L. & Guo, W.-Z. (2005). An improved fish-search algorithm for information retrieval. IEEE International Conference on Natural Language Processing and Knowledge Engineering (IEEE NLP-KE), pp. 523–528.

Luo, Y., Zhang, J. & Li, X. (2007, August 18–21). The optimization of PID controller parameters based on artificial fish swarm algorithm. IEEE International Conference on Automation and Logistics, (pp. 1058–1062). Jinan, China.

Luo, Y., Wei, W. & Wang, S.X. (2010, August 25–27). Optimization of PID controller parameters based on an improved artificial fish swarm algorithm. IEEE Third International Workshop on Advanced Computational Intelligence (IWACI) (pp. 328–332). Suzhou, Jiangsu, China.

Ma, Q. & Lei, X. (2010). Application of artificial fish school algorithm in UCAV path planning. IEEE Fifth International Conference on Bio-Inspired Computing: Theories and Applications (BIC-TA), pp. 555–559.

Ma, X. & Liu, Q. (2009, August 20–24). An artificial fish swarm algorithm for Steiner tree problem. IEEE International Conference on Fuzzy Systems (FUZZ-IEEE). Korea, pp. 59–63.

Ma, H., & Wang, Y. (2009). An artificial fish swarm algorithm based on chaos search. *Fifth International Conference on Natural Computation, 4*, 118–121.

Madeiro, S.S., Lima-Neto, F.B.D., Bastos-Filho, C.J.A. & Figueiredo, E.M.D.N. (2011). Density as the segregation mechanism in fish school search for multimodal optimization problems. In Tan, Y. (Ed.) *ICSI 2011, Part II, LNCS 6729*, (pp. 563–572). Berlin: Springer.

Min, H. & Wang, Z. (2010, December 14–18). Group escape behavior of multiple mobile robot system by mimicking fish schools. IEEE International Conference on Robotics and Biomimetics (ROBIO), Tianjin, China, pp. 320–326.

Neshat, M., Yazdani, D., Gholami, E., Masoumi, A., & Sargolzae, M. (2011). A new hybrid algorithm based on artificial fishes swarm optimization and k-means for cluster analysis. *International Journal of Computer Science Issues, 8*, 251–259.

Neshat, M., Adeli, A., Sepidnam, G., Sargolzaei, M., & Toosi, A. N. (2012a). A review of artificial fish swarm optimization methods and applications. *International Journal on Smart Sensing and Intelligent Systems, 5*, 107–148.

Neshat, M., Sepidnam, G., Sargolzaei, M. & Toosi, A.N. (2012b). Artificial fish swarm algorithm: a survey of the state-of-the-art, hybridization, combinatorial and indicative applications. *Artificial Intelligence Review,* doi:10.1007/s10462-012-9342-2.

Nie, H., Wang, B., Zhang, D. & Bai, B. (2010). The multi-stage transmission network planning based on chaotic artificial fish school algorithm. International Conference on E-Product E-Service and E-Entertainment (ICEEE), pp. 1–5.

Niu, D., Shen, W. & Sun, Y. (2010). RBF and artificial fish swarm algorithm for short-term forecast of stock indices. IEEE Second International Conference on Communication Systems, Networks and Applications (ICCSNA), pp. 139–142.

Peng, Y. (2011). An improved artificial fish swarm algorithm for optimal operation of cascade reservoirs. *Journal of Computers, 6*, 740–746.

Qi, A.-L., Ma, H.-W., & Liu, T. (2009). A weak signal detection method based on artificial fish swarm optimized matching pursuit. *World Congress on Computer Science and Information Engineering, 6*, 185–189.

Rocha, A.M.A.C. & Fernandes, E.M.G.P. (2011a). Mutation-based artificial fish swarm algorithm for bound constrained global optimization. In Simos, T.E., (Ed.) ICNAAM 2011, Vol. 1389, pp. 751–754.

Rocha, A.M.A.C. & Fernandes, E.M.G.P. (2011b, December 5–6). On hyperbolic penalty in the mutated artificial fish swarm algorithm in engineering problems. 16th Online World Conference on Soft Computing in Industrial Applications (WSC16). WWW, pp. 1–11.

Rocha, A.M.A.C., Fernandes, E.M.G.P. & Martins, T.F.M.C. (2011a). Novel fish swarm heuristics for bound constrained global optimization problems. In Murgante, B., Gervasi, O., Iglesias, A., Taniar, D. & Apduhan, B. (Eds.) *ICCSA 2011, Part III, LNCS 6784,* (PP. 185–199). Berlin: Springer.

Rocha, A. M. A. C., Martins, T. F. M. C., & Fernandes, E. M. G. P. (2011b). An augmented Lagrangian fish swarm based method for global optimization. *Journal of Computational and Applied Mathematics, 235*, 4611–4620.

Shen, W., Guo, X., Wu, C., & Wu, D. (2011). Forecasting stock indices using radial basis function neural networks optimized by artificial fish swarm algorithm. *Knowledge-Based Systems, 24*, 378–385.

Shi, H.-Y. & Shang, Z.-Q. (2010). Study on a solution of pursuit-evasion differential game based on artificial fish school algorithm. Chinese Control and Decision Conference (CCDC), pp. 2092–2096. IEEE.

Song, J., Sun, R.-Y., Zhang, Y.-J., Li, N.-N. & Gu, J.-H. (2008). The splicing method of images of rare point's feature based on artificial fish-swarm algorithm. International Conference on Advanced Computer Theory and Engineering (ICACTE), pp. 783–787.

Song, X., Wang, C., Wang, J. & Zhang, B. (2010). A hierarchical routing protocol based on AFSO algorithm for WSN. IEEE International Conference On Computer Design and Appliations (ICCDA), pp. V2-635–V2639.

Sun, T., Xie, X.-F., Sun, Y.-Q. & Li, S.-Y. (2009). Airplane route planning for plane-missile cooperation using improved fish-search algorithm. International Joint Conference on Artificial Intelligence (JCAI), pp. 853–856.

Sun, S., Zhang, J. & Liu, H. (2011, December 16–18). Key frame extraction based on artificial fish swarm algorithm and k-means. IEEE International Conference on Transportation, Mechanical, and Electrical Engineering (TMEE) (pp. 1650–1653). Changchun, China.

Tian, W., & Liu, J. (2009). An improved artificial fish swarm algorithm for multi robot task scheduling. *Fifth International Conference on Natural Computation, 4*, 127–130.

Tian, W. & Tian, Y. (2009). An improved artificial fish swarm algorithm for resource leveling. International Conference on Management and Service Science (MASS), pp. 1–4.

Tian, W., Ai, L., Tian, Y., & Liu, J. (2009a). A new optimization algorithm for fuzzy set design. *International Conference on Intelligent Human-Machine Systems and Cybernetics (IHMSC), 2*, 431–435.

Tian, W., Geng, Y., Liu, J. & Ai, L. (2009b). Optimal parameter algorithm for image segmentation. IEEE Second International Conference on Future Information Technology and Management Engineering (FITME), pp. 179–182.

Tsai, H.-C., & Lin, Y.-H. (2011). Modification of the fish swarm algorithm with particle swarm optimization formulation and communication behavior. *Applied Soft Computing, 11*, 5367–5374.

Turabieh, H. & Abdullah, S. (2011). A hybrid fish swarm optimisation algorithm for solving examination timetabling problems. In Coello, C.A.C. (Ed.) *LION 5, LNCS 6683*, (pp. 539–551). Berlin: Springer.

Wang, L. & Ma, L. (2011, August 12–14). A hybrid artificial fish swarm algorithm for bin-packing problem. IEEE International Conference on Electronic and Mechanical Engineering and Information Technology (EMEIT). pp. 27–29.

Wang, C.-J., & Xia, S.-X. (2010). Application of probabilistic causal-effect model based artificial fish-swarm algorithm for fault diagnosis in mine hoist. *Journal of Software, 5*, 474–481.

Wang, C.-R., Zhou, C.-L., & Ma, J.-W. (2005). An improved artificial fish-swarm algorithm and its application in feed-forward neural networks. *Fourth International Conference on Machine Learning and Cybernetics, Guangzhou, China, 18–21*(August), 2890–2894.

Wang, F., Xu, X. & Zhang, J. (2008). Strategy for aircraft sequencing based on artificial fish school algorithm. Control and Decision Conference (CCDC), pp. 861–864.

Wang, F., Xu, X. & Zhang, J. (2009). Application of artificial fish school and K-means clustering algorithms for stochastic GHP. Control and Decision Conference (CCDC), pp. 4280–4283.

Wang, Y., Liao, H. & Hu, H. (2012). Wireless sensor network deployment using an optimized artificial fish swarm algorithm. IEEE International Conference on Computer Science and Electronics Engineering (ICCSEE), pp. 90–94.

Wei, X.-X., Zeng, H.-W. & Zhou, Y.-Q. (2010). Hybrid artificial fish school algorithm for solving ill-conditioned linear systems of equations. IEEE International Conference on Intelligent Computing and Intelligent Systems (ICIS), pp. 290–294.

Wu, Y., Gao, X.-Z. & Zenger, K. (2011a). Knowledge-based artificial fish-swarm algorithm. 8th IFAC World Congress, 28 August–2 September, Milano, Italy, pp. 14705–14710. International Federation of Automatic Control (IFAC).

Wu, Y., Kiviluoto, S., Zenger, K., Gao, X. Z., & Huang, X. (2011b). Hybrid swarm algorithms for parameter identification of an actuator model in an electrical machine. *Advances in Acoustics and Vibration, 2011*, 1–12.

Xiao, L. (2010). A clustering algorithm based on artificial fish school. IEEE 2nd International Conference on Computer Engineering and Technology (ICCET), pp. V7-766–V7-76.

Xu, L. & Liu, S. (2010). Case retrieval strategies of tabu-based artificial fish swarm algorithm. IEEE Second International Conference on Computational Intelligence and Natural Computing (CINC), pp. 365–369.

Xu, H., Li, R., Guo, J., & Wang, H. (2009). An adaptive meta-cognitive artificial fish school algorithm. *International Forum on Information Technology and Applications (IFITA), 1*, 594–597.

Xue, Y., Du, H., & Jian, W. (2004). Optimum steelmaking charge plan using artificial fish swarm optimization algorithm. *IEEE International Conference on Systems, Man and Cybernetics, 5*, 4360–4364.

Yang, F., Tang, G. & Jin, H. (2011). Knowledge mining of traditional Chinese medicine constitution classification rules based on artificial fish school algorithm. IEEE 3rd International Conference on Communication Software and Networks (ICCSN), pp. 462–466.

Yazdani, D., Golyari, S. & Meybodi, M. R. (2010a). A new hybrid algorithm for optimization based on artificial fish swarm algorithm and cellular learning automata. IEEE 5th International Symposium on Telecommunications (IST), pp. 932–937.

Yazdani, D., Toosi, A.N. & Meybodi, M.R. (2010b). Fuzzy adaptive artificial fish swarm algorithm. *Advances in Artificial Intelligence, LNCS 6464*, (pp. 334–343). Berlin: Springer.

Yazdani, D., Akbarzadeh-Totonchi, M.R., Nasiri, B. & Meybodi, M.R. (2012, June 10–15). vA new artificial fish swarm algorithm for dynamic optimization problems. IEEE World Congress on Computational lnteliigence (WCCI). Brisbane, Australia, pp. 1–8.

Yu, G., & He, D.-X. (2011). Based on AFSA-tabu search algorithm combined QoS multicast routing algorithm. *Energy Procedia, 13*, 5746–5752.

Yu, Y., Tian, Y.-F. & Yin, Z.-F. (2005). Multiuser detector based on adaptive artificial fish school algorithm. ISCIT, pp. 1433–1437.

Yu, S., Wang, R., Xu, H., Wan, W., Gao, Y. & Jin, Y. (2011). WSN nodes deployment based on artificial fish school algorithm for traffic monitoring system. IEEE IET International Conference on Smart and Sustainable City (ICSSC), pp. 1–5.

Yu, H., Wei, J. & Li, J. (2012). Transformer fault diagnosis based on improved artificial fish swarm optimization algorithm and BP network. IEEE 2nd International Conference on Industrial Mechatronics and Automation (ICIMA), pp. 99–104.

Yuan, Y., Zhu, H., Zhang, M., Zhu, H., Wang, X., Wang, H., Chen, J. & Zhang, J. (2010). Reactive power optimization of distribution network based on improved artificial fish swarm algorithm. IEEE China International Conference on Electricity Distribution (CICED), pp. 1–5.

Zhang, M., Shao, C., Li, F., Gan, Y. & Sun, J. (2006a, June 25–28). Evolving neural network classifiers and feature subset using artificial fish swarm. IEEE International Conference on Mechatronics and Automation. Luoyang, China, pp. 1598–1602.

Zhang, M., Shao, C., Li, M. & Sun, J. (2006b, June 21–23). Mining classification rule with artificial fish swarm. 6th World Congress on Intelligent Control and Automation (pp. 5877–5881). Dalian, China.

Zhang, B., Mao, J. & Li, H. (2011, March 20–23). A hybrid algorithm for sensing coverage problem in wireless sensor netwoks. IEEE International Conference on Cyber Technology in Automation, Control, and Intelligent Systems (CYBER) (pp. 162–165). Kunming, China.

Zheng, T. & Li, J. (2010, July 6–9). Multi-robot task allocation and scheduling based on fish swarm algorithm. IEEE 8th World Congress on Intelligent Control and Automation (WCICA) (pp. 6681–6685). Jinan, China.

Zheng, G., & Lin, Z. (2012). A winner determination algorithm for combinatorial auctions based on hybrid artificial fish swarm algorithm. *Physics Procedia, 25*, 1666–1670.

Zhou, Y. & Huang, H. (2009). Hybrid artificial fish school algorithm based on mutation operator for solving nonlinear equations. IEEE International Workshop on Intelligent Systems and Applications (ISA), pp. 1–5.

Zhou, Y. & Liu, B. (2009). Two novel swarm intelligence clustering analysis methods. IEEE Fifth International Conference on Natural Computation (ICNC), pp. 497–501.

Zhu, K. & Jiang, M. (2009). An improved artificial fish swarm algorithm based on chaotic search and feedback strategy. International Conference on Computational Intelligence and Software Engineering (CISE), pp. 1–4.

Zhu, K. & Jiang, M. (2010, July 6–9). Quantum artificial fish swarm algorithm. IEEE 8th World Congress on Intelligent Control and Automation (WCICA) (pp. 1–5). Jinan, China.

Zhu, K., Jiang, M. & Cheng, Y. (2010). Niche artificial fish swarm algorithm based on quantum theory. IEEE 10th International Conference on Signal Processing (ICSP), pp. 1425–1428.

Zhu, W., Jiang, J., Song, C., & Bao, L. (2012). Clustering algorithm based on fuzzy C-means and artificial fish swarm. *Procedia Engineering, 29*, 3307–3311.

Chapter 10
Frog Inspired Algorithms

Abstract In this chapter, we present two frog inspired computational intelligence (CI) algorithms, namely, shuffled frog leaping algorithm (SFLA) and frog calling algorithm (FCA). We first provide a brief introduction in Sect. 10.1. Then, the fundamentals and performance of SFLA are introduced in Sect. 10.2. Next, Sect. 10.3 outlines some core working principles and preliminary experimental studies relative to FCA. Finally, Sect. 10.4 summarises this chapter.

10.1 Introduction

In this chapter, we will introduce two computational intelligence (CI) algorithms that are inspired by some interesting behaviours exhibited by frogs (Wang et al. 2008; Mills et al. 2010; Rock et al. 2009; Reilly and Jorgensen 2011). These two algorithms are called shuffled frog leaping algorithm (SFLA) and frog calling algorithm (FCA), respectively.

10.2 Shuffled Frog Leaping Algorithm

10.2.1 Fundamentals of Shuffled Frog Leaping Algorithm

Shuffled frog leaping algorithm (SFLA) was recently proposed in (Eusuff and Lansey 2003; Eusuff et al. 2006; Eusuff 2004) for solving problems with discrete decision variables. Inspired by natural memetics, SFLA is a population-based cooperative search metaphor combining the benefits of the genetic-based memetic algorithm (MA) and the social behaviour based particle swarm optimization (PSO). Such algorithms have been developed to arrive at near-optimum solutions to complex and large-scale optimization problems which cannot be solved by gradient-based mathematical programming techniques.

B. Xing and W.-J. Gao, *Innovative Computational Intelligence: A Rough Guide to 134 Clever Algorithms*, Intelligent Systems Reference Library 62, DOI: 10.1007/978-3-319-03404-1_10, © Springer International Publishing Switzerland 2014

In SFLA, a population of randomly generated P solutions forms an initial population, where each solution called a frog is represented by an n-dimensional vector. SFLA starts with the whole population partitioned into a number of parallel subsets referred to as memeplexes. Then each memeplex is considered as a different culture of frogs and permitted to evolve independently to search the space. Within each memeplex, the individual frogs hold their own ideas, which can be affected by the ideas of other frogs, and experience a memetic evolution. During the evolution, the frogs may change their memes by using the information from the memeplex best or the best individual of entire population. Incremental changes in memo-types correspond to a leaping step size and the new meme corresponds to the frog's new position. In each cycle, only the frog with the worst fitness in the current memeplex is improved by a process similar to PSO.

In order to implement SFLA, the following procedures need to be followed (Eusuff and Lansey 2003; Eusuff et al. 2006; Eusuff 2004):

- Step 0: Setting $im = 0$ and $iN = 0$, where the number of memeplexes will be counted by im, and the number of evolutionary steps is recorded by iN.
- Step 1: Setting $im = im + 1$.
- Step 2: Setting $iN = iN + 1$.
- Step 3: Constructing a submemeplex. The weights are allocated based on a triangular probability distribution which is defined by Eq. 10.1 (Eusuff and Lansey 2003):

$$p_j = \frac{2(n+1-j)}{n(n+1)}, \quad j = 1, 2, \ldots, n. \tag{10.1}$$

- Step 4: Improving the worst frog's location. In SFLA, the new position can be computed through Eq. 10.2 (Eusuff and Lansey 2003):

$$D_{(iq=q)} = D_W + d. \tag{10.2}$$

If $D_{(iq=q)}$ falls within the feasible space, then computing the new performance value $f_{(iq=q)}$; otherwise going to Step 5. If the new $f_{(iq=q)}$ is better than the previous $f_{(iq=q)}$, then replacing the old $D_{(iq=q)}$ with the new one and jumping to Step 7; otherwise, going to Step 5.

- Step 5: If previous step (i.e., Step 4) could not generate a better solution, then computing the step and the new position for frog based on the present global optimal solution.
- Step 6: Censorship. If the frog's new location is either unsuitable or no good than the old one, the spread of defective meme is terminated by stochastically generating a new frog at a suitable position to replace the frog whose new position was not possible to move towards an optimum value.
- Step 7: Upgrading the memeplex.
- Step 8: If $iN < N$, returning to Step 2.

- Step 9: If $im < m$, returning to Step 1; otherwise performing shuffling operation to create new memeplex sets.

10.2.2 Performance of SFLA

To verify the efficacy of SFLA, the New York City Water Supply Tunnel System case study was employed in (Eusuff and Lansey 2003). The simulation results showed that SFLA was capable to find previous best solutions for two example networks and a near optimal solution for the third case. In comparison with other CI techniques (e.g., genetic algorithm (GA), simulated annealing (SA), etc.), the SFLA converged within fewer iteration rounds which make it a versatile tool in dealing with optimization problems.

10.3 Emerging Frog Inspired Algorithm

In addition to the aforementioned SFLA, the characteristics of this interesting animal also motivate researchers to develop another frog inspired innovative CI algorithm.

10.3.1 Frog Calling Algorithm

10.3.1.1 Fundamentals of Frog Calling Algorithm

Frog calling algorithm (FCA) was originally proposed in (Mutazono et al. 2012) for dealing with power consumption issue in the context of wireless sensor networks. Inspired by Japanese tree frog calling (or satellite) behaviour, a self-organizing scheduling scheme was presented in (Mutazono et al. 2012) to achieve a energy-efficient data transmission. To fully utilize the FCA, the following three factors have to be considered (Mutazono et al. 2012):

- Factor 1: Territory. A frog will first check that if there is any calling frog in its own territory range, and then it will confirm that if the total number of calling frogs existing in the paddy field is still within an acceptable range. Once it is done with these, it will decide to produce calls or not.
- Factor 2: Number of competing frogs. A frog will evaluate its surroundings and compare itself with other calling frogs according to some criteria. If the probability for the frog to win is high, it will begin to call anyway.
- Factor 3: Body size. Once the weak calling frog detects its current condition, it will adopt sleep strategy to avoid competition.

10.3.1.2 Performance of FCA

Mutazono et al. (2012) tested the proposed FCA on a single-hop network. The preliminary computer simulation results demonstrated that proposed FCA method extends network lifetime by a factor of 6.7 in comparison with the method without sleep control strategy for a coverage ratio of 80 %.

10.4 Conclusions

In this chapter, we presented two frog inspired CI algorithms, namely, SFLA and FCA. Although they both are newly introduced CI methods, we have witnessed the following rapid spreading of at least one of them, i.e., SFLA.

First, several enhanced versions of SFLA can be found in the literature as outlined below:

- Binary SFLA (Gómez-González and Jurado 2011).
- Chaos-based SFLA (Li et al. 2008; Zhang et al. 2011).
- Clonal selection-based SFLA (Bhaduri 2009).
- Composite SFLA (Zhang et al. 2010).
- Discrete SFLA (Baghmisheh et al. 2011).
- Hybrid SFLA (Rahimi-Vahed and Mirzaei 2007; Rao and Lakshmi 2012; Niknam and Farsani 2010; Luo et al. 2009a; Farahani et al. 2010; Khorsandi et al. 2011; Niknam et al. 2012b).
- Improved SFLA (Malekpour et al. 2012; Zhang et al. 2008; Zhen et al. 2009; Jahani et al. 2011c; Li et al. 2012b).
- Modified SFLA (Huynh 2008; Nejad et al. 2010a, b; Narimani 2011; Elbeltagi et al. 2007; Jahani et al. 2010, 2011a ; Niknam et al. 2011b, c; Luo et al. 2009b; Roy and Chakrabarti 2011; Pu et al. 2011; Ahandani et al. 2011; Zhang et al. 2012).
- Multiobjective SFLA (Rahimi-Vahed et al. 2009; Liu et al. 2011a, 2012; Wang and Gong 2013; Li et al. 2010; Niknam et al. 2011a).
- Tribe-based SFLA (Niknam et al. 2012a).

Second, the SFLA has also been successfully applied to a variety of optimization problems as listed below:

- Bridge life cycle management (Elbehairy 2007; Elbehairy et al. 2006).
- Circuit design (Zhu and Zhi 2012).
- Controller design optimization (Huynh 2008).
- Data mining (Amiri et al. 2009; Liu et al. 2011a, 2012).
- Image processing (Bhaduri 2009; Wang et al. 2010).
- Laminate composite structures optimization (Rao and Lakshmi 2011).
- Manufacturing optimization (Rahimi-Vahed and Mirzaei 2007; Pakravesh and Shojaei 2011).

- Network virtualization (Liu et al. 2011b).
- Power system optimization (Rameshkhah et al. 2010a, b, 2011; Nejad et al. 2010a; Gómez-González and Jurado 2011; Narimani 2011; Sedighizadeh et al. 2011; Payam et al. 2011; Bijami et al. 2011; Jahani et al. 2010, Jahani et al. 2011a, b; Jalilzadeh et al. 2011; Ebrahimi et al. 2011; Yammani et al. 2012; Malekpour et al. 2012; Niknam and Farsani 2010; Nejad et al. 2010b; Niknam et al. 2011a, b, c, 2012b; Khorsandi et al. 2011; Roy and Chakrabarti 2011).
- Project management (Elbeltagi et al. 2007).
- Robot control (Pu et al. 2011).
- Scheduling optimization (Rahimi-Vahed et al. 2009; Tavakolan 2011; Rahimi-Vahed and Mirzaei 2008; Pan et al. 2011; Fang and Wang 2012; Li et al. 2012a; Wang and Fang 2011).
- Travelling salesman problem (Xue-Hui et al. 2008; Luo et al. 2009b).
- Water resource management (Eusuff and Lansey 2003; Eusuff 2004; Mora-Meliá et al. 2010; Chung 2007; Chung and Lansey 2009; Pasha and Lansey 2009; Seifollahi-Aghmiuni et al. 2011; Li et al. 2010).

Interested readers are referred to them as a starting point for a further exploration and exploitation of frog inspired algorithms.

References

Ahandani, M. A., Shirjoposh, N. P., & Banimahd, R. (2011). Three modified versions of differential evolution algorithm for continuous optimization. *Soft Computing, 15*, 803–830.

Amiri, B., Fathian, M., & Maroosi, A. (2009). Application of shuffled frog-leaping algorithm on clustering. *International Journal of Advanced Manufacturing Technology, 45*, 199–209.

Baghmisheh, M. T. V., Madani, K., & Navarbaf, A. (2011). A discrete shuffled frog optimization algorithm. *Artificial Intelligence Review, 36*, 267–284.

Bhaduri, A. (2009). Color image segmentation using clonal selection-based shuffled frog leaping algorithm. In *Proceedings of the International Conference on Advances in Recent Technologies in Communication and Computing (ARTCom)*. (pp. 517–520). IEEE.

Bijami, E., Abshari, R., Askari, J., Hosseinnia, S., & Farsangi, M. M. (2011). Optimal design of damping controllers for multi-machine power systems using metaheuristic techniques. *International Review of Electrical Engineering, 6*, 1883–1894.

Chung, G. (2007). *Water supply system management design and optimization under uncertainty*. PhD Thesis, University of Arizona.

Chung, G., & Lansey, K. (2009). Application of the shuffled frog leaping algorithm for the optimization of a general large-scale water supply system. *Water Resources Management, 23*, 797–823.

Ebrahimi, J., Hosseinian, S. H., & Gharehpetian, G. B. (2011). Unit commitment problem solution using shuffled frog leaping algorithm. *IEEE Transactions on Power Systems, 26*, 573–581.

Elbehairy, H., Elbeltagi, E., Hegazy, T., & Soudki, K. (2006). Comparison of two evolutionary algorithms for optimization of bridge deck repairs. *Computer-Aided Civil and Infrastructure Engineering, 21*, 561–572.

Elbehairy, H. (2007). *Ridge management system with integrated life cycle cost optimization*. PhD Thesis, University of Waterloo.

Elbeltagi, E., Hegazy, T., & Grierson, D. (2007). A modified shuffled frog-leaping optimization algorithm: Applications to project management. *Structure and Infrastructure Engineering, 3*, 53–60.

Eusuff, M. M., & Lansey, K. E. (2003). Optimization of water distribution network design using the shuffled frog leaping algorithm. *Journal of Water Resources Planning and Management, 129*, 210–225.

Eusuff, M. M. (2004). *Water resources decision making using meta-heuristic optimization methods*. PhD Thesis, University of Arizona.

Eusuff, M., Lansey, K., & Pasha, F. (2006). Shuffled frog-leaping algorithm: A memetic meta-heuristic for discrete optimization. *Engineering Optimization, 38*, 129–154.

Fang, C., & Wang, L. (2012). An effective shuffled frog-leaping algorithm for resource-constrained project scheduling problem. *Computers and Operations Research, 39*, 890–901.

Farahani, M., Movahhed, S. B. & Ghaderi, S. F. (2010, September 6–9). A hybrid meta-heuristic optimization algorithm based on SFLA. In *Proceedings of the 2nd International Conference on Engineering Optimization*. (pp. 1–8). Lisbon, Portugal.

Gómez-González, M., & Jurado, F. (2011). A binary shuffled frog-leaping algorithm for the optimal placement and sizing of photovoltaics grid-connected systems. *International Review of Electrical Engineering, 6*, 452–458.

Huynh, T.-H. (2008). A modified shuffled frog leaping algorithm for optimal tuning of multivariable PID controllers. In *IEEE International Conference on Industrial Technology (ICIT)*. (pp. 1–6). doi:10.1109/ICIT.2008.4608439.

Jahani, R., Nejad, H. C., Malekshah, A. S., & Shayanfar, H. A. (2010). A new advanced heuristic method for optimal placement of unified power flow controllers in electrical power systems. *International Review of Electrical Engineering, 5*, 2786–2794.

Jahani, R., Malekshah, A. S., Nejad, H. C., & Araskalaei, A. H. (2011a). Applying a new advanced intelligent algorithm for optimal distributed generation location and sizing in radial distribution systems. *Australian Journal of Basic and Applied Sciences, 5*, 642–649.

Jahani, R., Nejad, H. C., Araskalaei, A. H., & Hajinasiri, M. (2011b). Optimal DG allocation in distribution network using a new heuristic method. *Australian Journal of Basic and Applied Sciences, 5*, 635–641.

Jahani, R., Nejad, H. C., Khayat, O., Abadi, M. M., & Zadeh, H. G. (2011c). An improved shuffled frog leaping algorithm approach for unit commitment problem. *Australian Journal of Basic and Applied Sciences, 5*, 1379–1387.

Jalilzadeh, S., Noroozian, R., Sabouri, M., & Behzadpoor, S. (2011). PSS and SVC controller design using chaos, PSO and SFL algorithms to enhancing the power system stability. *Energy and Power Engineering, 3*, 87–95.

Khorsandi, A., Alimardani, A., Vahidi, B., & Hosseinian, S. H. (2011). Hybrid shuffled frog leaping algorithm and Nelder—Mead simplex search for optimal reactive power dispatch. *IET Generation, Transmission and Distribution, 5*, 249–256.

Li, Y., Zhou, J., Yang, J., Liu, L., Qin, H. & Yang, L. (2008). The chaos-based shuffled frog leaping algorithm and its application. In *Proceedings of the Fourth International Conference on Natural Computation (ICNC)*. (pp. 481–485). doi:10.1109/ICNC.2008.242.

Li, Y., Zhou, J., Zhang, Y., Qin, H., & Liu, L. (2010). Novel multiobjective shuffled frog leaping algorithm with application to reservoir flood control operation. *Journal of Water Resources Planning and Management, 136*, 217–226.

Li, J., Pan, Q., & Xie, S. (2012a). An effective shuffled frog-leaping algorithm for multi-objective flexible job shop scheduling problems. *Applied Mathematics and Computation, 218*, 9353–9371.

Li, X., Luo, J., Chen, M.-R., & Wang, N. (2012b). An improved shuffled frog-leaping algorithm with extremal optimisation for continuous optimisation. *Information Sciences, 192*, 143–151.

Liu, J., Li, Z., Hu, X. & Chen, Y. (2011a). Multiobjective optimizaiton shuffled frog-leaping biclustering. In *Proceedings of the IEEE International Conference on Bioinformatics and Biomedicine Workshop (BIBMW)*.

Liu, W., Li, S., Xiang, Y., & Tang, X. (2011b). Virtual network embedding based on shuffled frog leaping algorithm in TUNIE. *International Journal of Advancements in Computing Technology, 3*, 402–409.

Liu, J., Li, Z., Hu, X., Chen, Y., & Liu, F. (2012). Multi-objective dynamic population shuffled frog-leaping biclustering of microarray data. *BMC Genomics, 13*, 1–11.

Luo, J., Chen, M.-R. & Li, X. (2009a). A novel hybrid algorithm for global optimization based on EO and SFLA. In *Proceedings of the 4th IEEE Conference on Industrial Electronics and Applications (ICIEA)*. (pp. 1935–1939). IEEE.

Luo, X.-H., Yang, Y., & Li, X. (2009b). Modified shuffled frog-leaping algorithm to solve traveling salesman problem (in Chinese). *Journal on Communications, 30*, 130–135.

Malekpour, A. R., Tabatabaei, S., & Niknam, T. (2012). Probabilistic approach to multi-objective Volt/Var control of distribution system considering hybrid fuel cell and wind energy sources using improved shuffled frog leaping algorithm. *Renewable Energy, 39*, 228–240.

Mills, D. S., Marchant-Forde, J. N., McGreevy, P. D., Morton, D. B., Nicol, C. J., Phillips, C. J. C., et al. (Eds.). (2010). *The encyclopedia of applied animal behaviour and welfare*. UK: CAB International. ISBN 978-0-85199-724-7.

Mora-Meliá, D., Iglesias-Rey, P. L., Bosque-Chacón, G. & López-Jiménez, P. A. (2010, October 29–30) *Statistical analysis of water distribution networks design using shuffled frog leaping algorithm. Proceedings of the International Workshop on Environmental Hydraulics*, IWEH09. (pp. 327–331). Valencia, Spain.

Mutazono, A., Sugano, M., & Murata, M. (2012). Energy efficient self-organizing control for wireless sensor networks inspired by calling behavior of frogs. *Computer Communications, 35*, 661–669.

Narimani, M. R. (2011). A new modified shuffle frog leaping algorithm for non-smooth economic dispatch. *World Applied Sciences Journal, 12*, 803–814.

Nejad, H. C., Jahani, R. & Shayanfar, H. A. (2010a). Comparison of modified shuffled frog leaping algorithm and other algorithms for optimal distributed generation location and sizing. *International Review of Electrical Engineering* (*I.R.E.E.*), *5*, 2286–2292.

Nejad, H. C., Jahani, R., Shayanfar, H. A., & Olamaei, J. (2010b). Comparison of novel heuristic technique and other evolutionary methods for optimal unit commitment of power system. *International Review on Modelling and Simulations, 3*, 1476–1482.

Niknam, T., & Farsani, E. A. (2010). A hybrid self-adaptive particle swarm optimization and modified shuffled frog leaping algorithm for distribution feeder reconfiguration. *Engineering Applications of Artificial Intelligence, 23*, 1340–1349.

Niknam, T., Farsani, E. A., & Nayeripour, M. (2011a). An efficient multi-objective modified shuffled frog leaping algorithm for distribution feeder reconfiguration problem. *European Transactions on Electrical Power, 21*, 721–739.

Niknam, T., Firouzi, B. B., & Mojarrad, H. D. (2011b). A new evolutionary algorithm for non-linear economic dispatch. *Expert Systems with Applications, 38*, 13301–13309.

Niknam, T., Narimani, M. R., Jabbari, M., & Malekpour, A. R. (2011c). A modified shuffle frog leaping algorithm for multi-objective optimal power flow. *Energy, 36*, 6420–6432.

Niknam, T., Farsani, E. A., Nayeripour, M., & Firouzi, B. B. (2012a). A new tribe modified shuffled frog leaping algorithm for multi-objective distribution feeder reconfiguration considering distributed generator units. *European Transactions on Electrical Power, 22*, 308–333.

Niknam, T., Narimani, M. R., & Azizipanah-Abarghooee, R. (2012b). A new hybrid algorithm for optimal power flow considering prohibited zones and valve point effect. *Energy Conversion and Management, 58*, 197–206.

Pakravesh, H., & Shojaei, A. (2011). Optimization of industrial CSTR for vinyl acetate polymerization using novel shuffled frog leaping based hybrid algorithms and dynamic modeling. *Computers and Chemical Engineering, 35*, 2351–2365.

Pan, Q.-K., Wang, L., Gao, L., & Li, J. (2011). An effective shuffled frog-leaping algorithm for lot-streaming flow shop scheduling problem. *International Journal of Advanced Manufacturing Technology, 52*, 699–713.

Pasha, M. F. K., & Lansey, K. (2009). Water quality parameter estimation for water distribution systems. *Civil Engineering and Environmental Systems, 26*, 231–248.

Payam, M. S., Bijami, E., Abdollahi, M., & Dehkordi, A. S. (2011). Optimal coordination of directional overcurrent relay for power delivery system. *Australian Journal of Basic and Applied Sciences, 5*, 1949–1957.

Pu, H., Zhen, Z., & Wang, D. (2011). Modified shuffled frog leaping algorithm for optimization of UAV flight controller. *International Journal of Intelligent Computing and Cybernetics, 4*, 25–39.

Rahimi-Vahed, A., & Mirzaei, A. H. (2007). A hybrid multi-objective shuffled frog-leaping algorithm for a mixed-model assembly line sequencing problem. *Computers and Industrial Engineering, 53*, 642–666.

Rahimi-Vahed, A., & Mirzaei, A. H. (2008). Solving a bi-criteria permutation flow-shop problem using shuffled frog-leaping algorithm. *Soft Computing, 12*, 435–452.

Rahimi-Vahed, A., Dangchi, M., Rafiei, H., & Salimi, E. (2009). A novel hybrid multi-objective shuffled frog-leaping algorithm for a bi-criteria permutation flow shop scheduling problem. *International Journal of Advanced Manufacturing Technology, 41*, 1227–1239.

Rameshkhah, F., Abedi, M. & Hosseinian, H. (2010a). Comparison and combination of shuffled frog-leaping algorithm and k-means for clustering of VCAs in power system. *International Review of Electrical Engineering (I.R.E.E.),* 5, 194–204.

Rameshkhah, F., Abedi, M., & Hosseinian, S. H. (2010b). Clustering of voltage control areas in power system using shuffled frog-leaping algorithm. *Electrical Engineering, 92*, 269–282.

Rameshkhah, F., Abedi, M., & Hosseinian, S. H. (2011). Comparison of shuffled frog leaping algorithm and PSO in data clustering with constraint for grouping voltage control areas in power systems. *European Transactions on Electrical Power, 21*, 1763–1782.

Rao, A. R. M. & Lakshmi, K. (2012). Optimal design of stiffened laminate composite cylinder using a hybrid SFL algorithm. *Journal of Composite Materials*. doi:10.1177/0021998311435674.

Reilly, S. M., & Jorgensen, M. E. (2011). The evolution of jumping in frogs: Morphological evidence for the basal anuran locomotor condition and the radiation of locomotor systems in crown group anurans. *Journal of Morphology, 272*, 149–168.

Rock, M., Murphy, J. T., Rasiah, R., Seters, P. V., & Managi, S. (2009). A hard slog, not a leap frog: globalization and sustainability transitions in developing Asia. *Technological Forecasting and Social Change, 76*, 241–254.

Roy, P., & Chakrabarti, A. (2011). Modified shuffled frog leaping algorithm for solving economic load dispatch problem. *Energy and Power Engineering, 3*, 551–556.

Sedighizadeh, M., Sarvi, M., & Naderi, E. (2011). Multi objective optimal power flow with FACTS devices using shuffled frog leaping algorithm. *International Review of Electrical Engineering, 6*, 1794–1801.

Seifollahi-Aghmiuni, S., Haddad, O. B., Omid, M. H., & Mariño, M. A. (2011). Long-term efficiency of water networks with demand uncertainty. *Water Management, 164*, 147–159.

Tavakolan, M. (2011). *Development of construction projects scheduling with evolutionary algorithms.* PhD Thesis, Columbia University.

Wang, L., & Fang, C. (2011). An effective shuffled frog-leaping algorithm for multi-mode resource-constrained project scheduling problem. *Information Sciences, 181*, 4804–4822.

Wang, L. & Gong, Y. (2013). Multi-objective dynamic population shuffled frog leaping algorithm. In: Y. Tan, Y. Shi, & H. Mo (Eds.) *Advances in swarm intelligence, LNCS 7982*, (pp. 24–31). Berlin: Springer.

Wang, M., Zang, X.-Z., Fan, J.-Z., & Zhao, J. (2008). Biological jumping mechanism analysis and modeling for frog robot. *Journal of Bionic Engineering, 5*, 181–188.

Wang, N., Li, X., & Chen, X.-H. (2010). Fast three-dimensional Otsu thresholding with shuffled frog-leaping algorithm. *Pattern Recognition Letters, 31*, 1809–1815.

Xue-Hui, L., Yang, Y. & Li, X. (2008). Solving TSP with shuffled frog-leaping algorithm. In *Eighth international conference on intelligent systems design and applications (ISDA)*. (pp. 228–232). doi:10.1109/ISDA.2008.346.

Yammani, C., Maheswarapu, S., & Matam, S. (2012). Multiobjective optimization for optimal placement and size of DG using shuffled frog leaping algorithm. *Energy Procedia, 14*, 990–995.

Zhang, X., Hu, F., Tang, J., Zou, C. & Zhao, L. (2010). A kind of composite shuffled frog leaping algorithm. In *Sixth international conference on natural computation (ICNC)*. (vol. 5, pp. 2232–2235). doi:10.1109/ICNC.2010.5584419.

Zhang, X., Hu, F., Zou, C., & Zhao, L. (2011). The research of swarm intelligence algorithm based on chaotic frog behavior. *Energy Procedia, 13*, 1189–1196.

Zhang, X., Hu, X., Cui, G., Wang, Y., & Niu, Y. (2008, June 25–27). An improved shuffled frog leaping algorithm with cognitive behavior. In *The 7th World Congress on Intelligent Control and Automation (WCICA)*. Chongqing, China. (pp. 6197–6202). doi:10.1109/WCICA.2008.4592798.

Zhang, X., Zhang, Y., Shi, Y., Zhao, L., & Zou, C. (2012). Power control algorithm in cognitive radio system based on modified shuffled frog leaping algorithm. *International Journal of Electronics and Communications, 66*, 448–454.

Zhen, Z., Wang, D. & Liu, Y. (2009, May 18–21). Improved shuffled frog leaping algorithm for continuous optimization problem. IEEE Congress on Evolutionary Computation. (pp. 992–2995). Trondheim, Norway.

Zhu, A., & Zhi, L. (2012). Automatic test pattern generation based on shuffled frog leaping algorithm for sequential circuits. *Procedia Engineering, 29*, 856–860.

Chapter 11
Fruit Fly Optimization Algorithm

Abstract In this chapter, we present a novel optimization algorithm called fruit fly optimization algorithm (FFOA) which is inspired by the behaviour of fruit flies. We first describe the general knowledge of the foraging behaviour of fruit flies in Sect. 11.1. Then, the fundamentals and performance of FFOA are introduced in Sect. 11.2. Finally, Sect. 11.3 summarises this chapter.

11.1 Introduction

If you have been seeing small flies in your kitchen, they're probably fruit flies. In fact, the fruit flies can be viewed as the second smallest member among the model animals in the narrow sense which has only hundreds of neurons and has no brain (Shimada et al. 2005). During the summer, they are attracted to ripened or fermenting food through their sensing and perception characteristics, especially in osphresis and vision (Pan 2012; Touhara 2013). Inspired by the behaviour of real fruit flies, recently Pan (2012) proposed a new algorithm called fruit flies optimization algorithm (FFOA).

11.1.1 The Foraging Behaviour of Fruit Flies

Fruit flies are small flies and usually with red eyes. They are especially attracted to ripened foods in the kitchen. They can even smell food source from 40 km away (Pan 2012). In addition, the number of the fruit fly's eye (i.e., compound eye) are huge in which contains 760 unit eyes (Chapman 2013). Based on those characteristics, the fruit fly can exploit an extraordinarily wide range of food sources. Generally speaking, the food finding process of fruit fly is as follows (Pan 2012): firstly,

B. Xing and W.-J. Gao, *Innovative Computational Intelligence: A Rough Guide to 134 Clever Algorithms*, Intelligent Systems Reference Library 62,
DOI: 10.1007/978-3-319-03404-1_11,

it smells the food source by osphresis organ, and flies towards that location; then, after it gets close to the food location, the sensitive vision is also used for finding food and other fruit flies' flocking location; finally, it flies towards that direction.

11.2 Fruit Fly Optimization Algorithm

11.2.1 Fundamentals of Fruit Fly Optimization Algorithm

Fruit fly optimization algorithm (FFOA) was originally proposed in Pan (2011, 2012) that is based on the food foraging behaviour of fruit fly. Generally, the procedures of FFOA are described as follows (Pan 2012):

- Initialization phase: The fruit flies are randomly distributed in the search space (*InitX_axis* and *InitY_axis*) via Eqs. 11.1 and 11.2, respectively (Pan 2012):

$$X_i = X\_axis + RandomValue, \tag{11.1}$$

$$Y_i = Y\_axis + RandomValue, \tag{11.2}$$

  where the term "*RandomValue*" is a random vector that were sampled from a uniform distribution.
- Path construction phase: The distance and smell concentration value of each fruit fly can be defined via Eqs. 11.3 and 11.4, respectively (Pan 2012):

$$Dist_i = \sqrt{X_i^2 + Y_i^2}, \tag{11.3}$$

$$S_i = \frac{1}{Dist_i}, \tag{11.4}$$

  where $Dist_i$ is the distance between the *i*th individual and the food location, and S_i is the smell concentration judgment value which is the reciprocal of distance.

- Fitness function calculation phase. It can be defined via Eqs. 11.5 and 11.6, respectively (Pan 2012):

$$Smell_i = Function(S_i), \tag{11.5}$$

$$[bestSmell, bestIndex] = \max(Smell_i), \tag{11.6}$$

  where $Smell_i$ is the smell concentration of the individual fruit fly, *bestSmell* and *bestIndex* represent the largest elements and its indices along different dimensions of smell vectors, and $\max(Smell_i)$ is the maximal smell concentration among the fruit flies.

- Movement phase: The fruit fly keeps the best smell concentration value and will use vision to fly towards that location via Eqs. 11.7–11.9, respectively (Pan 2012):

$$Smellbest = bestSmell. \tag{11.7}$$

$$X\_axis = X(bestIndex). \tag{11.8}$$

$$Y\_axis = Y(bestIndex). \tag{11.9}$$

Overall, taking into account the key phases described above, the steps of implementing FFOA can be summarized as follows (Pan 2012):

- Step 1: Initialize the optimization problem and algorithm parameters.
- Step 2: Repeat till stopping criteria met. First, randomly select a location via distance and smell concentration judgment value. Second, calculate its fitness function $Function(S_i)$. Third, find out the fruit fly with maximal smell concentration among the fruit fly swarm. Fourth, rank the solutions and move to the best solution.
- Step 3: Post process and visualize results.

11.2.2 Performance of FFOA

In order to show how the FFOA performs, two functions (i.e., one minimum and one maximum) are tested in Pan (2012). Computational results showed that FFOA is capable to find the minimal value and the maximal value.

11.3 Conclusions

Nowadays, a number of algorithmic approaches based on the animals' foraging behaviour were developed and applied to variety of combinatorial optimization problems. Among others, FFOA is a new member. Two characteristics of fruit flies (osphresis and vision) are the building blocks of FFOA. The main advantages of FFOA include simple computational process, ease understanding, and easy implementation (Pan 2012). Although it is a newly introduced CI method, we have witnessed the following rapid spreading of FFOA:

First, several enhanced versions of FFOA can be found in literature as outlined below:

- Adaptive mutation FFOA (Han and Liu 2013).
- Binary FFOA (Wang et al. 2013).

- Modified FFOA (Liu et al. 2012).

Second, the FFOA has also been successfully applied to a variety of optimization problems as listed below:

- Autonomous surface vessels control (Abidin et al. 2012).
- Control optimization (Liu et al. 2012).
- Data mining (Chen et al. 2013; Tu et al. 2012).
- Multidimensional knapsack problem (Wang et al. 2013).
- Power load forecasting (Li et al. 2012, 2013, 2012).
- Traffic flow control (Zhu et al. 2013).

Interested readers please refer to them as a starting point for a further exploration and exploitation of FFOA.

References

Abidin, Z. Z., Hamzah, M. S. M., Arshad, M. R., & Ngah, U. K. (2012). A calibration framework for swarming ASVs' system design. *Indian Journal of Geo-Marine Sciences, 41*, 581–588.

Chapman, R. F. (2013). In S. J. Simpson, A. E. Douglas (Eds.) *The insects: structure and function*. New York: Cambridge University Press. ISBN 978-0-521-11389-2.

Chen, P.-W., Lin, W.-Y., Huang, T.-H., & Pan, W.-T. (2013). Using fruit fly optimization algorithm optimized grey model neural network to perform satisfaction analysis for e-business service. *Applied Mathematics and Information Sciences, 7*, 459–465.

Han, J.-Y., & Liu, C.-Z. (2013). Fruit fly optimization algorithm with adaptive mutation (in Chinese). *Application Research of Computers, 30*, 1–6. (in Chinese).

Li, H., Guo, S., Zhao, H., Su, C., & Wang, B. (2012). Annual electric load forecasting by a least squares support vector machine with a fruit fly optimization algorithm. *Energies, 5*, 4430–4445.

Li, H.-Z., Guo, S., Li, C.-J., & Sun, J.-Q. (2013). A hybrid annual power load forecasting model based on generalized regression neural network with fruit fly optimization algorithm. *Knowledge-Based Systems, 37*, 378–387.

Liu, Y., Wang, X. & Li, Y. (2012, July 6–8). A modified fruit-fly optimization algorithm aided PID controller designing. In *IEEE 10th World Congress on Intelligent Control and Automation*. (pp. 233–238). Beijing, China.

Pan, W.-T. (2011). *Fruit fly optimization algorithm* . Taiwan: Tsang Hai Book Publishing Co. ISBN 978-986-6184-70-3. (in Chinese).

Pan, W.-T. (2012). A new fruit fly optimization algorithm: Taking the financial distress model as an example. *Knowledge-Based Systems, 26*, 69–74.

Shimada, T., Kato, K., Kamikouchi, A., & Ito, K. (2005). Analysis of the distribution of the brain cells of the fruit fly by an automatic cell counting algorithm. *Physica A, 350*, 144–149.

Touhara, K. (Ed.) (2013). *Pheromone signaling: Methods and protocols*. London: Springer. ISBN 978-1-62703-618-4.

Tu, C.-S., Chang, C.-T., Chen, K–. K., & Lu, H.-A. (2012). A study on business performance with the combination of Z-score and FOAGRNN hybrid model. *African Journal of Business Management, 6*, 7788–7798.

Wang, L., Zheng, X.-L., & Wang, S.-Y. (2013). A novel binary fruit fly optimization algorithm for solving the multidimensional knapsack problem. *Knowledge-Based Systems, 48*, 17–23.

Zhu, W., Li, N., Shi, C., & Chen, B. (2013). SVR based on FOA and its application in traffic flow prediction. *Open Journal of Transportation Technologies, 2*, 6–9. (in Chinese).

Chapter 12
Group Search Optimizer Algorithm

Abstract In this chapter, we introduce a new optimization algorithm called group search optimizer (GrSO) which is inspired from the relationship of group foraging behaviours, i.e., producer-scrounger paradigm. We first describe the general knowledge of the producer-scrounger model in Sect. 12.1. Then, the fundamentals and performance of GrSO are introduced in Sect. 12.2. Finally, Sect. 12.3 summarises this chapter.

12.1 Introduction

The most easily recognized animals' behaviour is the foraging behaviour, i.e., searching for and exploiting food resources (Mills et al. 2010). Nowadays, several population-based algorithms are proposed based on the foraging theory, such as ant colony optimization (ACO) and particle swarm optimization (PSO). Recently, He et al. (2009) introduced a newly developed algorithm called group search optimizer (GrSO) algorithm which is inspired from the relationship of group foraging behaviours, i.e., producer-scrounger paradigm (Millor et al. 2006).

12.1.1 Producer-Scrounger Model

Generally speaking, the producer-scrounger (PS) model is a group-living foraging strategy in which the food will be founded by the discoverers (producers) and shared with others (scroungers) (Barnard and Sibly 1981; Brockmann and Barnard 1979). It assumed that individuals should specialize in either producing or scrounging at any one time (Parker 1984). In fact, it is the novel behaviour and adopted by many animals. To understand when and how such exploitative relationships will occur, several studies are made. For example Giraldeau and Lefebvre (1987) studied the PS model in pigeons, and (Biondolillo et al. 1997)

B. Xing and W.-J. Gao, *Innovative Computational Intelligence: A Rough Guide to 134 Clever Algorithms*, Intelligent Systems Reference Library 62,
DOI: 10.1007/978-3-319-03404-1_12,

tested the PS model between the zebra finches. In addition Vickery et al. (1991) proposed a newly model in which an information sharing mechanism (i.e., a third strategist: producer-scrounger opportunist) is incorporated.

12.2 Group Search Optimizer Algorithm

12.2.1 Fundamentals of Group Search Optimizer Algorithm

Group search optimizer (GrSO) algorithm was originally proposed in He et al. (2006). The population of GrSO is called group and each one inside is called member. Based on the PS model, the main steps of GrSO can be described as follows (Chen et al. 2012; Shen et al. 2009):

- Initializing phase: In GrSO, ith member at the kth searching iteration has a position $\mathbf{X}_i^k \in R^n$, a head angel $\boldsymbol{\varphi}_i^k = \left(\varphi_{i,1}^k, \ldots, \varphi_{i,n}^k\right) \in R^{n-1}$, and a head direction $\mathbf{D}_i^k(\boldsymbol{\varphi}_i^k) = \left(\mathbf{d}_{i,1}^k, \ldots, \mathbf{d}_{i,n}^k\right) \in R^n$ in an $n-$dimensional search space. In general, the distance can be calculated from $\boldsymbol{\varphi}_i^k$ via a polar to Cartesian coordinate transformation via Eqs. 12.1–12.3, respectively (Chen et al. 2012):

$$\mathbf{d}_{i,1}^k = \prod_{p=1}^{n-1} \cos\left(\varphi_{i,p}^k\right), \tag{12.1}$$

$$\mathbf{d}_{i,j}^k = \sin\left(\varphi_{i,(j-1)}^k\right) \cdot \prod_{p=j}^{n-1} \cos\left(\varphi_{i,p}^k\right), \quad \text{for } j = 2, \ldots, n-1, \tag{12.2}$$

$$\mathbf{d}_{i,n}^k = \sin\left(\varphi_{i,(n-1)}^k\right). \tag{12.3}$$

- Producing phase: The searching process of producer $\mathbf{X}_p$ at the kth iteration samples three points randomly via Eq. 12.4 (He et al. 2009):

$$\begin{cases} \mathbf{X}_z = \mathbf{X}_p^k + r_1 \cdot l_{\max} \cdot \mathbf{D}_p^k(\boldsymbol{\varphi}^k) \\ \mathbf{X}_r = \mathbf{X}_p^k + r_1 \cdot l_{\max} \cdot \mathbf{D}_p^k(\boldsymbol{\varphi}^k + \mathbf{r}_2 \cdot \theta_{\max}/2), \\ \mathbf{X}_l = \mathbf{X}_p^k + r_1 \cdot l_{\max} \cdot \mathbf{D}_p^k(\boldsymbol{\varphi}^k - \mathbf{r}_2 \cdot \theta_{\max}/2) \end{cases} \tag{12.4}$$

where $\mathbf{X}_p^k$ is the current position of pth individual in kth generation, $\mathbf{X}_z$, $\mathbf{X}_r$, and $\mathbf{X}_l$ are the positions which the pth individual found in zero degree, right and left direction of it, respectively, $r_1 \in R^1$ is a normally distributed random number with mean 0 and standard deviation 1, $\mathbf{r}_2 \in R^{n-1}$ is a uniformly distributed random sequence in the range $(0, 1)$, and $\theta_{\max}$ and $l_{\max}$ are the maximum pursuit angle and distance, respectively.

If the searching on three directions is ended, there are three states as follows (He et al. 2009):

When the new position has a better fitness value, the producer will move to the new point. The producer will keep its current position, however turn its head to a new angle via Eq. 12.5 (He et al. 2009):

$$\varphi^{k+1} = \varphi^{k} + \mathbf{r}_2 \cdot \alpha_{\max}, \tag{12.5}$$

where $\alpha_{\max} \in R^1$ is the maximum tuning angle.

When there is no better position can be found after α iterations, the producer will turn its head back to zero degree via Eq. 12.6 (He et al. 2009):

$$\varphi^{k+\alpha} = \varphi^{k}, \tag{12.6}$$

where $\alpha \in R^1$ is a pre-defined constant.

- Scrounging phase: After the determination of the producer, the scroungers will perform random walks by searching the opportunities to join the resources found by the producer via Eq. 12.7 (He et al. 2009):

$$\mathbf{X}_i^{k+1} = \mathbf{X}_i^{k} + \mathbf{r}_3 \cdot \left(\mathbf{X}_p^{k} + \mathbf{X}_i^{k}\right), \tag{12.7}$$

  where $\mathbf{r}_3$ is an uniform random sequence in the range $(0, 1)$, $\mathbf{X}_i^k$ and $\mathbf{X}_i^{k+1}$ are the positions of ith scrounger in t and $t+1$ iterations, respectively.
- Ranging phase: The inefficiency foragers will be selected as rangers that will perform a new searching process based on the random walks, i.e., generating a new random head angle (φ_i), choosing a random distance (l_i), and moving to the new position ($\mathbf{X}_i^{k+1}$), via Eqs. 12.8–12.10, respectively (He et al. 2009):

$$\varphi^{k+1} = \varphi_i^{k} + \mathbf{r}_2 \cdot \alpha_{\max}, \tag{12.8}$$

$$l_i = \alpha \cdot r_1 \cdot l_{\max}, \tag{12.9}$$

$$\mathbf{X}_i^{k+1} = \mathbf{X}_i^{k} + l_i \cdot \mathbf{D}_i^{k}\left(\varphi^{k+1}\right), \tag{12.10}$$

  where $r_1 \in R^1$ is a normally distributed random number with mean 0 and standard deviation 1, $\mathbf{r}_2 \in R^{n-1}$ is a uniformly distributed random sequence in the range $(0, 1)$, $l_{\max}$ is the maximum pursuit distance, $\alpha_{\max} \in R^1$ is the maximum tuning angle.

Taking into account the key phases described above, the steps of implementing the GrSO algorithm can be summarized as follows (He et al. 2006, 2009):

- Step 1: Defining the optimization problem, and initializing the optimization parameters.
- Step 2: Repeat till stopping criteria met.
- Step 3: Choose a member as producer.

- Step 4: The producer performs producing.
- Step 5: Choose scroungers.
- Step 6: Scroungers perform scrounging.
- Step 7: Dispersed the rest members to perform ranging.
- Step 8: Evaluate members.
- Step 9: Check if maximum iteration is reached, go to Step 2 for new beginning, if a specified termination criteria is satisfied, stop and return the best solution.

12.2.2 Performance of GrSO

In order to show how the GrSO algorithm performs, the founders have conducted a set of studies to convince us. First, in He et al. (2006), four benchmark functions are studied. Second, in He et al. (2009), an intensive study based on a set of 23 benchmark functions are illustrated. For comparison purposes, several traditional computational intelligence (CI) methods are employed, namely genetic algorithm (GA), PSO, evolutionary programming (EP), fast EP (FEP), evolution strategies (ES), and fast ES (FES). Experimental results showed that GrSO outperforms others in solving multimodal functions, while performing a similar performance for unimodal functions in terms of accuracy and convergence rate.

12.3 Conclusions

Based on the producer-scrounger model, we introduced an interesting algorithm called GrSO in which three types of members are involved: producers (i.e., seeking food resources), scroungers (i.e., joining resources founded by the producer), and rangers (i.e., performing random walks from their current positions). For simplification, He et al. (2009) assumed that there is only one producer at each searching iteration and the remaining members are divided into scroungers and rangers, respectively. In addition, three mechanisms are employed to perform better searching strategies, i.e., environment scanning (i.e., vision) for producers, area copying for scroungers and random walks for rangers. Also, it is worth mentioning that the GrSO algorithm is capable of handling a variety of optimization problems, especially for the large scale optimization problems. The differences between GrSO and other algorithms (such as ACO, EA, and PSO) please refer to He et al. (2009) for more details. Although it is a newly introduced CI method, we have witnessed the following rapid spreading of GrSO:

First, several enhanced versions of GrSO can be found in the literature as outlined below:

- Fast GrSO (Zhan et al. 2011; Qin et al. 2009).
- GrSO with multiple producers (Guo et al. 2012).
- Hybrid GrSO and extreme learning machine (Silva et al. 2011a).

- Hybrid GrSO with metropolis rule (Fang et al. 2010).
- Improved GrSO (Xie et al. 2009; Shen et al. 2009; Silva et al. 2011b; Chen et al. 2012).
- Multiobjective GrSO (Wang et al. 2012).

Second, the GrSO algorithm has also been successfully applied to a variety of optimization problems as listed below:

- Artificial neural network training (He and Li 2008; Silva et al. 2011a, b).
- Mechanical design optimization (Shen et al. 2009).
- Power system optimization (Wu et al. 2008; Zhan et al. 2011; Kang et al. 2011, 2012; Guo et al. 2012; Liao et al. 2012).
- Truss structure design optimization (Liu et al. 2008; Xie et al. 2009).

Interested readers are referred to them as a starting point for a further exploration and exploitation of the GrSO algorithm.

References

Barnard, C. J., & Sibly, R. M. (1981). Producers and scroungers: a general model at its application to captive flocks of house sparrows. *Animal Behaviour, 29*, 543–550.

Biondolillo, K., Stamp, C., Woods, J., & Smith, R. (1997). Working and scrounging by zebra finches in an operant task. *Behavioural Processes, 39*, 263–269.

Brockmann, H. J., & Barnard, C. J. (1979). Kleptoparasitism in birds. *Animal Behaviour, 27*, 487–514.

Chen, D., Wang, J., Zou, F., Hou, W., & Zhao, C. (2012). An improved group search optimizer with operation of quantum-behaved swarm and its application. *Applied Soft Computing, 12*, 712–725.

Fang, J., Cui, Z., Cai, X., & Zeng, J. A. (2010, July 17–19). *Hybrid group search optimizer with metropolis rule. The 2010 International Conference on Modelling, Identification and Control (ICMIC)* pp. (556–561). Japan: IEEE.

Giraldeau, L.-A., & Lefebvre, L. (1987). Scrounging prevents cultural transmission of food-finding behaviour in pigeons. *Animal Behaviour, 35*, 387–394.

Guo, C. X., Zhan, J. P., & Wu, Q. H. (2012). Dynamic economic emission dispatch based on group search optimizer with multiple producers. *Electric Power Systems Research, 86*, 8–16.

He, S., & Li, X. (2008). *Application of a group search optimization based artificial neural network to machine condition monitoring. IEEE International Conference on Emerging Technologies and Factory Automation (ETFA)* (pp. 1260–1266). IEEE.

He, S., Wu, Q. H., & Saunders, J. R. A. (2006, July 16–21). *Novel group search optimizer inspired by animal behavioural ecology. IEEE Congress on Evolutionary Computation (CEC) Sheraton Vancouver Wall Centre Hotel, Vancouver, BC*, (pp. 1272–1278). Canada: IEEE.

He, S., Wu, Q. H., & Saunders, J. R. (2009). Group search optimizer: an optimization algorithm inspired by animal searching behavior. *IEEE Transactions on Evolutionary Computation, 13*, 973–990.

Kang, Q., Lan, T., Yan, Y., An, J., & Wang, L. (2011). *Swarm-based optimal power flow considering generator fault in distribution systems. IEEE International Conference on Systems, Man, and Cybernetics (SMC)* (pp. 786–790). IEEE.

Kang, Q., Lan, T., Yan, Y., Wang, L., & Wu, Q. (2012). Group search optimizer based optimal location and capacity of distributed generations. *Neurocomputing, 78*, 55–63.

Liao, H., Chen, H., Wu, Q., Bazargan, M., & Ji, Z. (2012). Group search optimizer for power system economic dispatch. In: Y. Tan, Y. Shi, & Z. Ji. (Eds.), ICSI 2012, Part I, LNCS 7331, (pp. 253–260). Berlin: Springer.

Liu, F., Xu, X.-T., Li, L.-J., & Wu, Q. H. (2008). *The group search optimizer and its application on truss structure design. Fourth International Conference on Natural Computation (ICNC)* (pp. 688–692). IEEE.

Millor, J., Amé, J. M., Halloy, J., & Deneubourg, J. L. (2006). Individual discrimination capability and collective decision-making. *Journal of Theoretical Biology, 239*, 313–323.

Mills, D. S., Marchant-Forde, J. N., McGreevy, P. D., Morton, D. B., Nicol, C. J., Phillips, C. J. C., et al. (Eds.). (2010). *The encyclopedia of applied animal behaviour and welfare*. UK: CAB International. ISBN 978-0-85199-724-7.

Parker, G. A. (1984). Evolutionarily stable strategies. In: J. R. Krebs, & N. B. Davies, (Eds.), *Behavioural ecology: an evolutionary approach.* Oxford: Blackwell Scientific.

Qin, G., Liu, F., & Li, L. (2009). *A quick group search optimizer with passive congregation and its convergence analysis. International Conference on Computational Intelligence and Security (CIS)* (pp. 249–253). IEEE.

Shen, H., Zhu, Y., Niu, B., & Wu, Q. H. (2009). An improved group search optimizer for mechanical design optimization problems. *Progress in Natural Science, 19*, 91–97.

Silva, D. N. G., Pacifico, L. D. S., & Ludermir, T. B. (2011a). *An evolutionary extreme learning machine based on group search optimization IEEE Congress on Evolutionary Computation (CEC)* (pp. 574–580). IEEE.

Silva, D. N. G., Pacifico, L. D. S., & Ludermir, T. B. (2011b). *Improved group search optimizer based on cooperation among groups for feedforward networks training with weight decay. IEEE International Conference on Systems, Man, and Cybernetics (SMC)* (pp. 2133–2138). IEEE.

Vickery, W. L., Giraldeau, L.-A., Templeton, J. J., Kramer, D. L., & Chapman, C. A. (1991). Producer, scroungers, and group foraging. *The American Naturalist, 137*, 847–863.

Wang, L., Zhong, X., & Liu, M. (2012). A novel group search optimizer for multi-objective optimization. *Expert Systems with Applications, 39*, 2939–2946.

Wu, Q. H., Lu, Z., Li, M. S., & Ji, T. Y. (2008). *Optimal placement of FACTS devices by a group search optimizer with multiple producer. IEEE World Congress on Computational Intelligence (WCCI)* (pp. 1033–1039). IEEE.

Xie, H., Liu, F., & Li, L. (2009). *A topology optimization for truss based on improved group search optimizer. International Conference on Computational Intelligence and Security (CIS)* (pp. 244–148). IEEE.

Zhan, J. P., Yin, Y. J., Guo, C. X., & Wu, Q. H. (2011). *Integrated maintenance scheduling of generators and transmission lines based on fast group searching optimizer. IEEE Power and Energy Society General Meeting* (pp. 1–6). IEEE.

Chapter 13
Invasive Weed Optimization Algorithm

Abstract In this chapter, we present an interesting algorithm called invasive weed optimization (IWO) which is inspired from colonizing weeds. We first describe the general knowledge of the biological invasion in Sect. 13.1. Then, the fundamentals and performance of IWO are introduced in Sect. 13.2. Finally, Sect. 13.3 summarises this chapter.

13.1 Introduction

Weeds are one of the most robust and troublous plants in agriculture. When we were young, you may have heard that "the weeds always win". This is due to the weeds have some strong properties, such as adaptation, robustness, vigorousness, and invasion. Based on those properties, a novel numerical stochastic optimization algorithm called invasive weed optimization (IWO) is proposed by Mehrabian and Lucas (2006) which is based on the natural selection (survival of the fittest) in the biological world.

13.1.1 Biological Invasion

Generally speaking, biological invasion is a phenomenon in which the groups of individuals (such as weeds) migrate to new environments and compete with native populations (Shigesada and Kawasaki 1997). In fact, it is not a novel phenomenon however, it is one of the most important impacts on the earth's ecosystems (Jose et al. 2013). Also, it can be used as a fundamental framework in designing effective optimization algorithms (Falco et al. 2012).

B. Xing and W.-J. Gao, *Innovative Computational Intelligence: A Rough Guide to 134 Clever Algorithms*, Intelligent Systems Reference Library 62,
DOI: 10.1007/978-3-319-03404-1_13, © Springer International Publishing Switzerland 2014

13.2 Invasive Weed Optimization Algorithm

13.2.1 Fundamentals of Invasive Weed Optimization Algorithm

Invasive weed optimization (IWO) algorithm was originally proposed in Mehrabian and Lucas (2006). To implement the IWO algorithm, the following steps need to be performed (Mehrabian and Lucas 2006; Roshanaei et al. 2008):

- Initialization: a population of initial weeds $W = (w_1, w_2, \ldots, w_m)$, each representing one trial solution of the optimization problem at hand, is being dispread over the d-dimensional problem space with random positions.
- Reproduction: each member of the population is allowed to produce seeds depending on its own, as well as the colony's lowest and highest fitness to simulate the natural survival of the fittest process. Such that, the number of seed produced by a weed increases linearly from lowest possible seed for a weed with worst fitness to the maximum number of seeds for a plant with best fitness (which corresponds to the lowest objective function value for a minimization problem).
- Spatial distribution: the generated seeds are being randomly distributed over the d-dimensional search space by normally distributed random numbers with mean equal to zero; but varying variance parameter decreasing over the number of iteration. The reason for that is to guarantee that the produced seeds will be generated in a distant area but around the parent weed and decreases nonlinearly, which results in grouping the fitter plants are together and inappropriate plants are eliminated over times. Here, the standard deviation (σ) of the random function is made to decrease over the iterations from a previously defined initial value ($\sigma_{initial}$), to a final value (σ_{final}), is calculated in every time step via Eq. 13.1 (Mehrabian and Lucas 2006; Roshanaei et al. 2008):

$$\sigma_{iter} = \frac{(iter_{\max} - iter)^n}{(iter_{\max})^n}\left(\sigma_{initial} - \sigma_{final}\right) + \sigma_{final}, \tag{13.1}$$

  where $iter_{\max}$ is the maximum number of iterations, σ_{iter} is the standard deviation at the present time step and n is the non-linear modulation index usually set as 2.

- Competitive exclusion: due to fast reproduction, after passing some iteration the number of produced plants in a colony reaches to its maximum ($P_{\max}$). In this step, a competitive mechanism is activated for eliminating undesirable plants with poor fitness and allowing fitter plants to reproduce more sees as expected. This process continues until maximum iterations or some other stopping criteria are reached and the plant with the best fitness is selected as the optimal solution.

Taking into account the key phases described above, the steps of implementing the IWO algorithm can be summarized as follows (Ghosh et al. 2011; Roy et al. 2013; Li et al. 2011; Kundu et al. 2012; Mehrabian and Lucas 2006)

- Step 1: Initialize randomly generated weeds in the entire search space.
- Step 2: Evaluate fitness of the whole population members.
- Step 3: Allow each population member to produce a number of seeds with better population members produce more seeds (i.e., reproduction).
- Step 4: The generated seeds are distributed over the search space by normally distributed random numbers with mean equal to zero but varying variance (i.e., spatial dispersal).
- Step 5: When the weed population exceeds the upper limit, perform competitive exclusion.
- Step 6: Check the termination criteria.

13.2.2 Performance of IWO

In order to test the performance of IWO, a set of benchmark multidimensional functions are adopted in Mehrabian and Lucas (2006), such as Sphere function, Griewank function and Rastrigin function. Compared with other CI algorithms [such as genetic algorithm (GA), simulated annealing (SA), and particle swarm optimization (PSO)], computational results showed that IWO is capable of finding desired minima very fast.

13.3 Conclusions

Recently, there has been a considerable attention paid for employing nature inspired algorithms to solve optimization problems. Among others, IWO is a new member that motivated by a common phenomenon in agriculture, i.e., colonization of invasive weeds. Although it is a newly introduced CI method, we have witnessed the following rapid spreading of IWO:

First, several enhanced versions of IWO can be found in the literature as outlined below:

- Cooperative coevolutionary IWO (Hajimirsadeghi et al. 2009).
- Differential IWO (Basak et al. 2013).
- Differential IWO (Basak et al. 2013).
- Discrete IWO (Ghalenoei et al. 2009).
- Foraging weed colony optimization (Roy et al. 2010).
- Hybrid IWO and differential evolution algorithm (Roy et al. 2013).
- IWO for multiobjective optimization (Kundu et al. 2012).
- Modified IWO (Giri et al. 2010; Ghosh et al. 2011; Pahlavani et al. 2012; Basak et al. 2010).
- Non-dominated sorting IWO (Nikoofard et al. 2012).

Second, the IWO algorithm has also been successfully applied to a variety of optimization problems as listed below:

- Antenna design optimization (Roshanaei et al. 2008; Mallahzadeh et al. 2009; Basak et al. 2010; Li et al. 2011; Mallahzadeh and Taghikhani 2013).
- Communication scheme optimization (Hung et al. 2010).
- Control optimization (Ghosh et al. 2011).
- Data clustering (Mehrabian and Lucas 2006).
- Electricity market optimization (Hajimirsadeghi et al. 2009; Nikoofard et al. 2012).
- Feed-forward neural network training (Giri et al. 2010).
- Multimodal optimization (Roy et al. 2013).
- Multiple task allocation problem (Ghalenoei et al. 2009).
- Recommender system optimization (Rad and Lucas 2007).
- Solving nonlinear equations (Pourjafari and Mojallali 2012).
- Travel path optimization (Pahlavani et al. 2012).

Interested readers please refer to them as a starting point for a further exploration and exploitation of the IWO algorithm.

References

Basak, A., Pal, S., Das, S., Abraham, A., & Snasel, V. (2010, July 18–23). A modified invasive weed optimization algorithm for time-modulated linear antenna array synthesis. In *Proceedings of the IEEE World Congress on Computational Intelligence (WCCI), Barcelona* (pp. 372–379). CCIB, Barcelona: IEEE.

Basak, A., Maity, D., & Das, S. (2013). A differential invasive weed optimization algorithm for improved global numerical optimization. *Applied Mathematics and Computation, 219*, 6645–6668.

Falco, I. D., Cioppa, A. D., Maisto, D., Scafuri, U., & Tarantino, E. (2012). Biological invasion–inspired migration in distributed evolutionary algorithms. *Information Sciences, 207*, 50–65. http://dx.doi.org/10.1016/j.ins.2012.04.027.

Ghalenoei, M. R., Hajimirsadeghi, H., & Lucas, C. (2009, December 16–18). Discrete invasive weed optimization algorithm: Application to cooperative multiple task assignment of UAVs. In *Joint 48th IEEE Conference on Decision and Control and 28th Chinese Control Conference, Shanghai* (pp. 1665–1670). Shanghai: IEEE.

Ghosh, A., Das, S., Chowdhury, A., & Giri, R. (2011). An ecologically inspired direct search method for solving optimal control problems with Bézier parameterization. *Engineering Applications of Artificial Intelligence, 24*, 1195–1203.

Giri, R., Chowdhury, A., Ghosh, A., Das, S., Abraham, A., & Snasel, V. (2010, October 10–13). A modified invasive weed optimization algorithm for training of feed-forward neural networks. In *IEEE International Conference on Systems, Man, and Cybernetics (IEEE SMC), Istanbul* (pp. 3166–3173). Istanbul: IEEE.

Hajimirsadeghi, H., Ghazanfari, A., Rahimi-Kian, A., & Lucas, C. (2009). Cooperative coevolutionary invasive weed optimization and its application to nash equilibrium search in electricity markets. In *World Congress on Nature and Biologically Inspired Computing (NaBIC)*, (pp. 1532–1535). IEEE.

Hung, H.-L., Chao, C.-C., Cheng, C.-H., & Huang, Y.-F. (2010, October 10–13). Invasive weed optimization method based blind multiuser detection for MC-CDMA interference suppression over multipath fading channel. In *IEEE International Conference on Systems, Man, and Cybernetics (SMC), Istanbul* (pp. 2145–2150). Istanbul: IEEE.

Jose, S., Singh, H. P., Batish, D. R., & Kohli, R. K. (Eds.). (2013). *Invasive plant ecology*. Boca Raton: Taylor & Francis Group, LLC, ISBN 978-1-7398-8127-9.

Kundu, D., Suresh, K., Ghosh, S., Das, S., Panigrahi, B. K., & Das, S. (2012). Multi-objective optimization with artificial weed colonies. *Information Sciences, 181*, 2441–2454.

Li, Y., Yang, F., Ouyang, J., & Zhou, H. (2011). Yagi-Uda antenna optimization based on Invasive weed optimization method. *Electromagnetics, 31*, 571–577.

Mallahzadeh, A. R., & Taghikhani, P. (2013). Shaped elevation pattern synthesis for reflector antenna. *Electromagnetics, 33*, 40–50.

Mallahzadeh, A. R., Es'Haghi, S., & Hassani, H. R. (2009). Compact U-array MIMO antenna designs using IWO algorithm. *International Journal of RF and Microwave Computer-Aided Engineering, 19*, 568–576.

Mehrabian, A. R., & Lucas, C. (2006). A novel numerical optimization algorithm inspired from weed colonization. *Ecological Informatics, 1*, 355–366.

Nikoofard, A. H., Hajimirsadeghi, H., Rahimi-Kian, A., & Lucas, C. (2012). Multiobjective invasive weed optimization: application to analysis of pareto improvement models in electricity markets. *Applied Soft Computing, 12*, 100–112.

Pahlavani, P., Delavar, M. R., & Frank, A. U. (2012). Using a modified invasive weed optimization algorithm for a personalized urban multi-criteria path optimization problem. *International Journal of Applied Earth Observation and Geoinformation, 18*, 313–328.

Pourjafari, E., & Mojallali, H. (2012). Solving nonlinear equations systems with a new approach based on invasive weed optimization algorithm and clustering. *Swarm and Evolutionary Computation, 4*, 33–43.

Rad, H. S., & Lucas, C. (2007). A recommender system based on invasive weed optimization algorithm. *IEEE Congress on Evolutionary Computation (CEC),* (pp. 4297–4304). IEEE.

Roshanaei, M. M., Lucas, C., & Mehrabian, A. R. (2008). Adaptive beamforming using a novel numerical optimisation algorithm. *IET Microwaves, Antennas and Propagation, 3*, 765–773.

Roy, G. G., Chakroborty, P., Zhao, S.-Z., Das, S., & Suganthan, P. N. (2010, July 18–23). Artificial foraging weeds for global numerical optimization over continuous spaces. In *IEEE World Congress on Computational Intelligence (WCCI), Barcelona* (pp. 1189–1196). CCIB, Barcelona: IEEE.

Roy, S., Islam, S. M., Das, S., Ghosh, S., & Vasilakos, A. V. (2013). A simulated weed colony system with subregional differential evolution for multimodal optimization. *Engineering Optimization, 45*(4), 459–481. http://dx.doi.org/10.1080/0305215X.2012.678494.

Shigesada, N., & Kawasaki, K. (1997). *Biological invasions: Theory and practice*. USA: Oxford University Press.

Chapter 14
Music Inspired Algorithms

Abstract In this chapter, we introduce a set of music inspired algorithms, namely, harmony search (HS), melody search (MeS) algorithm, and method of musical composition (MMC) algorithm. We first describe the general knowledge of harmony in Sect. 14.1. Then, the fundamentals and performances of HS, MeS algorithm, and MMC algorithm are introduced in Sects. 14.2 and 14.3, respectively. Finally, Sect. 14.4 summarises this chapter.

14.1 Introduction

Everyone loves music. It is the one art form that is entirely defined by time. For example, it can be broadly divided into three groups, namely, classical music, Jazz, and rock (Jarrett and Day 2008). In addition, the fantastic thing of music is that it can be improvisational played. For example, when you go to a concert, you will find that two or three guitar players can improvise freely on the guitar based on their own trained habits (French 2012). It just like act of grabbing a few seemingly random notes, however, in the end you will admire to these excellent melodic ideas. Inspired by that, several music based algorithms are proposed recently.

14.1.1 Harmony

Generally speaking, harmony is one of the major building blocks when you build a musical bridge between your different melodic themes, the other two are rhythm and melody, respectively. It can be defined as any combination of notes that can be simultaneously played (Jarrett and Day 2008). The elementary study of harmony is about chord progression in which a series of chord are played in order (Yi and Goldsmith 2010). One good source for harmony is the melody itself. In fact, they are interacted with each other. In addition, different harmonies can give you a totally different feeling.

B. Xing and W.-J. Gao, *Innovative Computational Intelligence: A Rough Guide to 134 Clever Algorithms*, Intelligent Systems Reference Library 62,
DOI: 10.1007/978-3-319-03404-1_14,

14.2 Harmony Search Algorithm

14.2.1 Fundamentals of Harmony Search Algorithm

Harmony search (HS) algorithm was originally proposed by Geem et al. (2001). With the underlying fundamental of natural musical performance processes in which the musicians improvise their instruments' pitch by searching for the pleasing harmony (a perfect state), HS find the solutions through the determination of an objective function (i.e., the audience's aesthetics) in which a set of values (i.e., the musicians) assigned to each decision variable (i.e., the musical instrument's pitch). In general, the HS algorithm has three main operations: harmony memory (HM) consideration, pitch adjustment, and randomization (Geem et al. 2001). The HS algorithm is performed in several steps, outlined below (Geem et al. 2001):

- Preparation of harmony memory: The main building block of HS is the usage of HM, because multiple randomized solution vectors are stored in HM via Eq. 14.1 (Geem 2009):

$$\mathrm{HM} = \left[\begin{array}{cccc|c} D_1^1 & D_2^1 & \cdots & D_n^1 & f(\mathbf{D}^1) \\ D_1^2 & D_2^2 & \cdots & D_n^2 & f(\mathbf{D}^2) \\ \vdots & \vdots & \cdots & \vdots & \vdots \\ D_1^{HMS} & D_2^{HMS} & \cdots & D_n^{HMS} & f(\mathbf{D}^{HMS}) \end{array}\right], \tag{14.1}$$

  where D_i^j is the ith decision variable in the jth solution vector that has one discrete value out of a candidate set $\{D_i(1), D_i(2), \ldots, D_i(k), \ldots, D_i(K_i)\}$, $f(\mathbf{D}^j)$ is the objective function value for the jth solution vector, and HMS is the harmony memory size (i.e., the number of multiple vectors stored in the HM).
- Improvisation of new harmony: A new harmony vector $D_i^{new} = (D_1^{new}, D_2^{new}, \ldots, D_n^{new})$ is improvised by the following three rules (Geem 2009):

(1) Random selection: Based on this rule, one value is chosen out of the candidate set via Eq. 14.2 (Geem 2009):

$$D_i^{new} \leftarrow D_i(k),\ D_i(k) \in \{D_i(1), D_i(2), \ldots, D_i(K_i)\}. \tag{14.2}$$

(2) HM consideration: In memory consideration, one value is chosen out of the HM set with a probability of harmony memory consideration rate (HMCR) via Eq. 14.3 (Geem 2009):

$$D_i^{new} \leftarrow D_i(l),\ D_i(l) \in \{D_i^1, D_i^2, \ldots, D_i^{HMS}\}. \tag{14.3}$$

(3) Pitch adjustment: According to this rule, the obtained vale as in Eq. 14.3 is further changed into neighbouring values, with a probability of pitch adjusting rate (PAR) via Eq. 14.4 (Geem 2009):

$$D_i^{new} \leftarrow D_i(l \pm 1),\ D_i(l) \in \left\{D_i^1, D_i^2, \ldots, D_i^{HMS}\right\}. \tag{14.4}$$

Overall, these three rules are the core terms of the stochastic derivative of HS and can be summarized via Eq. 14.5 (Geem 2009):

$$\begin{aligned} \left.\frac{\partial f}{\partial D_i}\right|_{D_i = D_i(l)} &= \frac{1}{K_i} \cdot (1 - HMCR) + \frac{n(D_i(l))}{HMS} \cdot HMCR \cdot (1 - PAR) \\ &+ \frac{n(D_i(l \pm 1))}{HMS} \cdot HMCR \cdot PAR, \end{aligned} \tag{14.5}$$

where $\frac{1}{K_i} \cdot (1 - HMCR)$ denotes for the rate to choose a value $D_i(l)$ for the decision variable D_i by random selection, $\frac{n(D_i(l))}{HMS} \cdot HMCR \cdot (1 - PAR)$ chooses the rate by HM consideration, and $\frac{n(D_i(l \pm 1))}{HMS} \cdot HMCR \cdot PAR$ chooses the rate by pitch adjustment.

- Update of HM: Once the new vector $D_i^{new} = \left(D_1^{new}, D_2^{new}, \ldots, D_n^{new}\right)$ is completely generated, it will be compared with the other vectors that stored in HM. If it is better than the worst vector in HM with respect to the objective function, it will be updated (i.e., the new harmony is included in the HM and the existing worst harmony is excluded from the HM).

The optimization procedures of the HS algorithm are given as follows (Lee and Geem 2009; Geem et al. 2001):

- Step 1: Initialize the optimization problem and algorithm parameters.
- Step 2: Initialization of HM.
- Step 3: Improvise a new harmony from the HM.
- Step 4: Update the HM.
- Step 5: Repeat Steps 3 and 4 until the termination criterion is satisfied.

14.2.2 Performance of HS

In order to show how the HS algorithm performs, three problems are presented to demonstrate the searching ability of HS in Geem et al. (2001), i.e., travelling salesman problem, relatively simple constrained minimization problem, and water network design problem. The computational results showed that HS outperforms other existing heuristic methods [such as genetic algorithm (GA)] in two specific applications (i.e., relatively simple constrained minimization problem and water network design problem).

14.3 Emerging Music Inspired Algorithms

Although music inspired algorithm is a new member of computational intelligence (CI) family, a number of similar algorithms have been proposed in the literature. This section gives an overview to some of these algorithms which have been demonstrated to be very efficient and robust.

14.3.1 Melody Search Algorithm

14.3.1.1 Fundamentals of Melody Search Algorithm

Melody search (MeS) algorithm was originally proposed by Ashrafi and Dariane (2011). It is inspired by the basic concepts applied in HS, but unless the HS algorithm used a single HM, the MeS algorithm employed the procedure of the group improvisation [i.e., several memories called player memory (PM)] simultaneously for finding the best succession of pitches in a melody. Main steps of MeS are outlined as follows (Ashrafi and Dariane 2011):

- Step 1: Initializing the optimization problem and adopting algorithm parameters. In general, there are six major parameters defined in MeS, namely number of player memories (PMN), player memory size (PMS), maximum number of iterations (NI), maximum number of iterations for the initial phase (NII), bandwidth (bw), and player memory considering rate (PMCR).
- Step 2: Initial phase that includes two repeated procedures (i.e., improvise a new melody from each PM and update each PM) until the criterion for stopping this step (i.e., NII) is satisfied, is given as follows,

(1) Initialize PM is defined via Eq. 14.6 (Ashrafi and Dariane 2011):

$$MM = [PM_1, PM_2, \ldots, PM_{PMN}], \tag{14.6}$$

where *MM* denotes the melody memory in which a set of player memories are involved. The PM's matrixes are generated via Eqs. 14.7 and 14.8, respectively (Ashrafi and Dariane 2011):

$$PM_i = \left[\begin{array}{cccc|c} x_{i,1}^1 & x_{i,1}^2 & \cdots & x_{i,1}^D & Fit_i^1 \\ x_{i,2}^1 & x_{i,2}^2 & \cdots & x_{i,2}^D & Fit_i^2 \\ \vdots & \vdots & \cdots & \vdots & \vdots \\ x_{i,PMS}^1 & x_{i,PMS}^2 & \cdots & x_{i,PMS}^D & Fit_i^{PMS} \end{array}\right], \tag{14.7}$$

$$x_{i,j}^{k} = LB_k + r \cdot (UB_k - LB_k),\quad \text{for} \begin{cases} i = 1, 2, \ldots, PMN \\ j = 1, 2, \ldots, PMS \\ k = 1, 2, \ldots, D \end{cases}, \tag{14.8}$$

where D is the number of pitches of melodic line (i.e., decision variables), $[LB_k, UB_k]$ is the possible range of the searching dimension, and r is a real number uniformly distributed in $[0, 1]$.

(2) Improvise a new melody $X_{i,new} = \left(x_{i,new}^{1}, x_{i,new}^{2}, \ldots, x_{i,new}^{n}\right)$ from each PM according to three rules (Ashrafi and Dariane 2011):

Memory consideration: The value of each variable can be chosen from any value in the specified PM.

Pitch adjustment: Based on this rule, the value can be determined by a constant pitch bandwidth (bw) and a pitch adjusting rate (PAR) such as Eq. 14.9 (Ashrafi and Dariane 2011):

$$PAR_t = PAR_{\min} + \frac{PAR_{\max} - PAR_{\min}}{NI} \times t, \tag{14.9}$$

where PAR_t is the pitch adjusting rate of the ith iteration, $PAR_{\min}$ and $PAR_{\max}$ are the minimum and maximum adjusting rates, respectively, and NI is the maximum number of iterations.

Randomization: This rule is used to increase the diversity of the solutions.

(3) Update each PM.

- Step 3: Second phase that includes two repeated procedures until the NI is satisfied, namely,

(1) Improvise a new melody from each PM according to the possible range of pitches.
(2) Update each PM.
(3) Finally, determine the possible ranges of pitches for next improvisation (Just for randomization).

14.3.1.2 Performance of MeS

To evaluate the performance of MeS, five classical benchmark functions are tested in (Ashrafi and Dariane 2011). Compared with other CI methods [such as artificial bee colony (ABC), GA, HS, particle swarm optimization (PSO), and particle swarm and evolutionary algorithm (PS-EA)], the MeS is capable of finding better solutions.

14.3.2 Method of Musical Composition Algorithm

14.3.2.1 Fundamentals of Method of Musical Composition Algorithm

Method of musical composition (MMC) algorithm was originally proposed by Mora-Gutiérrez et al. (2012). The MMC algorithm used a dynamic creative system which means the composers exchange information among themselves and their environment to compose music. Normally, MMC involves four steps as follows (Mora-Gutiérrez et al. 2012):

- Initialization: In this step, the scores ($P_{*,*,i}$), which used as memory, are randomly generated via Eqs. 14.10 and 14.11, respectively (Mora-Gutiérrez et al. 2012):

$$P_{*,*,i} = \begin{pmatrix} x_{1,1} & x_{1,2} & \cdots & x_{1,n} \\ x_{2,1} & x_{2,2} & \cdots & x_{2,n} \\ \vdots & \vdots & \vdots & \vdots \\ x_{Ns,1} & x_{Ns,2} & \vdots & x_{Ns,n} \end{pmatrix}, \tag{14.10}$$

$$P_{*,*,i} = x_l^L + \left(rand \cdot \left(x_l^U - x_l^L\right)\right), \tag{14.11}$$

  where $P_{*,*,i}$ is the score of the ith composer, $x_{j,l}$ is the lth decision variable of jth tune, *rand* is a real number uniformly distributed in $[0, 1]$, and $\left(x_l^U - x_l^L\right)$ is the possible range of the searching dimension.
- Exchanging of information among agents: According to the interaction policy, i.e., "composer i exchange a tune with composer k if and only if there is a link between them and the worst tune of composer k is better than the worst tune of composer i". Two sub-phases (i.e., update of links among composers and exchange of information) are employed to exchange the information.
- Generating for each agent a new tune: Based on the composer's background and his innovative ideas, the new tune will be created. This phase includes two sub-phases, i.e., building the background of each composer ($KM_{*,*i}$) which includes the knowledge of composer i and the environment information that he perceived, and creating a new tune.
- The $P_{*,*,i}$ updating: Based on the value of objective function, the score will be updated.

14.3.2.2 Performance of MMC

To show the performance of MMC, 13 benchmark continuous optimization problems are performed in Mora-Gutiérrez et al. (2012). Compared with HS, improved HS, global-best HS, and self-adaptive HS, the experimental results showed that MMC improves the results obtained by the other methods, especially in the domain of multimodal functions.

14.4 Conclusions

In this chapter, we introduced a set of music inspired algorithms, namely, HS, MeS, and MMC. The former two are based on the idea of improvisation process by a skilled musician, while the last algorithm is inspired by the creative process of musical composition. Although the novelties of these music algorithms (e.g., HS) are still under debate (see (Weyland 2010) for details), we have witnessed the following rapid spreading of at least one of them, i.e., HS:

First, numerous enhanced versions of HS can be found in the literature as outlined below:

- Box-Muller HS (Fetanat et al. 2011).
- Chaotic differential HS (Coelho et al. 2010).
- Chaotic HS (Pan et al. 2011b; Alatas 2010).
- Coevolutionary differential evolution with HS (Wang and Li 2012).
- Differential HS (Wang and Li 2013; Qin and Forbes 2011b).
- Discrete HS (Gandhi et al. 2012; Pan et al. 2010b; Tasgetiren et al. 2012).
- Effective global best HS (Zou et al. 2011a).
- Efficient HS (Degertekin 2012).
- Global-best HS (Omran and Mahdavi 2008).
- Grouping HS (Landa-Torres et al. 2012; Askarzadeh and Rezazadeh 2011).
- Guided variable neighborhood embedded HS (Huang et al. 2009).
- Harmony fuzzy search algorithm (Alia et al. 2009a).
- Highly reliable HS (Taherinejad 2009).
- HS with dual-memory (Gao et al. 2012b).
- Hybrid clonal selection algorithm and HS (Wang et al. 2009).
- Hybrid differential evolution and HS (Mirkhani et al. 2013; Li and Wang 2009; Liao 2010; Duan et al. 2013; Gao et al. 2009).
- Hybrid global best HS and K-means algorithm (Cobos et al. 2010).
- Hybrid globalbest HS (Wang et al. 2010, 2011).
- Hybrid HS (Gao et al. 2012a; Gil-López et al. 2012; Wang et al. 2010).
- Hybrid HS and hill climbing (Al-Betar and Khader 2009).
- Hybrid HS and linear discriminate analysis (Moeinzadeh et al. 2009).
- Hybrid K-means and HS (Mahdavi and Abolhassani 2009; Forsati et al. 2008b).
- Hybrid modified subgradient and HS (Yaşar and Özyön 2011).
- Hybrid probabilistic neural networks and HS (Ameli et al. 2012).
- Hybrid swarm intelligence and HS (Pandi and Panigrahi 2011; Pandi et al. 2011).
- Improved discrete HS (Shi et al. 2011).
- Improved HS based on exponential distribution (Coelho and Mariani 2009).
- Intelligent tuned HS (Yadav et al. 2012).
- Learning automata-based HS (Enayatifar et al. 2013).
- Local-best HS with dynamic sub-harmony memories (Pan et al. 2011a).
- Mixed-discrete HS (Jaberipour and Khorram 2011).

- Modified HS (Kaveh and Nasr 2011; Zinati and Razfar 2012; Al-Betar and Khader 2012; Gao et al. 2008; Das et al. 2011; Mun and Cho 2012).
- Multiobjective HS (Sivasubramani and Swarup 2011a, b; Li et al. 2012).
- Novel global HS (Zou et al. 2010a, b, c, 2011b).
- Opposition-based HS (Chatterjee et al. 2012).
- Other hybrid HS (Jang et al. 2008; Yıldız 2008; Fesanghary et al. 2008; Zhao and Suganthan 2010).
- Other improved HS (Afshari et al. 2011; Fourie et al. 2010; Sirjani et al. 2011; Yadav et al. 2011; Kaveh and Abadi 2010; Geem and Williams 2008; Geem 2010, 2012; Mahdavi et al. 2007; Coelho and Bernert 2009; Chakraborty et al. 2009; Jaberipour and Khorram 2010b; Qin and Forbes 2011a; Al-Betar et al. 2012).
- Parallel HS (Lee and Zomaya 2009).
- Parameter-setting-free HS (Geem and Sim 2010).
- Particle-swarm enhanced HS (Geem 2009; Li et al. 2008; Zhao et al. 2011; Cheng et al. 2012).
- Quantum inspired HS (Layeb 2013).
- Self-adaptive global best HS (Kulluk et al. 2011; Pan et al. 2010a).
- Self-adaptive HS (Degertekin 2012; Wang and Huang 2010; Chang and Gu 2012).
- Social HS (Kaveh and Ahangaran 2012).

Second, the HS algorithm has been successfully applied to a variety of optimization problems as listed below:

- Adaptive parameter controlling (Nadi et al. 2010).
- Analog filter design optimization (Vural et al. 2013).
- Antenna design optimization (Guney and Onay 2011).
- Artificial neural network training (Kattan et al. 2010; Kattan and Abdullah 2011a, b; Kulluk et al. 2011, 2012).
- Communication networks optimization (Forsati et al. 2008a; Shi et al. 2011; Landa-Torres et al. 2012; Ser et al. 2012).
- Data mining (Mahdavi and Abolhassani 2009; Mahdavi et al. 2008; Forsati et al. 2008b; Moeinzadeh et al. 2009; Wang et al. 2009; Venkatesh et al. 2010; Cobos et al. 2010; Ramos et al. 2011).
- Engineering design optimization (Mohammadi et al. 2011; Gil-López et al. 2012; Lee and Geem 2005).
- Facility location optimization (Afshari et al. 2011; Kaveh and Nasr 2011).
- Fuel cell research (Askarzadeh and Rezazadeh 2011).
- Fuzzy-rough rule induction (Diao and Shen 2012).
- Ground motion records analysis (Kayhan et al. 2011).
- Image processing (Alia et al. 2009a, b, 2008, 2010; Fourie et al. 2010).
- Interaction parameter estimation problem (Merzougui et al. 2012).
- Knapsack problem (Zou et al. 2011b; Layeb 2013).
- Lot sizing problem (Piperagkas et al. 2012).

- Materials research (Mun and Geem 2009).
- Milling process optimization (Razfar et al. 2011; Zarei et al. 2009; Zinati and Razfar 2012).
- Music composition (Geem and Choi 2007).
- Orienteering problem (Geem et al. 2005c).
- Parameter-setting-free technique enhanced HS (Geem and Sim 2010).
- Power system optimization (Vasebi et al. 2007; Mukhopadhyay et al. 2008; Fesanghary and Ardehali 2009; Coelho and Mariani 2009; Coelho et al. 2010; Yaşar and Özyön 2011; Fetanat et al. 2011; Geem 2011; Pandi and Panigrahi 2011; Pandi et al. 2011; Sivasubramani and Swarup 2011a, b; Khorram and Jaberipour 2011; Boroujeni et al. 2011a, b, c, d; Sirjani et al. 2011; Khazali and Kalantar 2011; Shariatkhah et al. 2012; Ezhilarasi and Swarup 2012; Javadi et al. 2012; Chatterjee et al. 2012; Ameli et al. 2012; Wang and Li 2013; Zhang et al. 2013).
- Robot control optimization (Mirkhani et al. 2013).
- Scheduling optimization (Huang et al. 2009; Zou et al. 2010a; Wang et al. 2010, 2011; Pan et al. 2010b, 2011a, b; Yadav et al. 2011; Gao et al. 2012a; Ahmad et al. 2012; Geem 2007; Tasgetiren et al. 2012).
- Signal processing (Gandhi et al. 2012; Guo et al. 2012).
- Software design optimization (Alsewari and Zamli 2012a, b).
- Structure design optimization (Geem et al. 2005b; Geem and Hwangbo 2006; Degertekin 2008, 2012; Fesanghary et al. 2009, 2012; Kaveh and Talataha 2009; Kaveh and Abadi 2010; Hasançebi et al. 2010; Khajehzadeh et al. 2011; Bekdaş and Nigdeli 2011; Erdal et al. 2011; Lagaros and Papadrakakis 2012; Kaveh and Ahangaran 2012; Shahrouzi and Sazjini 2012; Miguel and Miguel 2012; Lee and Geem 2004; Ryu et al. 2007; Lee et al. 2011).
- Sudoku puzzle problem (Geem 2008a).
- Sum-of-ratios problem solving (Jaberipour and Khorram 2010a).
- Supply chain optimization (Wong and Guo 2010; Taleizadeh et al. 2011, 2012; Purnomo et al. 2012).
- System reliability optimization (Zou et al. 2010c, 2011a; Wang and Li 2012).
- Timetabling (Al-Betar and Khader 2009, 2012; Al-Betar et al. 2008, 2010).
- Transportation system optimization (Ceylan et al. 2008).
- Vehicle routing problem (Geem et al. 2005b).
- Water network optimization (Geem 2006a, b, 2008b, 2009; Ayvaz 2007, 2009; Mora-Meliá et al. 2009; Geem et al. 2011; Geem and Park 2006).

Interested readers please refer to them together with several excellent reviews [e.g., (Alia and Mandava 2011; Manjarres et al. 2013; Geem et al. 2008)] as a starting point for a further exploration and exploitation of these music inspired algorithms.

References

Afshari, S., Aminshahidy, B., & Pishvaie, M. R. (2011). Application of an improved harmony search algorithm in well placement optimization using streamline simulation. *Journal of Petroleum Science and Engineering, 78*, 664–678.

Ahmad, I., Mohammad, M. G., Salman, A. A., & Hamdan, S. A. (2012). Broadcast scheduling in packet radio networks using harmony search algorithm. *Expert Systems with Applications, 39*, 1526–1535.

Alatas, B. (2010). Chaotic harmony search algorithms. *Applied Mathematics and Computation, 216*, 2687–2699.

Al-Betar, M. A. & Khader, A. T. (2009, August 10–12). A hybrid harmony search for university course timetabling. In *Multidisciplinary International Conference on Scheduling: Theory and Applications (MISTA)*, Dublin, Ireland (pp. 157–179).

Al-Betar, M. A., & Khader, A. T. (2012). A harmony search algorithm for university course timetabling. *Annals of Operations Research, 194*, 3–31.

Al-Betar, M. A., Khader, A. T., & Gani, T. A. (2008). A harmony search algorithm for university course timetabling. In *7th International Conference on the Practice and Theory of Automated Timetabling*, Montreal, Canada (pp. 1–12).

Al-Betar, M. A., Khader, A. T., & Nadi, F. (2010, July 7–11). Selection mechanisms in memory consideration for examination timetabling with harmony search. In *Annual Conference on Genetic and Evolutionary Computation (GECCO)*, Portland, Oregon, USA (pp. 1203–1210).

Al-Betar, M. A., Doush, I. A., Khader, A. T., & Awadallah, M. A. (2012). Novel selection schemes for harmony search. *Applied Mathematics and Computation, 218*, 6095–6117.

Alia, O. M. D., & Mandava, R. (2011). The variants of the harmony search algorithm: an overview. *Artificial Intelligence Review, 36*, 49–68.

Alia, O. M. D., Mandava, R., Ramachandram, D., & Aziz, M. E. (2008). Dynamic fuzzy clustering using harmony search with application to image segmentation. In *IEEE International Symposium on Signal Processing and Information Technology (ISSPIT)* (pp. 538–543). IEEE.

Alia, O. M. D., Mandava, R., & Aziz, M. E. (2009a). A novel image segmentation algorithm based on harmony fuzzy search algorithm. In *International Conference of Soft Computing and Pattern Recognition (SOCPAR)* (pp. 335–340). IEEE.

Alia, O. M. D., Mandava, R., Ramachandram, D., & Aziz, M. E. (2009b). Harmony search-based cluster initialization for fuzzy C-means segmentation of MR images. In *IEEE Region 10 Conference TENCON* (pp. 1–6). IEEE.

Alia, O. M. D., Mandava, R., & Aziz, M. E. (2010). A hybrid harmony search algorithm to MRI brain segmentation. In *9th IEEE International Conference on Cognitive Informatics (ICCI)* (pp. 712–721). IEEE.

Alsewari, A. R. A., & Zamli, K. Z. (2012a). Design and implementation of a harmony-search-based variable-strength t-way testing strategy with constraints support. *Information and Software Technology, 54*, 553–568.

Alsewari, A. R. A., & Zamli, K. Z. (2012b). A harmony search based pairwise sampling strategy for combinatorial testing. *International Journal of the Physical Sciences, 7*, 1062–1072.

Ameli, M. T., Shivaie, M., & Moslehpour, S. (2012). Transmission network expansion planning based on hybridization model of neural networks and harmony search algorithm. *International Journal of Industrial Engineering Computations, 3*, 71–80.

Ashrafi, S. M. & Dariane, A. B. (2011, December 5–8). A novel and effective algorithm for numerical optimization: melody search (MS). In *11th International Conference on Hybrid Intelligent Systems (HIS)*, Melacca (pp. 109–114). IEEE.

Askarzadeh, A., & Rezazadeh, A. (2011). A grouping-based global harmony search algorithm for modeling of proton exchange membrane fuel cell. *International Journal of Hydrogen Energy, 36*, 5047–5053.

Ayvaz, M. T. (2007). Simultaneous determination of aquifer parameters and zone structures with fuzzy C-means clustering and meta-heuristic harmony search algorithm. *Advances in Water Resources, 30*, 2326–2338.

Ayvaz, M. T. (2009). Application of harmony search algorithm to the solution of groundwater management models. *Advances in Water Resources, 32*, 916–924.

Bekdaş, G., & Nigdeli, S. M. (2011). Estimating optimum parameters of tuned mass dampers using harmony search. *Engineering Structures, 33*, 2716–2723.

Boroujeni, S. M. S., Boroujeni, B. K., Abdollahi, M., & Delafkar, H. (2011a). Multi-area load frequency control using IP controller tuned by harmony search. *Australian Journal of Basic and Applied Sciences, 5*, 1224–1231.

Boroujeni, S. M. S., Boroujeni, B. K., Delafkar, H., Behzadipour, E., & Hemmati, R. (2011b). Harmony search algorithm for power system stabilizer tuning. *Indian Journal of Science and Technology, 4*, 1025–1030.

Boroujeni, S. M. S., Boroujeni, B. K., Delafkar, H., Behzadipour, E., & Hemmati, R. (2011c). Harmony search algorithm for STATCOM controllers tuning in a multi machine environment. *Indian Journal of Science and Technology, 4*, 1031–1035.

Boroujeni, S. M. S., Delafkar, H., Behzadipour, E., & Boro, A. S. (2011d). Reactive power planning for loss minimization based on harmony search algorithm. *International Journal of Natural and Engineering Sciences, 5*, 73–77.

Ceylan, H., Ceylan, H., Haldenbilen, S., & Baskan, O. (2008). Transport energy modeling with meta-heuristic harmony search algorithm, an application to Turkey. *Energy Policy, 36*, 2527–2535.

Chakraborty, P., Roy, G. G., Das, S., & Jain, D. (2009). An improved harmony search algorithm with differential mutation operator. *Fundamenta Informaticae, 95*, 1–26.

Chang, H., & Gu, X.-S. (2012). Multi-HM adaptive harmony search algorithm and its application to continuous function optimization. *Research Journal of Applied Sciences, Engineering and Technology, 4*, 100–103.

Chatterjee, A., Ghoshal, S. P., & Mukherjee, V. (2012). Solution of combined economic and emission dispatch problems of power systems by an opposition-based harmony search algorithm. *Electrical Power and Energy Systems, 39*, 9–20.

Cheng, Y. M., Li, L., Sun, Y. J., & Au, S. K. (2012). A coupled particle swarm and harmony search optimization algorithm for difficult geotechnical problems. *Structural and Multidisciplinary Optimization, 45*, 489–501.

Cobos, C., Andrade, J., Constain, W., Mendoza, M., & León, E. (2010, July 18–23). Web document clustering based on global-best harmony search, *k*-means, frequent term sets and Bayesian information criterion. In *Proceedings of the IEEE World Congress on Computational Intelligence (WCCI)*, CCIB, Barcelona, Spain (pp. 4637–4644). IEEE.

Coelho, L. D. S., & Bernert, D. L. D. A. (2009). An improved harmony search algorithm for synchronization of discrete-time chaotic systems. *Chaos, Solitons and Fractals, 41*, 2526–2532.

Coelho, L. D. S., & Mariani, V. C. (2009). An improved harmony search algorithm for power economic load dispatch. *Energy Conversion and Management, 50*, 2522–2526.

Coelho, L. D. S., Bernert, D. L. D. A., & Mariani, V. C. (2010, July 18–23). Chaotic differential harmony search algorithm applied to power economic dispatch of generators with multiple fuel options. In *IEEE World Congress on Computational Intelligence (WCCI)*, CCIB, Barcelona, Spain (pp. 1416–1420). IEEE.

Das, S., Mukhopadhyay, A., Roy, A., Abraham, A., & Panigrahi, B. K. (2011). Exploratory power of the harmony search algorithm: Analysis and improvements for global numerical optimization. *IEEE Transactions on Systems, Man, and Cybernetics—Part B: Cybernetics, 41*, 89–106.

Degertekin, S. O. (2008). Optimum design of steel frames using harmony search algorithm. *Structural and Multidisciplinary Optimization, 36*, 393–401.

Degertekin, S. O. (2012). Improved harmony search algorithms for sizing optimization of truss structures. *Computers and Structures, 92–93*, 229–241.

Diao, R. & Shen, Q. (2012, June 10–15). A harmony search based approach to hybrid fuzzy-rough rule induction. In *IEEE World Congress on Computational Intelligence (WCCI)*, Brisbane, Australia (pp. 1–8). IEEE.

Duan, Q., Liao, T. W., & Yi, H. Z. (2013). A comparative study of different local search application strategies in hybrid metaheuristics. *Applied Soft Computing, 13*, 1464–1477.

Enayatifar, R., Yousefi, M., Abdullah, A. H., & Darus, A. N. (2013). LAHS: a novel harmony search algorithm based on learning automata. In *Communications in Nonlinear Science and Numerical Simulation, 18*, 3481–3497. http://dx.doi.org/10.1016/j.cnsns.2013.04.028.

Erdal, F., Doğan, E., & Saka, M. P. (2011). Optimum design of cellular beams using harmony search and particle swarm optimizers. *Journal of Constructional Steel Research, 67*, 237–247.

Ezhilarasi, G. A. & Swarup, K. S. (2012). Network partitioning using harmony search and equivalencing for distributed computing. *Journal of Parallel and Distributed Computing*, *72*, 936–943. doi:10.1016/j.jpdc.2012.04.006.

Fesanghary, M., & Ardehali, M. M. (2009). A novel meta-heuristic optimization methodology for solving various types of economic dispatch problem. *Energy, 34*, 757–766.

Fesanghary, M., Mahdavi, M., Minary-Jolandan, M., & Alizadeh, Y. (2008). Hybridizing harmony search algorithm with sequential quadratic programming for engineering optimization problems. *Computer Methods in Applied Mechanics and Engineering, 197*, 3080–3091.

Fesanghary, M., Damangir, E., & Soleimani, I. (2009). Design optimization of shell and tube heat exchangers using global sensitivity analysis and harmony search algorithm. *Applied Thermal Engineering, 29*, 1026–1031.

Fesanghary, M., Asadi, S., & Geem, Z. W. (2012). Design of low-emission and energy-efficient residential buildings using a multi-objective optimization algorithm. *Building and Environment, 49*, 245–250.

Fetanat, A., Shafipour, G., & Ghanatir, F. (2011). Box-Muller harmony search for optimal coordination of directional overcurrent relays in power system. *Scientific Research and Essays, 6*, 4079–4090.

Forsati, R., Haghighat, A. T., & Mahdavi, M. (2008a). Harmony search based algorithms for bandwidth-delay-constrained least-cost multicast routing. *Computer Communications, 31*, 2505–2519.

Forsati, R., Mahdavi, M., Kangavari, M., & Safarkhani, B. (2008b). Web page clustering using harmony search optimization. In *Canadian Conference on Electrical and Computer Engineering (CCECE)* (pp. 001601–001604). IEEE.

Fourie, J., Mills, S., & Green, R. (2010). Harmony filter: a robust visual tracking system using the improved harmony search algorithm. *Image and Vision Computing, 28*, 1702–1716.

French, R. M. (2012). *Technology of the guitar*. New York, Springer Science + Business Media, ISBN 978-1-4614-1920-4.

Gandhi, T. K., Chakraborty, P., Roy, G. G., & Panigrahi, B. K. (2012). Discrete harmony search based expert model for epileptic seizure detection in electroencephalography. *Expert Systems with Applications, 39*, 4062–4065.

Gao, X. Z., Wang, X., & Ovaska, S. J. (2008). Modified harmony search methods for uni-modal and multi-modal optimization. In *Eighth International Conference on Hybrid Intelligent Systems* (pp. 65–72).

Gao, X.-Z., Wang, X., & Ovaska, S. J. (2009). Uni-modal and multi-modal optimization using modified harmony search methods. *International Journal of Innovative Computing, Information and Control, 5*, 2985–2996.

Gao, K.-Z., Pan, Q.-K., Li, J.-Q., & Wang, Y.-T. (2012a). A hybrid harmony search algorithm for the no-wait flow-shop scheduling problems. *Asia-Pacific Journal of Operational Research, 29*, 1–23.

Gao, X. Z., Wang, X., Zenger, K., & Wang, X. (2012b, October 14–17). A novel harmony search method with dual memory. In *IEEE International Conference on Systems, Man, and Cybernetics (SMC)*, COEX, Seoul, Korea (pp. 177–183). IEEE.

Geem, Z. W. (2005, June 25–29). School bus routing using harmony search. GECCO 2005, Washington, DC, USA (pp. 1–6). ACM.

Geem, Z. W. (2006a). Optimal cost design of water distribution networks using harmony search. *Engineering Optimization, 38*, 259–280.

Geem, Z. W. (2006b). Parameter estimation for the nonlinear Muskingum model using BFGS technique. *Journal of Irrigation and Drainage Engineering, 132*, 474–478.

Geem, Z. W. (2007). Optimal scheduling of multiple dam system using harmony search algorithm. In *Computational and Ambient Intelligence, LNCS 4507* (pp. 316–323). Berlin Heidelberg: Springer.

Geem, Z. W. (2008a). Harmony search algorithm for solving Sudoku. In B. Apolloni., R. J. Howlett., & L. Jain (Eds.), *Knowledge-Based Intelligent Information and Engineering Systems, LNCS 4692* (pp. 371–378). Berlin Heidelberg: Springer.

Geem, Z. W. (2008b). Novel derivative of harmony search algorithm for discrete design variables. *Applied Mathematics and Computation, 199*, 223–230.

Geem, Z. W. (2009). Particle-swarm harmony search for water network design. *Engineering Optimization, 41*, 297–311.

Geem, Z. W. (2010). State-of-the-art in the structure of harmony search algorithm. In *Recent Advances in Harmony Search Algorithm* (pp. 1–10). Berlin: Springer.

Geem, Z. W. (2011). Discussion on "Combined heat and power economic dispatch by harmony search algorithm" by A. Vasebi et al. *International Journal of Electrical Power and Energy Systems*, *29*(2007), 713–719. *Electrical Power and Energy Systems, 33*, 1348.

Geem, Z. W. (2012). Effects of initial memory and identical harmony in global optimization using harmony search algorithm. *Applied Mathematics and Computation, 218*, 11337–11343. http://dx.doi.org/10.1016/j.amc.2012.04.070.

Geem, Z. W. & Choi, J.-Y. (2007). Music composition using harmony search algorithm. In M. Giacobini (Ed.), *Applications of Evolutionary Computing* (pp. 593–600). Berlin Heidelberg: Springer.

Geem, Z. W. & Hwangbo, H. (2006). Application of harmony search to multi-objective optimization for satellite heat pipe design. In *US-Korea Conference on Science, Technology, and Entrepreneurship (UKC)*. Teaneck, Nj, USA (pp. 1–3).

Geem, Z. W. & Park, Y. (2006, April 16–18). Optimal layout for branched networks using harmony search. In *5th WSEAS International Conference on Applied Computer Science*, Hangzhou, China (pp. 364–367).

Geem, Z. W., & Sim, K.-B. (2010). Parameter-setting-free harmony search algorithm. *Applied Mathematics and Computation, 217*, 3881–3889.

Geem, Z. W. & Williams, J. C. (2008, March 24–26). Ecological optimization using harmony search. In *American Conference on Applied Mathematics*, Harvard, Massachusetts, USA (pp. 148–152). World Scientific and Engineering Academy and Society (WSEAS).

Geem, Z. W., Kim, J. H., & Loganathan, G. V. (2001). A new heuristic optimization algorithm: Harmony search. *Simulation, 76*, 60–68.

Geem, Z. W., Lee, K. S., & Park, Y. (2005a). Application of harmony search to vehicle routing. *American Journal of Applied Sciences, 2*, 1552–1557.

Geem, Z. W., Lee, K. S., & Tseng, C.-L. (2005b, June 25–29). Harmony search for structural design. In *GECCO'05*, Washington, DC, USA (pp. 651–652). ACM.

Geem, Z. W., Tseng, C.-L., & Park, Y. (2005c). Harmony search for generalized orienteering problem: best touring in China. In L. Wang., K. Chen K., & Y. Ong (Eds.), *ICNC 2005, LNCS 3612* (pp. 741–750). Berlin Heidelberg: Springer.

Geem, Z. W., Fesanghary, M., Choi, J.-Y., Saka, M. P., Williams, J. C., Ayvaz, M. T., Li, L., Ryu, S., & Vasebi, A. (2008). Recent advances in harmony search. In W. Kosiński (Ed.), *Advances in Evolutionary Algorithms, ISBN 978-953-7619-11-4, Chapter 7* (pp. 127–142). Vienna, Austria: I-Tech Education and Publishing.

Geem, Z. W., Kim, J.-H., & Jeong, S.-H. (2011). Cost efficient and practical design of water supply network using harmony search. *African Journal of Agricultural Research, 6*, 3110–3116.

Gil-López, S., Ser, J. D., Salcedo-Sanz, S., Pérez-Bellido, Á. M., Cabero, J. M. A., & Portilla-Figueras, J. A. (2012). A hybrid harmony search algorithm for the spread spectrum radar polyphase codes design problem. *Expert Systems with Applications, 39*, 11089–11093.

Guney, K., & Onay, M. (2011). Optimal synthesis of linear antenna arrays using a harmony search algorithm. *Expert Systems with Applications, 38*, 15455–15462.

Guo, P., Wang, J., Gao, X. Z., & Tanskanen, J. M. A. (2012, October 14–17). Epileptic EEG signal classification with marching pursuit based on harmony search method. In *IEEE International Conference on Systems, Man, and Cybernetics (SMC)*, COEX, Seoul, Korea (pp. 177–183). IEEE.

Hasançebi, O., Erdal, F., & Saka, M. P. (2010). Optimum design of geodesic steel domes under code provisions using metaheuristic techniques. *International Journal of Engineering and Applied Sciences, 2*, 88–103.

Huang, M., Dong, H.-Y., Wang, X.-W., Zheng, B.-L., & Ip, W. H. (2009, June 12–14). Guided variable neighborhood harmony search for integrated charge planning in primary steelmaking processes. In *GEC'09*, Shanghai, China (pp. 231–238). ACM.

Jaberipour, M., & Khorram, E. (2010a). Solving the sum-of-ratios problems by a harmony search algorithm. *Journal of Computational and Applied Mathematics, 234*, 733–742.

Jaberipour, M., & Khorram, E. (2010b). Two improved harmony search algorithms for solving engineering optimization problems. *Communications in Nonlinear Science and Numerical Simulation, 15*, 3316–3331.

Jaberipour, M., & Khorram, E. (2011). A new harmony search algorithm for solving mixed–discrete engineering optimization problems. *Engineering Optimization, 43*, 507–523.

Jang, W. S., Kang, H. I., & Lee, B. H. (2008). Hybrid simplex-harmony search method for optimization problems. In *IEEE Congress on Evolutionary Computation (CEC)* (pp. 4157–4164). IEEE.

Jarrett, S. & Day, H. (2008). *Music composition for dummies*. 111 River St. Hoboken, NJ, USA: Wiley Publishing, Inc., ISBN 978-0-470-22421-2.

Javadi, M. S., Sabramooz, S., & Javadinasab, A. (2012). Security constrained generation scheduling using harmony search optimization case study: Day-ahead heat and power scheduling. *Indian Journal of Science and Technology, 5*, 1812–1820.

Kattan, A. & Abdullah, R. (2011a). An enhanced parallel and distributed implementation of the harmony search based supervised training of artificial neural networks. In *Third International Conference on Computational Intelligence, Communication Systems and Networks (CICSyN)* (pp. 275–280). IEEE.

Kattan, A. & Abdullah, R. (2011b). A parallel and distributed implementation of the harmony search based supervised training of artificial neural networks. In *Proceedings of the Second International Conference on Intelligent Systems, Modelling and Simulation (ISMS)* (pp. 277–283). IEEE.

Kattan, A., Abdullah, R., & Salam, R. A. (2010). Harmony search based supervised training of artificial neural networks. In *International Conference on Intelligent Systems, Modelling and Simulation (ISMS)* (pp. 105–110). IEEE.

Kaveh, A., & Abadi, A. S. M. (2010). Cost optimization of a composite floor system using an improved harmony search algorithm. *Journal of Constructional Steel Research, 66*, 664–669.

Kaveh, A., & Ahangaran, M. (2012). Discrete cost optimization of composite floor system using social harmony search model. *Applied Soft Computing, 12*, 372–381.

Kaveh, A., & Nasr, H. (2011). Solving the conditional and unconditional *p*-center problem with modified harmony search: A real case study. *Scientia Iranica A, 18*, 867–877.

Kaveh, A., & Talataha, S. (2009). Particle swarm optimizer, ant colony strategy and harmony search scheme hybridized for optimization of truss structures. *Computers and Structures, 87*, 267–283.

Kayhan, A. H., Korkmaz, K. A., & Irfanoglu, A. (2011). Selecting and scaling real ground motion records using harmony search algorithm. *Soil Dynamics and Earthquake Engineering, 31*, 941–953.

Khajehzadeh, M., Taha, M. R., El-Shafie, A., & Eslami, M. (2011). Economic design of foundation using harmony search algorithm. *Australian Journal of Basic and Applied Sciences, 5*, 936–943.

Khazali, A. H., & Kalantar, M. (2011). Optimal reactive power dispatch based on harmony search algorithm. *Electrical Power and Energy Systems, 33*, 684–692.

Khorram, E., & Jaberipour, M. (2011). Harmony search algorithm for solving combined heat and power economic dispatch problems. *Energy Conversion and Management, 52*, 1550–1554.

Kulluk, S., Ozbakir, L., & Baykasoglu, A. (2011). Self-adaptive global best harmony search algorithm for training neural networks. *Procedia Computer Science, 3*, 282–286.

Kulluk, S., Ozbakir, L., & Baykasoglu, A. (2012). Training neural networks with harmony search algorithms for classification problems. *Engineering Applications of Artificial Intelligence, 25*, 11–19.

Lagaros, N. D., & Papadrakakis, M. (2012). Applied soft computing for optimum design of structures. *Structural and Multidisciplinary Optimization, 45*, 787–799.

Landa-Torres, I., Gil-Lopez, S., Salcedo-Sanz, S., Ser, J. D., & Portilla-Figueras, J. A. (2012). A novel grouping harmony search algorithm for the multiple-type access node location problem. *Expert Systems with Applications, 39*, 5262–5270.

Layeb, A. (2013). A hybrid quantum inspired harmony search algorithm for 0–1 optimization problems. *Journal of Computational and Applied Mathematics, 253*, 14–25.

Lee, K. S., & Geem, Z. W. (2004). A new structural optimization method based on the harmony search algorithm. *Computers and Structures, 82*, 781–798.

Lee, K. S., & Geem, Z. W. (2005). A new meta-heuristic algorithm for continuous engineering optimization: Harmony search theory and practice. *Computer Methods in Applied Mechanics and Engineering, 194*, 3902–3933.

Lee, Y. C. & Zomaya, A. Y. (2009). Interweaving heterogeneous metaheuristics using harmony search. In *IEEE International Symposium on Parallel and Distributed Processing (IPDPS)* (pp. 1–8). IEEE.

Lee, K. S., Han, S. W., & Geem, Z. W. (2011). Discrete size and discrete-continuous configuration optimization methods for truss structures using the harmony search algorithm. *International Journal of Optimization in Civil Engineering, 1*, 107–126.

Li, L.-P. & Wang, L. (2009, June 12–14). Hybrid algorithms based on harmony search and differential evolution for global optimization. In *GEC*, Shanghai, China (pp. 271–278).

Li, H.-Q., Li, L., Kim, T.-H., & Xie, S.-L. (2008). An improved PSO-based of harmony search for complicated optimization problems. *International Journal of Hybrid Information Technology, 1*, 91–98.

Li, Y., Chen, J., Liu, R., & Wu, J. (2012, June 10–15). A spectral clustering-based adaptive hybrid multi-objective harmony search algorithm for community detection. In *IEEE World Congress on Computational Intelligence (WCCI)*, Brisbane, Australia (pp. 1–8). IEEE.

Liao, T. W. (2010). Two hybrid differential evolution algorithms for engineering design optimization. *Applied Soft Computing, 10*, 1188–1199.

Mahdavi, M., & Abolhassani, H. (2009). Harmony K-means algorithm for document clustering. *Data Mining and Knowledge Discovery, 18*, 370–391.

Mahdavi, M., Fesanghary, M., & Damangir, E. (2007). An improved harmony search algorithm for solving optimization problems. *Applied Mathematics and Computation, 188*, 1567–1579.

Mahdavi, M., Chehreghani, M. H., Abolhassani, H., & Forsati, R. (2008). Novel meta-heuristic algorithms for clustering web documents. *Applied Mathematics and Computation, 201*, 441–451.

Manjarres, D., Landa-Torres, I., Gil-Lopez, S., Ser, J. D., Bilbao, M. N., Salcedo-Sanz, S., & Geem, Z. W. (2013). A survey on applications of the harmony search algorithm. *Engineering Applications of Artificial Intelligence, 26*, 1818–1831. http://dx.doi.org/10.1016/j.engappai.2013.05.008.

Merzougui, A., Hasseine, A., & Laiadi, D. (2012). Application of the harmony search algorithm to calculate the interaction parameters in liquid–liquid phase equilibrium modeling. *Fluid Phase Equilibria, 324*, 94–101.

Miguel, L. F. F., & Miguel, L. F. F. (2012). Shape and size optimization of truss structures considering dynamic constraints through modern metaheuristic algorithms. *Expert Systems with Applications, 39*, 9458–9467.

Mirkhani, M., Forsati, R., Shahri, A. M., & Moayedikia, A. (2013). A novel efficient algorithm for mobile robot localization. *Robotics and Autonomous Systems, 61*, 920–931. http://dx.doi.org/10.1016/j.robot.2013.04.009.

Moeinzadeh, H., Asgarian, E., Zanjani, M., Rezaee, A., & Seidi, M. (2009). Combination of harmony search and linear discriminate analysis to improve classification. In *Proceedings of the Third Asia International Conference on Modelling and Simulation (AMS)* (pp. 131–135). IEEE.

Mohammadi, M., Houshyar, A., Pahlavanhoseini, A., & Ghadimi, N. (2011). Using harmony search algorithm for optimization the component sizing of plug-in hybrid electric vehicle. *International Review of Electrical Engineering, 6*, 2990–2999.

Mora-Gutiérrez, R. A., Ramírez-Rodríguez, J., & Rincón-García, E. A. (2012). An optimization algorithm inspired by musical composition. *Artificial Intelligence Review*. doi:10.1007/s10462-011-9309-8.

Mora-Meliá, D., Iglesias-REY, P. L., Lopez-Patiño, G., & Fuertes-Miquel, V. S. (2009, October 29–30). Application of the harmony search algorithm to water distribution networks design. In G. Palau-Salvador (Ed.), *International Workshop on Environmental Hydraulics, IWEH09*, Valencia, Spain (pp. 265–271). CRC Press.

Mukhopadhyay, A., Roy, A., Das, S., Das, S., & Abraham, A. (2008). Population-variance and explorative power of harmony search: an analysis. In *Third International Conference on Digital Information Management (ICDIM)* (pp. 775–781). IEEE.

Mun, S., & Cho, Y.-H. (2012). Modified harmony search optimization for constrained design problems. *Expert Systems with Applications, 39*, 419–423.

Mun, S., & Geem, Z. W. (2009). Determination of viscoelastic and damage properties of hot mix asphalt concrete using a harmony search algorithm. *Mechanics of Materials, 41*, 339–353.

Nadi, F., Khader, A. T., & Al-Betar, M. A. (2010, July 7–11). Adaptive genetic algorithm using harmony search. In *Proceedings of the Annual Conference on Genetic and Evolutionary Computation (GECCO)*, Portland, Oregon, USA (pp. 819–820).

Omran, M. G. H., & Mahdavi, M. (2008). Global-best harmony search. *Applied Mathematics and Computation, 198*, 643–656.

Pan, Q.-K., Suganthan, P. N., Tasgetiren, M. F., & Liang, J. J. (2010a). A self-adaptive global best harmony search algorithm for continuous optimization problems. *Applied Mathematics and Computation, 216*, 830–848.

Pan, Q.-K., Tasgetiren, M. F., Suganthan, P. N., & Liang, Y.-C. (2010b, July 18–23). Solving lot-streaming flow shop scheduling problems using a discrete harmony algorithm. In *IEEE World Congress on Computational Intelligence (WCCI)*, CCIB, Barcelona, Spain (pp. 4134–4139). IEEE.

Pan, Q.-K., Suganthan, P. N., Liang, J. J., & Tasgetiren, M. F. (2011a). A local-best harmony search algorithm with dynamic sub-harmony memories for lot-streaming flow shop scheduling problem. *Expert Systems with Applications, 38*, 3252–3259.

Pan, Q.-K., Wang, L., & Gao, L. (2011b). A chaotic harmony search algorithm for the flow shop scheduling problem with limited buffers. *Applied Soft Computing, 11*, 5270–5280.

Pandi, V. R., & Panigrahi, B. K. (2011). Dynamic economic load dispatch using hybrid swarm intelligence based harmony search algorithm. *Expert Systems with Applications, 38*, 8509–8514.

Pandi, V. R., Panigrahi, B. K., Bansal, R. C., Das, S., & Mohapatra, A. (2011). Economic load dispatch using hybrid swarm intelligence based harmony search algorithm. *Electric Power Components and Systems, 39*, 751–767.

Piperagkas, G. S., Konstantaras, I., Skouri, K., & Parsopoulos, K. E. (2012). Solving the stochastic dynamic lot-sizing problem through nature-inspired heuristics. *Computers and Operations Research, 39*, 1555–1565.

Purnomo, H. D., Wee, H. M., & Praharsi, Y. (2012). Two inventory review policies on supply chain configuration problem. *Computers and Industrial Engineering, 63*, 448–455.

Qin, A. K. & Forbes, F. (2011a, July 12–16). Dynamic regional harmony search with opposition and local learning. In *Annual Conference on Genetic and Evolutionary Computation (GECCO)*, Dublin, Ireland (pp. 53–54).

Qin, A. K. & Forbes, F. (2011b, July 12–16). Harmony search with differential mutation based pitch adjustment. In *Annual Conference on Genetic and Evolutionary Computation (GECCO)*, Dublin, Ireland (pp. 545–552).

Ramos, C. C. O., Souza, A. N., Chiachia, G., Falcão, A. X., & Papa, J. P. (2011). A novel algorithm for feature selection using harmony search and its application for non-technical losses detection. *Computers and Electrical Engineering, 37*, 886–894.

Razfar, M. R., Zinati, R. F., & Haghshenas, M. (2011). Optimum surface roughness prediction in face milling by using neural network and harmony search algorithm. *International Journal of Advanced Manufacturing Technology, 52*, 487–495.

Ryu, S., Heyl, C. N., Duggal, A. S., & Geem, Z. W. (2007, June 10–15). Mooring cost optimization via harmony search. In *26th International Conference on Offshore Mechanics and Arctic Engineering (EMAE)*, San Diego, California, USA (pp. 1–8). ASME.

Ser, J. D., Matinmikko, M., Gil-López, S., & Mustonen, M. (2012). Centralized and distributed spectrum channel assignment in cognitive wireless networks: A harmony search approach. *Applied Soft Computing, 12*, 921–930.

Shahrouzi, M., & Sazjini, M. (2012). Refined harmony search for optimal scaling and selection of accelerograms. *Scientia Iranica, Transactions A: Civil Engineering, 19*, 218–224.

Shariatkhah, M.-H., Haghifam, M.-R., Salehi, J., & Moser, A. (2012). Duration based reconfiguration of electric distribution networks using dynamic programming and harmony search algorithm. *Electrical Power and Energy Systems, 41*, 1–10.

Shi, F., Xia, X., Chang, C., Xu, G., Qin, X., & Jia, Z. (2011). An application in frequency assignment based on improved discrete harmony search algorithm. *Procedia Engineering, 24*, 247–251.

Sirjani, R., Mohamed, A., & Shareef, H. (2011). Optimal capacitor placement in three-phase distribution systems using improved harmony search algorithm. *International Review of Electrical Engineering, 6*, 1783–1793.

Sivasubramani, S., & Swarup, K. S. (2011a). Environmental/economic dispatch using multi-objective harmony search algorithm. *Electric Power Systems Research, 81*, 1778–1785.

Sivasubramani, S., & Swarup, K. S. (2011b). Multi-objective harmony search algorithm for optimal power flow problem. *Electrical Power and Energy Systems, 33*, 745–752.

Taherinejad, N. (2009). Highly reliable harmony search algorithm. In *European Conference on Circuit Theory and Design (ECCTD)* (pp. 818–822). IEEE.

Taleizadeh, A. A., Niaki, S. T. A., & Barzinpour, F. (2011). Multiple-buyer multiple-vendor multi-product multi-constraint supply chain problem with stochastic demand and variable lead-time: A harmony search algorithm. *Applied Mathematics and Computation, 217*, 9234–9253.

Taleizadeh, A. A., Niaki, S. T. A., & Seyedjavadi, S. M. H. (2012). Multi-product multi-chance-constraint stochastic inventory control problem with dynamic demand and partial back-ordering a harmony search algorithm. *Journal of Manufacturing Systems, 31*, 204–213.

Tasgetiren, M. F., Bulut, O., & Fadiloglu, M. M. (2012, June 10–15). A discrete harmony search algorithm for the economic lot scheduling problem with power of two policy. In *IEEE World Congress on Computational Intelligence (WCCI)*, Brisbane, Australia (pp. 1–8). IEEE.

Vasebi, A., Fesanghary, M., & Bathaee, S. M. T. (2007). Combined heat and power economic dispatch by harmony search algorithm. *Electrical Power and Energy Systems, 29*, 713–719.

Venkatesh, S. K., Srikant, R., & Madhu, R. M. (2010, January 22–23). Feature selection and dominant feature selection for product reviews using meta-heuristic algorithms. In *Proceedings of the Compute'10*, Bangalore, Karnataka, India (pp. 1–4). ACM.

Vural, R. A., Bozkurt, U., & Yildirim, T. (2013). Analog active filter component selection with nature inspired metaheuristics. *International Journal of Electronics and Communications, 67*, 197–205.

Wang, C.-M., & Huang, Y.-F. (2010). Self-adaptive harmony search algorithm for optimization. *Expert Systems with Applications, 37*, 2826–2837.

Wang, L., & Li, L.-P. (2012). A coevolutionary differential evolution with harmony search for reliability–redundancy optimization. *Expert Systems with Applications, 39*, 5271–5278.

Wang, L., & Li, L.-P. (2013). An effective differential harmony search algorithm for the solving non-convex economic load dispatch problems. *Electrical Power and Energy Systems, 44*, 832–843.

Wang, X., Gao, X.-Z., & Ovaska, S. J. (2009). Fusion of clonal selection algorithm and harmony search method in optimization of fuzzy classification systems. *International Journal of Bio-Inspired Computation, 1*, 80–88.

Wang, L., Pan, Q.-K., & Tasgetiren, M. F. (2010). Minimizing the total flow time in a flow shop with blocking by using hybrid harmony search algorithms. *Expert Systems with Applications, 37*, 7929–7936.

Wang, L., Pan, Q.-K., & Tasgetiren, M. F. (2011). A hybrid harmony search algorithm for the blocking permutation flow shop scheduling problem. *Computers and Industrial Engineering, 61*, 76–83.

Weyland, D. (2010). A rigorous analysis of the harmony search algorithm: How the research community can be misled by a "novel" methodology. *International Journal of Applied Metaheuristic Computing, 1–2*, 50–60.

Wong, W. K., & Guo, Z. X. (2010). A hybrid intelligent model for medium-term sales forecasting in fashion retail supply chains using extreme learning machine and harmony search algorithm. *International Journal of Production Economics, 128*, 614–624.

Yadav, P., Kumar, R., Panda, S. K., & Chang, C. S. (2011). An improved harmony search algorithm for optimal scheduling of the diesel generators in oil rig platforms. *Energy Conversion and Management, 52*, 893–902.

Yadav, P., Kumar, R., Panda, S. K., & Chang, C. S. (2012). An intelligent tuned harmony search algorithm for optimisation. *Information Sciences, 196*, 47–72.

Yaşar, C., & Özyön, S. (2011). A new hybrid approach for nonconvex economic dispatch problem with valve-point effect. *Energy, 36*, 5838–5845.

Yi, L., & Goldsmith, J. (2010). Decision-theoretic harmony: A first step. *International Journal of Approximate Reasoning, 51*, 263–274.

Yildiz, A. R. (2008). Hybrid Taguchi-harmony search algorithm for solving engineering optimization problems. *International Journal of Industrial Engineering, 15*, 286–293.

Zarei, O., Fesanghary, M., Farshi, B., Saffar, R. J., & Razfar, M. R. (2009). Optimization of multi-pass face-milling via harmony search algorithm. *Journal of Materials Processing Technology, 209*, 2386–2392.

Zhang, Z.-N., Liu, Z.-L., Chen, Y., & Xie, Y.-B. (2013). Knowledge flow in engineering design: an ontological framework. *Proceedings of the Institution of Mechanical Engineers, Part C: Journal of Mechanical Engineering Science, 227*, 760–770.

Zhao, S.-Z. & Suganthan, P. N. (2010, July 18–23). Dynamic multi-swarm particle swarm optimizer with sub-regional harmony search. In *IEEE World Congress on Computational Intelligence (WCCI)*, CCIB, Barcelona, Spain (pp. 1983–1990). IEEE.

Zhao, S.-Z., Suganthan, P. N., Pan, Q.-K., & Tasgetiren, M. F. (2011). Dynamic multi-swarm particle swarm optimizer with harmony search. *Expert Systems with Applications, 38*, 3735–3742.

Zinati, R. F., & Razfar, M. R. (2012). Constrained optimum surface roughness prediction in turning of X20Cr13 by coupling novel modified harmony search-based neural network and modified harmony search algorithm. *International Journal of Advanced Manufacturing Technology, 58*, 93–107.

Zou, D., Gao, L., Li, S., Wu, J., & Wang, X. (2010a). A novel global harmony search algorithm for task assignment problem. *The Journal of Systems and Software, 83*, 1678–1688.

Zou, D., Gao, L., Wu, J., & Li, S. (2010b). Novel global harmony search algorithm for unconstrained problems. *Neurocomputing, 73*, 3308–3318.
Zou, D., Gao, L., Wu, J., Li, S., & Li, Y. (2010c). A novel global harmony search algorithm for reliability problems. *Computers and Industrial Engineering, 58*, 307–316.
Zou, D., Gao, L., Li, S., & Wu, J. (2011a). An effective global harmony search algorithm for reliability problems. *Expert Systems with Applications, 38*, 4642–4648.
Zou, D., Gao, L., Li, S., & Wu, J. (2011b). Solving 0–1 knapsack problem by a novel global harmony search algorithm. *Applied Soft Computing, 11*, 1556–1564.

Chapter 15
Imperialist Competitive Algorithm

Abstract In this chapter, we present a new optimization algorithm called imperialist competitive algorithm (ICA) which is inspired by the human socio-political evolution process. We first describe the general knowledge of the imperialism in Sect. 15.1. Then, the fundamentals and performance of ICA are introduced in Sect. 15.2. Finally, Sect. 15.3 summarises this chapter.

15.1 Introduction

In politics and history, in order to explain the extending control over weaker people or areas, the term "empire" is appeared. One of the famous examples is the Roman Empire from 27 BC to 476. The most powerful person is usually called imperialist and different imperialists will extent their authority and control of one state or people over another through the form of competition. Inspired by this human socio-political evolution process, a newly developed evolutionary algorithm called Imperialist competitive algorithm (ICA) is proposed by Atashpaz-Gargari and Lucas (2007).

15.1.1 Imperialism

Imperialism was greatly influenced by an economic theory known as mercantilism which inspired the government to extend their power and the rule beyond its own boundaries (Atashpaz-Gargari and Lucas 2007). In its initial forms, due to the limited supply of wealth, counties focused on building their empire with natural resources, such as gold and silver. However, nowadays the new imperialism focused on the new technological advances and developments. The ultimate goal is to increase the number of their colonies and spreading their empires over the world.

B. Xing and W.-J. Gao, *Innovative Computational Intelligence: A Rough Guide to 134 Clever Algorithms*, Intelligent Systems Reference Library 62,
DOI: 10.1007/978-3-319-03404-1_15,

15.2 Imperialist Competitive Algorithm

15.2.1 *Fundamentals of Imperialist Competitive Algorithm*

Imperialist competitive algorithm (ICA) was originally proposed by Atashpaz-Gargari and Lucas (2007). In ICA, the countries can be viewed as population individuals and basically divided into two groups based on their power, i.e., colonies and imperialists. Also, one empire is formed by one imperialist with its colonies. Furthermore, two operators called assimilation and revolution and one strategy called imperialistic competition are the main building blocks that employed in ICA. The implementation procedures are described as below (Atashpaz-Gargari and Lucas 2007):

- Initializing phase:

1. Preparation of initial populations. Each solution (i.e., *country*) that in form of an array can be defined via Eq. 15.1 (Atashpaz-Gargari and Lucas 2007):

$$country = [p_1, p_2, \ldots, p_{N_{\text{var}}}], \tag{15.1}$$

where $p_i s$ represent different variables which based on various socio-political characteristics (such as culture, language, and economical policy), and N_{var} denotes the total number of the characteristics (i.e., n–dimension of the problems) to be optimized.
2. Creating the cost function. In order to evaluate the cost of countries, the cost function can be defined via Eq. 15.2 (Atashpaz-Gargari and Lucas 2007):

$$\cos t = f(country) = f(p_1, p_2, \ldots, p_{N_{\text{var}}}). \tag{15.2}$$

3. Initializing the empires: In general, the initial size of populations (N_{pop}) involves two types of countries [i.e., colony (N_{col}) and imperialist (N_{imp})] which together form the empires. To form the initial empires proportionally, the normalized cost of an imperialist is defined via Eq. 15.3 (Atashpaz-Gargari and Lucas 2007):

$$NC_n = c_n - \max_i\{c_i\}, \tag{15.3}$$

where c_n is the cost of nth imperialist, NC_n denotes its normalized cost.

Normally, two methods can be used to divide colonies among imperialist: (1) from the imperialists' point of view which based on the power of each imperialist; (2) from the colonies' point of view which based on the relationship with the imperialist (i.e., the colonies should be possessed by the imperialist according to the power). Both methods are given via Eqs. 15.4 and 15.5, respectively (Atashpaz-Gargari and Lucas 2007):

$$Power_n = \left| \frac{NC_n}{\sum_{i=1}^{N_{imp}} NC_i} \right|, \tag{15.4}$$

$$NOC_n = round\{Power_n, N_{col}\}, \tag{15.5}$$

where $Power_n$ is the normalized power of each imperialist, N_{col} and N_{imp} represent the number of all colonies and imperialists, respectively, and NOC_n is the initial number of colonies of nth empire.

- Moving phase:

1. Assimilation strategy: According to this strategy, all colonies will move toward their relevant imperialist with x units via Eq. 15.6 (Atashpaz-Gargari and Lucas 2007):

$$x \sim U(0, \beta \times d), \tag{15.6}$$

where x is a random variable with uniform distribution, β is a number greater than 1, and d is the distance between a colony and an imperialist.

2. Revolution strategy: According to this strategy, a random amount of deviation to the direction of movement is incorporated via Eq. 15.7 (Atashpaz-Gargari and Lucas 2007):

$$\theta \sim U(-\gamma, \gamma), \tag{15.7}$$

where θ is a random variable with uniform distribution, and γ is a parameter that adjusts the deviation from the original direction.

- Exchanging phase: Based on the cost function, when the new position of a colony is better than that of the corresponding imperialist, the imperialist and the colony change their positions and the new location with lower cost becomes the imperialist.
- Imperialistic competition phase:

1. Calculating the total power of an empire: It is influenced by the power of imperialist country and the colonies of an empire via Eq. 15.8 (Atashpaz-Gargari and Lucas 2007):

$$TC_n = Cost(imperialist_n) + \xi\, mean\{Cost(colonies\, of\, empire_n)\}, \tag{15.8}$$

where TC_n is the total cost of the nth empire, and ξ is a positive number which is considered to be less than 1.

2. Imperialistic competition strategy: According to this strategy, all empires try to take the possession of the colonies of other empires and control them. To

modelled this strategy, the weakest colonies of the weakest empires will be chose to competition among all other empires in order to possess this colony. Based on TC_n, the normalized total cost is simply obtained via Eq. 15.9 (Atashpaz-Gargari and Lucas 2007):

$$NTC_n = TC_n - \max_i\{TC_i\}, \quad (15.9)$$

where NTC_n represents the total normalized cost of nth empire. Having NTC_n, the possession probability of each empire is evaluated via Eq. 15.10 (Atashpaz-Gargari and Lucas 2007):

$$P_{p_n} = \left| \frac{NTC_n}{\sum_{i=1}^{N_{imp}} NTC_i} \right|. \quad (15.10)$$

- Eliminating phase: When an empire loses all its colonies (i.e., their colonies will be divided among other empires), it is assumed to be collapsed and will be eliminated.
- Convergence phase: At the end, all the colonies will be under the control of the most powerful empire, which means all the colonies have the same positions and same costs and will be controlled by an imperialist with the same position and cost as themselves. In other words, there are no difference not only among colonies but also between colonies and imperialist.

Taking into account the key phases described above, the steps of implementing ICA can be summarized as follows (Atashpaz-Gargari and Lucas 2007):

- Step 1: Defining the optimization problem.
- Step 2: Generating initial empires by pick some random points on the function.
- Step 3: Move the colonies towards imperialist states in different directions (i.e., assimilation).
- Step 4: Random changes occur in the characteristics of some countries (i.e., revolution).
- Step 5: Position exchange between a colony and imperialist.
- Step 6: Compute the total cost of all empires.
- Step 7: Use imperialistic competition and pick the weakest colony from the weakest empire.
- Step 8: Eliminate the powerless empires.
- Step 9: Check if maximum iteration is reached, go to Step 3 for new beginning. If a specified termination criteria is satisfied stop and return the best solution.

15.2.2 *Performance of ICA*

In order to show how the ICA performs, four minimization problems are tested in Atashpaz-Gargari and Lucas (2007). Compared with genetic algorithm (GA) and particle swarm optimization (PSO), the results showed that ICA is capable of reaching the global minimum.

15.3 Conclusions

In this chapter, an optimization algorithm motivated by one of the socio-politically models (i.e., imperialistic competition) is introduced. The term "countries" are designed as the initial populations and categorized into colony and imperialist states, respectively. Both colonies and imperialist together form the empires. The basic idea behind ICA is to lead the search process toward the powerful imperialist or the optimum points based on their "power", i.e., the weakest empires will lost their colonies until there will be no colony in that. The objective of ICA is to find an ideal world in which there is no difference not only among colonies but also between colonies and imperialist. Although it is a newly introduced computational intelligence method, we have witnessed the following rapid spreading of ICA:

First, several enhanced versions of ICA can be found in the literature as outlined below:

- Bacterial foraging optimization based ICA (Acharya et al. 2010).
- Chaotic improved ICA (Talatahari et al. 2012a, b).
- Constrained genetic algorithm based ICA (Acharya et al. 2010).
- Fast ICA (Acharya et al. 2010).
- Hybrid artificial neural network and ICA (Taghavifar et al. 2013).
- Hybrid evolutionary algorithm and ICA (Ramezani et al. 2012).
- Modified ICA (Niknam et al. 2011).
- Multiobjective ICA (Bijami et al. 2011; Mohammadi et al. 2011).
- Quad countries algorithm (Soltani-Sarvestani et al. 2012).

Second, the ICA has also been successfully applied to a variety of optimization problems as listed below:

- Artificial neural network training (Moadi et al. 2011).
- Assembly line balancing (Bagher et al. 2011).
- Data mining (Ghanavati et al. 2011; Niknam et al. 2011).
- Facility location optimization (Mohammadi et al. 2011; Moadi et al. 2011).
- Power system optimization (Nejad and Jahani 2011; Bijami et al. 2011).
- Product mix-outsourcing problem (Nazari-Shirkouhi et al. 2010).
- Scheduling optimization (Forouharfard and Zandieh 2010; Ayough et al. 2012; Lian et al. 2012; Kayvanfar and Zandieh 2012).

- Soil compaction prediction (Taghavifar et al. 2013).
- Structure design optimization (Talatahari et al. 2012b).

Interested readers please refer to them as a starting point for a further exploration and exploitation of ICA.

References

Acharya, D. P., Panda, G., & Lakshmi, Y. V. S. (2010). Effects of finite register length on fast ICA, bacterial foraging optimization based ICA and constrained genetic algorithm based ICA algorithm. *Digital Signal Processing, 20*, 964–975.

Atashpaz-Gargari, E., & Lucas, C. (2007). Imperialist competitive algorithm: An algorithm for optimization inspired by imperialistic competition. In *IEEE Congress on Evolutionary Computation (CEC 2007)* (pp. 4661–4667). IEEE.

Ayough, A., Zandieh, M., & Farsijani, H. (2012). GA and ICA approaches to job rotation scheduling problem: Considering employee's boredom. *International Journal of Advanced Manufacturing Technology, 60*, 651–666.

Bagher, M., Zandieh, M., & Farsijani, H. (2011). Balancing of stochastic U-type assembly lines: An imperialist competitive algorithm. *International Journal of Advanced Manufacturing Technology, 54*, 271–285.

Bijami, E., Abshari, R., Askari, J., Hosseinnia, S., & Farsangi, M. M. (2011). Optimal design of damping controllers for multi-machine power systems using metaheuristic techniques. *International Review of Electrical Engineering, 6*, 1883–1894.

Forouharfard, S., & Zandieh, M. (2010). An imperialist competitive algorithm to schedule of receiving and shipping trucks in cross-docking systems. *International Journal of Advanced Manufacturing Technology, 51*, 1179–1193.

Ghanavati, M., Gholamian, M. R., Minaei, B., & Davoudi, M. (2011). An efficient cost function for imperialist competitive algorithm to find best clusters. *Journal of Theoretical and Applied Information Technology, 29*, 22–31.

Kayvanfar, V., & Zandieh, M. (2012). The economic lot scheduling problem with deteriorating items and shortage: An imperialist competitive algorithm. *International Journal of Advanced Manufacturing Technology*. doi:10.1007/s00170-011-3820-6.

Lian, K., Zhang, C., Shao, X., & Gao, L. (2012). Optimization of process planning with various flexibilities using an imperialist competitive algorithm. *International Journal of Advanced Manufacturing Technology, 59*, 815–828.

Moadi, S., Mohaymany, A. S., & Babaei, M. (2011). Application of imperialist competitive algorithm to the emergency medical services location problem. *International Journal of Artificial Intelligence and Applications (IJAIA), 2*, 137–147.

Mohammadi, M., Tavakkoli-Moghaddam, R., & Rostami, H. (2011). A multi-objective imperialist competitive algorithm for a capacitated hub covering location problem. *International Journal of Industrial Engineering Computations, 2*, 671–688.

Nazari-Shirkouhi, S., Eivazy, H., Ghodsi, R., Rezaie, K., & Atashpaz-Gargari, E. (2010). Solving the integrated product mix-outsourcing problem using the imperialist competitive algorithm. *Expert Systems with Applications, 37*, 7615–7626.

Nejad, H. C., & Jahani, R. (2011). A new approach to economic load dispatch of power system using imperialist competitive algorithm. *Australian Journal of Basic and Applied Sciences, 5*, 835–843.

Niknam, T., Fard, E. T., Ehrampoosh, S., & Rousta, A. (2011). A new hybrid imperialist competitive algorithm on data clustering. *Sādhanā, 36*, 293–315.

Ramezani, F., Lotfi, S., & Soltani-Sarvestani, M. A. (2012). A hybrid evolutionary imperialist competitive algorithm (HEICA). In J.-S. Pan, S.-M. Chen & N. T. Nguyen (Eds.) *ACIIDS 2012, Part I, LNAI 7196* (pp. 359–368). Berlin: Springer.

Soltani-Sarvestani, M. A., Lotfi, S., & Ramezani, F. (2012). Quad countries algorithm (QCA). In J.-S. Pan, S.-M. Chen & N. T. Nguyen (Eds.) *ACIIDS 2012, Part III, LNAI 7198* (pp. 119–129). Berlin: Springer.

Taghavifar, H., Mardani, A., & Taghavifar, L. (2013). A hybridized artificial neural network and imperialist competitive algorithm optimization approach for prediction of soil compaction in soil bin facility. *Measurement, 46*, 2288–2299.

Talatahari, S., Azar, B. F., Sheikholeslami, R., & Gandomi, A. H. (2012a). Imperialist competitive algorithm combined with chaos for global optimization. *Communications in Nonlinear Science and Numerical Simulation, 17*, 1312–1319.

Talatahari, S., Kaveh, A., & Sheikholeslami, R. (2012b). Chaotic imperialist competitive algorithm for optimum design of truss structures. *Structural and Multidisciplinary Optimization*. doi:10.1007/s00158-011-0754-4.

Chapter 16
Teaching–Learning-based Optimization Algorithm

Abstract In this chapter, we present an interesting algorithm called teaching–learning-based optimization (TLBO) which is inspired by the teaching and learning behaviour. We first describe the general knowledge of the teacher-student relationships in Sect. 16.1. Then, the fundamentals and performance of TLBO algorithm are introduced in Sect. 16.2. Finally, Sect. 16.3 summarises this chapter.

16.1 Introduction

Education is one of the most important things throughout our lifecycle. It would be difficult to find a school or a university in which the ability of teachers is not of significant relevance. The functions of the teachers may vary among cultures, such as they may provide instruction in literacy and numeracy, craftsmanship or vocational training, the arts, religion, civics, community roles, even the life skills. In addition, some experimental studies showed that student motivation and learning ability are closely linked to student–teacher relationships (Kaur et al. 2013). Inspired by this phenomenon, recently, Rao et al. (2011) proposed a new computational intelligence (CI) algorithm called teaching–learning-based optimization (TLBO) algorithm.

16.2 Teaching–Learning-based Optimization

16.2.1 Fundamentals of Teaching-Learning-based Optimization Algorithm

Teaching–learning-based optimization (TLBO) was originally proposed by Rao et al. (2011). The basic idea of TLBO is that the teacher is considered as the most knowledgeable person in a class who shares his/her knowledge with the students to

B. Xing and W.-J. Gao, *Innovative Computational Intelligence: A Rough Guide to 134 Clever Algorithms*, Intelligent Systems Reference Library 62,
DOI: 10.1007/978-3-319-03404-1_16,

improve the output (i.e., grades or marks) of the class. The quality of the learners is evaluated by the mean value of the student's grade in class. Furthermore, learners also learn from interaction between themselves, which also helps in their results.

To perceive the function basis of the TLBO algorithm, suppose there are two different teachers, T_1 and T_2. They both teach a subject with the same content to the same merit level learners in two different classes. The distribution of marks obtained by the learners of two different evaluated by the teachers is defined as the means of class-1 (M_1) and class-2 (M_2) achieved scores, respectively. Assume that curve-2 represents better mean for the results of the learners than curve-1 and so it can be said that teacher T_2 is better than teacher T_1 in terms of teaching. In addition, a normal distribution is assumed for the obtained grades after taking an exam by the teachers, though in practice it may have skewness. Typically, a normal distribution can be defined via Eq. 16.1 (Ross 1998):

$$f(x) = \frac{1}{\sigma\sqrt{2\pi}} e^{-(x-\mu)^2/(2\sigma^2)}, \tag{16.1}$$

where σ^2 is the variance, μ is the mean, and x is any value for which the normal distribution function is required.

Meanwhile, in terms of learner phase, for the marks obtained for learners in a class, we assumed the means of marks is M_A, and the best learner is mimicked as a teacher defined as T_A. His or her mission is trying to move mean M_A towards their own level according to his or her capability, thereby increasing the learners' level to new mean M_B. Like above mentioned, the quality of the students is judged from the mean value of the population. However, during the "leaner phase" the learners learn by interaction between each other. When teacher T_A convey knowledge among the learners and those level increases toward his or her own level, at which stage the students require a new teacher, of superior quality than themselves, i.e. in this case the new teacher is T_B. Hence, there will be a new curve-B with new teacher T_B.

Based on above procedures, the main processes of TLBO can be divided into two phases, i.e. the "teacher phase", where candidate solutions are randomly distributed over the search space and the best solution is determined among those then it shares the information with others; and the "learner phase", where the solutions put effort into passing the own information through the interaction to each other. Working principles of both phases are explained below (Rao et al. 2011):

- Teacher phase:

In the model, this phase produces a random ordered state of points called learners within the search space. Then a point is considered as the teacher, who is highly learned person and shares his or her knowledge with the learners, and others learn significant group information from the teacher. It is the first part of the algorithm where the mean of a class increases from M_A to M_B depending upon a good teacher. At this point, Rao et al. (2011) assumed a good teacher is one who brings his/her learners up to his/her level in terms of knowledge. However, in practice this is not possible and a teacher can only move the mean of a class up to some extent depending on the capability of the class. This follows a random process depending on many factors (Rao et al. 2011).

Let M_i be the mean and Ti be the teacher at any iteration i. Ti will try to move mean Mi towards its own level, so now the new mean will be Ti designated as M_{new}. The solution is updated according to the difference between the existing and the new mean given via Eq. 16.2 (Rao et al. 2011):

$$Difference\_Mean_i = r_i(M_{new} - T_F M_i), \tag{16.2}$$

where T_F is a teaching factor that decides the value of mean to be changed, and r_i is a random number in the range [0,1]. The value of T_F can be either 1 or 2, which is again a heuristic step and decided randomly with equal probability via Eq. 16.3 (Rao et al. 2011):

$$T_F = round[1 + rand(0,1)\{2-1\}]. \tag{16.3}$$

This difference modifies the existing solution via Eq. 16.4 (Rao et al. 2011):

$$X_{new,i} = X_{old,i} + Difference\_Mean_i. \tag{16.4}$$

- Learner phase:

It is the second part of the algorithm where learners increase their knowledge by interaction among themselves. So, a solution is randomly interacted to learn something new with other solutions in the population. In the light of this statement, a solution will learn new information if the other solutions have more knowledge than him or her. Mathematically the learning phenomenon of this phase is expressed by Eq. 16.5 (Rao et al. 2011):

$$\begin{aligned} X_{new,i} &= X_{old,i} + r_i(X_i - X_j), \quad \text{if} \quad f(X_i) < f(X_j) \\ X_{new,i} &= X_{old,i} + r_i(X_j - X_i), \quad \text{if} \quad f(X_j) < f(X_j). \end{aligned} \tag{16.5}$$

At any iteration i, considering two different learners X_i and X_j, where $i \neq j$.

Consequently, accept X_{new}, if it gives better function value. After a number of sequential teaching–learning cycles, where the teacher convey knowledge among the learners and those level increases toward his or her own level, the distribution of the randomness within the search space becomes smaller and smaller about to point considering as teacher. It means knowledge level of the whole class shows smoothness and the algorithm converges to a solution. Also, a termination criterion can be a predetermined maximum iteration number is reached.

Taking into account two key phases described above, the steps of implementing the TLBO algorithm can be summarized as follows (Rao et al. 2011; Zou et al. 2013):

- Step 1: Defining the optimization problem, and initializing the optimization parameters.

- Step 2: Initializing the population.
- Step 3: Starting teacher phase where the main activity is learners learning from their teacher.
- Step 4: Starting learner phase where the main activity is learners further tune their knowledge through the interaction with their peers.
- Step 5: Evaluating stopping criteria. Terminate the algorithm in the maximum generation number is reached; otherwise return to Step 3 and the algorithm continues.

16.2.2 Performance of TLBO

In order to show how the TLBO algorithm performs, five different constrained benchmark functions, four different benchmark mechanical design problems, and six real world mechanical design applications are tested in Rao et al. (2011). Computational results showed that TLBO is more effective and efficient compared with other CI algorithms.

16.3 Conclusions

In this chapter, we introduced a new efficient population based optimization method, i.e., TLBO, that is inspired by the effect of influence of a teacher on the output of learners in a class, which learners first acquire knowledge from a teacher (i.e., teacher phase) and then from classmates (i.e., learner phase).

In principle, population consists of learners in a class and design variables are courses offered. The output in TLBO algorithm is measured according to the results or grades of the learners which normally determined by the level of teacher. That means, a high quality teacher is usually considered as a highly learned person who trains learners so that they can have better results in terms of their marks or grades. Moreover, learners also learn from the interaction among themselves which also helps in improving their results. In many aspects, TLBO resembles evolutionary algorithms (Michalewicz 1996). For example, Črepinšek et al. (2012) pointed out three similarities:

- An initial population is randomly generated.
- Moving/learning towards teacher and classmates can be regarded as a special mutation operator.
- Selection is deterministic (i.e., two solutions are compared and the better one always survives), which is also used often in many other evolutionary algorithms such as evolutionary strategies.

Although the effectiveness of TLBO is still under debate [see (Črepinšek et al. 2012; Waghmare 2013) for details], we have witnessed the following rapid spreading of TLBO:

First, several enhanced version of TLBO can also be found in the literature as outlined below:

- Elitist TLBO (Rao and Patel 2012);
- Improved TLBO (Niknam et al. 2012a; Rao and Patel 2013a).
- Modified TLBO (Rao and Patel 2013b, c).
- Multiobjective TLBO (Nayak et al. 2012; Niknam et al. 2012c; Zou et al. 2013).

Second, the TLBO algorithm has also been successfully applied to a variety of optimization problems as listed below:

- Clustering method optimization (Naik et al. 2012).
- Continuous non-linear large scale optimization (Rao et al. 2012).
- Dynamic economic emission dispatch (Niknam et al. 2012c).
- Heat exchanger optimization (Rao and Patel 2013b).
- Machining process parameters' optimization (Pawar and Rao 2012; Rao and Kalyankar 2013a).
- Mechanical design optimization (Rao and Savsani 2012).
- Micro-grid operation (Niknam et al. 2012a).
- Multi-pass turning process parameter optimization (Rao and Kalyankar 2013b).
- Optimal location of automatic voltage regulators (Niknam et al. 2012b).
- Power flow optimization (Nayak et al. 2012).
- Truss structure optimization (Degertekin and Hayalioglu 2013).
- Two-stage thermoelectric cooler optimization (Rao and Patel 2013c).
- Unconstrained and constrained optimization (Rao et al. 2012).
- Used products pre-sorting system optimization (Xing and Gao 2014).

Interested readers please refer to them as a starting point for a further exploration and exploitation of the TLBO algorithm.

References

Črepinšek, M., Liu, S.-H., & Mernik, L. (2012). A note on teaching–learning-based optimization algorithm. *Information Sciences, 212*, 79–93.

Degertekin, S. O., & Hayalioglu, M. S. (2013). Sizing truss structures using teaching-learning-based optimization. *Computers and Structures*, *119*, 177–188. http://dx.doi.org/10.1016/j.compstruc.2012.12.011.

Kaur, B., Anthony, G., Ohtani, M., & Clarke, D. (Eds.). (2013). *Student voice in mathematics classrooms around the world*. P.O. Box 21858, 3001 AW Rotterdam, The Netherlands: Sense Publishers. ISBN 978-94-6209-348-5.

Michalewicz, Z. (1996). *Genetic algorithms + data structures = evolution programs* (3rd ed.). Berlin, Heidelberg: Springer.

Naik, A., Satapathy, S. C., & Parvathi, K. (2012). Improvement of initial cluster center of C-means using teaching learning based optimization. *Procedia Technology, 6*, 428–435.

Nayak, M. R., Nayak, C. K., & Rout, P. K. (2012). Application of multi-objective teaching learning based optimization algorithm to optimal power flow problem. *Procedia Technology, 6*, 255–264.

Niknam, T., Azizipanah-Abarghooee, R., & Narimani, M. R. (2012a). An efficient scenario-based stochastic programming framework for multi-objective optimal micro-grid operation. *Applied Energy, 99*, 455–470.

Niknam, T., Azizipanah-Abarghooee, R., & Narimani, M. R. (2012b). A new multi objective optimization approach based on TLBO for location of automatic voltage regulators in distribution systems. *Engineering Applications of Artificial Intelligence, 25*, 1577–1588.

Niknam, T., Golestaneh, F., & Sadeghi, M. S. (2012c). θ-Multiobjective teaching–learning-based optimization for dynamic economic emission dispatch. *IEEE Systems Journal, 6*(2), 341–352.

Pawar, P. J., & Rao, R. V. (2013). Parameter optimization of machining processes using teaching–learning-based optimization algorithm. *International Journal of Advanced Manufacturing Technology, 67*, 995–1006. doi:10.1007/s00170-012-4524-2.

Rao, R. V., & Kalyankar, V. D. (2013a). Parameter optimization of modern machining processes using teaching–learning-based optimization algorithm. *Engineering Applications of Artificial Intelligence, 26*, 524–531.

Rao, R. V., & Kalyankar, V. D. (2013b). Multi-pass turning process parameter optimization using teaching–learning-based optimization algorithm. *Scientia Iranica, Transactions D: Computer SCience & Engineering and Electrical Engineering, 20*, 967–974. doi:10.1016/j.scient.2013.01.002.

Rao, R. V., & Patel, V. (2012). An elitist teaching–learning-based optimization algorithm for solving complex constrained optimization problems. *International Journal of Industrial Engineering Computations, 3*, 535–560.

Rao, R. V., & Patel, V. (2013a). An improved teaching-learning-based optimization algorithm for solving unconstrained optimization problems. *Scientia Iranica D, 20*, 710–720. doi:10.1016/j.scient.2012.12.005.

Rao, R. V., & Patel, V. (2013b). Multi-objective optimization of heat exchangers using a modified teaching–learning-based optimization algorithm. *Applied Mathematical Modelling, 37*, 1147–1162.

Rao, R. V., & Patel, V. (2013c). Multi-objective optimization of two stage thermoelectric cooler using a modified teaching–learning-based optimization algorithm. *Engineering Applications of Artificial Intelligence, 26*, 430–445.

Rao, R. V., & Savsani, V. J. (2012). *Mechanical design optimization using advanced optimization techniques*. London: Springer. ISBN 978-1-4471-2747-5.

Rao, R. V., Savsani, V. J., & Vakharia, D. P. (2011). Teaching–learning-based optimization: A novel method for constrained mechanical design optimization problems. *Computer-Aided Design, 43*, 303–315.

Rao, R. V., Savsani, V. J., & Balic, J. (2012a). Teaching–learning-based optimization algorithm for unconstrained and constrained real-parameter optimization problems. *Engineering Optimization, 44*(12), 1447–1462.

Rao, R. V., Savsani, V. J., & Vakharia, D. P. (2012b). Teaching–learning-based optimization: An optimization method for continuous non-linear large scale problems. *Information Sciences, 183*, 1–15.

Ross, S. (1998). *A first course in probability* (5th ed.). Upper Saddle River, New Jersey: Prentice-Hall Inc.

Waghmare, G. (2013). Comments on "A note on teaching–learning-based optimization algorithm". *Information Sciences, 229*, 159–169.

Xing, B., & Gao, W.-J. (2014). *Computational intelligence in remanufacturing*. 701 E. Chocolate Avenue, Suite 200, Hershey PA 17033: IGI Global. ISBN 978-1-4666-4908-8.

Zou, F., Wang, L., Hei, X., Chen, D., & Wang, B. (2013). Multi-objective optimization using teaching-learning-based optimization algorithm. *Engineering Applications of Artificial Intelligence, 26*, 1291–1300. http://dx.doi.org/10.1016/j.engappai.2012.11.006.

Chapter 17
Emerging Biology-based CI Algorithms

Abstract In this chapter, a group of (more specifically 56 in total) emerging biology-based computational intelligence (CI) algorithms are introduced. We first, in Sect. 17.1, describe the organizational structure of this chapter. Then, from Sects. 17.2 to 17.57, each section is dedicated to a specific algorithm which falls within this category, respectively. The fundamentals of each algorithm and their corresponding performances compared with other CI algorithms can be found in each associated section. Finally, the conclusions drawn in Sect. 17.58 closes this chapter.

17.1 Introduction

Several novel biology-based algorithms were detailed in previous chapters. In particular, Chap. 2 detailed the bacteria inspired algorithms, Chap. 3 was dedicated to the bat inspired algorithms, Chap. 4 discussed the bee inspired algorithms, Chap. 5 introduced the biogeography-based optimization algorithm, Chap. 6 was devoted to the cat swarm optimization algorithm, Chap. 7 explained the cuckoo inspired algorithms, Chap. 8 focused on the luminous insect inspired algorithms, Chap. 9 concentrated on the fish inspired algorithms, Chap. 10 targeted on the frog inspired algorithms, Chap. 11 studied the fruit fly optimization algorithm, Chap. 12 addressed the group search optimizer algorithm, Chap. 13 worked on the invasive weed optimization algorithm, Chap. 14 covered the music inspired algorithms, Chap. 15 talked about the imperialist competition algorithm, and Chap. 16 described the teaching-learning-based optimization algorithm. Apart from those quasi-mature biology principles inspired CI methods, there are some emerging algorithms also fall within this category. This chapter collects 56 of them that are currently scattered in the literature and organizes them as follows:

- Section 17.2: Amoeboid Organism Algorithm.
- Section 17.3: Artificial Searching Swarm Algorithm.

B. Xing and W.-J. Gao, *Innovative Computational Intelligence: A Rough Guide to 134 Clever Algorithms*, Intelligent Systems Reference Library 62,
DOI: 10.1007/978-3-319-03404-1_17,

- Section 17.4: Artificial Tribe Algorithm.
- Section 17.5: Backtracking Search Algorithm.
- Section 17.6: Bar Systems.
- Section 17.7: Bean Optimization Algorithm.
- Section 17.8: Bionic Optimization.
- Section 17.9: Blind, Naked Mole-Rats.
- Section 17.10: Brain Storm Optimization Algorithm.
- Section 17.11: Clonal Selection Algorithm.
- Section 17.12: Cockroach Swarm Optimization Algorithm.
- Section 17.13: Collective Animal Behaviour.
- Section 17.14: Cultural Algorithm.
- Section 17.15: Differential Search.
- Section 17.16: Dove Swarm Optimization.
- Section 17.17: Eagle Strategy.
- Section 17.18: Fireworks Optimization Algorithm.
- Section 17.19: FlockbyLeader.
- Section 17.20: Flocking-based Algorithm.
- Section 17.21: Flower Pollinating Algorithm.
- Section 17.22: Goose Optimization Algorithm.
- Section 17.23: Great Deluge Algorithm.
- Section 17.24: Grenade Explosion Method.
- Section 17.25: Group Leaders Optimization Algorithm.
- Section 17.26: Harmony Elements Algorithm.
- Section 17.27: Human Group Formation.
- Section 17.28: Hunting Search.
- Section 17.29: Krill Herd.
- Section 17.30: League Championship Algorithm.
- Section 17.31: Membrane Algorithm.
- Section 17.32: Migrating Birds Optimization.
- Section 17.33: Mine Blast Algorithm.
- Section 17.34: Monkey Search Algorithm.
- Section 17.35: Mosquito Host-Seeking Algorithm.
- Section 17.36: Oriented Search Algorithm.
- Section 17.37: Paddy Field Algorithm.
- Section 17.38: Photosynthetic Algorithm.
- Section 17.39: Population Migration Algorithm.
- Section 17.40: Roach Infestation Optimization.
- Section 17.41: Saplings Growing Up Algorithm.
- Section 17.42: Seeker Optimization Algorithm.
- Section 17.43: Self-Organizing Migrating Algorithm.
- Section 17.44: Sheep Flock Heredity Model.
- Section 17.45: Simple Optimization.
- Section 17.46: Slime Mould Algorithm.
- Section 17.47: Social Emotional Optimization Algorithm.
- Section 17.48: Social Spider Optimization Algorithm.

- Section 17.49: Society and Civilization Algorithm.
- Section 17.50: Stem Cells Optimization Algorithm.
- Section 17.51: Stochastic Focusing Search Algorithm.
- Section 17.52: Swallow Swarm Optimization.
- Section 17.53: Termite-hill Algorithm.
- Section 17.54: Unconscious Search.
- Section 17.55: Wisdom of Artificial Crowds.
- Section 17.56: Wolf Colony Algorithm.
- Section 17.57: Wolf Pack Search.

The effectiveness of theses newly developed algorithms are validated through the testing on a wide range of benchmark functions and engineering design problems, and also a detailed comparison with various traditional performance leading CI algorithms such as particle swarm optimization (PSO), genetic algorithm (GA), differential evolution (DE), evolutionary algorithm (EA), fuzzy system (FS), ant colony optimization (ACO), and simulated annealing (SA).

17.2 Amoeboid Organism Algorithm

In this section, we will introduce an emerging CI algorithm that is derived from the amoeboid related studies (Reece et al. 2011).

17.2.1 Fundamentals of Amoeboid Organism Algorithm

Amoeboid organism algorithm (AOA) was recently proposed in Zhang et al. (2007, 2013a) and Nakagaki et al. (2000). To implement AOA for find the shortest path problem, the following steps need to be performed (Zhang et al. 2007, 2013a; Nakagaki et al. 2000):

- Step 1: Removing the edges with conductivity equals to zero.
- Step 2: Calculating the pressure of each node based on each node's current conductivity and length which can be obtained through Eq. 17.1 (Zhang et al. 2007, 2013a; Nakagaki et al. 2000):

$$\sum_i \frac{D_{ij}}{L_{ij}}\left(p_i - p_j\right) = \begin{cases} -1 & j = 1 \\ 1 & j = 2 \\ 0 & \text{otherwise} \end{cases} . \tag{17.1}$$

- Step 3: Using the pressure of each node acquired via Step 2 to compute each node's conductivity based on Eq. 17.2 (Zhang et al. 2007, 2013a; Nakagaki et al. 2000:

$$Q_{ij} = \frac{D_{ij}}{L_{ij}}(p_i - p_j), \tag{17.2}$$

where p_i represents the pressure at the node N_i, D_{ij} denotes the conductivity of the edge M_{ij}, and Q_{ij} is used to express the flux through tube M_{ij} from N_i to N_j.

- Step 4: Evaluating the value of each edge's conductivity. If it equals to 1, moving to Step 5; otherwise, jumping to Step 7.
- Step 5: Calculating the next time flux and conductivity based on the current flux and conductivity value via Eq. 17.3 (Zhang et al. 2007, 2013a; Nakagaki et al. 2000):

$$\begin{cases} \sum_i Q_{i1} + I_0 = 0 \\ \sum_i Q_{i1} - I_0 = 0 \\ \frac{d}{dt} D_{ij} = f\left(\left|Q_{ij}\right|\right) - rD_{ij} \end{cases}. \tag{17.3}$$

- Step 6: Returning to Step 1.
- Step 7: Outputting the solution and terminating the algorithm.

17.2.2 Performance of AOA

Six benchmark test problems with various dimensions were employed in Zhang et al. (2013a) to test the performance of the proposed AOA. From the simulation results it can be observed that AOA was able to find the optimal solutions for all cases, in particular, AOA offers better results that are reported so far in the literature.

17.3 Artificial Searching Swarm Algorithm

In this section, we will introduce an emerging CI algorithm that is based on the simulation of the natural biology system.

17.3.1 Fundamentals of Artificial Searching Swarm Algorithm

Artificial searching swarm algorithm (ASSA) was recently proposed in Chen (2009), Chen et al. (2009a, b, c, 2010a). The procedures of implementing ASSA are outlined as below (Chen 2009):

- Step 1: Setting up the parameters, generating the initial population, and evaluating the fitness value.
- Step 2: Dealing with the individual swarm member in turn as follows: Moving toward the calling peer by one step if a signal is received from such peer; otherwise implementing the reconnaissance mechanism. Sending a signal to other peers if a better is found; otherwise moving one step randomly.
- Step 3: Calculating the fitness value and comparing it with the best value found so far.
- Step 4: Checking whether the terminating criterion is met. If yes, stopping the algorithm; otherwise, going back to Step 2.

17.3.2 Performance of ASSA

Chen (2009) tested the ASSA on a typical optimal design optimization problem for the purpose of verifying its effectiveness. The preliminary experimental results showed that ASSA outperforms GA and offers better solution quality. Chen (2009) claimed at the end of the study that the small swarm size will help ASSA to achieve a good searching capability.

17.4 Artificial Tribe Algorithm

In this section, we will introduce an emerging CI algorithm that is inspired by the natural tribe's survival mechanism (Magstadt 2013).

17.4.1 Fundamentals of Artificial Tribe Algorithm

Artificial tribe algorithm (ATA) was recently proposed in Chen et al. (2012). The basic inspiration of ATA is renewing the tribe through the strategies of propagation and migration, and relocating the tribe by moving to a better living environment if the current one is getting worse. The two unique characteristics of ATA make it different to other popular swarm intelligence techniques: First, if the present living condition is good, the tribe will tend to propagate, through propagation strategy, the nest generation which is similar to the feature found in genetic algorithm; Second, on the contrary, if the current living situation is bad, the tribe will intend to relocate, by using migration strategy, to another place. Once they are settled, the tribe will continue to propagate. This feature of ATA and the position changing policy used in PSO are alike. Built upon the aforementioned concepts, the running flow of ATA can be described as follows (Onwubolu 2006; Chen et al. 2006, 2012; Coelho and Bernert 2009):

- Step 1: Setting parameters, initializing the tribe, and computing the fitness value.
- Step 2: Adding one to iteration counter, evaluating the current living condition of the tribe, and making decisions according to a simple rule (i.e., if living condition is good, then propagation; otherwise, migration).
- Step 3: Calculating the fitness value.
- Step 4: Determining whether the terminating criteria is met (if so, then stopping the iteration; otherwise, returning to Step 2).

17.4.2 Performance of ATA

Seven benchmark test functions were employed in Chen et al. (2012) to test the performance of the proposed ATA. From the simulation results it can be observed that the tribe size is an important factor for a successful implementation of ATA. In general, the larger size we set for a tribe, the better performance we can obtain but with the cost of a reduced ATA's efficiency. On the other hand, the ATA is able to run fast with a small tribe size but which unfortunately results in low population diversity.

17.5 Backtracking Search Algorithm

In this section, we will introduce an emerging CI algorithm that simulates the movement exhibited by an migrating organism, namely, Brownian-like random-walk (Bolstad 2012; Durrett 1984; Shlesinger et al. 1999).

17.5.1 Fundamentals of Backtracking Search Algorithm

Backtracking search algorithm (BSA) was originally proposed in Civicioglu (2013). The motivation of developing BSA is to design simpler and more effective search algorithms. Therefore, unlike many other optimization algorithms, BSA has only one controlling variable and its initial value also does affect the BSA's overall problem-solving ability. To implement BSA, the following five processes need to be performed (Civicioglu 2013):

- Process 1: Initialization. In BSA, the initial population P can be defined through Eq. 17.4 (Civicioglu 2013):

$$P_{i,j} \sim U\left(low_j, up_j\right), \quad i = 1, 2, \ldots, N \text{ and } j = 1, 2, \ldots, D, \tag{17.4}$$

where the population size and problem dimension are denoted by N and D, respectively, U represents a uniform distribution, and P_i stands for a target individual in the population P.

- Process 2: Selection-I. In BSA, the historical population *oldP* is determined at this stage for computing the search direction. The initial historical population is computed through Eq. 17.5 (Civicioglu 2013):

$$oldP_{i,j} \sim U\left(low_j, up_j\right), \quad i = 1, 2, \ldots, N \text{ and } j = 1, 2, \ldots, D. \tag{17.5}$$

At the start of each iteration, an *oldP* redefining mechanism is introduced in BSA through the if-then rule defined by Eq. 17.6 (Civicioglu 2013):

$$\text{if } a < b \text{ then } oldp := P|a, b \sim U(0, 1), \tag{17.6}$$

where := denotes the updating operation.

- Process 3: Mutation. At this stage, the initial form of the trial population *Mutant* is created by Eq. 17.7 (Civicioglu 2013):

$$Mutant = P + F \cdot (oldP - P). \tag{17.7}$$

- Process 4: Crossover. The final form of the trial population T is generated at this stage.
- Process 5: Selection-II. At this step, a set of T_{iS} which have better fitness values than the corresponding P_{iS} are utilized to renew the P_{iS} according to a greedy selection mechanism.

17.5.2 Performance of BSA

To verify the proposed BSA, Civicioglu (2013) employed 3 test function sets in which the Set-1 involves 50 widely recognized benchmark functions, the Set-2 contains 25 benchmark problems that used in CEC 2005, and the Set-3 consists of three real-world cases used in CEC 2011. Through a detailed comparison and analysis, the results showed that BSA can solve a greater number of benchmark problems and can offer statistically better outcomes than its competitors.

17.6 Bar Systems Algorithm

In this section, we will introduce an emerging CI algorithm that is based on a common phenomenon observed from human social life (Ramachandran 2012a, b, c; Carlson 2013).

17.6.1 Fundamentals of Bar Systems Algorithm

Bar systems (BSs) algorithm was recently proposed in Acebo and Rosa (2008). The BSs algorithm was inspired by the social behaviour of the staffs or bartenders, and can be enclosed in the broader class of swarm intelligence. In the bar, bartenders have to act in a highly dynamic, asynchronous and time-critical environment, and no obvious greedy strategy (such as serving first the best customer, serving first the nearest customer or serving first the customer who has arrived first) gives good results (Acebo and Rosa 2008). Thus, the multi-agent system provides a good framework to rise to the challenge of developing a new class of adaptive and robustness systems.

In general, the crucial step in BSs algorithm is the choice of the task which the agent has to execute for the next time step. In BSs, acting as bartenders, agents operate concurrency into the environment in a synchronous manner; execute the task where they should pour the drinks. After an initial phase, the "bartenders" make their decisions according to the different problem-dependent properties (e.g., weight, speed, location, response time, maximum load, etc.), instead of making decisions randomly. Over time, if an agent is unable to adapt the environment to the preconditions of the task (such as the cost for agent to execute the task in the current state of the environment) or if it is unable to carry the task out by itself then it will be eliminated. Briefly, the BSs algorithm can be defined as a quadruple (E, T, A, F) where (Acebo and Rosa 2008):

- E is a (physical or virtual) environment. The state of the environment at each moment is determined by a set of state variables (V_E). One of those variables is usually the time, due to the major objective of bartenders is to keep the customers waiting for a shorter time. The set of all possible states of the environment is defined as S which is the set of all the possible simultaneous instantiations of the set of state variables (V_E).
- $T = \{t_1, t_2, \ldots, t_M\}$ is a set of tasks to be accomplished by the agents within the environment. Each task (t_i) has associated: $pre(t_i)$ denotes a set of preconditions over V_E which determine whether the task (t_i) can be done; $imp(t_i)$ stands for a non-negative real value which reflects the importance of the task (t_i); and $urg(t_i)$ denotes a function of V_E which indicates the urgency of task (t_i) in the current state of the environment. It will usually be a non-decreasing function of time.
- $A = \{a_1, a_2, \ldots, a_N\}$ is a set of agents situated into the environment. Each agent (a_i) can have different objective (e.g., weight, speed, location, response time, maximum load, etc.). A cost, $\text{cost}(a_i, t_i)$, is associated with each agent. If an agent is unable to adapt the environment to the preconditions of the task or if it is unable to carry the task out by itself, then the $\text{cost}(a_i, t_i)$ will be defined as infinite. In general, this cost can be divided in two parts: the cost for a_i to make the environment fulfil the preconditions of task (t_i), usually this can include the cost of stop doing his current tasks; and the cost for a_i to actually execute t_j.
- $F : S \times A \times T \to \mathbf{R}$ is the function which reflects the degree to which agents are "attracted" by tasks. Overall, given a state of the environment, an agent and a

task, $F(s, a_i, t_i)$, must be defined in a way such that it increases with $imp(t_i)$ and $urg(t_i)$ and it decreases with $\text{cost}(a_i, t_i)$.

17.6.2 Performance of BSs

At the end of their work, Acebo and Rosa (2008) tested the applicability and efficiency of the proposed BSs algorithm on a NP-hard problem in which a group of loading robots in a commercial harbour has to be well scheduled so that all required containers are transported to the targeted ship while keeping the transportation cost as low as possible. The experiments results indicated that BSs can provide much better results than other greedy algorithms.

17.7 Bean Optimization Algorithm

In this section, we will introduce an emerging CI algorithm that is based on some studies of bean (Marcus 2013; Sizer and Whitney 2014; Maathuis 2013; Reece et al. 2011).

17.7.1 Fundamentals of Bean Optimization Algorithm

Bean optimization algorithm (BeOA) was recently proposed in Zhang et al. (2008b, 2013b, c). It has shown good performance in solving some difficult optimization problems such as travelling salesman problem (Zhang et al. 2012a; Li 2010) and scheduling problem (Zhang et al. 2010; Wang and Cheng 2010).

Just like other CI algorithms, a potential solution of problem space is firstly encoded into BeOA representation of search space. Situation of each individual bean can thus be expressed as vector like $X = \{x_1, x_2, x_3, \ldots, x_n\}$ indicating the current state of each bean, where n is determined by the scale of problem to be resolved. The environment in which the beans are sown is mainly the solution space and the states of other beans. The basic equation of implementing BeOA is shown in Eq. 17.8 (Zhang et al. 2012a):

$$X[i] = \begin{cases} X[i] & \text{if } X[i] \text{ is a father bean} \\ X_{mb} + \text{Distribution}(X_{mb}) \cdot A & \text{if X[i] is not a father bean} \end{cases}, \tag{17.8}$$

where $X[i]$ is the position of bean i, X_{mb} is the position of the father bean. $\text{Distribution}(X_{mb})$ is the random variable with a certain distribution of father bean in order to get the positions of its descendants. Parameter A can be set according to the range of the problem to be resolved.

In addition, when the descendant beans finished locating, their fitness values are to be evaluated. The beans with most optimal fitness value will be selected as the candidates of father beans in the next generation. The candidates of father beans should also satisfy the condition that the distance between every two father beans should be larger than the distance threshold. This condition assures that the father beans can have a fine distribution to avoid premature convergence and enhance the performance of the BeOA for global optimization. If all the conditions can be satisfied, the candidate can be set as the father bean for next generation.

17.7.2 Performance of BeOA

In general, the BeOA shares many common points inspired from models of the natural evolution of species. For example, they are population-based algorithms that use operators inspired by population genetics to explore the search space (the most typical genetic operators are reproduction, mutation, and crossover). In addition, they update the population and search for the optimum with random techniques. Differences among the different biology-based CI algorithms concern the particular representations chosen for the individuals and the way genetic operators are implemented. For example, unlike GA, BeOA does not use genetic operators like mutation, they update themselves with distance threshold.

17.8 Bionic Optimization Algorithm

In this section, we will introduce an emerging CI algorithm that is based on studies related to the bionic research (Levin 2013a, b, c, d, e, f).

17.8.1 Fundamentals of Bionic Optimization Algorithm

Bionic optimization (BO) algorithm was recently proposed in Song et al. (2013) for dealing with turbine layout optimization problem in a wind farm. The core concept of BO is to treat each turbine as an individual bion, attempting to be repositioned where its own power outcomes can be increased. There are several BO related studies available in the literature (Zang et al. 2010; Steinbuch 2011; Wei 2011). In Song et al. (2013), the authors defined the BO as a two-stage optimization process in which the Steps 1–6 are included in the Stage 1 and the Stage 2 contains the Steps 7–11. The detailed descriptions about each corresponding step are provided as below (Song et al. 2013):

- Step 1: When a turbine is being added to an existing wind farm, an evaluation function will be employed to assess each discretized points for the newly introduced turbine. In Song et al. (2013), the evaluation function is defined by Eq. 17.9:

$$E(\mathbf{x}) = -\frac{P(\mathbf{x})}{P_{\max}} + \sum_{i=1}^{N} D(\|\mathbf{x} - \mathbf{x}_i\|). \tag{17.9}$$

Calculating $u(\mathbf{x})$ through Eq. 17.10 to obtain the flow field for empty layout (Song et al. 2013).

$$P(\mathbf{x}) = F(u'(\mathbf{x})) = F(u(\mathbf{x})[1 - \beta c(\mathbf{x})]). \tag{17.10}$$

- Step 2: Computing the evaluation values for all the discretized points through Eq. 17.11 (Song et al. 2013):

$$E(\mathbf{x}) = -\frac{P(\mathbf{x})}{P_{\max}} + \sum_{i=1}^{N} D(\|\mathbf{x} - \mathbf{x}_i\|). \tag{17.11}$$

- Step 3: Adding a turbine at the point where the evaluation value is the least.
- Step 4: Terminating the Stage 1 if the turbine numbers pass a specified boundary.
- Step 5: Through the particle model mechanism, simulating the wake flow for all turbines and computing $c(\mathbf{x})$ through Eq. 17.12 (Song et al. 2013):

$$P(\mathbf{x}) = F(u'(\mathbf{x})) = F(u(\mathbf{x})[1 - \beta c(\mathbf{x})]). \tag{17.12}$$

- Step 6: Going back to Step 2.
- Step 7: Since the wake flow created by the later added turbines could still influence the former existing turbines, there is a necessity to further optimize the layout. At this step, one turbine with the same order as in the adding process will be removed.
- Step 8: Calculating the wake flow through the particle model mechanism.
- Step 9: Computing the evaluation values for all points.
- Step 10: Re-adding a turbine into the layout at the point with the least evaluation value.
- Step 11: Going back to Step 7.

17.8.2 Performance of BO

In BO, the layout adjustment strategy within each step is controlled by the evaluation function without any randomness which make BO require much less computational time in comparison with other CI algorithm, e.g., GA and PSO. Through several case studies such as flat terrain scenario, complex terrain scenario, and grid dependency of time cost context, Song et al. (2013) claimed at the end of their study that, for the considered cases, the BO produced better solution quality, in particular for complex terrain case.

17.9 Blind, Naked Mole-Rats Algorithm

In this section, we will introduce an emerging CI algorithm that is based on the blind naked mole-rats' social behaviour in looking for food resources and preventing the whole colony from the potential invasions (Mills et al. 2010).

17.9.1 Fundamentals of Blind, Naked Mole-Rats Algorithm

Blind, naked mole-rats (BNMR) algorithm was recently proposed by Taherdangkoo et al. (2012a). For the purpose of simplification, the BNMR algorithm does not distinguish the soldier moles from the employed moles, i.e., these two types of moles are simply placed in one single group which is called employed moles in BNMR. To implement BNMR algorithm, the following steps need to be performed (Taherdangkoo et al. 2012a):

- First, randomly generating the initial population of the blind naked mole-rats colony across the whole problem space. In BNMR, the number of the population is designed twice as much as the food resources where each of the food resources denotes a response for target problem space. According to Taherdangkoo et al. (2012a), some parameters can be defined by Eq. 17.13:

$$x_i = x_i^{\min} + \beta\left(x_i^{\max} - x_i^{\min}\right), \quad i = 1, \ldots, S, \tag{17.13}$$

  where x_i denotes the ith food source, β represents a random variable which falls within $[0, 1]$, and S is the total number of food sources.
- In addition, the underground temperature is also taken into account as defined by Eq. 17.14 (Taherdangkoo et al. 2012a):

$$\begin{aligned} &H(x) = \rho(x)C(x)\tfrac{\Delta T(x,t)}{\Delta t} \\ &\rho C = f_s(\rho C)_s + f_a(\rho C)_a + f_w(\rho C)_w, \\ &f_s + f_a + f_w = 1 \end{aligned} \tag{17.14}$$

  where $H(x)$ stands for the soil temperature which changes with the depth x, $\rho(x)$ and $C(x)$ denotes the soil's thermal properties (e.g., the density and the specific heat capacity). Although $\rho(x)$ and $C(x)$ are variables vary with the changing of environment, in BNMR, they are treated as constant which falls within $[2, 4]$, $\Delta T(x, t)/\Delta t$ is the rate of the soil temperature varying with the time, f stands for the volumetric contribution of each element in the compound, and the three subscripts (i.e., s, a, and w) indicate the soil components (e.g., sand, air, and water).
- During the search of neighbours for food sources, the attenuation coefficient A has to be updated in each iteration. The Eq. 17.15 is used to express such fact (Taherdangkoo et al. 2012a):

$$A_i^t = A_i^{t-1}\left[1 - \exp\left(\frac{-\alpha t}{T}\right)\right], \tag{17.15}$$

where α denotes a random number which falls within $[0, 1]$ (in BNMR, a fixed value of $\alpha = 0.95$ is employed for simplicity), and t represents the iteration step.

- Then, for each food source, two employed moles will be dispatched. The acquired food sources are grouped by queen mole according to the probability of P which is calculated via Eq. 17.16 (Taherdangkoo et al. 2012a):

$$P_i = \frac{Fitness_i = FS_i \times R_i}{\sum_{j=1}^{N} Fitness_j}, \tag{17.16}$$

where $Fitness_i$ is assessed by its employed moles, FS_i is relative to the best food sources, R_i represents the route to the food source, and N stands for the food sources number.

- Finally, BNMR algorithm also takes the colony defence into account which is calculated through Eq. 17.17 (Taherdangkoo et al. 2012a):

$$B_i^t = \zeta \times B_i^{t-1}, \tag{17.17}$$

where ζ is a user defined coefficient ($\zeta \geq 1$), and B_i^t denotes the number of eliminated points for the ith food source during the tth iteration.

17.9.2 Performance of BNMR

In order to show how the BNMR algorithm performs, Taherdangkoo et al. (2012a) used 24 benchmark test functions such as Shifted Sphere function, Shifted Rotated High Conditioned Elliptic Function, Shifted Rosenbrock's Function, Shifted Rotated Griewank's Function without Bounds, and Shifted Rastrigin's Function. Compared with other CI techniques (e.g., GA, PSO, SA, etc.), the BNMR algorithm has better convergence than its competitive algorithms which demonstrates that BNMR is capable of getting out of local minimum in the problem space and reaching the global minimum.

17.10 Brain Storm Optimization Algorithm

In this section, we will introduce an emerging CI algorithm that is based on the outputs of human brain related research (Gross 2014; Wilson 2013; Taylor 2012).

17.10.1 Fundamentals of Brain Storm Optimization Algorithm

Brain storm optimization algorithm (BSOA) was recently proposed by Shi (2011a). Since the human beings are among on of the most intelligent social animals on earth, the BSO was engineered to have the ability of both convergence and divergence. The process of brainstorming (or brainwaving) is often utilized in dealing with a set of complicated problems which are not always solvable for an individual person. A detailed description about the natural human being brain storm process can be found in Shi (2011b). A typical brain storm process generally follows the eight steps (Shi 2011b; Xue et al. 2012; Zhan et al. 2012; Zhou et al. 2012; Krishnanand et al. 2013):

- Step 1: Getting together a brainstorming group of people with as diverse background as possible.
- Step 2: Generating many ideas according to the four principles (i.e., suspend judgment, anything goes, cross-fertilize, and go for quantity) of idea generation guidance.
- Step 3: Having several customers act as the owners of the problem to pick up a couple of ideas as better ideas for solving the targeted problem.
- Step 4: Using the fact that the ideas (selected in Step 3) enjoy a higher chosen probability than their competitor ideas as an evidence to generate more ideas based again on the four principles.
- Step 5: Having the customers to select several better ideas again as they did in Step 3.
- Step 6: Picking up an object randomly and using the intrinsic characteristics of the object as the indication to create more ideas (still based on the four principles of idea generation guidance).
- Step 7: Letting the customers choose several better ideas as they did in Step 3.
- Step 8: Obtaining a fairly good enough problem solution at the end of the brain storm process.

Although the three-round brain storm process, participated by a group of real human beings, can not last for too long, in a computer simulation environment, we can set the round of idea generation to a very large number as we desire.

17.10.2 Performance of BSOA

To test the performance of BSOA, Shi (2011b) chose ten benchmark functions (among them, five are unimodal functions, while the other five are multimodal functions). The simulation results indicated that BSOA algorithm performed reasonably well.

17.11 Clonal Selection Algorithm

In this section, we will introduce an emerging CI algorithm that is based on Darwin's evolutionary theory and clone related studies (Gamlin 2009; Mayfield 2013; Woodward 2008; Steinitz 2014).

17.11.1 Fundamentals of Clonal Selection Algorithm

Clonal selection algorithm (CSA) was recently proposed in Castro and Zuben (2000). There are several CSA related variants and applications can be found in the literature (Castro and Zuben 2002; Campelo et al. 2005; Gao et al. 2013; Wang et al. 2009; Batista et al. 2009; Ding and Li 2009; Riff et al. 2013). Interested readers are referred to two excellent reviews (Brownlee 2007; Ulutas and Kulturel-Konak 2011) for updated information. To implement CSA, the following steps need to be performed (Castro and Zuben 2000):

- Step 1: Creating a set of candidate solutions (denoted by P), composing of the subset of memory cells (represented by M), and adding to the remaining population (P_r), i.e., $P = P_r + M$.
- Step 2: According to an affinity measure, choosing the n best individuals of the population, named P_n.
- Step 3: Cloning the population of these n best individuals and giving rise to a intermediate population clones, called C. The clone size is regarded as an increasing function of the affinity with the antigen.
- Step 4: Submitting the population of clones to a hypermutation mechanism. A maturated antibody population is then generated and denoted by C^*.
- Step 5: Reselecting the improved individuals from C^* to compose the memory set, i.e., M.
- Step 6: Replacing d andibodies by novel ones (introduced through diversity strategy). In CSA, the replacement probability of lower affinity cells is in general high.

17.11.2 Performance of CSA

To verify the CSA, three problem sets are considered in Castro and Zuben (2000), namely, binary character recognition task, multimodal optimization problem, and the classic travelling salesman problem. In comparison with GA, the simulation results demonstrated that CSA is a very promising CI algorithm which has showed a fine tractability regarding the computational cost.

17.12 Cockroach Swarm Optimization Algorithm

In this section, we will introduce an emerging CI algorithm that is based on the behaviours observed through cockroach studies (Bell et al. 2007; Lihoreau et al. 2010; Lihoreau et al. 2012; Chapman 2013; Bater 2007; Reece et al. 2011).

17.12.1 Fundamentals of Cockroach Swarm Optimization Algorithm

Cockroach swarm optimization algorithm (CSOA) was recently proposed in Chen and Tang (2010, 2011; Cheng et al. 2010). The basic concept of CSOA is that located in the D-dimensional search space R^D, there is a swarm of cockroaches which contains N cockroach individuals. The ith individual denotes a D-dimensional vector $\mathbf{X}(i) = (x_{i1}, x_{i2}, \ldots, x_{iD})$ for $(i = 1, 2, \ldots, N)$, the location of each individual is a potential solution to the targeted problem. The model of CSOA consists of three behaviours, namely, chase-swarming, dispersing, and ruthless which are explained as below (Chen and Tang 2010, 2011; Cheng et al. 2010).

- Chase-swarm behaviour: Each individual cockroach $\mathbf{X}(i)$ will run after (within its visual range) a cockroach $\mathbf{P}(i)$ which carries the local optimum. This behaviour is modelled as Eq. 17.18 (Chen and Tang 2010, 2011; Cheng et al. 2010):

$$\mathbf{X}'(i) = \begin{cases} \mathbf{X}(i) + step \cdot rand \cdot [\mathbf{P}(i) - \mathbf{X}(i)] & \text{if } \mathbf{X}(i) \neq \mathbf{P}(i) \\ \mathbf{X}(i) + step \cdot rand \cdot [\mathbf{P}_g - \mathbf{X}(i)] & \text{if } \mathbf{X}(i) = \mathbf{P}(i) \end{cases}, \tag{17.18}$$

  where $\mathbf{P}_g = Opt_i\{\mathbf{X}(i),\ i = 1, \ldots, N\}$ denotes the global optimum individual cockroach, $\mathbf{P}(i) = Opt_j\{\mathbf{X}(j) \| \mathbf{X}(i) - \mathbf{X}(j) \| \leq visual,\ i = 1, \ldots, N \text{ and } j = 1, \ldots, N\}$, *step* represents a fixed value, and *rand* stands for a random number within the interval of [0, 1].
- Dispersing behaviour: During a certain time interval, each individual cockroach will be randomly dispersed for the purpose of keeping the diversity of the current swarm. This behaviour is modelled through Eq. 17.19 (Chen and Tang 2010, 2011; Cheng et al. 2010):

$$\mathbf{X}'(i) = \mathbf{X}(i) + rand(1, D), \quad i = 1, \ldots, N, \tag{17.19}$$

  where $rand(1, D)$ is a D-dimensional random vector which falls within a certain interval.
- Ruthless behaviour: At a certain time interval, the cockroach which carries the current best value substitute another cockroach in a randomly selection manner.

This behaviour is modelled through Eq. 17.20 (Chen and Tang 2010, 2011; Cheng et al. 2010):

$$\mathbf{X}(k) = \mathbf{P}_g, \tag{17.20}$$

where k is a random integer within the interval of $[1, N]$.

Built on these three behaviours, the working procedure of the CSOA algorithm can be classified into the following steps (Chen and Tang 2010, 2011; Cheng et al. 2010):

- Step 1: Setting parameters and initializing population;
- Step 2: Search $\mathbf{P}(i)$ and $\mathbf{P}_g$;
- Step 3: Performing chase-swarming and updating $\mathbf{P}_g$;
- Step 4: Executing dispersing behaviour and updating $\mathbf{P}_g$;
- Step 5: Running ruthless behaviour;
- Step 6: Checking stopping criterion. If yes, generate output; otherwise, go back to step 2.

17.12.2 Performance of CSOA

In Chen and Tang (2011), the authors made an attempt to employ CSOA to solve vehicle routing problem (VRP), more specifically, the VRP with time windows (VRPTW for short). In general the VRPTW can be stated as follows: Products are to be delivered to a group of customers by a fleet of vehicles from a central depot. The locations of the depot and the customers are known. The object is to find a suitable route which minimizes the total travel distance or cost subject to the constraints listed below.

- Each customer is visited only once by exactly one vehicle;
- Each vehicle has the fixed starting and ending point (i.e., the depot);
- The vehicles are capacitated which means the total demand of any route should not exceed the maximum capacity of an assigned vehicle;
- The visit to a customer is time restrict, i.e., each customer can only be served during a certain time period.

To test the effectiveness of the CSOA for focal problem, Chen and Tang (2011) conducted a study on VRP and VRPTW separately. The experimental results were compared with PSO and the improved PSO. Through the comparison, the authors claimed that CSOA is able to explore the optimum with higher optimal rate and shorter time.

17.13 Collective Animal Behaviour Algorithm

In this section, we will introduce a new CI algorithm which inspired by the collective decision-making mechanisms among the animal groups (Sulis 1997; Tollefsen 2006; Nicolis et al. 2003; Schutter et al. 2001; You et al. 2009; Couzin 2009; Aleksiev et al. 2008; Stradner et al. 2013; Zhang et al. 2012b, Niizato and Gunji 2011; Oca et al. 2011; Eckstein et al. 2012; Petit and Bon 2010).

17.13.1 Fundamentals of Collective Animal Behaviour Algorithm

Collective animal behaviour (CAB) algorithm was originally proposed by Cuevas et al. (2013). In CAB, each animal position is viewed as a solution within the search space. Also, a set of rules that model the collective animal behaviours will be employed in the proposed algorithm. The main steps of CAB are outlined below (Cuevas et al. 2013):

- Initializing the population. Generate a set $\mathbf{A}$ of N_p animal positions ($\mathbf{A} = \{\mathbf{a}_1, \mathbf{a}_2, \ldots, \mathbf{a}_{N_p}\}$) randomly in the D-dimensional search space as defined by Eq. 17.21 (Cuevas et al. 2013):

$$\begin{gathered} a_{j,i} = a_j^{low} + rand(0,1) \cdot \left(a_j^{high} - a_j^{low}\right) \\ j = 1, 2, \ldots, D;\ i = 1, 2, \ldots, N_p. \end{gathered} , \tag{17.21}$$

  where a_j^{low} and a_j^{high} represent the lower bound and upper bound, respectively, and $a_{j,i}$ is the jth parameter of the ith individual.
- Calculating and sorting the fitness value for each position. According to the fitness function, the best position (B) which is chosen from the new individual set $\mathbf{X} = \{\mathbf{x}_1, \mathbf{x}_2, \ldots, \mathbf{x}_{N_p}\}$ will be stored in a memory that includes two different elements as expressed in Eq. 17.22 (Cuevas et al. 2013):

$$\begin{cases} \mathbf{M}_g: & \begin{array}{l}\text{for maintaining the best found positions}\\ \text{in each generation}\end{array} \\ \mathbf{M}_h: & \begin{array}{l}\text{for storing the best history positions}\\ \text{during the complete evolutionary process}\end{array} \end{cases} . \tag{17.22}$$

- Keep the position of the best individuals. In this operation, the first B elements of the new animal position set $\mathbf{A}(\{\mathbf{a}_1, \mathbf{a}_2, \ldots, \mathbf{a}_B\})$ are generated. This behaviour rule is modelled via Eq. 17.23 (Cuevas et al. 2013):

$$\mathbf{a}_l = \mathbf{m}_h^l + \mathbf{v}, \tag{17.23}$$

where $l \in \{1, 2, \ldots, B\}$ while $\mathbf{m}_h^l$ represents the historic memory $\mathbf{M}_h$, and $\mathbf{v}$ is a random vector holding an appropriate small length.

- Move from or to nearby neighbours. This operation can be defined by Eq. 17.24 (Cuevas et al. 2013):

$$\mathbf{a}_i = \begin{cases} \mathbf{x}_i \pm r \cdot \left(\mathbf{m}_h^{\text{nearest}} - \mathbf{x}_i\right) & \text{with probability } H \\ \mathbf{x}_i \pm r \cdot \left(\mathbf{m}_g^{\text{nearest}} - \mathbf{x}_i\right) & \text{with probability } (1 - H) \end{cases}, \tag{17.24}$$

where $i \in \{B+1, B+2, \ldots, N_p\}$, $\mathbf{m}_h^{\text{nearest}}$ and $\mathbf{m}_g^{\text{nearest}}$ represent the nearest elements of $\mathbf{M}_h$ and $\mathbf{M}_g$ to $\mathbf{x}_i$, respectively, and r is a random number between $[-1, 1]$.

- Move randomly. This rule is defined by Eq. 17.25 (Cuevas et al. 2013):

$$\mathbf{a}_i = \begin{cases} \mathbf{r} & \text{with probability } P \\ \mathbf{x}_i & \text{with probability } (1 - P) \end{cases}, \tag{17.25}$$

where $i \in \{B+1, B+2, \ldots, N_p\}$, and $\mathbf{r}$ is a random vector defined within the search space.

- Updating the memory. The updating procedure is as follows (Cuevas et al. 2013):
  Two memory elements are merged together as shown in Eq. 17.26 (Cuevas et al. 2013):

$$\mathbf{M}_U\left(\mathbf{M}_U = \mathbf{M}_h \cup \mathbf{M}_g\right). \tag{17.26}$$

Based on the parameter (ρ), the elements of the memory $\mathbf{M}_U$ is calculated. The ρ value is computed via Eq. 17.27 (Cuevas et al. 2013):

$$\rho = \frac{\prod_{j=1}^{D}\left(a_j^{high} - a_j^{low}\right)}{10 \cdot D}, \tag{17.27}$$

where a_j^{low} and a_j^{high} represent the pre-specified lower bound and the upper bound, respectively, within a D-dimensional space.

- Optimal determination. It is defined by Eq. 17.28 (Cuevas et al. 2013):

$$\text{Th} = \frac{Max_{fitness}(\mathbf{M}_h)}{6}, \tag{17.28}$$

where Th represents a threshold value that decide which elements will be considered as a significant local minimum, and $Max_{fitness}(\mathbf{M}_h)$ represent the best fitness value among $\mathbf{M}_h$ elements.

17.13.2 Performance of CAB

In order to evaluate the performance of CAB, a set of multimodal benchmark functions were adopted in Cuevas et al. (2013), namely, Deb's function, Deb's decreasing function, Roots function, two dimensional multimodal function, Rastringin's function, Shubert function, Griewank function, and modified Griewank function. Compared with other CI algorithms, computational results showed that CAB outperforms the other algorithms in terms of the solution quality.

17.14 Cultural Algorithm

In this section, we will introduce an CI algorithm that is based on the human social evolution (Mayfield 2013).

17.14.1 Fundamentals of Cultural Algorithm

Cultural algorithm (CA) was originally proposed in Reynolds (1994, 1999). There are several variants and application can be found in the literature (Digalakis and Margaritis 2002; Alexiou and Vlamos 2012; Ochoa-Zezzatti et al. 2012; Srinivasan and Ramakrishnan 2012; Silva et al. 2012). In CA, the evolution process can be viewed as a dual-inheritance system in which two search spaces (i.e., the population space and the belief space) are included.

In general, the population space is used to represent a set of behavioural traits associated with each individual. On the other hand, the belief space is used to describe different domains of knowledge that the population has of the search space and it can be delivered into distinct categories, such as normative knowledge, domain specific knowledge, situational knowledge, temporal knowledge, and spatial knowledge. In other words, the belief space is used to store the information on the solution of the problem.

Furthermore, at each iteration, two functions (i.e., acceptance function and influence function) and two operators (i.e., crossover and mutation) are employed to maintain the CA algorithm. The acceptance function is used to decide which knowledge sources influence individuals. On the other hand, the influence function is used to determine which individuals and their behaviours can impact the belief space knowledge. Also, the crossover and mutation operators are used to support the population space that control the beliefs' changes in individuals.

The main steps of CA can be outlined as follows (Reynolds 1994):

- Step 1: Generate the initial population.
- Step 2: Initialize the belief space. In CA, if only two knowledge components, i.e., situational knowledge component and normative knowledge component are employed, the belief space can be defined by Eq. 17.29 (Reynolds 1994, 1999):

$$B(t) = (S(t), N(t)), \tag{17.29}$$

  where the situational knowledge component is represented by $S(t)$, and $N(t)$ denotes the normative knowledge component.
- Step 3: Evaluate the initial population.
- Step 4: Iterative procedure. First, update the belief space (with the individuals accepted). Second, apply the variation operators (under the influence of the belief space). Third, evaluate each child. Fourth, perform selection.
- Step 5: Check termination criteria.

17.14.2 Performance of CA

To verify CA, a set of studies are conducted in Reynolds (1994). The experiments results demonstrated that CA is indeed a very promising solver for dealing with optimization problems.

17.15 Differential Search Algorithm

In this section, we will introduce an emerging CI algorithm that simulates the movement exhibited by an migrating organism, namely, Brownian-like random-walk (Bolstad 2012; Durrett 1984; Shlesinger et al. 1999).

17.15.1 Fundamentals of Differential Search Algorithm

Differential search (DS) algorithm was originally proposed in Civicioglu (2012). To implement DS, the following features need to be considered (Civicioglu 2012; Sulaiman 2013):

- Feature 1: In DS, a set of artificial organisms making up a super-organism, namely, $Superorganism_g$, $g = \{1, 2, \ldots, \text{maxgeneration}\}$ in which the number of organisms is equivalent to the size of the problem (i.e., $x_{i,j}$, $j = \{1, 2, \ldots, D\}$).
- Feature 2: In DS, a member of a super-organism (i.e., an artificial organism) in its initial position can be defined through Eq. 17.30 (Civicioglu 2012):

$$x_{i,j} = rand \cdot \left(up_j - low_j\right) + low_j, \tag{17.30}$$

where $X_i = \left[x_{i,j}\right]$ represents a group of artificial organism, and the artificial super-organism can thus be expressed by $Superorganism_g = [X_i]$.

- Feature 3: In DS, the movement style for an artificial super-organism finding a stopover site is modelled by Brownian-like random walk. Several randomly chosen individuals within an artificial super-organism move forward to the targets of donor which equals to $\left[X_{random\_shuffling(i)}\right]$ for the purpose of discovering stopover sites which is generated through Eq. 17.31 (Civicioglu 2012):

$$StopoverSite = Superorganism + Scale \cdot (donor - Superorganism). \tag{17.31}$$

- Feature 4: In DS, in order to generate the scale value, a gamma-random number creator (i.e., *randg*) controlled by an uniform-random number creator (i.e., *rand*) and both falling within the range of $[0, 1]$ are employed.
- Feature 5: In DS, the numbers of individual artificial organism to join the stopover site search process are decided in an random manner.
- Feature 6: In DS, if a more fertile stopover site is discovered, a group of artificial organisms will move to the newly founded place, while the artificial super-organism will keep searching.
- Feature 7: There are only two controlling variables (i.e., p_1 and p_2) are used in DS. Through conducting a set of detailed tests, Civicioglu (2012) suggested the following values (see Eq. 17.32) can provide the best solutions for the respective problems.

$$p_1 = p_2 = 0.3 \cdot rand. \tag{17.32}$$

17.15.2 Performance of DS

To verify the proposed DS, Civicioglu (2012) employed two test function sets in which the Test Set-1 consists of 40 benchmark functions (e.g., Shubert function, Stepint function, Trid function, etc.) and the Test Set-2 is composed of 12 benchmark test functions which include such as Shifted Sphere function, Shifted Schwefel's function, Shifted Rastrigin Function, and Shifted Rosenbrock function. In comparison with other 8 widely used optimization algorithms through the use of statistical approaches, the experimental results demonstrated that DS is a very attractive solver for numerical optimization problems. At the end of the study, Civicioglu (2012) further applied DS to the problem of transforming the geocentric cartesian coordinates into geodetic coordinates. Compared with the other 9 classical methodologies and 8 CI algorithms which have been previously reported in dealing with the same problem, the results also confirmed the practicability and high level of accuracy of DS.

17.16 Dove Swarm Optimization Algorithm

In this section, we will introduce an emerging CI algorithm that is based on foraging behaviours observed from a dove swarm (Mills et al. 2010).

17.16.1 Fundamentals of Dove Swarm Optimization Algorithm

Dove swarm optimization (DSO) algorithm was recently proposed in Su et al. (2009). The basic working principles of DSO are listed as follows (Su et al. 2009):

- Step 1: Initializing the number of doves and deploying the doves on the 2-dimensional artificial ground.
- Step 2: Setting the number of epochs ($e = 0$), and the degree of satiety, $f_j^e = 0$ for $j = 1, \ldots, M \times N$. Initializing the multi-dimensional sense organ vector, $\vec{w}_j$ for $j = 1, \ldots, M \times N$.
- Step 3: Computing the total amount of the satiety degrees in the flock, $T(e) = \sum_{j=1}^{M \times N} f_j^e$.
- Step 4: Presenting an input pattern (i.e., piece of artificial crumb) $\vec{x}_k$ to the $M \times N$ doves.
- Step 5: Locating the dove b_f closest to the crumb $\vec{x}_k$ according to the minimum-distance criterion shown in Eq. 17.33 (Su et al. 2009):

$$b_f = \arg \min_j \left\| \vec{x}_k - \vec{w}_j(k) \right\|, \quad \text{for } j = 1, \ldots, M \times N, \tag{17.33}$$

  The dove with the artificial sense organ vector which is the most similar to the artificial crumb, $\vec{x}_k$, is claimed to be the winner.

- Step 6: Updating each dove's satiety degree through Eq. 17.34 (Su et al. 2009):

$$f_j^e(new) = \frac{\left\| \vec{x}_k - \vec{w}_{b_f}(k) \right\|}{\left\| \vec{x}_k - \vec{w}_j(k) \right\|} + \lambda f_j^e(old), \quad \text{for } j = 1, \ldots, M \times N. \tag{17.34}$$

- Step 7: Selecting the dove, b_f, with the highest satiety degree based on the following criterion expressed as Eq. 17.35 (Su et al. 2009):

$$b_s = \arg \max_{1 \le j \le M \times N} f_j^e. \tag{17.35}$$

- Step 8: Updating the sense organ vectors and the position vectors via Eqs. 17.36 and 17.37, respectively (Su et al. 2009):

$$\vec{w}_j(k+1) = \begin{cases} \vec{w}_{b_f}(k) + \eta_w\left(\vec{x}_k - \vec{w}_{b_f}(k)\right) & \text{for } j = b_f \\ \vec{w}_j(k) & \text{for } j \neq b_f \end{cases}, \tag{17.36}$$

$$\vec{p}_j(k+1) = \vec{p}_j(k) + \eta_p \beta\left(\vec{p}_{b_s}(k) - \vec{p}_j(k)\right), \quad \text{for } j = 1, \ldots, M \times N. \tag{17.37}$$

- Step 9: Returning to Step 4 until all patterns are processes.
- Step 10: Stopping the whole training procedure if the following criterion (see Eq. 17.38) is met (Su et al. 2009):

$$\left| \sum_{j=1}^{M \times N} f_j^e - T(e) \right| \leq \varepsilon. \tag{17.38}$$

Otherwise, increasing the number of epochs by one ($e = e + 1$), and go back to Step 3 until the pre-defined limit for the number of epochs is met. The satisfaction of the criterion given above means that the total amount of satiety degree has converged to some extent.

17.16.2 Performance of DSO

In general there are two main obstacles encountered in data clustering: the geometric shapes of the clusters are full of variability, and the cluster numbers are not often known a priori. In order to determine the optimal number of clusters, Su et al. (2009) employed DSO to perform data projection task, i.e., projecting high-dimensional data onto a low-dimensional space to facilitate visual inspection of the data. This process allows us to visualize high-dimensional data as a 2-dimensional scatter plot. The basic idea in their work can be described as follows (Su et al. 2009): In a data set, each data pattern, $\vec{x}$, is regarded as a piece of artificial crumb and these artificial crumbs (i.e., data patterns) will be sequentially tossed to a flock of doves on a two-dimensional artificial ground. The flock of doves adjusts its physical movements to seek these artificial crumbs. Individual members of the flock can profit from discoveries of all of the other members of the flock during the foraging procedure because an individual is usually influenced by the success of the best individual of the flock and thus has a desire to imitate the behaviour of the best individual. Gradually, the flock of the doves will be divided into several groups based on the distributions of the artificial crumbs. Those formed groups will naturally correspond to the hidden data structure in the data set. By viewing the distributions of the doves on the 2-dimensional artificial ground, we may quickly find out the number of clusters inherent in the data set. However, many practical data sets have high-dimensional data points. Therefore, the aforementioned idea has to be generalized so that it can process high-dimensional data. In the real world, each dove has a pair of eyes to find out where crumbs are, but in the artificial world, a virtual dove does not have the capability to perceive a piece of multi-dimensional artificial crumb that is located around it. In order to cope with issue, Su et al. (2009) equipped each dove with functionalities, i.e., a multi-dimensional artificial sense organ represented as a sense organ vector, $\vec{w}$, which has the same dimensionality as a data pattern, $\vec{x}$, and a 2-dimensional position vector, $\vec{p}$, which represents its position on the 2-dimensional artificial ground. In addition to these two vectors, $\vec{w}$ and $\vec{p}$, a parameter called the satiety parameter is also attached to each dove. This special parameter endows a dove with the ability of expressing its present satiety status with respect to the food, that is, a dove with a low degree of satiety will have a strong desire to change its present foraging policy and be more willing to imitate the behaviour of the dove which performs the best among the flock.

To test the performance of DSO, five (two artificial and three real) data sets were selected in the study. These data sets include Two-Ellipse, Chromosomes, Iris, Breast Cancer, and 20-Dimensional Non-Overlapping. The projection capability of DSO was compared with the other popular projection algorithms, e.g., Sammon's algorithm. For DSO, the maximum number of epochs for every data set (excluding Iris and 20-Dimensional data sets) were set to be 5, while for the Iris and 20-Dimensional data sets, were set to be 10 and 20, respectively. The case studies showed that DSO can fulfil the projection task. Meanwhile, the performance of DSO is not so sensitive to the size of dove swarm.

17.17 Eagle Strategy

In this section, we will introduce an emerging strategy or search method that is based on the eagle search (hunting) behaviour.

17.17.1 Fundamentals of Eagle Strategy

Eagle strategy (ES) algorithm was proposed in Yang and Deb (2010, 2012) and Gandomi et al. (2012). It is a two-stage method, i.e., exploring the search space globally using Lévy flight random walks and then employing an intensive local search mechanism for optimization, such as hill-climbing and the downhill simplex method. The main steps of ES can be described as follows (Yang and Deb 2010, 2012; Gandomi et al. 2012):

- Step 1: Initialize the population and parameters.
- Step 2: Iterative procedure. First, perform random search in the global search space defined by Eq. 17.39 (Yang and Deb 2010):

$$\text{Lévy} \sim u = t^{-\lambda}, \quad (1 < \lambda \leq 3), \tag{17.39}$$

where $\lambda = 3$ corresponds to Brownian motion, while $\lambda = 1$ has a characteristics of stochastic tunnelling.

Second, evaluate the objective functions. Third, make an intensive local search with a hypersphere via any optimization technique such as downhill simplex (i.e., Nelder-Mead) method. Fourth, calculate the fitness and keep the best solutions. Fifth, increase the iteration counter. Sixth, calculate means and standard deviations.

- Step 3: Post process results and visualization.

17.17.2 Performance of ES

To evaluate the efficiency of ES, the Ackley function is adopted in Yang and Deb (2010). Compared with other CI algorithms (such as PSO and GA), the results showed that ES outperforms the others in finding the global optima with the success rates of 100 %. As all CI algorithms require a balance between the exploration and exploitation, this strategy can be combined into any algorithms [such as firefly algorithm (Yang and Deb 2010) and DE (Gandomi et al. 2012)] to improve the computational results.

17.18 Fireworks Optimization Algorithm

In this section, we will introduce an emerging CI algorithm that is derived from the explosion process of fireworks, an explosive devices invented by our clever ancestor, which can produce striking display of light and sound (Lancaster et al. 1998).

17.18.1 Fundamentals of Fireworks Optimization Algorithm

Fireworks optimization algorithm (FOA) was recently proposed in Tan and Zhu (2010). The basic idea was when we need to find a point x_j satisfying $f(x_i) = y$, a set of fireworks will be continuously fired in the potential search space until an agent (i.e., a spark in fireworks context) gets to or reasonably close to the candidate point x_j. Based on this understanding, to implement FOA algorithm, the following steps need to be performed (Janecek and Tan 2011; Pei et al. 2012; Tan and Zhu 2010):

- Step 1: Fireworks explosion process designing. Since the number of sparks and their coverage in the sky determines whether an explosion is good or not, Tan and Zhu (2010) first defined the number of sparks created by each firework x_j through Eq. 17.40:

$$s_i = m \cdot \frac{y_{\max} - f(x_i) + \xi}{\sum_{i=1}^{n} [y_{\max} - f(x_i)] + \xi}, \tag{17.40}$$

where m is a parameter used to control the total number of sparks created by the n fireworks, $y_{\max} = \max(f(x_i))$ (for $i = 1, 2, \ldots, n$) stands for the maximum value of the objective function among the $y_{\max}$ fireworks, and ξ represents a small constant which is used to avoid zero-division-error. Meanwhile, in order to get rid of the overwhelming effects of the splendid fireworks, bounds s_i are also defined by Eq. 17.41 (Tan and Zhu 2010):

$$\hat{s}_i = \begin{cases} round(a \cdot m) & \text{if } s_i < am \\ round(b \cdot m) & \text{if } s_i > bm,\ a < b < 1 \\ round(s_i) & \text{otherwise} \end{cases}, \tag{17.41}$$

where a and b are constant parameters.

Next, Tan and Zhu (2010) also designed the explosion amplitude via Eq. 17.42:

$$A_i = \hat{A} \cdot \frac{f(x_i) - y_{\min} + \xi}{\sum_{i=1}^{n} [f(x_i) - y_{\min}] + \xi}, \tag{17.42}$$

where $\hat{A}$ represents the maximum amplitude of an explosion, and $y_{\min} = \min(f(x_i))$ (for $i = 1, 2, \ldots, n$) denotes the minimum value of the objective function among the n fireworks.

Finally, the directions of the generated sparks are computed using Eq. 17.43 (Tan and Zhu 2010):

$$z = round(d \cdot rand(0, 1)), \tag{17.43}$$

where d denotes the dimensionality of the location x, and $rand(0, 1)$ represents an uniformly distributed number within $[0, 1]$.

- Step 2: In order to obtain a good implementation of FOA, the locations of where we want the fireworks to be fired need to be chosen properly. According to Tan and Zhu (2010), the general distance between a location x and other locations can be expressed as Eq. 17.44:

$$R(x_i) = \sum_{j \in K} d(x_i, x_j) = \sum_{j \in K} \left\| x_i - x_j \right\|, \tag{17.44}$$

  where K denotes a group of current locations of all fireworks and sparks. The selection probability of a location x_i is then defined via Eq. 17.45 (Tan and Zhu 2010):

$$p(x_i) = \frac{R(x_i)}{\sum_{j \in K} R(x_j)}. \tag{17.45}$$

17.18.2 Performance of FOA

To validate the performance of the proposed FOA, 9 benchmark test functions were chosen by Tan and Zhu (2010) and the comparisons were conducted among the FOA, the standard PSO, and the clonal PSO. The experiment results indicated that the FA clearly outperforms the other algorithms in both optimization accuracy and convergence speed.

17.19 FlockbyLeader Algorithm

In this section, we will introduce an emerging CI algorithm that is based on the leadership pattern found in flocks of pigeon birds (Couzin et al. 2005; Giraldeau et al. 1994).

17.19.1 Fundamentals of FlockbyLeader Algorithm

The FlockbyLeader algorithm was proposed by Bellaachia and Bari (2012) in which the recently discovered leadership dynamic mechanisms in pigeon flocks are incorporated in the normal flocking model [i.e., Craig Reynolds' Model (Reynolds 1987)]. In every iteration, the algorithm starts by finding flock leaders. The main steps are illustrated as follows (Bellaachia and Bari 2012):

- Calculating fitness value of each flock leader (L_i) according to the objective function (i.e., $d_{\max}^{L_i}$). It will be defined by Eq. 17.46 (Bellaachia and Bari 2012):

$$d_{\max}^{L_i} = \max_{o \in kNB_t(x_i)} \{\rho(x_i, o)\}, \tag{17.46}$$

  where $kNB_t(x_i)$ is the k-neighbourhood of x_i at iteration t, $d_{\max}^{L_i}$ as radius associated with leader L_i at iteration t, x_i is a node in the feature graph, and $\rho(x_i, o)$ is the given distance function between objects x_i and o.
- Ranking the LeaderAgent (A_i). This procedure is defined by Eqs. 17.47–17.49, respectively (Bellaachia and Bari 2012):

$$Rank_t(A_i) = Log\left(\frac{|N_{i,t}|}{|N_t|} * 10\right) * ARF_t(A_i), \tag{17.47}$$

$$ARF_t(A_i) = \frac{|DR\_kNB_t(x_i)|}{|DR\_kNB_t(x_i)| + |D\_kNB_t(x_i)|}, \tag{17.48}$$

$$\begin{cases} \text{if } ARF_t(A_i) \geq 0.5, & \text{then } x_i \text{ is a flockleader} \\ \text{if } ARF_t(A_i) < 0.5, & \text{then } x_i \text{ is a follower} \end{cases}, \tag{17.49}$$

where $DR\_kNB_t(x_i)$ represents the dynamic reverse k-neighbourhood of x_i at iteration t, $ARF_t(A_i)$ is the dynamic agent role factor of the agent A_i at iteration t, $|N_{i,t}|$ is the number of the neighbours A_i at iteration t, and $|N_t|$ is the number of unvisited nodes at iteration t.
- Performing the flocking behaviour.
- Updating the FindFlockLeaders (G_f).

17.19.2 Performance of FlockbyLeader

To test the efficiency of the proposed algorithm, two large datasets that one is consists of 100 news articles collected from cyberspace, and the other one is the iris plant dataset were adopted by Bellaachia and Bari (2012). Compared with other CI algorithms, the proposed algorithm is significant improve the results.

17.20 Flocking-based Algorithm

In this section, we will introduce an emerging CI algorithm that is derived from the emergent collective behaviour found in social animal or insects (Lemasson et al. 2009; Ballerini et al. 2008; Luo et al. 2010; Kwasnicka et al. 2011).

17.20.1 Fundamentals of Flocking-based Algorithm

Flocking-based algorithm (FBA) was originally proposed in Cui et al. (2006), Picarougne et al. (2007) and Luo et al. (2010). The basic flocking model is composed of three simple steering rules (see below) that need to be executed at each instance over time, for each individual agent.

- Rule 1: Separation. Steering to avoid collision with other boids nearby.
- Rule 2: Alignment. Steering toward the average heading and speed of the neighboring flock mates.
- Rule 3: Cohesion. Steering to the average position of the neighboring flock mates.
- In the proposed algorithm, a fourth rule is added as below:
- Rule 4: Feature similarity and dissimilarity rule. Steering the motion of the boids with the similarity among targeted objects.

All these four rules can be formally express by the following equations (Cui et al. 2006):

- The function of separation rule is to act as an active boid trying to pull away before crashing into each other. The mathematical implementation of this rule is thus can be described by Eq. 17.50 (Cui et al. 2006):

$$d(P_x, P_b) \leq d_2 \Rightarrow \vec{v}_{sr} = \sum_x^n \frac{\overline{\vec{v}_x + \vec{v}_b}}{d(P_x, P_b)}, \tag{17.50}$$

where v_{sr} is velocity driven by Rule 1, d_2 is the distance pre-defined, v_b and v_x are the velocities of boids B and X.

- The function of alignment rule is to act as the active boid trying to align its velocity vector with the average velocity vector of the flock in its local neighbourhood. The degree of locality of this rule is determined by the sensor range of the active flock boid. This rule can be presented in a mathematical way through Eq. 17.51 (Cui et al. 2006):

$$d(P_x, P_b) \leq d_1 \cap d(P_x, P_b) \geq d_2 \Rightarrow \vec{v}_{ar} = \frac{1}{n} \sum_x^n \vec{v}_x, \tag{17.51}$$

where v_{cr} is velocity driven by Rule 3, d_1 and d_2 are pre-defined distance, and $(P_x - P_b)$ calculates a directional vector point.

- The flock boid tries to stay with the other boids that share the similar features with it. The strength of the attracting force is proportional to the distance (between the boids) and the similarity (between the boids' feature values) which can be expressed as Eq. 17.52 (Cui et al. 2006):

$$v_{ds} = \sum_x^n (S(B, X) \times d(P_x, P_b)), \tag{17.52}$$

where v_{ds} is the velocity driven by feature similarity, $S(B, X)$ is the similarity value between the features of boids B and X.

- The flock boid attempts to stay away from other boids with dissimilar features. The strength of the repulsion force is inversely proportional to the distance (between the boids) and the similarity value (between the boids' features) which are defined by Eq. 17.53 (Cui et al. 2006):

$$v_{dd} = \sum_x^n \frac{1}{S(B, X) \times d(P_x, P_b)}, \tag{17.53}$$

where v_{dd} is the velocity driven by feature dissimilarity. To get comprehensive flocking behavior, the actions of all the rules are weighted and summed to obtain a net velocity vector required for the active flock boid using Eq. 17.54 (Cui et al. 2006):

$$v = w_{sr} v_{sr} + w_{ar} v_{ar} + w_{cr} v_{cr} + w_{ds} v_{ds} + w_{dd} v_{dd}, \tag{17.54}$$

where v is the boid's velocity in the virtual space, and w_{sr}, w_{ar}, w_{cr}, w_{ds}, w_{dd} are pre-defined weight values.

17.20.2 Performance of FBA

Document clustering is an essential operation used in unsupervised document organization, automatic topic extraction, and information retrieval. It provides a structure for organizing large bodies of data (in text form) for efficient browsing and searching. Cui et al. (2006) utilized FBA for document clustering analysis. A synthetic data set and a real document collection (including 100 news articles collected from the Internet) were used in their study. In the synthesis data set, four data types were included with each containing 200 2-dimensional (x, y) data objects. Parameters x and y are distributed according to Normal distribution $N(\mu, \sigma)$; while for the real document collection data set, 100 news articles collected from the Internet at different time stages were categorized by human experts and manually clustered into 12 categories such as Airline safety, Iran Nuclear, Storm Irene, Volcano, and Amphetamine. In order to reduce the impact of the length variations of different documents, Cui et al. (2006) further normalized each file vector to make it in unit length. Each term stands one dimension in the document vector space. The total number of terms in the 100 stripped test files is thus 4,790 (i.e., 4,790 dimensions). The experimental studies were carried out on the synthetic and the real document collection data sets, respectively, among FBA and other popular clustering algorithms such as ant clustering algorithm and K-means algorithm. The final testing results illustrated that the FBA can have better performance with fewer iterations in comparison with the K-means and ant clustering algorithm. In the meantime, the clustering results generated by FBA were easy to be visualized and recognized even by an untrained human user.

17.21 Flower Pollinating Algorithm

In this section, we will introduce an emerging CI algorithm that is derived from the findings related to pollination studies (Acquaah 2012; Alonso et al. 2012)

17.21.1 Fundamentals of Flower Pollinating Algorithm

Flower pollinating algorithm (FPA) was originally proposed in Yang (2012). To implement FPA, the following four rules need to be followed (Yang 2012; Yang et al. 2013):

- Rule 1: Treating the biotic and cross-pollination as a global pollination process, and pollen-carrying pollinators following Lévy flights. In FPA, this rule can be defined by Eq. 17.55 (Yang 2012; Yang et al. 2013):

$$\mathbf{x}_i^{t+1} = \mathbf{x}_i^t + \gamma L(\lambda)\left(\mathbf{x}_i^t - \mathbf{g}_*\right), \quad (17.55)$$

where $\mathbf{x}_i^t$ denotes the pollen i or solution vector $\mathbf{x}_i$ at the tth iteration, $\mathbf{g}_*$ stands for the best solution found so far among all solutions at the current generation.

- Rule 2: For local pollination, abiotic and self-pollination are employed.
- Rule 3: Insects can play the role of pollinators for developing flower constancy. In FPA, the value of flower constancy is set equivalent to a probability called reproduction which is proportional to the similarity of two flowers involved. For modelling the local pollination, both Rule 2 and Rule can be expressed as Eq. 17.56 (Yang 2012; Yang et al. 2013):

$$\mathbf{x}_i^{t+1} = \mathbf{x}_i^t + \varepsilon\left(\mathbf{x}_j^t - \mathbf{x}_k^t\right), \quad (17.56)$$

where the pollen from different flowers of the same plant species is denoted by $\mathbf{x}_j^t$ and $\mathbf{x}_k^t$, respectively.

- Rule 4: Controlling the interaction or switching between the local and global pollination through a switch probability parameter p which falls within the range of $[0, 1]$. In FPA, a slightly biased mechanism is added here for local pollination.

17.21.2 Performance of FPA

The FPA was originally developed in Yang (2012) for dealing with single objective optimization problems. Ideally, it would be great that a new algorithm can be verified on all available test function. Nevertheless, this is quite a time-consuming job. Therefore, Yang (2012) selected a set of benchmark testing functions to check the effectiveness of FPA. The preliminary experimental results demonstrated that FPA is indeed a very effective optimization algorithm.

17.22 Goose Optimization Algorithm

In this section, we will introduce an emerging CI algorithm that is based on the characteristics of Canada geese flight (Hagler 2013) and the PSO algorithm.

17.22.1 Fundamentals of Goose Optimization Algorithm

Goose optimization algorithm (GOA) was proposed by Liu et al. (2006). Since then, this and similar ideas have attracted a steadily increasing amount of researchers, such as Sun and Lei (2009), Cao et al. (2012) and Dai et al. 2013). The main steps of GOA are described as follows (Sun and Lei 2009):

- Step 1: Initialize the population.
- Step 2: Calculate each goose's current fitness and ascertain each goose's individual optimum (*pfbest*) and its corresponding position (*pbest*).
- Step 3: Update each goose's local optimum ($pbest_i$)
- Step 4: Sort the population according to each goose's historical individual optimum ($pfbest_i$) in every generation and receive the sorted population (*spop*).
- Step 5: Replace the *i*th goose's global optimal with the $(i-1)$th goose's individual optimum of the sorted population.
- Step 6: Improve the velocity-location as defined by Eqs. 17.57 and 17.58, respectively (Sun and Lei 2009):

$$v_{id}^{k+1} = \omega \cdot v_{id}^{k} + \alpha\left(spop_{id}^{k} - x_{id}^{k}\right) + \beta\left(pbest_{(i-1)d}^{k} - x_{id}^{k}\right), \tag{17.57}$$

$$x_{id}^{k+1} = x_{id}^{k} + v_{id}^{k+1}, \tag{17.58}$$

  where $\alpha(spop_{id}^{k} - x_{id}^{k})$ can be regarded as a crossover operation between the *i*th goose of the current population and the *i*th goose of the stored population, $\beta(pbest_{(i-1)d}^{k} - x_{id}^{k})$ can be viewed as a crossover operation between the foregoing acquired goose position and the $(i-1)$th goose position of the stored population, and $\omega \cdot v_{id}^{k}$ can be perceived as a mutation operation by which the crossed geese are disturbed randomly.
- Step 7: Rank the solutions and store the current best as optimal fitness value as defined by Eq. 17.59 (Sun and Lei 2009):

$$\begin{cases} \text{if } f\left(x_{temp}\right) - f(x_i) < 0, & \text{then } x_{temp} \\ \text{if } f\left(x_{temp}\right) - f(x_i) > 0, & \text{then } x_i \end{cases}, \tag{17.59}$$

  where x_{temp} is the new goose position that is generated by mutation operator.
- Step 8: Check the termination criteria.

17.22.2 Performance of GOA

To test the efficiency of GOA, a set of travelling salesman benchmark problems were adopted in Sun and Lei (2009). Compared with other CI algorithms (such as GA, SA), computational results showed that GOA outperforms the others in terms of convergence speed and the quality of the solutions.

17.23 Great Deluge Algorithm

In this section, we will introduce a new CI algorithm that is based flood related research (Samuels et al. 2009).

17.23.1 Fundamentals of Great Deluge Algorithm

Great deluge algorithm (GDA) was originally proposed by Dueck (1993). There are several GDA related variants and applications can be found in the literature (Burke et al. 2004; Ravi 2004; Weigert et al. 2006; Sacco et al. 2006; AL-Milli 2010; Nahas et al. 2010; Ghatei et al. 2012; Abdullah et al. 2009). In order to implement GDA, the following facts need to be taken into account (Weigert et al. 2006; Dueck 1993):

- Normally, every place within in the search space can be reached at the beginning of the GDA.
- With the time advances, the landscape of the search space will be divided into several islands according to Eq. 17.60 (Weigert et al. 2006; Dueck 1993):

$$\begin{aligned} p_i &= \Theta(L_i - C_i) \\ L_i &= L_{i-1} - \Delta L \end{aligned}, \tag{17.60}$$

  where the water level is denoted by L, and the rain quantity is represented by ΔL.
- When the GDA is used to deal with the minimization problem, it can be renamed to great drought algorithm, through not technically necessary. In such situation, ΔL will actually refer to the water evaporation quantity. The "walker" in the original GDA will have be replaced by an artificial "fish" which continuously search for a place with sufficient water.
- The water level and rain quantity are controlling variables which play a key role in GDA. While the probability of satisfaction is independent of ΔC which depends only on the absolute value of the objective function C.

17.23.2 Performance of GDA

In order to evaluate the performance of GDA, two typical travelling salesman problems, i.e., the 442-city problem and the 532-city problem were selected in Dueck (1993). The experimental results demonstrated that GDA has the ability of finding the equally good results reported in the literature, but with much easier implementation effort. By further testing GDA on much harder problem such as chip placement case, the GDA generated better results than other known methods (including the results obtained by SA).

17.24 Grenade Explosion Method

In this section, we will introduce a new CI algorithm that is inspired by the mechanism of grenade explosion. In general, there are three types of grenade, i.e., explosive grenades, chemical grenades, and gas grenades (Adams 2004). Although it is a small bomb that is hurled by hand, it is particularly effective in knocking out enemy positions.

17.24.1 Fundamentals of Grenade Explosion Method

Grenade explosion method (GEM) was proposed in Ahrari et al. (2009) and Ahrari and Atai (2010). The core idea behind GEM is when grenade explodes, the thrown pieces of shrapnel destruct the objects near the explosion location. The main procedures of GEM are listed as follows (Ahrari et al. 2009; Ahrari and Atai 2010):

- Initializing the population. The initial grenades (N_g) are generated in random locations in an n-dimension search space $\vec{X}_i \in [-1, 1]^n$, $\left(i = 1, \ldots, N_g\right)$.
- Generate a point (X') around the jth grenade through Eq. 17.61 (Ahrari et al. 2009; Ahrari and Atai 2010):

$$X_j' = \{X_m + sign(r_m) \cdot |r_m|^p \cdot L_e\}, \quad j = 1, 2, \ldots, N_q, \tag{17.61}$$

  where $X = \{X_m\}$, $m = 1, 2, \ldots, n$ is the current location in the n-dimension search space, r_m is a uniformly distributed random number in $[-1, 1]$, L_e is the length of explosion along each coordinate, and p is a constant that defined as Eq. 17.62 (Ahrari et al. 2009; Ahrari and Atai 2010):

$$p = \max\left\{1, n \cdot \frac{\log(R_t/L_e)}{\log(T_w)}\right\}, \tag{17.62}$$

  where T_w is the probability that a produced piece of shrapnel collides an object in n-dimension hyper-box which circumscribes the grenade's territory, and R_t is the territory radius.

If X' is outside the feasible space, transport it to a new location inside the feasible region (i.e., $[-1, 1]^n$) as defined by Eq. 17.63 (Ahrari et al. 2009; Ahrari and Atai 2010):

$$\text{if} \quad X'_j \notin [-1,1]^n \quad \Rightarrow \quad \left(B'_j = \frac{X'_j}{|\text{Largest component of } x'_j \text{ in value}|} \right)$$

$$\rightarrow \quad B''_j = r'_j \cdot \left(B'_j - X \right) + X \tag{17.63}$$

$$\begin{cases} j = 1 \text{ to } N_q \text{ (Shrapnel Number)} \\ 0 < r'_j < +1 \text{ (Random Number)} \end{cases},$$

where X'_j is the collision location outside the feasible space, B''_j is the new location inside the feasible space, and N_q is the number of shrapnel pieces.

- Evaluate the distance between each grenade based on the territory radius (R_t). If X' is a distance of at least R_t apart from the location of grenades $(1, 2, \ldots, i-1)$, then X' is accepted.
- Calculate the fitness of the new generated points around the jth grenade. If the fitness of the best point is better than current location of the jth grenade, move the grenade to the location of the best point.
- Reduce R_t. For increasing the ability of global investigation, the territory radius will be reduced according to Eq. 17.64 (Ahrari et al. 2009; Ahrari and Atai 2010):

$$R_t = \frac{R_{t-initial}}{(R_{rd})^{(iteration\,No/total\,No\,of\,iterations)}}, \tag{17.64}$$

where R_{rd} is user defined (set before the algorithm starts).

Also, the length of explosion (L_e) is reduced via Eq. 17.65 (Ahrari et al. 2009; Ahrari and Atai 2010):

$$L_e = (L_{e-initial})^m (R_t)^{1-m}, \quad 0 \le m \le 1, \tag{17.65}$$

where m can be constant during the algorithm, or reduced from a higher value to a lower one.

17.24.2 Performance of GEM

To demonstrate the efficiency of GEM, a set of optimization benchmark functions such as De Jong's function, Goldstein and Price function, Branin function, Martin and Gaddy function, Rosenbrock function, Schwefel function, and Hyper Sphere function were employed in Ahrari and Atai (2010). Compared with other CI methods (e.g., GA, ACO), computational results showed that GEM can perform well in finding all global minima.

17.25 Group Leaders Optimization Algorithm

In this section, we will introduce a new CI algorithm that inspired by the influence of the leaders in social groups and cooperative co-evolutionary mechanism (Creel 1997; Theiner et al. 2010; Mosser and Packer 2009).

17.25.1 Fundamentals of Group Leaders Optimization Algorithm

Group leaders optimization algorithm (GLOA) was proposed by Daskin and Kais (2011). In order to implement GLOA, the following procedure need to be followed (Daskin and Kais 2011):

- Step 1: Generate p number of population for each group randomly.
- Step 2: Calculate fitness values for all members in all groups.
- Step 3: Determine the leaders for each group.
- Step 4: Mutation and recombination.
- Step 5: Parameter transfer from other groups (one way crossover).
- Step 6: Repeat Steps 3–5 until a termination criterion is satisfied.

17.25.2 Performance of GLOA

To demonstrate the efficiency of GLOA, a set of single and multi-dimensional optimization functions were adopted in Daskin and Kais (2011), namely Beale function, Easom function, Goldstein-Price's function, Shubert's function, Rosenbrock's Banana function, Griewank's function, Ackley's function, Sphere function, and Rastrigin function. Computational results showed that GLOA is very flexible and rarely gets trapped in local minima.

17.26 Harmony Elements Algorithm

In this section, we will introduce an emerging CI algorithm that is inspired by the human life model in traditional Chinese medicine and graph theory.

17.26.1 Fundamentals of Harmony Elements Algorithm

Harmony elements algorithm (HEA) or five-element string algorithm was recently proposed in Cui et al. (2008, 2009) and Rao et al. (2009). The five-elements theory posits wood, fire, earth, metal, and water as the basic elements of the material world, such as people, companies, games, plants, music, art and so on. In terms of traditional Chinese medicine, this theory is used to interpret the relationship between the physiology and pathology of the human body and the natural environment. In other words, they are metaphors for describing how things interact and relate with each other. To implement HEA, the following steps need to be followed (Cui et al. 2008, 2009):

- Step 1: Random initialization: Stochastically creating $2N$ five-element strings as candidate solutions, then grouping the candidate solutions into two string vectors (two element matrices) where the first one is denoted by $Q_{\min}$ and the second one is represented by $Q_{\max}$. The searching range for the ith component of the system state $\underline{x}$ is $[u_{\min}, u_{\max}]$.
- Step 2: $2N$ string cycles generation. By applying $\lambda[\,]$ to $Q_{\min}$ and $Q_{\max}$, respectively, ten string vectors can be created by Eq. 17.66 (Cui et al. 2009):

$$\begin{aligned} Q_i &= \lambda^{(i-1)}[Q_{\min}], \quad i = 1, 2, 3, 4, 5 \\ Q_i &= \lambda^{(i-6)}[Q_{\max}], \quad i = 6, 7, 8, 9, 10. \end{aligned} \tag{17.66}$$

- Step 3: Ranking the strings. Fitness checking and best-worst string vectors generation.
- Step 4: Best element selection and worst element removal. Performing packed-rolling operation and worst elements excising operation.
- Step 5: Checking whether the stopping criterion is met. If yes, terminating the HEA and outputting the results; otherwise, return to Step 1.

17.26.2 Performance of HEA

To verify the proposed HEA, Cui et al. (2009) employed 3 benchmark test functions, namely, Rosenbrock function, Rastrigin function, and Griewank function. In comparison with other CI algorithms (e.g., GA), the experimental results demonstrated that HEA's excellent global searching ability with very attractive speed and impressive solution quality. All these make HEA a quite promising optimization algorithm.

17.27 Human Group Formation Algorithm

In this section, we will introduce an emerging CI algorithm that is derived from a common phenomenon of individuals classification observed from human society (Frank 1998; Magstadt 2013; Ramachandran 2012a, b, c; Mayfield 2013; Howell 2014).

17.27.1 Fundamentals of Human Group Formation Algorithm

Human group formation (HGF) algorithm was recently proposed in Thammano and Moolwong (2010). The key concept of this algorithm is about the behaviour of in-group members that try to unite with their own group as much as possible, and at the same time maintain social distance from the out-group members. To implement HGF algorithm, the following steps need to be performed (Thammano and Moolwong 2010):

- Step 1: Cluster centres representation refers to the number of classes, number of available input patterns, and number, type, and scale of the features available to the clustering algorithm. At first, there are a total of Q clusters, which is equal to the number of target output classes.
- Step 2: Accuracy selection is usually measured by a distance function defined on pairs of patterns as shown in Eq. 17.67 (Thammano and Moolwong 2010):

$$\begin{aligned} &\text{Accuracy} = \frac{\sum_{i=1}^{P} A_i}{P} \\ &A_i = \begin{cases} 1, & \text{if } J \in Y_i \\ 0, & \text{otherwise} \end{cases}, \\ &J = \arg_j \min(d_j(X_i)), d_j(X_i) = \left\| X_i - z_j \right\| \end{aligned} \tag{17.67}$$

  where P denotes the total number of patterns in the training data set; J represents the index of a cluster whose reference pattern is the closest match to the incoming input pattern X_i; Y_i stands for the target output of the ith input pattern; z_j refers to the centre of the jth cluster; and $d_j(X_i)$ states the Euclidean distance between the input pattern X_i and the centre of the jth cluster.

- Step 3: The grouping/formation step can be performed in a way that in-group member try to unite with their own group and maintain social distance from the non-members as much as possible, update the centre value of each cluster (Z_j) by using Eq. 17.68 (Thammano and Moolwong 2010):

$$\begin{aligned} &Z_{jk}^{new} = Z_{jk}^{old} + \Delta Z_{jk} \\ &\Delta Z_{jk} = \sum_{m \in q} \eta_{jm} \beta_j \delta_{jm} (Z_{mk} - Z_{jk}) - \sum_{n \notin q} \eta_{jn} \beta_j \delta_{jn} (Z_{nk} - Z_{jk}), \end{aligned} \tag{17.68}$$

where k $(k = 1, 2, 3, \ldots, k)$ is the number of features in the input pattern; q is the class to which the jth cluster belongs; $\eta_{jm} = e^{-\left[\left(Z_{jk}-Z_{mk}\right)/\sigma\right]^2}$ and $\eta_{jn} = e^{-\left[\left(Z_{jk}-Z_{nk}\right)/\sigma\right]^2}$ have values between 0 and 1which determine the influence of mth and nth clusters on the jth cluster. In general, the further apart mth and nth clusters are from the jth cluster, the lower the values of η_{jm} and η_{jn}; β_j is the velocity of the jth cluster with respect to its own ability to move in the search space; and δ_{jm} is the parameter to prevent clusters of the same class from being too close to one another and normally with respect to two factors: (1) the distance between the jth cluster and the mth cluster, and (2) the territorial boundary of the clusters (T). If the distance between the jth cluster and the mth cluster is less than T, the value of δ_{jm} will be decreased by a predefined amount. After each centre is updated, if the accuracy is higher, save this new center value and then continue updating the next cluster centre; if it is lower, discard the new center value and return to the previous centre; and if it does not change, save the new center value and decrease the value of β_j by a predefined amount.

- Step 4: Cluster validity analysis is the assessment of clustering procedure's output. The cluster which satisfies the Eq. 17.69 will be deleted (Thammano and Moolwong 2010):

$$-\frac{1}{2\log_2\left(\frac{n_j}{p}\right)}\left(\frac{n_j^q}{n_j}\right)\left(\frac{\sum_{\forall X_i^j \in q}\|X_i^j - z_j\|}{n_j}\right) < \rho, \tag{17.69}$$

where n_j is the number of input patterns in the jth cluster; n_j^q is the number of input patterns in the jth cluster whose target outputs (Y) are q; X_i^j is the ith input pattern in the jth cluster; and ρ is the vigilance parameter.

- Step 5: Recalculating the accuracy of the model according to Eq. 17.67 (Thammano and Moolwong 2010):
- Step 6: For each remaining cluster, if the distance between the new centre updated in step 3 and the previous centre is less than 0.0001 $(\|Z_{jk}^{new} - Z_{jk}^{old}\| < 0.0001)$, randomly pick k small numbers between -0.1 and 0.1, and then add them to the centre value of the cluster. The purpose of this step is to prevent the premature convergence of the proposed algorithm to sub-optimal solutions.
- Step 7: Terminating process is to check the end condition, if it is satisfied, stop the loop; if not, examine the following conditions: (1) if the accuracy of the model improves over the previous iteration, randomly select one input pattern from the training data set of each target output class that still has error. Then go to step 2; and (2) if the accuracy does not improve, randomly select the input patterns, a number equal to the number of clusters deleted in step 4, from the training data set of each target output class. Then go to step 2.

17.27.2 Performance of HGF

To test the performance of HGF, Thammano and Moolwong (2010) employed 16 data sets (4 artificial and 12 real-world). The experimental results were compared with the fuzzy neural network, the radial basis function network, and the learning vector quantization network. The performance comparisons demonstrated that the validity of the proposed HGF algorithm.

17.28 Hunting Search Algorithm

In this section, we will introduce an emerging CI algorithm that is inspired by the group hunting of animals, such as African wild dogs (Gusset and Macdonald 2010), rodents (Ebensperger 2001), and wolves (Muro et al. 2011). Although these hunters have difference behavioural patterns during the hutting process, they are share a natural phenomenon in which all of them look for a prey in a group.

17.28.1 Fundamentals of Hunting Search Algorithm

Hunting search (HuS) algorithm was recently proposed in Oftadeh et al. (2010). To implement HuS algorithm, the following steps need to be performed (Oftadeh et al. 2010):

- Step 1: Initialize the optimization problem and algorithm parameters [such as hunting group size (*HGS*), maximum movement toward the leader (*MML*), and hunting group consideration rate (*HGCR*)].
- Step 2: Initialize the hunting group (HG) based on the number of hunters (*HGS*).
- Step 3: Moving toward the leader. The new hunters' positions are generated via Eq. 17.70 (Oftadeh et al. 2010):

$$x_i' = x_i + rand \cdot MML \cdot \left(x_i^L - x_i\right), \tag{17.70}$$

  where $x' = \left(x_1', x_2', \ldots, x_N'\right)$ represents the new hunters' positions, *MML* is the maximum movement toward the leader, *rand* is a uniform random number which varies between 0 and 1, and x_i^L is the position value of the leader for the *i*th variable.

- Step 4: Position correction-cooperation between members. The updating rule of the real value correction and digital value correction are given by Eqs. 17.71 and 17.72 respectively (Oftadeh et al. 2010):

$$x_i^{j'} \leftarrow \begin{cases} x_i^{j'} \in \{x_i^1, x_i^2, \ldots, x_i^{HGS}\} & \text{with probability } HGCR \\ x_i^{j'} = x_i^j \pm R_a & \text{with probability } (1 - HGCR) \\ i = 1, \ldots, N; & \\ i = 1, \ldots, N; & \end{cases}, \tag{17.71}$$

$$xd_{ik}^{j'} \leftarrow \begin{cases} d_{ik}^{j'} \in \{d_{ik}^1, d_{ik}^2, \ldots, d_{ik}^{HGS}\} & \text{with probability } HGCR \\ d_{ik}^{j'} = d_{ik}^j \pm a & \text{with probability } (1 - HGCR) \\ i = 1, \ldots, N; & \\ j = 1, \ldots, HGS & \\ k = 1, \ldots, M \text{ (number of digits in each variable)} & \end{cases}, \tag{17.72}$$

where $HGCR$ is the probability of choosing one value from the hunting group stored in the HG, $(1 - HGCR)$ is the probability of doing a position correction, a can be any number between 1 and 9, and R_a is an arbitrary distance radius for the continuous design variable as defined by Eq. 17.73 (Oftadeh et al. 2010):

$$R_a(it) = R_{a_{\min}}(\max(x_i) - \min(x_i)) \exp\left(\frac{Ln\left(\frac{R_{a_{\min}}}{R_{a_{\max}}}\right) \cdot it}{itm}\right), \tag{17.73}$$

where it is the iteration number, $\max(x_i)$ and $\min(x_i)$ are the maximum or minimum possible value of variable x_i, respectively, $R_{a_{\max}}$ and $R_{a_{\min}}$ are the maximum and minimum of relative search radius of the hunter, respectively, and itm is the maximum number of iterations in the optimization process.

- Step 5: Reorganizing the hunting group. The rule for members' recognition is defined by Eq. 17.74 (Oftadeh et al. 2010):

$$x_i' = x_i^L \pm rand \cdot (\max(x_i) - \min(x_i)) \cdot \alpha \exp(-\beta \cdot EN), \tag{17.74}$$

where x_i^L is the position value of the leader for the ith variable, $rand$ is a uniform random number which varies between 0 and 1, $\max(x_i)$ and $\min(x_i)$ are the maximum and minimum possible values of variable x_i, respectively, EN counts the number of times that the group has been trapped until this step, and α and β are positive real values.

- Step 6: Termination. Repeat Steps 3–5 until the termination criterion is satisfied.

17.28.2 Performance of HuS

In order to show how the HuS algorithm performs, different unconstrained and constrained standard benchmark test functions were adopted in Oftadeh et al. (2010). Compared with other CI techniques (e.g., EA, GA, PSO, ACO, etc.), the performance of HuS algorithm is very competitive.

17.29 Krill Herd Algorithm

In this section, we will introduce an emerging CI algorithm that is based on the krill swarm related studies (Verdy and Flierl 2008; Brierley and Cox 2010; Goffredo and Dubinsky 2014).

17.29.1 Fundamentals of Krill Herd Algorithm

Krill herd (KH) algorithm was recently proposed in Gandomi and Alavi (2012). In order to implement the KH algorithm, the following steps need to be followed (Gandomi and Alavi 2012; Wang et al. 2013):

- Step 1: Defining the simple boundaries, algorithm parameters, and so on.
- Step 2: Initialization. Stochastically creating the initial population within the search space.
- Step 3: Fitness evaluation. Evaluating each individual krill based on its position.
- Step 4: Calculating motion conditions. In KH algorithm, the motion caused by the presence of other individual krill is computed via Eq. 17.75 (Gandomi and Alavi 2012):

$$N_i^{new} = N^{\max}\alpha_i + \omega_n N_i^{old}, \tag{17.75}$$

where α_i equals to $\alpha_i^{local} + \alpha_i^{\text{target}}$, and $N^{\max}$ represents the maximum induced speed, N_i^{old} denotes the last induced motion. Meanwhile, the foraging motion for the ith krill individual is defined by Eq. 17.76 (Gandomi and Alavi 2012):

$$F_i = V_f\beta_i + \omega_f F_i^{old}, \tag{17.76}$$

where β_i is equivalent to $\beta_i^{food} + \beta_i^{best}$, and the foraging speed is denoted by V_f. Finally, the physical diffusion motion of the krill is treated as a random process. This motion can be expressed in Eq. 17.77 (Gandomi and Alavi 2012):

$$D_i = D^{\max}\delta, \tag{17.77}$$

where $D^{\max}$ denotes the maximum diffusion speed, and δ represents a random direction vector. All three motions can be defined by using the following Lagrangian model (see Eq. 17.78) (Gandomi and Alavi 2012):

$$\frac{dX_i}{dt} = N_i + F_i + D_i, \tag{17.78}$$

where N_i denotes the motion induced by other individual krills.

- Step 5: Implementing the genetic operators.
- Step 6: Updating the position of each individual krill within the search space.

- Step 7: Checking whether the stopping condition is met. If not, returning to Step 3; otherwise, terminating the algorithm.

17.29.2 Performance of KH

In order to show how the KH algorithm performs, 20 benchmark test functions such as Sphere function, Goldstein and Price function, Griewank function, and Ackley function are employed in Gandomi and Alavi (2012). Compared with other CI techniques, the performance of KH algorithm is very competitive.

17.30 League Championship Algorithm

In this section, we will introduce an emerging CI algorithm that is based on some interesting findings relative to sports science (Smolin and Grosvenor 2010; Abernethy et al. 2013).

17.30.1 Fundamentals of League Championship Algorithm

League championship algorithm (LCA) was recently proposed in Kashan (2009). In order to model an artificial championship environment, there are the following 6 idealization rules employed in LCA (Kashan 2009, 2011; Kashan and Karimi 2010):

- Rule 1: In LCA, playing strength is defined as the capability of one team defeating the other team.
- Rule 2: The game results is not predictable even if the team's playing strength is know perfectly.
- Rule 3: The winning probability of a team *i* over the other team *j* is assumed to be the same, no matter from which team's viewpoint.
- Rule 4: In the basic version of LCA, tie is not taken into account which means win or loss will be the only game result option.
- Rule 5: Teams only concentrate on the forthcoming math and with no interest of the other distant future game.
- Rule 6: When team *i* defeats the other team *j*, any strength of assisting team *i* in winning will have a dual weakness in causing team *j* to lose.

In order to implement LCA algorithm, the following modules need to be well designed (Kashan 2009, 2011; Kashan and Karimi 2010):

- Module 1: Creating the league timetable. In LCA, an important step is to simulate a real championship environment by establishing a schedule which forms a "virtual season". For instance, a single round-robin schedule mechanism can be employed for ensuring that each team plays against every other team once in each virtual season.
- Module 2: Confirming the winner or loser. Using the playing strength criterion, the winner or loser in LCA is identified in a random manner. Based on the abovementioned Rule 1, the expected chance of winning for team i (or j) can be defined as Eq. 17.79 (Kashan 2011):

$$p_i^t = \frac{f\left(X_j^t\right) - \hat{f}}{f\left(X_j^t\right) + f(X_i^t) - 2\hat{f}}. \tag{17.79}$$

- Module 3: Deploying a suitable mixture of team members. Since the strengths and weaknesses of the each individual team member are not the same, it is often important for coach to generate a good team members mixture by taking various constraint into account. In LCA, a similar process is also performed through an artificial analysis mechanism, more specifically, an artificial SWOT (denoting strengths, weaknesses, opportunities, and threats) analysis is utilized for generating a suitable focus strategy. Based on a thorough analysis, in order to get a new formation of team, the random number of changes made in B_i^t (i.e., best team formation for team i at week t) can be computed through Eq. 17.80 (Kashan 2011):

$$q_i^t = \left\lceil \frac{\ln\left(1 - \left(1 - (1 - p_c)^{n - q_0 + 1}\right)r\right)}{\ln(1 - p_c)} \right\rceil + q_0 - 1, \quad q_i^t \in \{q_0, q_0 + 1, \ldots, n\}, \tag{17.80}$$

where r denotes a random number which falls within the range of $[0, 1]$, and $p_c < 1$, $p_c \neq 0$ represents a controlling variable.

17.30.2 Performance of LCA

To verify the capability of LCA, Kashan (2009) employed 5 benchmark test functions which include such as Sphere function, Rosenbrock function, Rastrigin function, Ackley function, and Schwefel function. In comparison with other CI techniques (e.g., PSO), the simulation results proved that LCA is a dependable method which can converge very fast to the global optimal.

17.31 Membrane Algorithm

In this section, we will introduce an emerging CI algorithm that is based on some studies relative to biological membrane (Reece et al. 2011; Yeagle 2005) and some of its basic features inspired membrane computing (Păun 2000, 2002; Gheorghe et al. 2012; Xiao et al. 2013; Maroosi and Muniyandi 2013; Muniyandi and Zin 2013; Kim 2012; Gofman 2012; Nabil et al. 2012; Zhang et al. 2011; Murphy 2010; Aman 2009; Sedwards 2009; Păun 2007; Nguyen et al. 2008; Woodworth 2007; Ishdorj 2006; Zaharie and Ciobanu 2006; Ciobanu et al. 2003).

17.31.1 Fundamentals of Membrane Algorithm

Membrane algorithm (MA), an approach built on membrane system or P-system diagram (Păun 2000, 2002), was initially proposed in Nishida (2005):

- Component 1: A set of regions which are normally divided by nested membranes.
- Component 2: Each individual region contains a sub-algorithm and several tentative solutions of the targeted optimization problem.
- Component 3: Solution transferring strategy between adjacent regions.

Once the initial settings are done, the following steps need to be performed for implementing MA algorithm (Nishida 2005):

- Step 1: Simultaneously updating the solutions by using the sub-algorithm existing in each individual region.
- Step 2: Sending the best and worst solutions to all the adjacent inner and outer regions, respectively. This mechanism is performed for each region.
- Step 3: Repeating the solutions updating and transferring procedure until a stopping criterion is met.
- Step 4: Outputting the best solution found in the innermost region.

17.31.2 Performance of MA

Nishida (2005) employed the classic travelling salesman problem as a benchmark for verifying the performance of MA. The simulation results demonstrated that the performance of MA is very attractive. As Nishida (2005) commented in the work: On one hand, since other CI algorithms such as GA and SA can be used to play the role of sub-algorithm, an MA is likely to be able to avoid the local optimal. On the other hand, since different sub-algorithms are separated by membranes and the communications happen only among adjacent regions, MA can be easily implemented in other types of computing systems such as parallel, distributed, and grid computing. All these merits make MA a promising candidate in defeating "No Free Lunch Theorem (Wolpert and Macready 1997)".

17.32 Migrating Birds Optimization Algorithm

In this section, we will introduce an emerging CI algorithm that is inspired by the v-flight formation of the migrating birds. It gets this name because of the similarity of the shape the birds that, through one bird leading the flock and two lines of other birds following it, make to the letter "V" (Shettleworth 2010). In addition, the v-formation is one example of the fluid dynamics at work (Hagler 2013). Also, it is believed as a very efficient way for long distance flying due to it is possible to save energy and it can help birds avoid collisions (Badgerow and Hainsworth 1981; Cutts and Speakman 1994; Lissaman and Shollenberger 1970).

17.32.1 Fundamentals of Migrating Birds Optimization Algorithm

Migrating birds optimization (MBO) algorithm was proposed by Duman et al. (2012). In MBO, it is assumed that after flying fro some time, when the leader birds gets tired, it goes to the end of the line and one of the birds following it takes the leader position. As a result, MBO is capable of finding more areas of the feasible solution space by looking at the neighbour solutions. The main steps of MBO are outlined below (Duman et al. 2012):

- Step 1: Initializing the population and parameters.
- Step 2: Repeating the following procedure till stopping criteria met. First, randomly select a leading bird (i). Second, calculate its fitness function. Third, randomly select a neighbour bird among k available neighbour birds (e.g., j). Fourth, if $(F_i > F_j)$, then replace the j by the new solution. Fifth, improve each solution (s_r) in the flock (except leader) by evaluating neighbours' wing-tip spacing (WTS) through Eq. 17.81 (Duman et al. 2012):

$$WTS_{opt} = -0.05b, \tag{17.81}$$

  where WTS_{opt} represents the optimum WTS, and b is the wing span. Sixth, calculate fitness and keep the best solutions. Seventh, rank the solutions and store the current best as optimal fitness value.
- Step 3: Posting process and visualizing results.

17.32.2 Performance of MBO

To test the performance of the MBO algorithm, a series of quadratic assignment problems were taken as the benchmarks (Duman et al. 2012). Compared with other CI algorithm (e.g., TS, SA, and GA), MBO obtained very successful results.

17.33 Mine Blast Algorithm

In this section, we will introduce a new CI algorithm that is based on the observation of the mine bombs (a notorious invention by human) explosion in real world. Just like the volcano or earthquake (Rose 2008), with such force, the mine bombs will be blasted into billions of tiny pieces. In addition, the thrown pieces of shrapnel remain bore with other mine bombs near the explosion area resulting in their explosion.

17.33.1 Fundamentals of Mine Blast Algorithm

Mine blast algorithm (MBA) was proposed by Sadollah et al. (2012, 2013). The main steps of MBA are listed as follows (Sadollah et al. 2012, 2013):

- Step 0: Initializing the population. The initial population is generated by a first shot (X_0^f) explosion producing a number of individuals. The first shot point value is updated via Eq. 17.82 (Sadollah et al. 2012, 2013):

$$X_0^{new} = LB + rand \cdot (UB - LB), \tag{17.82}$$

  where LB and UB are the lower and upper bonds of the problem, respectively, and X_0^{new} is the new generated first shot point.
- Step 1: Initializing the parameters.
- Step 2: Check the condition of exploration constant (μ).
- Step 3: If condition of exploration constant is satisfied, calculate the distance of shrapnel pieces and their location, otherwise, go to Step 10. The calculating equations are given by Eq. 17.83 (Sadollah et al. 2012, 2013):

$$\begin{aligned} d_{n+1}^f &= d_n^f \cdot (|randn|)^2, \quad n = 0, 1, 2, \ldots \\ X_{e(n+1)}^f &= d_{n+1}^f \cdot \cos(\theta), \quad n = 0, 1, 2, \ldots \end{aligned}, \tag{17.83}$$

  where $X_{e(n+1)}^f$ is the location of exploding mine bomb, d_{n+1}^f is the distance of the thrown shrapnel pieces in each iteration, and *randn* is normally distributed pseudorandom number (obtained using *randn* function in MATLAB).

- Step 4: Calculate the direction of shrapnel pieces through Eq. 17.84 (Sadollah et al. 2012, 2013):

$$m_{n+1}^{f} = \frac{F_{n+1}^{f} - F_{n}^{f}}{X_{n+1}^{f} - X_{n}^{f}}, \quad n = 0, 1, 2, 3, \ldots, \tag{17.84}$$

where F is the function value of the X, and m_{n+1}^{f} is the direction of shrapnel pieces.

- Step 5: Generate the shrapnel pieces and compute their improved locations via Eq. 17.85 (Sadollah et al. 2012, 2013):

$$X_{n+1}^{f} = X_{e(n+1)}^{f} + \exp\left(-\sqrt{\frac{m_{n+1}^{f}}{d_{n+1}^{f}}}\right) \cdot X_{n}^{f}, \quad n = 0, 1, 2, 3, \ldots, \tag{17.85}$$

where $X_{e(n+1)}^{f}$, d_{n+1}^{f}, and m_{n+1}^{f} are the location of exploding mine bomb collided by shrapnel, the distance of shrapnel and the direction (slope) of the thrown shrapnel in each iteration, respectively.

- Step 6: Check the constraints for generated shrapnel pieces.
- Step 7: Save the best shrapnel piece as the best temporal solution.
- Step 8: Does the shrapnel piece have the lower function value than the best temporal solution?
- Step 9: If true, exchange the position of the shrapnel with the best temporal solution. Otherwise, go to Step 10.
- Step 10: Calculate the distance of shrapnel pieces and their locations, then return to Step 4. The calculating equations are given by Eq. 17.86 (Sadollah et al. 2012, 2013):

$$\begin{aligned} d_{n+1}^{f} &= \sqrt{\left(X_{n+1}^{f} - X_{n}^{f}\right)^2 + \left(F_{n+1}^{f} - F_{n}^{f}\right)^2}, \quad n = 0, 1, 2, \ldots \\ X_{e(n+1)}^{f} &= d_{n}^{f} \cdot rand \cdot \cos(\theta), \quad n = 0, 1, 2, \ldots \end{aligned}, \tag{17.86}$$

where $X_{e(n+1)}^{f}$ is the location of exploding mine bomb, *rand* is a uniformly distributed random number, and θ is the angle of the shrapnel which is calculated through Eq. 17.87 (Sadollah et al. 2012, 2013):

$$\theta = 360/N_s, \tag{17.87}$$

where N_s is the number of shrapnel pieces which are produced by the mine bomb explosion.

- Step 11: Reduce the distance of the shrapnel pieces according to Eq. 17.88 (Sadollah et al. 2012, 2013):

$$d_n^f = \frac{d_{n+1}^f}{\exp(k/\alpha)}, \quad n = 1, 2, 3, \ldots, \tag{17.88}$$

where α and k are the reduction constant which is user parameter and depends on the complexity of the problem and iteration number, respectively.

- Step 12: Check the convergence criteria. If the stopping criterion is satisfied, the algorithm will be stopped. Otherwise, return to Step 2.

17.33.2 Performance of MBA

To test the efficiency of MBA, five well-known truss structures problems were adopted in Sadollah et al. (2012), namely, 10-bar truss, 15-bar truss, 52-bar truss, 25-bar truss, and 72-bar truss. Compared with other CI algorithms (e.g., PSO), computational results showed that MBA clearly outperforms the others in terms of convergence speed and computational cost.

17.34 Monkey Search Algorithm

In this section, we will introduce an emerging CI algorithm that is inspired by the monkey foraging behaviour (King et al. 2011; Mills et al. 2010; Sueur et al. 2010; Lee and Quessy 2003; Taffe and Taffe 2011).

17.34.1 Fundamentals of Monkey Search Algorithm

Monkey search algorithm (MSA) was proposed by Mucherino and Seref (2007). In MSA, the food is viewed as the desirable solutions and the branches of the trees are illustrated as perturbations between two neighbouring feasible solutions. In addition, at each iteration, the starting solution is viewed as the root of a branch and the new neighbour solution is given at the tip of the same branch. The height of the trees (i.e., the functional distance between the two solutions, h_t) is determined by the random perturbation. Also, it is assumed that when the monkeys look for food, they will also learn which branches lead to better food resources. The main steps of MSA are described as follows (Mucherino and Seref 2007):

- Step 1: Initialize populations and parameters.
- Step 2: Repeat till stopping criteria met. First, randomly select a branch of a tree (n_w) as root. Second, calculate its fitness function. Third, perform the perturbations process to generate a new solution at the tip of the same branch as follows (Mucherino and Seref 2007): (1) Random changes to X_{cur}, as in the SA methods; (2) Crossover operator applied for generating a child solution from the

parents X_{cur} and X_{best}, as in GA; (3) The mean solution built from X_{cur} and X_{best}, inspired by ACO; (4) Directions that lead X_{cur} to X_{best}, as in directional evolution; (5) Creating solutions from X_{cur} and X_{best} and introducing random notes, as in harmony search.
- Step 3: Check the termination criteria.

In addition, for avoiding local optima, the predetermined number of n_m best solutions (i.e., the memory bank) are updated by each successive tree.

17.34.2 Performance of MSA

To test the performance of MSA, two global optimization problem of finding stable conformations of clusters of atoms' energy functions (i.e., Lennard Jones potential energy and Morse potential energy) were adopted in Mucherino and Seref (2007). In addition, a protein folding problem (i.e., the tube model) is also considered as test function. Compared with other CI algorithms (such as SA), computational results showed that the proposed algorithm outperforms others in terms of the quality of solutions.

17.35 Mosquito Host-Seeking Algorithm

In this section, we will introduce an emerging CI algorithm that is inspired by the host-seeking behaviour of mosquitoes (Levin 2013d).

17.35.1 Fundamentals of Mosquito Host-Seeking Algorithm

Mosquito host-seeking algorithm (MHSA) was proposed by Feng et al. (2009). According to the observation of mosquito host-seeking behaviour, there is only female mosquitoes search the host to attract (Woodward 2008). Recently, based on mosquito host-seeking behaviour, Cummins et al. (2012) proposed a new mathematical model to describe the effect of spatial heterogeneity. Furthermore, it is possible to design artificial female mosquitoes that, by seeking towards the hosts which emit an odor (e.g., CO_2), find the shortest path between the two nodes corresponding to the nest and to the food source. The following description, which is developed using the travelling salesman problem as a running example, is completed for implementing the proposed algorithm.

- Step 1: Initializing the population and parameters.

- Step 2: Calculate the distance between $(u_{ij}(t))$ between an artificial mosquito and the host at time t in parallel as defined by Eq. 17.89 (Feng et al. 2009):

$$u_{ij}(t) = \exp\left(-c_{ij}(t) r_{ij}(t) x_{ij}(t)\right), \tag{17.89}$$

where x_{ij} is the sex value of each artificial mosquito (m_{ij}) as defined by Eq. 17.90 (Feng et al. 2009):

$$\begin{cases} x_{ij} = 1, & m_{ij} \text{ is female} \\ x_{ij} = 0, & m_{ij} \text{ is male} \end{cases}, \tag{17.90}$$

and c_{ij} represents the relative strength of the artificial mosquito. It can be defined as Eq. 17.91 (Feng et al. 2009):

$$\begin{cases} t = 0, & c_{ij} = \max_{i,j} d_{ij} - d_{ij} \\ t > 0, & c_{ij} \in [0, 1] \end{cases}, \tag{17.91}$$

where c_{ij} represents the distance between city pair (C_i, C_j), and d_{ij} is defined by Eq. 17.92 (Feng et al. 2009):

$$d_{ij} = \sqrt{(x_i - x_j)^2 + (y_i - y_j)^2}. \tag{17.92}$$

The utility of all artificial mosquitoes will be summarized as Eq. 17.93 (Feng et al. 2009):

$$J(t) = \sum_{i=1}^{n} \sum_{j=1}^{n} u_{ij}(t). \tag{17.93}$$

- Step 3: Calculating the motion of artificial mosquitoes via Eqs. 17.94–17.96, respectively (Feng et al. 2009):

$$\frac{du_{ij}(t)}{dt} = \psi_1(t) + \psi_2(t), \tag{17.94}$$

$$\psi_1(t) = -u_{ij}(t) + \gamma v_{ij}(t), \tag{17.95}$$

$$\begin{aligned} \psi_2(t) = {} & \left[-\lambda_1 - \lambda_2 \frac{\partial J(t)}{\partial u_{ij}(t)} - \lambda_3 \frac{\partial P(t)}{\partial u_{ij}(t)} - \lambda_4 \frac{\partial Q(t)}{\partial u_{ij}(t)} \right] \\ & \times \left\{ \left[\frac{\partial u_{ij}(t)}{\partial r_{ij}(t)} \right]^2 + \left[\frac{\partial u_{ij}(t)}{\partial c_{ij}(t)} \right]^2 \right\} \end{aligned}, \tag{17.96}$$

where $\gamma > 1$, and $v_{ij}(t)$ is a piecewise linear function of $u_{ij}(t)$ defined by Eq. 17.97 (Feng et al. 2009):

$$v_{ij}(t) = \begin{cases} 0 & \text{if } u_{ij}(t) < 0 \\ u_{ij}(t) & \text{if } 0 \le u_{ij}(t) < 1\,, \\ 1 & \text{if } u_{ij}(t) > 1 \end{cases} \tag{17.97}$$

and $P(t)$ and $Q(t)$ represent the attraction functions as defined by Eqs. 17.98 and 17.99, respectively (Feng et al. 2009):

$$P(t) = \varepsilon^2 \ln \sum_{i=1}^{n} \sum_{j=1}^{n} \exp\left[-u_{ij}^2(t) \big/ 2\varepsilon^2\right] - \varepsilon^2 \ln nn, \tag{17.98}$$

$$Q(t) = \sum_{i=1}^{n} \left| \sum_{j=1}^{n} r_{ij}(t) x_{ij}(t) - 2 \right|^2 - \sum_{i,j} \int_0^{u_{ij}} \left\{ [1 + \exp(-10x)]^{-1} - 0.5 \right\} dx, \tag{17.99}$$

where $0 < \varepsilon < 1$, and $r_{ij}(t)$ is defined by Eq. 17.100 (Feng et al. 2009):

$$\begin{cases} r_{ij} = 1, & Z \text{ pass } p_{ij} \\ r_{ij} = 0, & Z \text{ not pass } p_{ij} \end{cases}, \tag{17.100}$$

where Z is the shortest path through n cities, and p_{ij} is the path between C_i and C_j.

The general hybrid attraction function for artificial mosquito can be defined by Eq. 17.101 (Feng et al. 2009):

$$E_{ij}(t) = -\lambda_1 u_{ij}(t) - \lambda_2 J(t) - \lambda_3 P(t) - \lambda_4 Q(t), \tag{17.101}$$

where $E_{ij}(t) =$ is the hybrid attrition function, and $0 < \lambda_1, \lambda_2, \lambda_3, \lambda_4 < 1$.

- Step 4: For each artificial mosquito, calculate the value of $dr_{ij}(t)$ and $dc_{ij}(t)$ in order to increase the problem's dynamically according to Eqs. 17.102 and 17.103, respectively (Feng et al. 2009):

$$\frac{dr_{ij}(t)}{dt} = -\lambda_1 \frac{\partial u_{ij}(t)}{\partial r_{ij}(t)} - \lambda_2 \frac{\partial J(t)}{\partial r_{ij}(t)} - \lambda_3 \frac{\partial P(t)}{\partial r_{ij}(t)} - \lambda_4 \frac{\partial Q(t)}{\partial r_{ij}(t)}, \tag{17.102}$$

$$\frac{dc_{ij}(t)}{dt} = -\lambda_1 \frac{\partial u_{ij}(t)}{\partial c_{ij}(t)} - \lambda_2 \frac{\partial J(t)}{\partial c_{ij}(t)} - \lambda_3 \frac{\partial P(t)}{\partial c_{ij}(t)} - \lambda_4 \frac{\partial Q(t)}{\partial c_{ij}(t)}, \tag{17.103}$$

where $\frac{\partial Q(t)}{\partial u_{ij}(t)} = -\left\{ \left[1 + \exp\left(-10(t) u_{ij}(t)\right)\right]^{-1} - 0.5 \right\}$.

- Step 5: Updating the both value of $(r_{ij}(t))$ and $(c_{ij}(t))$ in parallel through Eqs. 17.104 and 17.105 (Feng et al. 2009):

$$r_{ij}(t+1) = r_{ij}(t) + \frac{dr_{ij}(t)}{dt}, \tag{17.104}$$

$$c_{ij}(t+1) = c_{ij}(t) + \frac{dc_{ij}(t)}{dt}. \tag{17.105}$$

- Step 6: If all $\frac{du_{ij}(t)}{dt} = 0$, then finish successfully; otherwise, go to Step 2.

17.35.2 Performance of MHSA

The travelling salesman problem has attracted many researchers from different fields. Recently, Feng et al. (2009) used MHSA for finding near-optimum solutions to the travelling salesman problem. Computational results showed that the proposed algorithm performs well and easy to jump into local optimal. Also, it is easy to adapt to a wide range of the travelling salesman problem.

17.36 Oriented Search Algorithm

In this section, we will introduce an emerging algorithm that is inspired by the human helping behaviour. Helping behaviours are activities where people intend to assist other person to solve the problems, such as to relieve distress (Stukas and Clary 2012).

17.36.1 Fundamentals of Oriented Search Algorithm

Oriented search algorithm (OSA) was proposed by Zhang et al. (2008a) that mimics the helping behaviour of a little girl when she lost her way in deep forest. In OSA, the optimal solution of the objective function is the lost girl who can transmit information for help in order to be found immediately. The main steps of OSA are illustrated as follows (Zhang et al. 2008a):

- Step 1: Initialization the population and parameters. For example, the objective function for each search individuals ($f(x_{0ji})$) and the position of search individuals (x_{0ji}) are defined by Eqs. 17.106–17.108, respectively (Zhang et al. 2008a):

$$x_{0ji} = X_{\text{min}i} + (X_{\text{max}i} - X_{\text{min}i}) * random_{ji}(0,1), \tag{17.106}$$

$$x_{0ji} = c_i, \quad \text{where } c_i \in [X_{\text{min}i}, X_{\text{max}i}], \tag{17.107}$$

$$x_{0ji} = c_i, \quad \text{where } c_i \in [X_{\text{min}i}, X_{\text{max}i}], \tag{17.108}$$

where a_{ji} and c_i are constants.

- Step 2: Exploration walks procedure. First, generate the strategy of updating Δx_{tji} via Eqs. 17.109 and 17.110, respectively (Zhang et al. 2008a):

$$x_{tji} = x_{0ji} + \Delta x_{tji}, \tag{17.109}$$

$$\Delta x_{tji} = \left(x_{tji\_global} \cdot \left(1 + w \cdot randn_{tji}(0,1) - x_{tji}\right) \cdot random(0,1)\right), \tag{17.110}$$

where x_{tji_global} denotes the current optimal position of the objective function, and w is a variable which can adjust the variable trend of oriented-neighbour-space.

Second, explore new position of the current search individual (x_{tji}). Third, evaluate the quality of the objective function $(f_{tj} = f(x_{tji}))$. Fourth, if $f_{tj} \leq f_{(t-1)j}$, then $x_{0ji} = x_{tji}$. Fifth, update the current position of the objective function optimal solution (x_{tji_global}). Sixth, check the termination criteria.

- Step 3: Posting process and visualizing results.

17.36.2 Performance of OSA

To test the performance of OSA, a reactive power optimization problem was adopted in Zhang et al. (2008a). Compared with other algorithms, computational results showed that OSA has better convergence property and precision. Also, it is capable of escaping from the local optima.

17.37 Paddy Field Algorithm

In this section, we will introduce an emerging CI algorithm that is based on the concept of sowing is carried out in accordance with individual fitness value and neighbour numbers of seed so that they will grow towards the best environment (optimal solution) (Maathuis 2013; Acquaah 2012).

17.37.1 Fundamentals of Paddy Field Algorithm

Paddy field algorithm (PFA) was recently proposed in Premaratne et al. (2009). In general, situation of each individual seed or plant can be illustrated as vector $X = (x_1, x_2, \ldots, x_k)$ and the fitness or objective function of X is denoted by $Y = f(X)$. Depending on the nature of the parameter space each dimension of the seed can be bonded such that Eq. 17.111 holds (Premaratne et al. 2009; Wang et al. 2011):

$$x_j \in \left[(x_j)_{\min}, (x_j)_{\max} \right]. \tag{17.111}$$

To implement PFA algorithm, the following steps need to be performed (Premaratne et al. 2009):

- Sowing behaviour: The algorithm operates by initially scattering seeds (initial population p_0) at random in an uneven field. The values of seed dimensions are uniformly distributed depending on the bounds of the parameter space.
- Selection behaviour: When the seeds produce plants, the best plants are selected depending on a threshold (y_t), which can be used to determine the number of seeds of a plant. The reason for having a threshold is to control the population of the system. That means, a plant will be selected to the next iteration only if its fitness value (y) is greater than the threshold as defined by Eq. 17.112 (Premaratne et al. 2009).

$$y \geq y_t. \tag{17.112}$$

- Seeding behaviour: In this stage, each plant develops a number of seeds proportional to its health. The total quantity of seeds (s) produced by any plant would be a function of the plant fitness function and the maximum number of seeds $(q_{\max})$ as defined by Eq. 17.113 (Premaratne et al. 2009):

$$s = \phi[f(x), q_{\max}], \tag{17.113}$$

In general, the fitness function is depending on its fitness in proportion to the fittest plant of the population $(y_{\max})$ as shown in Eq. 17.114 (Premaratne et al. 2009):

$$s = q_{\max} \left[\frac{y - y_t}{y_{\max} - y_t} \right]. \tag{17.114}$$

- Pollination behaviour: In any paddy field, the strong ones (best solution) have greater opportunity to pass their seeds to future generations via pollination behaviour. This behaviour is a major factor either via animals or through wind. High population density would increase the chance of pollination for pollen carried by the wind. That means, the plant with more neighbours (i.e. neighbourhood function N) will be better pollinated. The number of viable seeds (s_v) produced by a plant can be expressed as Eq. 17.115 (Premaratne et al. 2009):

$$s_v = N\phi[f(x), q_{\max}], \quad 0 \leq N \leq 1. \tag{17.115}$$

In order to satisfy this condition, a sphere of radius (a) is used. For two plants x_j and x_k, the perimeter formula (see Eq. 17.116) (Premaratne et al. 2009)

$$n(x_j, x_k) = \|x_j - x_k\| - a, \tag{17.116}$$

is used. If the two are within the sphere, then $n < 0$. From this for a particular plant, the number of neighbours (v_j) can be determined. Once this is done, the pollination factor for that plant can be obtained from Eq. 17.117 (Premaratne et al. 2009),

$$N_j = e^{\left[\frac{v_j}{v_{\max}} - 1\right]}, \tag{17.117}$$

where $v_{\max}$ is maximum neighbour number of the plant.

- Dispersion behaviour: In order to prevent getting stuck in local minima, the seeds of each plant are dispersed and then the cycle stars again from the selection stage. In PFA, when dispersing, the dimension values take a Gaussian distribution which could provide a faster convergence in local search. The new seed will land on a location in the parameter space given by Eq. 17.118 (Premaratne et al. 2009):

$$X_{seed}^{i+1} = F\left(x^i, \sigma\right), \tag{17.118}$$

where σ is the coefficient of dispersion, which can determine the dispersion degree of produced seeds.

17.37.2 Performance of PFA

The difference between PFA and other nature-inspired algorithms (such as evolutionary algorithms) is PFA uses pollination and dispersal between individuals as mainly operators. In addition, unlike the basic version of PSO, in PFA, the random numbers are not generated by applying the uniform distribution function. Instead, the Gaussian probability distribution function is applied. This offers the advantage of enhanced search capability while maintaining adequate exploitation capability (Premaratne et al. 2009).

17.38 Photosynthetic Algorithm

The process of which the carbon atoms in CO_2 are incorporated into glucose, $C_6H_{12}O_6$, in green plants is normally referred to as photosynthesis (Whitten et al. 2014). It is often regarded as one of the key biological process in the biosphere (Dubinsky 2013; Carpentier 2011). Normally, oxygenic photosynthesis can be found occurring in cyanobacteria, algae and land plants (Dubinsky 2013). Although the actual process is quite complex, the following net equation (see Eq. 17.119) can be used to simply describe the phenomenon of photosynthesis (Whitten et al. 2014; Reece et al. 2011; Jelinek 2013; Hobbs et al. 2013):

$$6CO_2 + 6H_2O \xrightarrow[\text{chlorophyll}]{\text{sunlight}} C_6H_{12}O_6 + 6O_2, \tag{17.119}$$

where chlorophyll contains magnesium ions which are bond to porphyrin rings and it is critical substance for photosynthesis.

In nature, ribulose-1, 5-bisphosphate carboxylase/oxygenase (Rubisco for short) is the most abundant protein on Eearth, comprising almost the half of the protein in leaves. Basically Rubisco catalyses the carboxylation of ribulose-1, 5-bishosphate (RuBP), generating two molecules of 3-phosphoglycerate (3-PGA). Rubisco is a vary useful bifunctional enzyme that fixes the liberated CO_2 in the chloroplasts of photosynthetic organism through its carboxylase activity (Dubinsky 2013). This irreversible first step of photosynthesis is therefore the entering point for carbon into the biosphere (Carpentier 2011).

On our planet, most of the energy required to develop and sustain life is supplied by the capture of sunlight by photosynthetic organisms (Carpentier 2011). Photosynthesis is thus often treated as the source of global food, feed, fibre, and timber production as well as biomass-based bio-energy. The renewability is the main characteristic of each of these products of photosynthesis. For instance the main products photosynthesis are starch and sucrose where the latter is also the main form of carbon translocated from leaves to other organs in plants (Dubinsky 2013). Since photosynthesis is, in itself, a multidisciplinary research area which involves such as agriculture, environmental sciences, forestry, plant genetics, photobiology, photophysics, plant physiology, and biochemistry, the detailed explanation of many of its general and fundamental research methods and recent advances is out of the scope of the present book, interested readers are referred to the corresponding studies, e.g., (Acquaah 2012; Carpentier 2011; Dubinsky 2013; Maathuis 2013), for more relative information.

For the rest of this section, we will introduce an emerging CI algorithm which is based on the findings extracted from photosynthesis research.

17.38.1 Fundamentals of Photosynthetic Algorithm

Motivated by the principle of Benson-Calvin cycle Phase-1 and the reaction that happens in the chloroplast subcellular compartment for photorespiration, photosynthetic algorithm (PA) was originally proposed in Murase (2000). To perform the PA, the following calculation processes need to be followed (Murase 2000):

- First, randomly generating the intensity of light.
- Second, evaluating the fixation rate of CO_2 via the following equation (also refer to as the stimulation function in the PA algorithm) based on the light intensity (Murase 2000). This is a unique characteristic of the PA algorithm. Such stimulation often happens as a result of randomly changed light intensity which in turn adjusts the influential degree on the elements of RuBP [i.e., ribulose-1, 5-bishosphate (Carpentier 2011)] by photorespiration as shown in Eq. 17.120.

$$C = \frac{V_{max}}{1 + A/L}, \tag{17.120}$$

where the CO_2 fixation rate is denoted by C, V_{max} represents the maximum CO_2 fixation rate, A stands for the affinity of CO_2, and L is used to express the light intensity.

- Third, based on the fixation rate obtained from the stage above, one of two cycles, either Benson-Calvin or photorespiration will be selected at this stage. For both cycles, Murase (2000) utilized 16-bit strings which shuffles based on carbon molecules recombination rule in photosynthetic pathways.
- Then after certain rounds of iterations, an amount of GAPs, i.e., glyceraldehyde-3-phosphate (Dubinsky 2013), are generated for representing intermediate knowledge strings in the PA algorithm. Each GAP is composed of 16 bits. The fitness of these GAPs will be evaluated at this stage. The best fit GAP will remain as a DHAP [i.e., di-hydroxyacetone phosphate (Carpentier 2011)] which is referred to as the current estimated value.

Taking into account the fundamental process described above, the steps of implementing PA can be summarized as follows (Murase 2000; Alatas 2011; Yang 2005):

- Step 1: Initializing the following problem parameters such as $f(x)$ (the object function), x_i (the decision variable), N (the number of decision variables), and the boundary of constraints.
- Step 2: Initializing the following problem parameters such as DHAPs, and CO_2 fixation parameters (e.g., affinity A, maximum fixation rate V_{max}, and light intensity L).
- Step 3: Calculating CO_2 concentration, determining O_2/CO_2 concentration ration, and setting Benson-Calvin/photorespiration frequency ratio.
- Step 4: Evaluating if the stopping criteria are met. If yes, the algorithm stops; otherwise, go to the next step.
- Step 5: Depending the fixation rate of CO_2, the 16-bit strings are shuffled in either Benson-Calvin or photorespiration cycle.
- Step 6: Comparing the fitness value where the poor results will be removed and the desired DHAP strings and results will be remained.
- Step 7: Updating the light intensity and the next round of iteration of the PA algorithm starts.

17.38.2 Performance of PA

In order to verify the proposed PA, the finite element inverse analysis problem was employed in Murase (2000). The prediction of the elastic moduli of the finite element model via PA was quite satisfied. The overall performance demonstrated by this preliminary application make PA a very attractive optimization algorithm.

17.39 Population Migration Algorithm

In this section, we will introduce an emerging CI algorithm that is inspired by the population migrating mechanism (Ramachandran 2012a, b, c).

17.39.1 Fundamentals of Population Migration Algorithm

Population migration algorithm (PMA) was originally proposed in Zhou and Mao (2003). There are several variants and application can be found in the literature (Zhang et al. 2009; Zhang and Zhou 2009; Wang et al. 2010; Lu and Liu 2011; Zhao and Liu 2009, 2011). To implement PMA, the following components need to be considered (Zhang and Zhou 2009; Zhou and Mao 2003):

- Component 1: In PMA, the social-cooperation strategy of PMA can be defined by Eq. 17.121 (Zhang and Zhou 2009):

$$\alpha^t = [x_{best}, 1, \text{population migration}, x_{best}]. \tag{17.121}$$

- Component 2: In PMA, the self-adaptation strategy can be divided into two parts, namely, population flow and population proliferation. Mathematically, the self-adaptation mechanism can be defined as Eq. 17.122 (Zhang and Zhou 2009):

$$\beta^t = \left[\left(\text{pop}_{flow}, \text{pop}_{proliferation}\right)\text{shrinkage the beneficial region}\right]. \tag{17.122}$$

- Component 3: Competition, i.e., population updating strategy can be described as Eq. 17.123 (Zhang and Zhou 2009):

$$\gamma^t = [\mu = \lambda, (\mu, \lambda), \text{ record and update } x_{best} \text{ and } f(x_{best})]. \tag{17.123}$$

17.39.2 Performance of PMA

To verify the proposed PMA, a set of experimental studies were conducted in Zhou and Mao (2003). The simulation results demonstrated that PMA is a very attractive optimization problem solver.

17.40 Roach Infestation Optimization

In this section, we will introduce an emerging CI algorithm that is based on the collective behaviour of some insects, e.g., roach (Bater 2007; Chapman 2013).

17.40.1 Fundamentals of Roach Infestation Optimization Algorithm

Roach infestation optimization (RIO) algorithm was proposed by Havens et al. (2008) that inspired by the recent observation of cockroaches' social (both collective and individual) behaviours. Typically, there are three types of behaviour are employed in RIO, i.e., search behaviour, social behaviour, and hungry (foraging) behaviour. Each one is outlined as follows (Havens et al. 2008):

- Search behaviour (*Find_Darkness*): As RIO is a cockroach-inspired PSO, this behaviour is defined as Eq. 17.124 (Havens et al. 2008):

$$\vec{v}_i = C_0\vec{v}_i + C_{\max}\vec{R}_1 .* \left(\vec{p}_i - \vec{x}_i\right), \tag{17.124}$$

  where $\vec{v}_i$ is the velocity of the ith agent (cockroach), $\vec{x}_i$ is the current location, $\vec{p}_i$ is the best location found by the ith agent, $\{C_0, C_{\max}\}$ are parameters, $\vec{R}_1$ is a vector of uniform random numbers and $.*$ is element-by-element vector multiplication.
- Social behaviour (*Find_Friends*): The agents will socialize and share their information by setting the darkest local location $(\vec{l})$ as shown in Eq. 17.125 (Havens et al. 2008):

$$\vec{l}_i = \vec{l}_j = \arg\min_k\left\{F\left(\vec{p}_k\right)\right\}, \quad k = \{i, j\}, \tag{17.125}$$

  where $\vec{l}_i$ is the group best solution, $\{i, j\}$ are the indices of the two socializing cockroaches, and $\vec{p}_k$ is darkest known location for the individual cockroach agent (i.e., the best personal location).
- Hunger behaviour (*Find_Food*): To model this behaviour, a parameter called hunger counter ($hunger_i$) is employed. The main procedure deals with the hungry degree checking of the agents which based on a threshed (t_{hunger}). If not (i.e., $hunger_i < t_{hunger}$), update the agent's velocity as shown in Eq. 17.125 (Havens et al. 2008):

$$\begin{aligned}\vec{v}_i &= C_0\vec{v}_i + C_{\max}\vec{R}_1 .* \left(\vec{p}_i - \vec{x}_i\right) + C_{\max}\vec{R}_2 .* \left(\vec{l}_i - \vec{x}_i\right)\\ \vec{x}_i &= \vec{x}_i + \vec{v}_i,\end{aligned} \tag{17.126}$$

  otherwise, the agent will be transported to other food location $(\vec{b})$ randomly. Also, this piece of food is randomly relocated.

17.40.2 Performance of RIO

To check the efficiency of RIO, a number of numerical examples were adopted in Havens et al. (2008), such as Sphere function, Rastrigin function, Rosenbrock function, Ackley function, Griewank function, Michalewicz function, Easom function, and Hump function. Compared with other CI algorithms (e.g., PSO), the proposed algorithm is more effective for optimizing highly-modal function.

17.41 Saplings Growing Up Algorithm

In this section, we will introduce an emerging CI algorithm that is based on the behaviours observed from saplings (Schnell and Priyadarshan 2012; Tidball and Krasny 2014; Reece et al. 2011).

17.41.1 Fundamentals of Saplings Growing Up Algorithm

Saplings growing up algorithm (SGuA) was originally proposed in Karci (2007a, b, c) and Karci and Alatas (2006). To implement SGuA, the following steps need to be performed (Karci 2007a, b, c; Karci and Alatas 2006):

- Sowing phase: In sowing phase, a uniform population method and divide-and-generate paradigm was proposed for initial population generating. Initially, two saplings are set where $S_0 = \{u_1, u_2, \ldots, u_n\}$ and $S_1 = \{l_1, l_2, \ldots, l_n\}$, n is the length of sampling and this case the dividing factor (k) is considered as $k = 1$, $u_i$and l_i are upper and lower bounds for corresponding variables. Then the factor k is determined. For $k = 2$ and two extra S_2 and S_3 are divided from S_0 and S_1 as shown in Eqs. 17.127 and 17.128, respectively (Karci 2007a, b, c; Karci and Alatas 2006):

$$S_2 = \{r_1, r_2, \ldots, r_{n/2}, r_{n/2+1}, r_{n/2+2}, \ldots, r_n\}, \tag{17.127}$$

$$S_3 = \{r \cdot l_1, r \cdot l_2, \ldots, r \cdot l_{n/2}, r \cdot u_{n/2+1}, r \cdot u_{n/2+2}, \ldots, r \cdot u_n\}, \tag{17.128}$$

  where r is a random number such as $0 \leq r \leq 1$. Let us consider the population P size as $|P|$ and the number of elements in the set of generated saplings S as $|S|$. So if $|S| < |P|$, then the value of k is increased by 1, and $2^3 - 2 = 8 - 2 = 6$ saplings can be derived from S_0 and S_1, which are not in S, since S_0 and S_1 are divided into three parts. The remaining saplings in the garden will be obtained by applying same method with increasing the value of k. This process goes on until $|S| \geq |P|$. Hereafter, the first $|P|$ elements of the set S are taken as saplings' population.

- Growing up phase: This phase contains three operators: mating, branching, and vaccinating operators.
  The aim of mating to generate a new sapling from currently existing saplings (global search) by interchanging current exist information between temporary solutions. In general, the distance between two saplings affects the mating process' taking place or not, and it depends on the distance between current pair (i.e., it has greater probability for near saplings and has small probability for saplings far away to each other). Let $P_m(S_1, S_2)$ can be computed in the following two ways as shown in Eqs. 17.129 and 17.130, respectively (Karci 2007a, b, c; Karci and Alatas 2006):

$$P_m(S_1, S_2) = 1 - \frac{\left(\sum_{i=1}^{n} \left(s_{1,i} - s_{2,i}\right)^2\right)^{\frac{1}{2}}}{R}, \tag{17.129}$$

$$P_m(S_1, S_2) = \frac{\left(\sum_{i=1}^{n} \left(s_{1,i} - s_{2,i}\right)^2\right)^{\frac{1}{2}}}{R} \left(1 - \frac{\left(\sum_{i=1}^{n} \left(s_{1,i} - s_{2,i}\right)^2\right)^{\frac{1}{2}}}{R}\right), \tag{17.130}$$

  where $R = \left(\sum_{i=1}^{n} (u_i - l_i)^2\right)^{\frac{1}{2}}$, u_i and l_i are upper and lower bounds for corresponding variables.
  Branching: each sapling consists of branches, and initially each sapling contains no branches $\left(P\left(s_{1,j}|s_{1,i}\right) = 1\right)$ and it is a body. In order to grow up a branch on any point (i.e., a new sapling) from currently exist saplings, the author used probabilistic method for determination of branch position depending on the currently exist branches position. It aims at embedding/removing new knowledge into/from the current solutions set. Let $S_1 = s_{1,1}s_{1,2}\ldots s_{1,i}\ldots s_{1,n}$ be a sapling. If a branch occurs in point $s_{1,i}$, then the probability of this pint could be calculated in two ways listed below (see Eqs. 17.131 and 17.132, respectively) (Karci 2007a, b, c; Karci and Alatas 2006): linear and non-linear. The distance between $s_{1,i}$ and $s_{1,j}$ (where $i \neq j$) can be considered as $|j - i|$ or $|i - j|$.

$$\text{linear case: } P\left(s_{1,j}|s_{1,i}\right) = 1 - \frac{1}{(|j - i|)^2}, \quad i \neq j, \tag{17.131}$$

$$\text{non-linear case: } P\left(s_{1,j}|s_{1,i}\right) = 1 - \frac{1}{e^{(|j-i|)^2}}, \quad i \neq j. \tag{17.132}$$

Vaccinating: aims to generate new saplings from currently exist saplings which are similar; since the dissimilarity of saplings affects the success of vaccinating process (i.e., vaccinating success is proportional to the dissimilarity of both saplings). In SGuA, if $\text{Dis}(S_1, S_2) \geq$ threshold, the dissimilarity of saplings is computed in the following two ways shown in Eqs. 17.133 and 17.134, respectively (Karci 2007a, b, c; Karci and Alatas 2006):

$$S_1 = \begin{cases} s_{1,i} & \text{if } s_{1,i} = s_{2,i} \\ random(1) & \text{if } s_{1,i} \neq s_{2,i} \end{cases}, \tag{17.133}$$

$$S_2 = \begin{cases} s_{2,i} & \text{if } s_{2,i} = s_{1,i} \\ random(1) & \text{if } s_{2,i} \neq s_{1,i} \end{cases}, \tag{17.134}$$

where S_1 and S_2 are obtained as consequence of applying vaccinating process to S_1 and S_2; $random(1)$ generates a random number which is 0 or 1.

The initial value of threshold depends on the problem solvers. For example, G and H are saplings and the similarity of S_1 and S_2 is Dis $(S_1, S_2) = \sum_{i=1}^{n} |s_{1,i} - s_{2,i}|/u_i - l_i$. If Dis $(S_1, S_2) \geq n \cdot \varepsilon$, where ε is a user-defined constant $(0 < \varepsilon < 1)$, then S_1 and S_2 are vaccinated through Eqs. 17.135 and 17.136, respectively (Karci 2007a, b, c; Karci and Alatas 2006):

$$S_1 = \begin{cases} s_{1,i} & \text{if } \frac{|s_{1,i} - s_{2,i}|}{u_i - l_i} \leq \varepsilon \\ s_{2,i} & \text{if } \frac{|s_{1,i} - s_{2,i}|}{u_i - l_i} > \varepsilon \end{cases}, \tag{17.135}$$

$$S_2 = \begin{cases} s_{2,i} & \text{if } \frac{|s_{1,i} - s_{2,i}|}{u_i - l_i} \leq \varepsilon \\ s_{1,i} & \text{if } \frac{|s_{1,i} - s_{2,i}|}{u_i - l_i} > \varepsilon \end{cases}. \tag{17.136}$$

In fact, the vaccinating process is opposite to mating process, since vaccinating operator uses dissimilarity in the garden. Thus, the vaccinating operator can also compute the distance between saplings and then compute the probability of saplings as defined by Eq. 17.137 (Karci 2007a, b, c; Karci and Alatas 2006):

$$P_v(S_1, S_2) = \frac{\left(\sum_{i=1}^{n} (s_{1,i} - s_{2,i})^2\right)^{1/2}}{R}, \tag{17.137}$$

where $P_v(S_1, S_2)$ is the probability of S_1 and S_2 to be vaccinated.

17.41.2 Performance of SGuA

Compare with GA, the SGuA has some unique characteristics (Karci 2007b): (1) it uses objective function for determination of quality of saplings in contrast to GA due to the difficulty of defining fitness function; (2) the SGuA uses less parameter determined by the user (only one for vaccinating operator) and obtained better results with less time steps with respect to GA; (3) it uses similarity and dissimilarity properties in to current solutions set with property of new information not adding in neighbor points which GA did; (4) GA is a global search method but SGuA contains both local and global search steps. Furthermore, one of the unique features of the PA is that the operator within SGuA can be applied in two different ways: sequentially and separately. Those processes allow the SGuA more flexible and faster than others.

17.42 Seeker Optimization Algorithm

In this section, we will introduce an emerging CI algorithm that is inspired by the act of human searching behaviour.

17.42.1 Fundamentals of Seeker Optimization Algorithm

Seeker optimization algorithm (SeOA) was originally proposed in Dai et al. (2006, 2007). There are several variants and applications can be found in the literature (Dai et al. 2009a, b, 2010a, b; Shaw et al. 2011)

- Step 1: Initialization. Creating S positions which are described as Eq. 17.138 (Dai et al. 2007):

$$\{x_i(t)|x_i(t) = (x_{i1}, x_{i2}, \ldots, x_{iD}); \quad i = 1, 2, \ldots, S; \quad t = 0\}. \tag{17.138}$$

  The positions are randomly and uniformly distributed in the parametric space.

- Step 2: Computing and evaluating each seeker's fitness value.
- Step 3: Performing searching strategy. Giving search variables which includes centre position vector, searching direction, searching radius, and trust degree.
- Step 4: Updating position. The new position of each seeker can be computed through Eq. 17.139 (Dai et al. 2007):

$$\vec{x}(t+1) = \vec{c} + \vec{d} \cdot \vec{r} \cdot \sqrt{-\log(\vec{\mu})}. \tag{17.139}$$

- Step 5: Checking whether the stopping criterion is met. If yes, terminating the algorithm; otherwise, return to Step 3.

17.42.2 Performance of SeOA

To verify the proposed SeOA, 6 typical testing functions with varying complexities and number of variables were employed in Dai et al. (2007). These functions included such as Goldstein and Price function, De Jong's function 2, and Griewangk's function. In comparison with other CI algorithms (e.g., GA, PSO), SeOA outperformed its competitors in all cases.

17.43 Self-organising Migrating Algorithm

In this section, we will introduce an emerging CI algorithm that is based on the competitive-cooperative behaviour of intelligent creatures.

17.43.1 *Fundamentals of Self-organising Migrating Algorithm*

Self-organising migrating algorithm (SOMA) was first proposed in Zelinka and Lampinen (2000). There are several variants and applications can be found in the literature (Nolle et al. 2005; Senkerik et al. 2010; Zelinka et al. 2009; Davendra and Zelinka 2009; Davendra et al. 2013). Two evolutionary operators (i.e., mutation and crossover) are employed in SOMA to maintain the perturbation of individuals and movement of an element. The main steps of SOMA are outlined as follows (Davendra et al. 2013):

- Step 1: Initializing the parameters.
- Step 2: Creation the populations. The population is generated via Eqs. 17.140 and 17.141, respectively (Davendra et al. 2013):

$$P = \left\{ X_1^t, X_2^t, \ldots, X_\beta^t \right\}, \tag{17.140}$$

$$X_i^t = \left(x_{i,1}^t, x_{i,2}^t, \ldots, x_{i,n}^t \right), \quad \text{where } x_{i,j}^t : \begin{cases} i = 1, \ldots, \beta \\ j = 1, \ldots, N \end{cases}, \tag{17.141}$$

  where β is the number of individuals, $x_{i,j}^t$ represents the element in each individual, and N is the dimension of the problem.

- Step 3: Iterative procedure. First, Evaluate the cost function of each individual as follows (Davendra et al. 2013):

$$C_i^t = f\left(X_i^t\right), \; i = 1, \ldots, \beta. \tag{17.142}$$

Second, create the perturbation matrix $(\mathbf{A}_{i,j}(PRT))$ for each element $(x_{i,j}^t)$ in an individual (X_i^t) as defined by Eq. 17.143 (Davendra et al. 2013):

$$\mathbf{A}_{i,j} = \begin{cases} 1 & \text{if } rand(\,) < PRT \\ 0 & \text{otherwise} \end{cases}, \begin{cases} i = 1, 2, \ldots, \beta \\ j = 1, 2, \ldots, N \end{cases}, \tag{17.143}$$

where $PRT \in [0, 1]$ is a parameter that to achieve perturbation.

Third, all individuals perform their run towards the selected individual (leader), which has the best fitness for the migration as follows (Davendra et al. 2013):

$$x_{i,j}^{t} = x_{i,j}^{t-1} + \left(x_{L,j}^{t-1} - x_{i,j}^{t-1}\right) s\mathbf{A}_{i,j}, \qquad (17.144)$$

where $x_{i,j}^{t}$ is new candidate individual, $x_{i,j}^{t-1}$ is the original individual, $x_{L,j}^{t-1}$ is the leader individual, $s \in [0, \text{path length}]$, and $\mathbf{A}_{i,j}$ is perturbation matrix.
Fourth, Calculate the cost function and keep the best solutions.

- Step 4: Post process and visualize results.

17.43.2 Performance of SOMA

To test the performance of SOMA, a set of experimental studies are conducted in Zelinka and Lampinen (2000). Computational results showed that SOMA is very competitive.

17.44 Sheep Flock Heredity Model Algorithm

In this section, we will introduce an emerging CI algorithm that is based on some observations from sheep herd (Mills et al. 2010).

17.44.1 Fundamentals of Sheep Flock Heredity Model Algorithm

Sheep flock heredity model (SFHM) algorithm was originally proposed in Nara et al. (1999) and Kim and Ahn (2001). There are several applications can be found in the literature (Chandrasekaran et al. 2006; Subbaiah et al. 2009; Anandaraman 2011; Venkumar and Sekar 2012; Anandaraman et al. 2012; Mukherjee et al. 2012). The natural evolution phenomenon of sheep flocks can be associated to the genetic operations of string which we can define the following two kinds of genetic operation: (1) Traditional genetic operations between two strings; and (2) Genetic operations between sub-strings within one string. This kind of genetic operation is referred to as the "multi-stage genetic operation". In summary, to implement SFHM algorithm, the following steps need to be performed (Chandrasekaran et al. 2006; Nara et al. 1999; Mukherjee et al. 2012; Kim and Ahn 2001):

- Step 0: Initializing the population of artificial sheep herd.
- Step 1: Selecting the parent, setting the probability of sub-chromosome level crossover, performing the sub-chromosome level of crossover.
- Step 2: Selecting two sequences from population, setting crossover probability, performing chromosome level of crossover.
- Step 3: Checking the termination condition.

17.44.2 Performance of SFHM

To verify the proposed SFHM, Kim and Ahn (2001) tested it on a 23 generators' maintenance scheduling problem. It was assumed in the study that the system load will increase by 2 % per annual. The simulation results showed that the proposed SFHM outperform its competitor algorithm with a better solution quality and almost twice of the calculation times' reduction.

17.45 Simple Optimization Algorithm

In this section, we will introduce an emerging CI algorithm that is based on two very simple mechanisms, namely, exploration and exploitation.

17.45.1 Fundamentals of Simple Optimization Algorithm

Simple optimization (SPOT) algorithm, a population-based approach, was recently proposed in Hasançebi and Azad (2012). Briefly, the SPOT algorithm can be summarized as follows (Hasançebi and Azad 2012):

- Step 1: Generating a population of stochastically proposed candidate solutions and initiating the exploration procedure.
- Step 2: Evaluating the members of population according to an objective function.
- Step 3: Determining the best candidate solution among the whole population group.
- Step 4: Calculating the standard deviation of each column of population.
- Step 5: Starting the exploitation process and creating a set of new candidate solutions through Eq. 17.145 (Hasançebi and Azad 2012):

$$x_{new(i)} = x_{best(i)} + \lambda_2 \cdot R_{(i)}, \tag{17.145}$$

  where $\lambda_2 = 0.5\lambda_1$, in comparison with the exploration stage.
- Step 6: Evaluating the newly created candidate solutions.
- Step 7: Replacing the worst population members with the better new ones.
- Step 8: Determining the best candidate solution within the population.
- Step 9: Calculating the standard deviation again for each column of population.
- Step 10: Activating the exploration procedure and creating a set of new candidate solutions through Eq. 17.146 (Hasançebi and Azad 2012):

$$x_{new(i)} = x_{best(i)} + \lambda_1 \cdot R_{(i)}, \tag{17.146}$$

where λ_1 stands for a positive constant, and $R_{(i)}$ represents a normal distributed random number with a mean zero and a stand deviation of $\sigma_{R(i)}$.

- Step 11: Evaluating the newly created candidate solutions.
- Step 12: Checking whether the stopping criterion is met. If not, repeating the algorithm from Step 3; otherwise, terminates the algorithm.

17.45.2 Performance of SPOT

Hasançebi and Azad (2012) employed two well-known benchmark engineering optimization problems, namely, welded beam and pressure vessel design optimization, to verify the proposed SPOT algorithm. Compared with the best available methods found in the literature, the results obtained by SPOT is very attractive.

17.46 Slime Mold Algorithm

In this section, we will introduce an emerging CI algorithm that is based on slime mold related studies (Newell 1978).

17.46.1 Fundamentals of Slime Mold Algorithm

Slime mold algorithm (SMA) was recently proposed in Li et al. (2011). There are several slime mold related applications can be found in the literature (Umedachi et al. 2010; Tero et al. 2010; Adamatzky and Oliveira 2011; Li et al. 2011; Shann 2008) To implement SSOA, the following steps need to be performed (Li et al. 2011):

- Step 1: Initializing slime mold around the one or more food sources.
- Step 2: The slime mold will stream towards a newly introduced food source based on a gradient descent rule. The total field φ is computed via Eq. 17.147 (Li et al. 2011):

$$\varphi = \sum_{i=1}^{N} (f_i - \bar{f}) \cdot \ln\left(|\mathbf{x} - \mathbf{x}_i|^2\right), \tag{17.147}$$

  where the number of discovered food sources is denoted by N, f_i is the remaining nutrient amount found at the ith food source, and $\bar{f}$ is the mean of all f_i.

- Step 3: Once slime mold arrives at a newly discovered food resource, such food resource will be linked to the network. Nutrient values of the new connected food resources will gradually decay.

- Step 4: Checking whether the stopping criterion is met. If yes, the algorithm terminates; otherwise, i.e., unconnected food sources still existing, return to Step 2.

17.46.2 Performance of SMA

Based on the proposed SMA, Li et al. (2011) presented two self-organizing routing protocols for wireless sensor networks by considering both efficiency and robustness. The simulation conducted in their study proved that the proposed protocol is effective in building network connectivities, with a trade-off between efficiency and robustness.

17.47 Social Emotional Optimization Algorithm

In this section, we will introduce an emerging CI algorithm which is inspired from the human society. As we know, group decisions are very important for us and they have been studied for millennia. They are range from small-scale decisions, e.g., some advices taken by groups of relatives, friends or colleagues, to large-scale decisions, e.g., nation-wide democratic electrons and international agreements (Conradt and List 2009).

17.47.1 Fundamentals of Social Emotional Optimization Algorithm

Social emotional optimization algorithm (SEOA) was originally proposed in Xu et al. (2010), Wu et al. (2011), Wei et al. (2010), Cui and Cai (2010), Chen et al. (2010b) and Cui et al. (2010, 2011). Each person is viewed as a solution. Through cooperation and competition mechanisms, the personal social status will be increased and the best one will win and output as the final solution. The main steps of SEOA are outlined as follows:

- Step 1: Initializing all individuals randomly in the search space. In the fist step, all individuals' emotion indexes are set to 1 as shown in Eq. 17.148 (Cui et al. 2012):

$$\vec{x}_j(1) = \vec{x}_j(0) \oplus \text{Manner}_1, \tag{17.148}$$

where $\vec{x}_j(1)$ represents the social position of the jth individual in the initialization period, $\oplus$ means the operation. The movement phase of Manner_1 is defined by Eq. 17.149 (Cui et al. 2012):

$$\text{Manner}_1 = -k_1 \cdot rand_1 \cdot \sum_{w=1}^{L} \left(\vec{x}_w(0) - \vec{x}_j(0) \right), \tag{17.149}$$

where k_1 is a parameter used to control the emotion changing size, *rand* is a random number with uniform distribution, L represents the worst individuals that are selected to provide a remainder for the jth individual to avoid the wrong behaviour.

- Step 2: Computing the fitness value of each individual according to the objective function.
- Step 3: For the jth individual, determining the value $\vec{X}_{j,best}(0)$.
- Step 4: For all population, determining the value $Status\vec{s}_{best}(0)$.
- Step 5: Determining three emotional index via Eq. 17.150 (Cui et al. 2012):

$$\begin{cases} \vec{x}_j(t+1) = \vec{x}_j(t) \oplus \text{Manner}_2 & \text{If } BI_j(t+1) < TH_1 \\ \vec{x}_j(t+1) = \vec{x}_j(t) \oplus \text{Manner}_3 & \text{If } TH_1 \leq BI_j(t+1) < TH_2 \\ \vec{x}_j(t+1) = \vec{x}_j(t) \oplus \text{Manner}_4 & \text{otherwise} \end{cases}, \tag{17.150}$$

where TH_1 and TH_2 are two thresholds aiming to restrict the different behaviour manner.

- Step 6: Determining different decisions according to Eqs. 17.151–17.153, respectively (Cui et al. 2012):

$$\text{Manner}_2 = k_3 \cdot rand_3 \cdot \left(\vec{X}_{j,best}(t) - \vec{x}_j(t) \right) + k_2 \cdot rand_2 \cdot \left(Status\vec{s}_{best}(t) - \vec{x}_j(t) \right), \tag{17.151}$$

$$\begin{aligned} \text{Manner}_3 = {} & k_3 \cdot rand_3 \cdot \left(\vec{X}_{j,best}(t) - \vec{x}_j(t) \right) \\ & + k_2 \cdot rand_2 \cdot \left(Status\vec{s}_{best}(t) - \vec{x}_j(t) \right) \\ & - k_1 \cdot rand_1 \cdot \sum_{w=1}^{L} \left(\vec{x}_w(0) - \vec{x}_j(0) \right), \end{aligned} \tag{17.152}$$

$$\text{Manner}_4 = k_3 \cdot rand_3 \cdot \left(\vec{X}_{j,best}(t) - \vec{x}_j(t) \right) - k_1 \cdot rand_1 \cdot \sum_{w=1}^{L} \left(\vec{x}_w(0) - \vec{x}_j(0) \right), \tag{17.153}$$

where $Status\vec{s}_{best}(t)$ represents the best society status position obtained from all people previously, and $\vec{X}_{j,best}(t)$ denotes the best status value obtained by the jth individual previously. Both can be defined by Eqs. 17.154 and 17.155, respectively (Cui et al. 2012):

$$\vec{Status}_{best}(t) = \arg\min\left\{f\left(\vec{x}_w(h)|1 \leq h < t\right)\right\}, \tag{17.154}$$

$$\vec{X}_{j,best}(t) = \arg\min\left\{f\left(\vec{x}_j(h)|1 \leq h < t\right)\right\}. \tag{17.155}$$

- Step 7: Making mutation operation.
- Step 8: Checking termination criteria, if it is satisfied, output the best solution; otherwise, go to Step 3.

17.47.2 Performance of SEOA

To test the performance of SEOA, a set of benchmark functions were adopted in Cui et al. (2012). Compared with other CI algorithms, SEOA has a remarkable superior performance in terms of accuracy and convergence speed.

17.48 Social Spider Optimization Algorithm

In this section, we will introduce an emerging CI algorithm that is based on some findings regarding the spider colony (Bater 2007; Chapman 2013; Levin 2013a, b, c, d, e, f).

17.48.1 Fundamentals of Social Spider Optimization Algorithm

Spider algorithm has been around for a while for dealing with like search engine optimization (Whitehouse and Lubin 1999; Jonassen 2006; Du et al. 2005). Social spider optimization algorithm (SSOA) was recently proposed in Cuevas et al. (2013). To implement SSOA, the following steps need to be performed (Cuevas et al. 2013):

- Step 1: Setting the total number of n-dimensional colony members as N and defining the number of male N_{male} and female N_{female} spiders in the entire colony **S** based on Eq. 17.156 (Cuevas et al. 2013):

$$\begin{aligned} N_{male} &= N - N_{female} \\ N_{female} &= floor[(0.9 - rand \cdot 0.25) \cdot N], \end{aligned} \tag{17.156}$$

where *rand* stands for a random number which falls within the range of $[0, 1]$, and $floor(\cdot)$ indicates the mapping between a real and an integer numbers.

- Step 2: Initializing stochastically the female and male members and computing the mating radius according to Eq. 17.157 (Cuevas et al. 2013):

$$r = \frac{\sum_{j=1}^{n} \left(p_j^{high} - p_j^{low} \right)}{2 \cdot n}. \tag{17.157}$$

- Step 3: Calculating the weight of each spider in colony **S** through Eq. 17.158 (Cuevas et al. 2013):

$$w_i = \frac{J(\mathbf{s}_i) - worst_{\mathbf{s}}}{best_{\mathbf{s}} - worst_{\mathbf{s}}}, \tag{17.158}$$

where $J(\mathbf{s}_i)$ denotes the fitness value acquired through the evaluation of the spider position $\mathbf{s}_i$ with regard to the objective function $J(\cdot)$.

- Step 4: Moving female spiders according to the female cooperative operator modelled as shown in Eq. 17.159 (Cuevas et al. 2013):

$$\mathbf{f}_i^{k+1} = \begin{cases} \mathbf{f}_i^k + \alpha \cdot Vibc_i \cdot \left(\mathbf{s}_c - \mathbf{f}_i^k\right) + \beta \cdot Vibb_i \cdot \left(\mathbf{s}_b - \mathbf{f}_i^k\right) \\ \quad +\delta \cdot \left(rand - \frac{1}{2}\right) \text{ with probability } PF \\ \mathbf{f}_i^k - \alpha \cdot Vibc_i \cdot \left(\mathbf{s}_c - \mathbf{f}_i^k\right) - \beta \cdot Vibb_i \cdot \left(\mathbf{s}_b - \mathbf{f}_i^k\right) \\ \quad +\delta \cdot \left(rand - \frac{1}{2}\right) \text{ with probability } 1 - PF \end{cases}, \tag{17.159}$$

where α, β, δ, and *rand* are random numbers which fall within the range of $[0, 1]$.

- Step 5: Similarly moving male spiders according to the male cooperative operator expressed as Eq. 17.160 (Cuevas et al. 2013):

$$\mathbf{m}_i^{k+1} = \begin{cases} \mathbf{m}_i^k + \alpha \cdot Vibf_i \cdot \left(\mathbf{s}_f - \mathbf{m}_i^k\right) + \delta \cdot \left(rand - \frac{1}{2}\right) & \text{if } w_{N_{female}+i} > w_{N_{female}+m} \\ \mathbf{m}_i^k + \alpha \cdot \left(\frac{\sum_{h=1}^{N_{male}} \mathbf{m}_h^k \cdot w_{N_{female}+h}}{\sum_{h=1}^{N_{male}} w_{N_{female}+h}} - \mathbf{m}_i^k \right) & \text{if } w_{N_{female}+i} \le w_{N_{female}+m} \end{cases}, \tag{17.160}$$

where $\mathbf{s}_f$ indicates the nearest female spider to the male individual.

- Step 6: Performing the mating operation.
- Step 7: Checking whether the stopping criterion is satisfied. If yes, the algorithm terminates; otherwise, return to Step 3.

17.48.2 Performance of SSOA

Different to other evolutionary algorithms, in SSOA, each individual spider is modelled by taking its gender into account. This design allows incorporating computational mechanisms to avoid critical flaws and incorrect exploration-exploitation trade-off. In order to show how the SSOA performs, Cuevas et al. (2013) collected a comprehensive set of 19 benchmark test function from the literature. In comparison with other CI algorithms (e.g., PSO), the experimental results confirmed an acceptable performance of the SSOA in terms of the solution quality.

17.49 Society and Civilization Algorithm

In this section, we will introduce an emerging CI algorithm that is based on some findings regarding the human social behaviour (Irvine 2013; Müller 2013; Gross 2014; Bhugra et al. 2013; Adair 2007; Chen and Lee 2008; Savage 2012; Magstadt 2013; Chalmers 2010).

17.49.1 Fundamentals of Society and Civilization Algorithm

Society and civilization algorithm (SCA) was recently proposed in Ray and Liew (2003). To implement US algorithm, the following steps need to be performed (Ray and Liew 2003):

- In SCA, for each individual, c indicates the constraint satisfaction vector denoted by $c = [c_1, c_2, \ldots, c_s]$ as shown in Eq. 17.161 (Ray and Liew 2003):

$$c_i = \begin{cases} 0 & \begin{array}{c}\text{if } i\text{th constraint satisfied}\\ i = 1, 2, \ldots, s\end{array} \\ -g_i(\mathbf{x}) & \begin{array}{c}\text{if } i\text{th constraint violated}\\ i = 1, 2, \ldots, q\end{array} \\ h_i(\mathbf{x}) - \delta & \begin{array}{c}\text{if } i\text{th constraint violated}\\ i = q+1, q+2, \ldots, q+r\end{array} \\ \delta - h_i(\mathbf{x}) & \begin{array}{c}\text{if } i\text{th constraint violated}\\ i = q+r+1, q+r+2, \ldots, s\end{array} \end{cases}. \tag{17.161}$$

- Creating N random individuals standing for a civilization.
- Evaluating each individual according to the computed objective function value and constraints.

- Building societies so that $\bigcup_{i=1}^{K(t)} Soc_i(t) = Civ(t)$ and $Soc_i(t) \cap Soc_j(t) = \phi$, for $i \neq j$.
- Identifying society leaders so that $Soc\_L_i(t) \subset Soc_i(t)$, $i = 1, 2, \ldots, K(t)$.
- Performing move function *Move*( ).

17.49.2 Performance of SCA

In order to show how the SCA algorithm performs, Ray and Liew (2003) employed 4 well-studied benchmark engineering design optimization problems such as welded beam design, spring design, speed reducer design, and the three-bar truss design. Compared with the best results obtained from the literature, the proposed SCA performed consistently well on all cases. Meanwhile, the algorithm exhibits a robust convergence for all selected problems which make it a very attractive optimization algorithm.

17.50 Stem Cells Optimization Algorithm

During the past three decades, our understanding of embryonic cells has increased dramatically. The humbling beginning started thirty years ago when embryonic stem cells were first cultured from mouse embryos. It was only 15 years later, we were able to derive human embryonic stem cells from human embryos that were donated from early blastocysts which are no more need for in vitro fertilization (Sell 2013). Throughout all creature's life, the balance between cell birth and death is largely regulated by complex genetic systems in response to various growth and death signals. During this dynamic procedure, stem cells are present within most if not all multicellular organisms and are crucial for developing, tissue repairing, as well as aging and cancer (Resende and Ulrich 2013; Sell 2013). Briefly, stems cells can be defined as biological cells which have the ability of self-renewal and the capability in differentiating into various cell types (Sell 2013). They are thus regarded as one of the most important biological components which is essential to the proper growth and development in the process of embryogenesis. Since the detailed information regarding the stem cells is out of the scope of the present book, interested readers are referred to the corresponding studies, e.g., (Resende and Ulrich 2013; Sell 2013), for more recent advancement in this field.

For the rest of this section, we will introduced an emerging CI algorithm which is based on the findings of some important characteristics of stem cells.

17.50.1 Fundamentals of Stem Cells Optimization Algorithm

Stem cells optimization algorithm (SCOA) was originally proposed in Taherdangkoo et al. (2011, 2012b). To perform the SCOA algorithm, the following procedure needs to be followed (Taherdangkoo et al. 2012b):

- First, dividing the problem space into sections. The process can be accomplished totally in a random manner;
- Second, generating the initial population randomly and uniformly distributed in the whole search space of the target problem. At this stage, similar to most optimization algorithms, a variable matrix needs to be established for the purpose of obtaining a feedback with respect to problem variables. In SCOA, the key stem cells features are used to form the initial variable matrix. Such features may include liver cells, intestinal cells, blood cells, neurons, heart muscle cells, pancreatic islets cells, etc. Basically, the initial matrix can be express as Eq. 17.162 (Taherdangkoo et al. 2012b):

$$\text{Population} = \begin{bmatrix} X_1 \\ X_2 \\ \dots \\ X_N \end{bmatrix}, \tag{17.162}$$

  where $X_i = \text{Stem Cells} = [SC_1, SC_2, \ldots, SC_N]$; $i = 1, 2, \ldots, N$.
  In SCOA, some initialized parameters are defined as follows: M represents the maximum of stem cells; P stands for population size $(10 < P \leq M)$; $C_{Optimum}$ indicates the best of stem cell in each iteration; χ denotes the penalty parameter which is used to stop the growth of stem cell; and sc^i is the ith stem cell in the population.
- Third, the cost of each stem cell is obtained a criterion function which is determined based on the nature of the target problem. In SCOA, two types of memory, namely, local- and global-memory, are defined for each cell in which the local-memory is used to store the cost of each stem cell, and the global-memory stores the best cost among all locally stored cost values;
- Then, a self-renewal process will be performed which involves only the best cells of each area. At this stage, the information of each area's best cells will be shared and the cell that possesses the best cost will thus be chosen. In SCOA, such cell is designed to play a more important role than other cells. Briefly, the stem cells' self-renewal operation is computed through Eq. 17.163 (Taherdangkoo et al. 2012b):

$$SC_{Optimum}(t+1) = \zeta SC_{Optimum}(t), \tag{17.163}$$

  where the iteration number is denoted by t, $SC_{Optimum}$ represents the best stem cell found in each iteration, and ζ is a random number which falls within $[0, 1]$.

- Next, the above mentioned procedure will continue until the SCOA arrives at the goal of getting the best cell while keeping the value of cost function as low as possible. This is acquired via Eq. 17.164 (Taherdangkoo et al. 2012b):

$$x_{ij}(t+1) = \mu_{ij} + \varphi\big(\mu_{ij}(t) - \mu_{kj}(t)\big), \tag{17.164}$$

  where the ith stem cell position for the solution dimension j is represented by x_{ij}, the iteration number is denoted by t, two randomly selected stem cells for the solution dimension j are denoted by μ_{ij} and μ_{kj}, respectively, and $\varphi(\tau)$ (if $\mu_{ij}(t) - \mu_{kj}(t) = \tau$) generates a random variable falls within $[-1, 1]$.
- Finally, the best stem cell is selected when it has the most power relative to other cells. The comparative power can be computed via Eq. 17.165 (Taherdangkoo et al. 2012b):

$$\varsigma = \frac{SC_{Optimum}}{\sum_{i=1}^{N} SC_i}, \tag{17.165}$$

  where ς stands for stem cells' comparative power, $SC_{Optimum}$ denotes the stem cells selected in terms of cost, and N represents the final population size, i.e., when the best solution is obtained and the SCOA algorithm terminates.

17.50.2 Performance of SCOA

Similar to other optimization algorithms, SCOA is also a population-based algorithm. But the difference between SCOA and other CI algorithms lies in that it employs minimal constraints and thus has a simpler implementation. The converging speed of SCOA is faster than other algorithms in virtue of its simplicity and its capability of escaping from local minima (Taherdangkoo et al. 2012b).

17.51 Stochastic Focusing Search Algorithm

In this section, we will introduce an emerging CI algorithm that is based on the behaviours observed from human randomized search.

17.51.1 Fundamentals of Stochastic Focusing Searching Algorithm

Stochastic focusing searching (SFS) algorithm was recently proposed in Zheng et al. (2009). In order to implement SFS, the following steps need to be considered (Zheng et al. 2009; Wang et al. 2008):

- In SFS, the targeted optimization problems are regarded as minimization problems, and the particles are controlled according to Eqs. 17.166–17.168, respectively (Zheng et al. 2009).

$$\vec{v}_i(t) = \begin{cases} Rand(\,)\cdot[R_{ti} - \vec{x}_i(t-1)] & \text{if } fun[\vec{x}_i(t-1)] \geq fun[\vec{x}_i(t-2)] \\ \vec{v}_i(t-1) & \text{if } fun[\vec{x}_i(t-1)] < fun[\vec{x}_i(t-2)] \end{cases}, \tag{17.166}$$

$$\vec{x}_i(t) = \vec{v}_i(t) + \vec{x}_i(t-1), \tag{17.167}$$

$$\vec{x}_i(t) = \vec{x}_i(t-1) \quad \text{if } fun[\vec{x}_i(t)] \geq fun[\vec{x}_i(t-1)], \tag{17.168}$$

where $fun[\vec{x}_i(t)]$ denotes the objective function value of $\vec{x}_i(t)$, and R_{ti} represents a random chosen position in the neighbour space R_t of $\vec{g}_{best}$.
- It can be observed that each individual particle search in a decreasing R_t. Therefore, an appropriate selection of $w$is crucial not only to the convergence of particles, but also to the avoidance of local optimal. Accordingly w can be defined by Eq. 17.169 (Zheng et al. 2009):

$$w = \left(\frac{G-t}{G}\right)^{\delta}, \tag{17.169}$$

where the maximum generation is denoted by G, and δ denotes a positive number.

17.51.2 Performance of SFS

To evaluate the proposed SFS, Zheng et al. (2009) employed a test suite of benchmark functions collected from the literature. The testing function group consists of a diverse set of problems such as functions 1–5 are unimodal, function 8–13 are multimodal function, and function 14–23 are low-dimensional function. In comparison with other CI algorithms (e.g., DE, PSO), the experimental results demonstrated that SFS posses a better global searching capability and faster converging speed for most of the testing cases.

17.52 Swallow Swarm Optimization Algorithm

In this section, we will introduce an emerging CI algorithm that is based on the social behaviour of swallow swarm.

17.52.1 Fundamentals of Swallow Swarm Optimization Algorithm

Swallow swarm optimization (SSO) algorithm was originally proposed in Neshat et al. (2013). The algorithm shares some similarities with other population-based CI algorithms, e.g., PSO, but with several key differences. Basically, there are three types of swallows in the proposed SSO algorithm (Neshat et al. 2013):

- Explorer swallow (e_i): The major population of swallow colony is composed of explorer swallows. Their main task is to search the target problem space. After arriving at a potential solution point, an explorer swallow will use a special to guide the group moving toward such place. If the place is indeed the best one found in the problem space, this swallow will play a role of head leader (HL_i). However, if the place is a good but not the best position in comparison with other locations, this swallow will be selected as a local leader (LL_i). Otherwise, each explorer swallow e_i will make a random movement according to Eqs. 17.170–17.181, respectively (Neshat et al. 2013):

$$V_{HL_{i+1}} = V_{HL_i} + \alpha_{HL} rand(\,)(e_{best} - e_i) + \beta_{HL} rand(\,)(HL_i - e_i), \tag{17.170}$$

$$\alpha_{HL} = \{\text{if } (e_i = 0 \| e_{best} = 0) \rightarrow 1.5\}, \tag{17.171}$$

$$\alpha_{HL} = \begin{cases} \text{if } (e_i < e_{best}) \&\& (e_i < HL_i) & \rightarrow \dfrac{rand(\,) \cdot e_i}{e_i \cdot e_{best}} \quad (\text{where } e_i, e_{best} \neq 0) \\ \text{if } (e_i < e_{best}) \&\& (e_i > HL_i) & \rightarrow \dfrac{2 \cdot rand(\,) \cdot e_{best}}{1/(2 \cdot e_i)} \quad (\text{where } e_i \neq 0) \\ \text{if } (e_i > e_{best}) & \rightarrow \dfrac{e_{best}}{1/(2 \cdot rand(\,))} \end{cases}, \tag{17.172}$$

$$\beta_{HL} = \{\text{if } (e_i = 0 \| e_{best} = 0) \rightarrow 1.5\}, \tag{17.173}$$

$$\beta_{HL} = \begin{cases} \text{if } (e_i < e_{best}) \&\& (e_i < HL_i) & \rightarrow \dfrac{rand(\,) \cdot e_i}{e_i \cdot HL_i} \quad (\text{where } e_i, HL_i \neq 0) \\ \text{if } (e_i < e_{best}) \&\& (e_i > HL_i) & \rightarrow \dfrac{2 \cdot rand(\,) \cdot HL_i}{1/(2 \cdot e_i)} \quad (\text{where } e_i \neq 0) \\ \text{if } (e_i > e_{best}) & \rightarrow \dfrac{HL_i}{1/(2 \cdot rand(\,))} \end{cases}, \tag{17.174}$$

$$V_{LL_{i+1}} = V_{LL_i} + \alpha_{LL} rand(\,)(e_{best} - e_i) + \beta_{LL} rand(\,)(LL_i - e_i), \tag{17.175}$$

$$\alpha_{LL} = \{\text{if } (e_i = 0 \| e_{best} = 0) \rightarrow 2\}, \tag{17.176}$$

$$\alpha_{LL} = \begin{cases} \text{if } (e_i < e_{best})\&\&(e_i < LL_i) & \rightarrow \frac{rand() \cdot e_i}{e_i \cdot e_{best}} \quad (\text{where } e_i, e_{best} \neq 0) \\ \text{if } (e_i < e_{best})\&\&(e_i > LL_i) & \rightarrow \frac{2 \cdot rand() \cdot e_{best}}{1/(2 \cdot e_i)} \quad (\text{where } e_i \neq 0) \\ \text{if } (e_i > e_{best}) & \rightarrow \frac{e_{best}}{1/(2 \cdot rand())} \end{cases}, \tag{17.177}$$

$$\beta_{LL} = \{\text{if } (e_i = 0 \| e_{best} = 0) \rightarrow 2\}, \tag{17.178}$$

$$\beta_{LL} = \begin{cases} \text{if } (e_i < e_{best})\&\&(e_i < LL_i) & \rightarrow \frac{rand() \cdot e_i}{e_i \cdot LL_i} \quad (\text{where } e_i, LL_i \neq 0) \\ \text{if } (e_i < e_{best})\&\&(e_i > LL_i) & \rightarrow \frac{2 \cdot rand() \cdot LL_i}{1/(2 \cdot e_i)} \quad (\text{where } e_i \neq 0) \\ \text{if } (e_i > e_{best}) & \rightarrow \frac{LL_i}{1/(2 \cdot rand())} \end{cases}, \tag{17.179}$$

$$V_{i+1} = V_{HL_{i+1}} + V_{LL_{i+1}}, \tag{17.180}$$

$$e_{i+1} = e_i + V_{i+1}. \tag{17.181}$$

Vector V_{HL_i} has an important impact on explorer swallow's behaviour. Each explorer swallow e_i uses the nearest swallow LL_i for the purpose of calculating the vector of V_{LL_i}

.

- Aimless swallow (o_i): When the search begins, these swallows doe not occupy a good position. Their task is thus doing random search which means their positions are not affected by HL_i and LL_i. The role of an aimless swallow is more like a scout. If an aimless swallow o_i gets a better solution during its searching, it will replace its position with the nearest explorer swallow's position and then continue with search. This process is defined by Eq. 17.182 (Neshat et al. 2013):

$$o_{i+1} = o_i + \left[rand((-1, 1)) \cdot \frac{rand(\min_s, \max_s)}{1 + rand()} \right]. \tag{17.182}$$

- Leader swallow (l_i): Unlike the PSO which has only one leader particle, in SSO, there may have n_l leader swallows that are distributed or gathered in the problem space. The best leader is called leader head, while the others are called local leader. They are candidate desirable solutions that we are looking for.

17.52.2 Performance of SSO

In order to show how the SSO algorithm performs, Neshat et al. (2013) employed 16 benchmark test functions such as Sphere function, Rosenbrock function, Quadric function, Rastrigin function, Rotated Rastrigin function, and Rotated Ackley function. In comparison with other CI techniques (e.g., variant PSO, etc.), the SSO algorithm is one of the best optimization approaches in the category of swarm intelligence. It thus has the ability in solving optimization problems encountered in different scenarios.

17.53 Termite-Hill Algorithm

In this section, we will introduce an emerging CI algorithm that is based on hill building behaviour observed from real termites (Keller and Gordon 2009).

17.53.1 Fundamentals of Termite-Hill Algorithm

Termite-hill algorithm (ThA) was originally proposed (Zungeru et al. 2012) for dealing with wireless sensor networks routing problem. The key components of ThA are detailed as below (Zungeru et al. 2012):

- Component 1: Pheromone table. Pheromone plays an important role in leveraging an effective and efficient communication in real world (Touhara 2013). Similarly in ThA, the pheromone is updated through Eq. 17.183 (Zungeru et al. 2012):

$$T'_{r,s} = T_{r,s} + \gamma, \tag{17.183}$$

  where γ can be expressed as Eq. 17.184 (Zungeru et al. 2012):

$$\gamma = \frac{N}{E - \left(\frac{E_{\min} - N_j}{E_{av} - N_j}\right)}, \tag{17.184}$$

  where E denotes the initial energy of the nodes.
- Component 2: Route selection. In ThA, each of the routing tables carried by the nodes is initialized with a uniform probability distribution defined as Eq. 17.185 (Zungeru et al. 2012):

$$P_{s,d} = \frac{1}{N_k}, \tag{17.185}$$

  where N_k denotes the number of nodes in the network.

17.53.2 Performance of ThA

To verify the proposed ThA, Zungeru et al. (2012) conducted a series of experimental studies. The simulation results demonstrated that ThA is a very competitive routing algorithm in terms of energy consumption, throughput, and network lifetime.

17.54 Unconscious Search Algorithm

In this section, we will introduce an emerging CI algorithm that is based on some findings from human brain and psychology studies (Irvine 2013; Chalmers 2010; Müller 2013; Gross 2014; Bhugra et al. 2013; Şen 2014).

17.54.1 Fundamentals of Unconscious Search Algorithm

Unconscious search (US) algorithm was recently proposed in Ardjmand and Amin-Naseri (2012). The key concept of US is to use some common features found between optimization and psychoanalysis to design search algorithm. To implement US algorithm, the following steps need to be performed (Ardjmand and Amin-Naseri 2012):

- Initially, a group of suitable solutions $P = \left(P_1, P_2, \ldots, P_{|MM|}\right)$ is created. $|MM|$ is the size of the measurement matrix (MM) where the assorted group of best feasible solutions. In US, MM can be expressed as Eq. 17.186 (Ardjmand and Amin-Naseri 2012):

$$MM = \left\{P_q \middle| C(P_q) < C(P_{q+1}), \quad q = 1, 2, \ldots, |MM|\right\}, \tag{17.186}$$

  where a translation function is employed in US to map the value of the objective function of any solution P_q into a range of $(\alpha, 1-\alpha)$ for α falls within $(0, 1)$ and $P_q \in MM$. In US, the translation function f_{t_l} is defined according to Eq. 17.187 (Ardjmand and Amin-Naseri 2012):

$$f_{t_l}\left(C(P_q)\right) = \frac{1}{1 + e^{a\left(C(P_q)\right)+b}}, \quad P_q \in MM, \tag{17.187}$$

  where f_{t_l} is a sigmoid function, and a and b are variables of f_{t_l} which will be updated in each iteration.
- In US, a displacement memory, Π, is employed to remember the displace pattern in the solutions which can be calculated through Eqs. 17.188 and 17.189, respectively (Ardjmand and Amin-Naseri 2012):

$$\Pi = \{(\Pi_{jI}, \Pi_{jE}) | j = 1, 2, \ldots, |X|\}, \tag{17.188}$$

$$\begin{array}{l}\Pi_{jI} = \{\pi_{jIi} | i = 1, 2, \ldots, n; j = 1, 2, \ldots, |X|\} \\ \Pi_{jE} = \{\pi_{jEi} | i = 1, 2, \ldots, n; j = 1, 2, \ldots, |X|\}\end{array}, \tag{17.189}$$

where the number of decision parameters is denoted by n, and π_{jIi} and π_{jEi} are computed via Eqs. 17.190 and 17.191, respectively (Ardjmand and Amin-Naseri 2012):

$$\begin{array}{l}\pi_{jIi} = \left(\sum_{MS} f_{t_I}(C(P_q))\right) \\ P_q \in MM; P_q(i) \in X_j; j = 1, 2, \ldots, |X|. \\ q = 1, 2, \ldots, |MM|; i = 1, 2, \ldots, n\end{array} \tag{17.190}$$

$$\pi_{jEi} = \left(\sum_{MS} h\right); \quad j = 1, 2, \ldots, |X|; \quad i = 1, 2, \ldots, n, \tag{17.191}$$

where $MS \in \mathbb{Z}^+ - \{0\}$ represents the memory size.

- By using the displacement memory, Π, a new solution, S_1, will be created. In US, the ith solution component $S_1(i)$ will be allocated with a possible range X_j in solution space with a probability defined as Eq. 17.192 (Ardjmand and Amin-Naseri 2012):

$$\text{Prob}\{S_1(i) \in X_j\} = \frac{\frac{\pi_{jIi}}{1+(\pi_{jEi})^{\beta}}}{\sum_{j=1}^{|X|} \frac{\pi_{jIi}}{1+(\pi_{jEi})^{\beta}}}, \tag{17.192}$$

where the probability function is denoted by Prob, and β represents a predetermined constant.

- Once a displacement-free solution is reached, in order to remove the condensational resistance pattern, a condensational memory, Π', is introduced in US and it is defined as Eq. 17.193 (Ardjmand and Amin-Naseri 2012):

$$\begin{array}{l}\Pi' = \{(\Pi_i^+, \Pi_i^-) | i = 1, 2, \ldots, n\} \\ \Pi_i^+ = \left\{(\pi_{iI}^+, \pi_{iE}^+)^T \middle| i = 1, 2, \ldots, n\right\} \\ \Pi_i^- = \left\{(\pi_{iI}^-, \pi_{iE}^-)^T \middle| i = 1, 2, \ldots, n\right\}\end{array}, \tag{17.193}$$

where $\pi_{iI}^+ = \sum_{MS} f_{t_I}(C(P_q))$, $\pi_{iE}^+ = \sum_{MS} h$, $\pi_{iI}^- = \sum_{MS} f_{t_I}(C(P_q))$, and $\pi_{iE}^- = \sum_{MS} h$

17.54.2 Performance of US

Overall, US is a multi-start CI algorithm which contains three phases, namely, construction, construction review, and local search. To test the performance of US algorithm, Ardjmand and Amin-Naseri (2012) employed one bounded and six unbounded benchmark test functions and engineering design optimization problems which include such as Rosenbrock function, Goldsten and Price function, Wood function, and pressure vessel design problem. Compared with other CI algorithms, the results obtained by US is very competitive. The parameter analysis carried out in (Ardjmand and Amin-Naseri 2012) demonstrated that US is a robust and easy-to-use approach in dealing with hard optimization problems.

17.55 Wisdom of Artificial Crowds Algorithm

In this section, we will introduce an emerging CI algorithm that is based on some findings regarding the human collective intelligence (Irvine 2013; Chalmers 2010; Müller 2013; Gross 2014; Bhugra et al. 2013; Adair 2007; Chen and Lee 2008; Savage 2012).

17.55.1 Fundamentals of Wisdom of Artificial Crowds Algorithm

Wisdom of artificial crowds (WoAC) algorithm was recently proposed in Ashby and Yampolskiy (2011), Port and Yampolskiy (2012) and Yampolskiy et al. (2012). The WoAC is a post-processing algorithm in which independently decision-making artificial agents aggregate their personal solutions to reach an agreement about which answer is superior to all other solutions presented in the population (Yampolskiy et al. 2012). To implement WoAC algorithm, the following steps need to be performed (Yampolskiy et al. 2012):

- Setting up an automatic aggregation mechanism which collecting the individual solutions and producing a common solution that reflects frequent local structures of individual solutions.
- After establishing an agreement matrix, in order to transform agreements between solutions and costs, a nonlinear monotonic transformation function is applied as shown in Eq. 17.193 (Yampolskiy et al. 2012):

$$c_{ij} = 1 - I_{a_{ij}}^{-1}(b_1, b_2), \tag{17.194}$$

where $I_{a_{ij}}^{-1}(b_1, b_2)$ represents the inverse regularized beta function.

17.55.2 Performance of WoAC

To test the performance of WoAC algorithm, Yampolskiy et al. (2012) employed the classic travelling salesman problem as a benchmark. Compared with other CI algorithms (e.g., GA), the results obtained by WoAC is very competitive, in particular with an average of 3–10 % solutions quality improvement.

17.56 Wolf Colony Algorithm

As one of the largest species of the genus Canis, Northern gray wolves exhibit some of the most complex intra-specific social behaviour within the carnivores (Macdonald et al. 2004). Such behaviours include such as living in social units (i.e., packs), hunting socially, participating in group care of young offspring, and group defences of food and territory (Muro et al. 2011). According to Fuller et al. (2003) and Vucetich et al. (2004) in real world environment, living and foraging in form of packs is commonly observed when the prey base is composed of large ungulates, when the risk of losing food to scavengers is high, and when territorial defence is critical. Wolves hunt large ungulates, e.g., moose (Sand et al. 2006), in pack of two or more animals (Fuller et al. 2003) for the purpose of, e.g., reducing foraging cost (Packer and Caro 1997). In Muro et al. (2011), the authors employed two simple decentralized rules [via conventional CI approach, i.e., multi-agent system (MAS)] to regenerate the main features of the wolf pack hunting behaviour. The rules developed in their study are (1) moving towards the prey until a minimum safe distance to the prey is acquired, and (2) moving away from the other wolves (under the situation of close enough to the prey) that are adjacent to the safe distance to the prey. The detailed information regarding the wolf pack hunting behaviour is out of the scope of present book. For the rest of this section, we will introduce an emerging CI algorithm that is based on the hunting behaviours observed from a wolf colony.

17.56.1 Fundamentals of Wolf Colony Algorithm

Wolf colony algorithm (WCA) was originally proposed in Liu et al. (2011). Let D represents the dimension of the search space, n denotes the individual number, X_i stands for the position of the ith artificial wolf, then we have Eq. 17.197 (Liu et al. 2011):

$$X_i = (X_{i1}, \ldots, X_{id}, \ldots, X_{iD}), \tag{17.195}$$

where $1 \le i \le n$, and $1 \le d \le D$.

The WCA algorithm mimics several behaviours that are commonly found in a wolf colony (Liu et al. 2011).

- Searching behaviour: q artificial wolves are initialized to detect the possible quarry activities for the purpose of increasing the probability of discovering the quarry. Suppose that q scout wolves are the wolves that are closest to the quarry, maxdh denotes the maximum searching number, XX_i is the location of the ith scout wolf (totally h locations are created around the candidate wolf), and the jth searching position is denoted by Y_j, then we have Eq. 17.195 (Liu et al. 2011):

$$Y_j = XX_i + randn \cdot stepa, \tag{17.196}$$

  where a uniformly distributed random number (falling within $[-1, 1]$) is denoted by *randn*, *stepa* represents the searching step. The searching behaviour will be terminated under the following situations, i.e., the searching number is greater than maxdh or the current location is better than the optimal searching location.
- Besieging behaviour: Once a quarry is discovered by scout wolves, howl is normally used to notify other wolves about the position of the quarry. Let the location of a quarry in the dth searching space after the $k$thiteration is denoted by G_d^k, and X_{id}^k stands for the position of the ith artificial wolf, then we have Eq. 17.197 (Liu et al. 2011):

$$X_{id}^{k+1} = X_{id}^k + rand \cdot stepb \cdot \left(G_d^k - X_{id}^k\right), \tag{17.197}$$

  where *rand* represents a uniformly distributed random number (falling within $[0, 1]$), *stepb* denotes the searching step, the iteration number is represented by k, and the range of $[XMIN_d, XMAX_d]$ is used to stand for the dth position.
- Food assignment behaviour: Assigning food to the strongest wolves first, and then to other wolves is often observed in a colony of wolves. Based on this observation, in WCA, the wolves (denoted by$m$) with the worst performances will be replaced by newly generated m artificial wolves which are randomly distributed within the wolf colony. This mechanism can assist the WCA algorithm in avoiding the local optimum.

Taking into account the fundamental behaviours described above, the steps of implementing WCA can be summarized as follows (Liu et al. 2011):

- Step 1: Initializing the following parameters such as n (the individual number), *maxk* (the maximum iteration number), q (the number of the searching artificial wolf), h (the searching direction), *maxdh* (the maximum searching number), *stepa* (the searching step), *stepb* (the besieging step), m (the number of the worst artificial wolves), and the ith $(1 \le i \le n)$ artificial wolf's (X_i) position.
- Step 2: Forming the group of searching wolves (q optimal artificial wolves) and each member of searching wolves moves according to $Y_j = XX_i + randn \cdot stepa$.
- Step 3: Choosing the best location of the searching artificial wolves as the quarry's position. Updating each artificial wolf's position based on $X_{id}^{k+1} =$

$X_{id}^k + rand \cdot stepb \ \cdot (G_d^k - X_{id}^k)$. If $X_{id} < XMIN_d$, then set $X_{id} = XMIN_d$; if $X_{id} \geq XMAX_d$, set $X_{id} = XMAX_d$.

- Step 4: Updating the wolf colony following the food assignment behaviour, i.e., replacing the worst *m* artificial wolves with *m* newly generated artificial wolves.
- Step 5: Evaluating the stopping criteria. If the circulation steps of WCA equals the predetermined maximum iteration number, the algorithm stops and outputs the current best position of artificial wolves; otherwise, WCA continues to run (i.e., returning to Step 2).

17.56.2 Performance of WCA

In order to test the performance of WCA, Liu et al. (2011) employed 5 benchmark test functions such as Sphere function, Rosenbrock function, Schwefel function, and Rastrigin function. In comparison with other CI techniques (e.g., PSO and GA), WCA showed a good convergence and a strong global searching capability.

17.57 Wolf Pack Search Algorithm

In this section, we will introduce another wolf (in particular, wolf pack search behaviour) inspired CI algorithm (Cordoni 2009, Heylighen 1992).

17.57.1 Fundamentals of Wolf Pack Search Algorithm

Wolf pack search (WPS) algorithm was originally proposed by Yang et al. (2007). Briefly, WPS works as follows (Yang et al. 2007):

- Initializing step.
- Initializing a pack of wolves in a random manner.
- Comparing and determining the best wolf *GBest* and its fitness *GBFit*.
- Circulating and updating the Eq. 17.198 (Yang et al. 2007):

$$wolf_{new} = wolf + step \cdot (GBest - wolf)/|GBest - wolf|. \tag{17.198}$$

- If the fitness of $wolf_{new}$ is better than *GBFit*, replacing *GBest* and *GBFit* with $wolf_{new}$ and its corresponding fitness value, respectively.

17.57.2 Performance of WPS

In Yang et al. (2007), the WPS algorithm was hybridized with honeybee mating optimization algorithm to form WPS-MBO. By testing it on classical travelling salesman problem and a set of benchmark functions (e.g., Rosenbrock function, Schwefel function, and generalized Rastrigin function), the WPS-MBO showed a very attractive performance.

17.58 Conclusions

In this chapter, 56 emerging biology-based CI methodologies are discussed. Although most of them are still in their infancy, their usefulness has been demonstrated throughout the preliminary corresponding studies. Interested readers are referred to them as a starting point for a further exploration and exploitation of these innovative CI algorithms.

References

Abdullah, S., Turabieh, H., & Mccollum, B. (2009). A hybridization of electromagnetic-like mechanism and great deluge for examination timetabling problems. *Hybrid Metaheuristics, LNCS* (Vol. 5818, pp. 60–72). Berlin: Springer.

Abernethy, B., Kippers, V., Hanrahan, S. J., Pandy, M. G., Mcmanus, A. M., & Mackinnon, L. (2013). *Biophysical foundations of human movement,*, Champaign: Human Kinetics. ISBN 978-1-4504-3165-1.

Acebo, E. D., & Rosa, J. L. D. L. (2008, April 1–4). Introducing bar systems: A class of swarm intelligence optimization algorithms. In *AISB 2008 Symposium on Swarm Intelligence Algorithms and Applications*, University of Aberdeen (pp. 18–23). The Society for the Study of Artificial Intelligence and Simulation of Behaviour.

Acquaah, G. (2012). *Principles of plant genetics and breeding*. River Street: Wiley. ISBN 978-0-470-66476-6.

Adair, J. (2007). *Develop your leadership skills*. London: Kogan Page Limited. ISBN 0-7494-4919-5.

Adamatzky, A., & Oliveira, P. P. B. D. (2011). Brazilian highways from slime mold's point of view. *Kybernetes, 40*, 1373–1394.

Adams, S. (2004). *World War I*. London: Dorling Kindersley Limited. ISBN 1-4053-0298-4.

Ahrari, A., & Atai, A. A. (2010). Grenade explosion method: A novel tool for optimization of multimodal functions. *Applied Soft Computing, 10*, 1132–1140.

Ahrari, A., Shariat-Panahi, M., & Atai, A. A. (2009). GEM: a novel evolutionary optimization method with improved neighborhood search. *Applied Mathematics and Computation, 210*, 379–386.

Al-Milli, N. R. (2010). Hybrid genetic algorithms with great deluge for course timetabling. *International Journal of Computer Science and Network Security, 10*, 283–288.

Alatas, B. (2011). Photosynthetic algorithm approaches for bioinformatics. *Expert Systems with Applications, 38*, 10541–10546.

Aleksiev, A. S., Longdon, B., Christmas, M. J., Sendova-Franks, A. B., & Franks, N. R. (2008). Individual and collective choice: Parallel prospecting and mining in ants. *Naturwissenschaften, 95*, 301–305.

Alexiou, A., & Vlamos, P. (2012). A cultural algorithm for the representation of mitochondrial population. *Advances in Artificial Intelligence, 2012*, 1–7.

Alonso, C., Herrera, C. M., & Ashman, T.-L. (2012). A piece of the puzzle: A method for comparing pollination quality and quantity across multiple species and reproductive events. *New Phytologist, 193*, 532–542.

Aman, B. (2009). *Spatial dynamic structures and mobility in computation.* Unpublished Doctoral Thesis, Romania Academy.

Anandaraman, C. (2011). An improved sheep flock heredity algorithm for job shop scheduling and flow shop scheduling. *International Journal of Industrial Engineering Computations, 2*, 749–764.

Anandaraman, C., Sankar, A. M., & Natarajan, R. (2012). Evolutionary approaches for scheduling a flexible manufacturing system with automated guided vehicles and robots. *International Journal of Industrial Engineering Computations, 3*, 627–648.

Ardjmand, E., & Amin-naseri, M. R. (2012). Unconscious search: A new structured search algorithm for solving continuous engineering optimization problems based on the theory of psychoanalysis. In Y. Tan, Y. Shi, & Z. Ji (Eds.), *ICSI 2012, Part I, LNCS* (Vol. 7331, pp. 233–242). Berlin: Springer.

Ashby, L. H. & Yampolskiy, R. V. (2011). Genetic algorithm and wisdom of artificial crowds algorithm applied to light up. In *16th International Conference on Computer Games (GAMES 2011)*, (pp. 27–32). IEEE.

Badgerow, J. P., & Hainsworth, F. R. (1981). Energy savings through formation flight? A re-examination of the vee formation. *Journal of Theoretical Biology, 93*, 41–52.

Ballerini, M., Cabibbo, N., Candelier, R., Cavagna, A., Cisbani, E., Giardina, I., et al. (2008). Empirical investigation of starling flock: A benchmark study in collective animal behaviour. *Animal Behaviour, 76*, 201–215.

Bater, L. (2007). *Incredible insects: Answers to questions about miniature marvels.* Vero Beach: Rourke Publishing LLC. Post Office Box 3328. ISBN 978-1-60044-348-0.

Batista, L. D. S., Guimarães, F. G., & Ramírez, J. A. (2009). A distributed clonal selection algorithm for optimization in electromagnetics. *IEEE Transactions on Magnetics, 45*, 1598–1601.

Bell, W. J., Roth, L. M., & Nalepa, C. A. (2007). *Cockroaches: Ecology, behavior, and natural history.* Maryland: The Johns Hopkins University Press. ISBN 978-0-8018-8616-4.

Bellaachia, A., & Bari, A. (2012). Flock by leader: A novel machine learning biologically inspired clustering algorithm. In Y. Tan, Y. Shi, & Z. Ji (Eds.), *ICSI 2012, Part I, LNCS* (Vol. 7332, pp. 117–126). Berlin: Springer.

Bhugra, D., Ruiz, P., & Gupta, S. (2013). *Leadership in psychiatry.* Hoboken: Wiley. ISBN 978-1-119-95291-6.

Bolstad, T. M. (2012). *Brownian motion.* Department of Physics and Technology, University of Bergen.

Brierley, A. S., & Cox, M. J. (2010). Shapes of krill swarms and fish schools emerge as aggregation members avoid predators and access oxygen. *Current Biology, 20*, 1758–1762.

Brownlee, J. (2007). Clonal selection algorithms. CIS Technical Report, 070209A, 1–13.

Burke, E., Bykov, Y., Newall, J., & Petrovic, S. (2004). A time-predefined local search approach to exam timetabling problems. *IIE Transactions, 3*, 509–528.

Campelo, F., Guimarães, F. G., Igarashi, H., & Ramírez, J. A. (2005). A clonal selection algorithm for optimization in electromagnetics. *IEEE Transactions on Magnetics, 41*, 1736–1739.

Cao, C.-H., Wang, L.-M., Han, C.-Y., Zhao, D.-Z., & Zhang, B. (2012). Geese PSO optimization in geometric constraint solving. *Information Technology Journal, 11*, 504.

Carlson, N. R. (2013). *Physiology of behavior.* New Jersey: Pearson Education, Inc. ISBN 978-0-205-23948-1.

Carpentier, R. (2011). *Photosynthesis research protocols*. New York: Springer. ISBN 978-1-60671-924-6.

Castro, L. N. D., & Zuben, F. J. V. (2000, July). The clonal selecton algorithm with engineering applications. In *Workshop on Artificial Immune Systems and Their Applications*, Las Vegas, USA, pp. 1–7.

Castro, L. N. D., & Zuben, F. J. V. (2002). Learning and optimization using the clonal selection principle. *IEEE Transactions on Evolutionary Computation, 6*, 239–251.

Chalmers, D. J. (2010). *The character of consciousness*. USA: Oxford University Press. ISBN 978-0-195-31111-2.

Chandrasekaran, M., Asokan, P., Kumanan, S., & Balamurugan, T. (2006). Sheep flocks heredity model algorithm for solving job shop scheduling problems. *International Journal of Applied Management and Technology, 4*, 79–100.

Chapman, R. F. (2013). *The insects: structure and function*. In S. J. Simpson & A. E. Douglas (Eds.). New York: Cambridge University Press. ISBN 978-0-521-11389-2.

Chen, C.-C., & Lee, Y.-T. (Eds.). (2008). *Leadership and management in China: Philosophies, theories, and practices*. Cambridge: Cambridge University Press. ISBN 978-0-511-40909-7.

Chen, K., Li, T., & Cao, T. (2006). Tribe-PSO: A novel global optimization algorithm and its application in molecular docking. *Chemometrics and Intelligent Laboratory Systems, 82*, 248–259.

Chen, T. (2009). A simulative bionic intelligent optimization algorithm: Artificial searching swarm algorithm and its performance analysis. In *IEEE International Joint Conference on Computational Sciences and Optimization* (*CSO*) (pp. 864–866).

Chen, T., Liu, Z., Shu, Q., & Zhang, L. (2009a). On the analysis of performance of the improved artificial searching swarm algorithm. In *IEEE 2nd International Conference on Intelligent Networks and Intelligent Systems* (*ICINIS*) (pp. 502–506).

Chen, T., Pang, L., Du, J., Liu, Z., & Zhang, L. (2009b). Artificial searching swarm algorithm for solving constrained optimization problems. In *IEEE International Conference on Intelligent Computing and Intelligent Systems* (*ICIS*) (pp. 562–565) .

Chen, T., Wang, Y., & Li, J. (2012). Artificial tribe algorithm and its performance analysis. *Journal of Software, 7*, 651–656.

Chen, T., Wang, Y., Pang, L., Liu, Z., & Zhang, L. (2010a). An improved artificial searching swarm algortihm and its performance analysis. In *IEEE 2nd International Conference on Computer Modeling and Simulation* (*ICCMS*) (pp. 260–263).

Chen, T., Zhang, L., Liu, Z., Pang, L., & Shu, Q. (2009c). On the analysis of performance of the artificial searching swarm algorithm. In *IEEE 5h International Conference on Natural Computation* (*ICNC*) (pp. 365–368).

Chen, Y., Cui, Z., & Zeng, J. (2010b, July 7–9). Structural optimization of lennard-jones clusters by hybrid social cognitive optimization algorithm. In F. Sun, Y. Wang, J. Lu, B. Zhang, W. Kinsner, & L. A. Zadeh (Eds.), In *9th International Conference on Cognitive Informatics (ICCI)* (pp. 204–208). IEEE. Beijing, China.

Chen, Z., & Tang, H. (2010). Cockroach swarm optimization. In *IEEE 2nd International Conference on Computer Engineering and Technology* (*ICCET*) (pp. 652–655).

Chen, Z., & Tang, H. (2011). Cockroach swarm optimization for vehicle routing problems. *Energy Procedia, 13*, 30–35.

Cheng, L., Xu, Y.-H., Zhang, H.-B., Qian, Z.-L., & Feng, G. (2010). New bionics optimization algorithm: food truck-cockroach swarm optimization algorithm (in Chinese). *Computer Engineering, 36*, 208–209.

Ciobanu, G., Desai, R., & Kumar, A. (2003). Membrane systems and distributed computing. In G. Păun (Ed.), *WMC-CdeA 2002, LNCS* (Vol. 2597, pp. 187–202). Berlin: Springer.

Civicioglu, P. (2012). Transforming geocentric cartesian coordinates to geodetic coordinates by using differential search algorithm. *Computers and Geosciences, 46*, 229–247.

Civicioglu, P. (2013). Backtracking search optimization algorithm for numerical optimization problems. *Applied Mathematics and Computation, 219*, 8121–8144.

Coelho, L. D. S., & Bernert, D. L. D. A. (2009). PID control design for chaotic synchronization using a tribes optimization approach. *Chaos, Solitons and Fractals, 42*, 634–640.

Conradt, L., & List, C. (2009). Group decisions in humans and animals: A survey. *Philosophical Transaction of the Royal Society B, 364*, 719–742.

Cordoni, G. (2009). Social play in captive wolves (*Canis lupus*): Not only an immature affair. *Behaviour, 146*, 1363–1385.

Couzin, I. D. (2009). Collective cognition in animal groups. *Trends in Cognitive Sciences, 13*, 36–43.

Couzin, I. D., Krause, J., Franks, N. R., & Levin, S. A. (2005). Effective leadership and decision-making making in animal groups on the move. *Nature, 434*, 513–516.

Creel, S. (1997). Cooperative hunting and group size: Assumptions and currencies. *Animal Behaviour, 54*, 1319–1324.

Cuevas, E., Cienfuegos, M., Zaldívar, D., & Pérez-Cisneros, M. (2013). A swarm optimization algorithm inspired in the behavior of the social-spider. *Expert Systems with Applications*. doi: http://dx.doi.org/10.1016/j.eswa.2013.05.041.

Cuevas, E., Zaldívar, D., & Pérez-Cisneros, M. (2013b). A swarm optimization algorithm for multimodal functions and its application in multicircle detection. *Mathematical Problems in Engineering, 2013*, 1–22.

Cui, X., Gao, J., & Potok, T. E. (2006). A flocking based algorithm for document clustering analysis. *Journal of Systems Architecture, 52*, 505–515.

Cui, Y., Guo, R., & Guo, D. (2009). A naïve five-element string algorithm. *Journal of Software, 4*, 925–934.

Cui, Y. H., Guo, R., Rao, R. V., & Savsani, V. J. (2008, December 15–17) Harmony element algorithm: A naive initial searching range. In *International Conference on Advances in Mechanical Engineering*, S.V. (pp. 1–6). National Institute of Technology, Gujarat, India.

Cui, Z., & Cai, X. (2010, July 7–9). Using social cognitive optimization algorithm to solve nonlinear equations. In F. Sun, Y. Wang, J. Lu, B. Zhang, W. Kinsner & L. A. Zadeh (Eds.), In *9th International Conference on Cognitive Informatics* (*ICCI*) (pp. 199–203). Beijing, China. IEEE.

Cui, Z., Cai, X., & Shi, Z. (2011). Social emotional optimization algorithm with group decision. *Scientific Research and Essays, 6*, 4848–4855.

Cui, Z., Shi, Z., & Zeng, J. (2010). Using social emotional optimization algorithm to direct orbits of chaotic systems. In B. K. Panigrahi, S. Das, P. N. Suganthan & S. S. Dash (Eds.), *Swarm, Evolutionary, and Memetic Computing, LNCS* (Vol. 6466, pp. 389–395). Berlin: Springer.

Cui, Z., Xu, Y., & Zeng, J. (2012). Social emotional optimization algorithm with random emotional selection strategy. In R. Parpinelli (Ed.), *Theory and New Applications of Swarm Intelligence, Chap. 3* (pp. 33–50). Croatia: InTech. ISBN 978-953-51-0364-6.

Cummins, B., Cortez, R., Foppa, I. M., Walbeck, J., & Hyman, J. M. (2012). A spatial model of mosquito host-seeking behavior. *PLoS Computational Biology, 8*, 1–13.

Cutts, C. J., & Speakman, J. R. (1994). Energy savings in formation flight of pink-footed geese. *Journal of Experimental Biology, 189*, 251–261.

Dai, C., Chen, W., & Zhu, Y. (2006, November). Seeker optimization algorithm. In *IEEE International Conference on Computational Intelligence and Security* (pp. 225–229). Guangzhou, China.

Dai, C., Chen, W., Song, Y., & Zhu, Y. (2010a). Seeker optimization algorithm: A novel stochastic search algorithm for global numerical optimization. *Journal of Systems Engineering and Electronics, 21*, 300–311.

Dai, C., Chen, W., & Zhu, Y. (2010b). Seeker optimization algorithm for digital IIR filter design. *IEEE Transactions on Industrial Electronics, 57*, 1710–1718.

Dai, C., Chen, W., Zhu, Y., & Zhang, X. (2009a). Reactive power dispatch considering voltage stability with seeker optimization algorithm. *Electric Power Systems Research, 79*, 1462–1471.

Dai, C., Chen, W., Zhu, Y., & Zhang, X. (2009b). Seeker optimization algorithm for optimal reactive power dispatch. *IEEE Transactions on Power Systems, 24*, 1218–1231.

Dai, C., Zhu, Y., & Chen, W. (2007). Seeker optimization algorithm. In Y. Wang, Y. Cheung & H. Liu (Eds.), *CIS 2006, LNAI* (Vol. 4456. pp. 167–176). Berlin: Springer.

Dai, S., Zhuang, P., & Xiang, W. (2013). GSO: An improved PSO based on geese flight theory. In Y. Tan, Y. Shi, & H. Mo (Eds.), *Advances in Swarm Intelligence, ICSI 2013, Part I, LNCS* (Vol. 7928, pp. 87–95). Berlin: Springer.

Daskin, A., & Kais, S. (2011). Group leaders optimization algorithm. *Molecular Physics, 109*, 761–772.

Davendra, D., & Zelinka, I. (2009). Optimization of quadratic assignment problem using self-organinsing migrating algorithm. *Computing and Informatics, 28*, 169–180.

Davendra, D., Zelinka, I., Bialic-Davendra, M., Senkerik, R., & Jasek, R. (2013). Discrete self-organising migrating algorithm for flow-shop scheduling with no-wait makespan. *Mathematical and Computer Modelling, 57*, 100–110.

Digalakis, J. G., & Margaritis, K. G. (2002). A multipopulation cultural algorithm for the electrical generator scheduling problem. *Mathematics and Computers in Simulation, 60*, 293–301.

Ding, S., & Li, S. (2009). Clonal selection algorithm for feature selection and parameters optimization for support vector machines. In *IEEE 2nd International Symposium on Knowledge Acquisition and Modeling* (pp. 17–20).

Du, Y., Li, H., Pei, Z., & Peng, H. (2005). Intelligent spider's algorithm of search engine based on keyword. *ECTI Transactions on Computer and Information Theory, 1*, 40–49.

Dubinsky, Z. (Ed.). (2013). *Photosynthesis*. InTech: Croatia. ISBN 978-953-51-1161-0.

Dueck, G. (1993). New optimization heuristics: The great deluge algorithm and the record-to-record travel. *Journal of Computational Physics, 104*, 86–92.

Duman, E., Uysal, M., & Alkaya, A. F. (2012). Migrating birds optimization: A new metaheuristic approach and its performance on quadratic assignment problem. *Information Sciences, 217*, 65–77.

Durrett, R. (1984). *Brownian motion and martingales in analysis*. Belmont: Wadsworth Advanced Books and Software, A Division of Wadsworth, Inc. ISBN 0-534-03065-3.

Ebensperger, L. A. (2001). A review of the evolutionary causes of rodent group-living. *Acta Theriologica, 46*, 115–144.

Eckstein, M. P., Das, K., Pham, B. T., Peterson, M. F., Abbey, C. K., Sy, J. L., et al. (2012). Neural decoding of collective wisdom with multi-brain computing. *NeuroImage, 59*, 94–108.

Feng, X., Lau, F. C. M., & Gao, D. (2009). A new bio-inspired approach to the traveling salesman problem. In J. Zhou (Ed.), *Complex 2009, Part II, LNICST*, (Vol. 5, pp. 1310–1321). Institute for Computer Sciences, Social Informatics and Telecommunications Engineering.

Frank, S. A. (1998). *Foundations of social evolution*. New Jersey: Princeton University Press. ISBN 0-691-05933-0.

Fuller, T. K., Mech, L. D., & Cockrane, J. F. (2003). Wolf population dynamics. In L. D. Mech & L. Boitani (Eds.), *Wolves: Behavior, Ecology and Conservation* (pp. 161–191). Chicago: University of Chicago Press.

Gamlin, L. (2009). *Evolution*. New York: Dorling Kindersley Limited. ISBN 978-0-7566-5028-5.

Gandomi, A. H., & Alavi, A. H. (2012). Krill herd: A new bio-inspired optimization algorithm. *Communications in Nonlinear Science and Numerical Simulation, 17*, 4831–1845.

Gandomi, A. H., Yang, X.-S., Talatahari, S., & Deb, S. (2012). Coupled eagle strategy and differential evolution for unconstrained and constrained global optimization. *Computers and Mathematics with Applications, 63*, 191–200.

Gao, S., Chai, H., Chen, B., & Yang, G. (2013). Hybrid gravitational search and clonal selection algorithm for global optimization. In Y. Tan, Y. Shi & H. Mo (Eds.), *Advances in Swarm Intelligence, LNCS* (Vol. 7929, pp. 1–10). Hybrid gravitational search and clonal selection algorithm for global optimization. Berlin: Springer.

Ghatei, S., Khajei, R. P., Maman, M. S., & Meybodi, M. R. (2012). A modified PSO using great deluge algorithm for optimization. *Journal of Basic and Applied Scientific Research, 2*, 1362–1367.

Gheorghe, M., Păun, G., Rozenberg, G., Salomaa, A., & Verlan, S. (Eds.). (2012). *Membrane computing*. Berlin: Springer. ISBN 978-3-642-28023-8.

Giraldeau, L.-A., Soos, C., & Beauchamp, G. (1994). A test of the producer-scrounger foraging game in captive flocks of spice finches, *Lonchura punctulata*. *Behavioral Ecology and Sociobiology, 34*, 251–256.

Goffredo, S., & Dubinsky, Z. (2014). *The Mediterranean Sea: Its history and present challenges*. New York: Springer. ISBN 978-94-007-6703-4.

Gofman, Y. (2012). *Computational studies of the interactions of biologically active peptides with membrane*. Unpublished Doctoral Thesis, Universität Hamburg.

Gross, R. (2014). *Psychology: The science of mind and behaviour*. London: Hodder Education. ISBN 978-1-4441-0831-6.

Gusset, M., & Macdonald, D. W. (2010). Group size effects in cooperatively breeding African wild dogs. *Animal Behaviour, 79*, 425–428.

Hagler, G. (2013). *Modelinig ships and space craft*. Berlin: Springer. ISBN 978-1-4614-4595-1.

Hasançebi, O., & Azad, S. K. (2012). An efficient metaheuristic algorithm for engineering optimization: SPOT. *International Journal of Optimization in Civil Engineering, 2*, 479–487.

Havens, T. C., Spain, C. J., Salmon, N. G., & Keller, J. M. (2008, September 21–23). Roach infestation optimization. In *IEEE Swarm Intelligence Symposium* (pp. 1–7). St. Louis MO, USA.

Heylighen, F. (1992). Evolution, selfishness and cooperation. *Journal of Ideas, 2*, 70–76.

Hobbs, R. J., Higgs, E. S., & Hall, C. M. (2013). *Novel ecosystems: Intervening in the new ecological world order*. Hoboken: Wiley. ISBN 978-1-118-35422-3.

Howell, D. C. (2014). *Fundamental statistics for the behavioral sciences*. Belmont: Cengage Learning. ISBN 978-1-285-07691-1.

Irvine, E. (2013). *Consciousness as a scientific concept: A philosophy of science perspective*. Dordrecht: Springer. ISBN 978-94-007-5172-9.

Ishdorj, T.-O. (2006). *Membrane computing, neural inspirations, gene assembly in Ciliates*. Unpublished Doctoral Thesis, University of Seville.

Janecek, A., & Tan, Y. (2011). Swarm intelligence for non-negative matrix factorization. *International Journal of Swarm Intelligence Research, 2*, 12–34.

Jelinek, R. (2013). *Biomimetics: A molecular perspective*. Berlin/Boston: Walter de Gruyter. ISBN 978-3-11-028117-0.

Jonassen, T. M. (2006, June). Symbolic dynamics, the spider algorithm and finding certain real zeros of polynomials of high degree. In *8th International Mathematica Symposium* (pp. 1–16). Avignon.

Karci, A. (2007a). Human being properties of saplings growing up algorithm. In *International Conference on Computational Cybernetics* (*ICCC*) (pp. 227–232). IEEE.

Karci, A. (2007b). Natural inspired computational intelligence method: saplings growing up algorithm. In *International Conference on Computational Cybernetics (ICCC)*, pp. 221–226. IEEE.

Karci, A. (2007c). Theory of saplings growing up algorithm. In *Adaptive and Natural Computing Algorithms, LNCS* (Vol. 4431, pp. 450–460). Berlin: Springer.

Karci, A., & Alatas, B. (2006). Thinking capability of saplings growing up algorithm. In *Intelligent Data Engineering and Automated Learning* (*IDEAL 2006*), *LNCS* (Vol. 4224, pp. 386–393). Berlin: Springer.

Kashan, A. H. (2009). League championship algorithm: a new algorithm for numerical function optimization. In *IEEE International Conference of Soft Computing and Pattern Recognition* (*SoCPAR*) (pp. 43–48).

Kashan, A. H. (2011). An efficient algorithm for constrained global optimization and application to mechanical engineering design: League championship algorithm (LCA). *Computer-Aided Design, 43*, 1769–1792.

Kashan, A. H., & Karimi, B. (2010, 18–23 July). A new algorithm for constrained optimization inspired by the sport league championships. In *World Congress on Computational Intelligence* (*WCCI*) (pp. 487–494). CCIB, Barcelona, Spain.

Keller, L., & Gordon, É. (2009). *The lives of ants (translated by James Grieve)*. Oxford: Oxford University Press Inc. ISBN 978–0–19–954186–7.

Kim, H., & Ahn, B. (2001). A new evolutionary algorithm based on sheep flocks heredity model. In *Conference on Communications, Computers and Signal Processing* (pp. 514–517). IEEE.

Kim, Y.-B. (2012). *Distributed algorithms in membrane systems*. Unpublished Doctoral Thesis, University of Auckland.

King, A. J., Sueur, C., Huchard, E., & Cowlishaw, G. (2011). A rule-of-thumb based on social affiliation explains collective movements in desert baboons. *Animal Behaviour, 82*, 1337–1345.

Krishnanand, K. R., Hasani, S. M. F., & Panigrahi, B. K. (2013). Optimal power flow solution using self-evolving brain–storming inclusive teaching-learning-based algorithm. In Y. Tan, Y. Shi, & H. Mo (Eds.), *ICSI 2013, Part I, LNCS* (Vol. 7928, pp. 338–345). Berlin: Springer.

Kwasnicka, H., Markowska-Kaczmar, U., & Mikosik, M. (2011). Flocking behaviour in simple ecosystems as a result of artificial evolution. *Applied Soft Computing, 11*, 982–990.

Lancaster, R., Butler, R. E. A., Lancaster, J. M., & Shimizu, T. (1998). *Fireworks: Principles and practice*. New York: Chemical Publishing Co., Inc. ISBN 0-8206-0354-6.

Lee, D., & Quessy, S. (2003). Visual search is facilitated by scene and sequence familiarity in rhesus monkeys. *Vision Research, 43*, 1455–1463.

Lemasson, B. H., Anderson, J. J., & Goodwin, R. A. (2009). Collective motion in animal groups from a neurobiological perspective: The adaptive benefits of dynamic sensory loads and selective attention. *Journal of Theoretical Biology, 261*, 501–510.

Levin, S. A. (2013a). *Encyclopedia of biodiversity*. Oxford: Academic Press, Elsevier Inc. ISBN 978-0-12-384719-5.

Levin, S. A. (2013b). *Encyclopedia of biodiversity*. Oxford: Academic Press, Elsevier Inc. ISBN 978-0-12-384719-5.

Levin, S. A. (2013c). *Encyclopedia of biodiversity*. Oxford: Academic Press, Elsevier Inc. ISBN 978-0-12-384719-5.

Levin, S. A. (2013d). *Encyclopedia of biodiversity*. Oxford: Academic Press, Elsevier Inc. ISBN 978-0-12-384719-5.

Levin, S. A. (2013e). *Encyclopedia of biodiversity*. Oxford: Academic Press, Elsevier Inc. ISBN 978-0-12-384719-5.

Levin, S. A. (2013f). *Encyclopedia of biodiversity,*. London: Academic Press, Elsevier Inc. ISBN 978-0-12-384719-5.

Li, K., Torres, C. E., Thomas, K., Rossi, L. F., & Shen, C.-C. (2011). Slime mold inspired routing protocols for wireless sensor networks. *Swarm Intelligence, 5*, 183–223.

Li, Y. (2010, October 22–24). Solving TSP by an ACO-and-BOA-based hybrid algorithm. In *IEEE International Conference on Computer Application and System Modeling* (*ICCASM*), (Vol. 12, pp. 189–192).

Lihoreau, M., Costa, J. T., & Rivault, C. (2012). The social biology of domiciliary cockroaches: colony structure, kin recognition and collective decisions. *Insectes Sociaux*. doi: 10.1007/s00040-012-0234-x.

Lihoreau, M., Deneubourg, J.-L., & Rivault, C. (2010). Collective foraging decision in a gregarious insect. *Behavioral Ecology and Sociobiology, 64*, 1577–1587.

Lissaman, P. B. S., & Shollenberger, C. A. (1970). Formation flight of birds. *Science, 168*, 1003–1005.

Liu, C., Yan, X., Liu, C., & Wu, H. (2011). The wolf colony algorithm and its application. *Chinese Journal of Electronics, 20*, 212–216.

Liu, J. Y., Guo, M. Z., & Deng, C. (2006). Geese PSO: An efficient improvement to particle swarm optimization. *Computer Science, 33*, 166–168.

Lu, Y., & Liu, X. (2011). A new population migration algorithm based on the chaos theory. In *IEEE 2nd International Symposium on Intelligence Information Processing and Trusted Computing* (*IPTC*) (pp. 147–10).

Luo, X., Li, S., & Guan, X. (2010). Flocking algorithm with multi-target tracking for multi-agent systems. *Pattern Recognition Letters, 31*, 800–805.

Maathuis, F. J. M. (2013). *Plant mineral nutrients: Methods and protocols*. New York: Springer. ISBN 978-1-62703-151-6.

Macdonald, D. W., Creel, S., & Mills, M. G. L. (2004). Society: Canid society. In D. W. Macdonald & C. Sillero-Zubiri (Eds.), *Biology and Conservation of Wild Carnivores* (pp. 85–106). Oxford: Oxford University Press.

Magstadt, T. M. (2013). *Understanding politics: ideas, institutions, and issues*. Cengage Learning: Belmont. ISBN 978-1-111-83256-8.

Marcus, J. B. (2013). *Culinary nutrition: The science and practice of healthy cooking*. Waltham: Elsevier. ISBN 978-0-12-391882-6.

Maroosi, A., & Muniyandi, R. C. (2013). Membrane computing inspired genetic algorithm on multi-core processors. *Journal of Computer Science, 9*, 264–270.

Mayfield, J. E. (2013). *The engine of complexity: Evolution as computation*. New York: Columbia University Press. ISBN 978-0-231-16304-0.

Mills, D. S., Marchant-Forde, J. N., McGreevy, P. D., Morton, D. B., Nicol, C. J., Phillips, C. J. C., et al. (Eds.). (2010). *The encyclopedia of applied animal behaviour and welfare*. Wallingford: CAB International. ISBN 978-0-85199-724-7.

Mosser, A., & Packer, C. (2009). Group territoriality and the benefits of sociality in the African lion, *Panthera leo*. *Animal Behaviour, 78*, 359–370.

Mucherino, A., & Seref, O. (2007). Monkey search: A novel metaheuristic search for global optimization. In *AIP Conference Proceedings* (Vol. 953, pp. 162–173).

Mukherjee, R., Chakraborty, S., & Samanta, S. (2012). Selection of wire electrical discharge machining process parameters using non-traditional optimization algorithms. *Applied Soft Computing, 12*, 2506–2516.

Müller, V. C. (Ed.). (2013). *Philosophy and theory of artificial intelligence*. Berlin: Springer. ISBN 978-3-642-31673-9.

Muniyandi, R. C., & Zin, A. M. (2013). Membrane computing as the paradigm for modeling system biology. *Journal of Computer Science, 9*, 122–127.

Murase, H. (2000). Finite element inverse analysis using a photosynthetic algorithm. *Computers and Electronics in Agriculture, 29*, 115–123.

Muro, C., Escobedo, R., Spector, L., & Coppinger, R. P. (2011). Wolf-pack (Canis lupus) hunting strategies emerge from simple rules in computational simulations. *Behavioural Processes, 88*, 192–197.

Murphy, N. (2010). *Uniformity conditions for membrane system uncovering complexity below P*. Unpublished Doctoral Thesis, National University of Ireland Maynooth.

Nabil, E., Badr, A., & Farag, I. (2012). *A membrane-immune algorithm for solving the multiple 0-1 knapsack problem* (pp. 3–15). LVII: Informatica.

Nahas, N., Nourelfath, M., & Aït-Kadi, D. (2010). Iterated great deluge for the dynamic facility layout problem. Canada: Interuniversity Research Centre on Enterprise Networks, Logistics and Transportation, Report No.: CIRRELT-2010-20.

Nakagaki, T., Yamada, H., & Tóth, Á. (2000). Maze-solving by an amoeboid organism. *Nature, 407*, 470.

Nara, K., Takeyama, T., & Kim, H. (1999). A new evolutionary algorithm based on sheep flocks heredity model and its application to scheduling problem. In *IEEE International Conference on Systems, Man, and Cybernetics* (*SMC*) (pp. VI-503–VI-508).

Neshat, M., Sepidnam, G., & Sargolzaei, M. (2013). Swallow swarm optimization algorithm: A new method to optimization. *Neural Computing & Application, 23*, 429–454. doi: 10.1007/s00521-012-0939-9.

Newell, P. C. (1978). Genetics of the cellular slime molds. *Annual Review of Genetics, 12*, 69–93.

Nguyen, V., Kearney, D., & Gioiosa, G. (2008). An implementation of membrane computing using reconfigurable hardware. *Computing and Informatics, 27*, 551–569.

Nicolis, S. C., Detrain, C., Demolin, D., & Deneubourg, J. L. (2003). Optimality of collective choices: a stochastic approach. *Bulletin of Mathematical Biology, 65*, 795–808.

Niizato, T., & Gunji, Y.-P. (2011). Metric–topological interaction model of collective behavior. *Ecological Modelling, 222*, 3041–3049.

Nishida, T. Y. (2005, July 18–21). Membrane algorithm: An approximate algorithm for NP-complete optimization problems exploiting P-systems. In R. Freund, G. Lojka, M. Oswald, & G. Păun (Eds.), In *6th International workshop on membrane computing* (*WMC*) (pp. 26–43). Vienna, Austria. Institute of Computer Languages, Faculty of Informatics, Vienna University of Technology.

Nolle, L., Zelinka, I., Hopgood, A. A., & Goodyear, A. (2005). Comparison of an self-organizing migration algorithm with simulated annealing and differential evolution for automated waveform tuning. *Advanced Engineering Software, 36*, 645–653.

Oca, M. A. M. D., Ferrante, E., Scheidler, A., Pinciroli, C., Birattari, M., & Dorigo, M. (2011). Majority-rule opinion dynamics with differential latency: A mechanism for self-organized collective decision-making. *Swarm Intelligence, 5*, 305–327.

Ochoa-Zezzatti, A., Bustillos, S., Jaramillo, R., & Ruíz, V. (2012). Improving practice sports in a largest city using a cultural algorithm. *International Journal of Combinatorial Optimization Problems and Informatics, 3*, 14–20.

Oftadeh, R., Mahjoob, M. J., & Shariatpanahi, M. (2010). A novel meta-heuristic optimization algorithm inspired by group hunting of animals: Hunting search. *Computers and Mathematics with Applications, 60*, 2087–2098.

Onwubolu, G. C. (2006). Performance-based optimization of multi-pass face milling operations using Tribes. *International Journal of Machine Tools and Manufacture, 46*, 717–727.

Packer, C., & Caro, T. M. (1997). Foraging costs in social carnivores. *Animal Behaviour, 54*, 1317–1318.

Păun, G. (2000). Computing with membranes. *Journal of Computer and System Sciences, 61*, 108–143.

Păun, G. (2002). A guide to membrane computing. *Theoretical Computer Science, 287*, 73–100.

Păun, G. (2007). Tracing some open problems in membrane computing. *Romanian Journal of Information Science and Technology, 10*, 303–314.

Pei, Y., Zheng, S., Tan, Y., & Takagi, H. (2012, October 14–17). An empirical study on influence of approximation approaches on enhancing fireworks algorithm. In *IEEE International Conference on Systems, Man, and Cybernetics* (*IEEE SMC 2012*) (pp. 1322–1327). Seoul, Korea.

Petit, O., & Bon, R. (2010). Decision-making processes: The case of collective movements. *Behavioural Processes, 84*, 635–647.

Picarougne, F., Azzag, H., Venturini, G., & Guinot, C. (2007). A new approach of data clustering using a flock of agents. *Evolutionary Computation, 15*, 345–367.

Port, A. C., & Yampolskiy, R. V. (2012). Using a GA and wisdom of artificial crowds to solve solitaire battleship puzzles. In *IEEE 17th International Conference on Computer Games* (*CGAMES 2012*) (pp. 25–29).

Premaratne, U., Samarabandu, J., & Sidhu, T. (2009, December 28–31). A new biologically inspired optimization algorithm. In *IEEE 4th International Conference on Industrial and Information Systems* (*ICIIS*) (pp. 279–284). Sri Lanka.

Ramachandran, V. S. (2012a). *Encyclopedia of human behavior*. London: Elsevier. ISBN 978-0-12-375000-6.

Ramachandran, V. S. (2012b). *Encyclopedia of human behavior*. London: Elsevier. ISBN 978-0-12-375000-6.

Ramachandran, V. S. (2012c). *Encyclopedia of human behavior*. London: Elsevier. ISBN 978-0-12-375000-6.

Rao, R. V., Vakharia, D. P., & Savsani, V. J. (2009). Mechanical engineering design optimisation using modified harmony elements algorithm. *International Journal of Design Engineering, 2*, 116–135.

Ravi, V. (2004). Optimization of complex system reliability by a modified great deluge algorithm. *Asia-Pacific Journal of Operational Research, 21*, 487–497.

Ray, T., & Liew, K. M. (2003). Society and civilization: an optimization algorithm based on the simulation of social behavior. *IEEE Transactions on Evolutionary Computation, 7*, 386–396.

Reece, J. B., Urry, L. A., Cain, M. L., Wasserman, S. A., Minorsky, P. V., & Jackson, R. B. (2011). *Campbell biology*. San Francisco: Pearson Education, Inc. ISBN 978-0-321-55823-7.

Resende, R. R., & Ulrich, H. (2013). *Trends in stem cell proliferation and cancer research*. Dordrecht: Springer. ISBN 978-94-007-6210-7.

Reynolds, C. W. (1987). Flocks, herds, and schools: A distributed behavioral model. *Computer Graphics, 21*, 25–34.

Reynolds, R. G. (1994). An introduction to cultural algorithms. In A. V. Sebald & L. J. Fogel (Eds.) *The 3rd Annual Conference on Evolutionary Programming* (pp. 131–139). World Scientific Publishing.

Reynolds, R. G. (1999). Cultural algorithms: theory and application In D. Corne, M. Dorigo & Glover, F. (Eds.), *New Ideas in Optimization*. NY: McGraw-Hill.

Riff, M. C., Montero, E., & Neveu, B. (2013). Reducing calibration effort for clonal selection based algorithms. *Knowledge-Based Systems, 41*, 54–67.

Rose, S. V. (2008). *Volcano and earthquake*. New York: Dorling Kindersley Limited. ISBN 978-0-7566-3780-4.

Sacco, W. F., Oliveira, C. R. E. D., & Pereira, C. M. N. A. (2006). Two stochastic optimization algorithms applied to nuclear reactor core design. *Progress in Nuclear Energy, 48*, 525–539.

Sadollah, A., Bahreininejad, A., Eskandar, H., & Hamdi, M. (2012). Mine blast algorithm for optimization of truss structures with discrete variables. *Computers and Structures, 102–103*, 49–63.

Sadollah, A., Bahreininejad, A., Eskandar, H., & Hamdi, M. (2013). Mine blast algorithm: A new population based algorithm for solving constrained engineering optimization problems. *Applied Soft Computing, 13*, 2592–2612.

Samuels, P., Huntington, S., Allsop, W., & Harrop, J. (2009). *Flood risk management: Research and practice*. London: Taylor & Francis Group. ISBN 978-0-415-48507-4.

Sand, H., Wikenros, C., Wabakken, P., & Liberg, O. (2006). Effects of hunting group size, snow depth and age on the success of wolves hunting moose. *Animal Behaviour, 72*, 781–789.

Savage, N. (2012). Gaining wisdom from crowds. *Communications of the ACM, 55*, 13–15.

Schnell, R. J., & Priyadarshan, P. M. (2012). *Genomics of tree crops*. New York: Springer. ISBN 978-1-4614-0919-9.

Schutter, G. D., Theraulaz, G., & Deneubourg, J.-L. (2001). Animal–robots collective intelligence. *Annals of Mathematics and Artificial Intelligence, 31*, 223–238.

Sedwards, S. (2009). *A natural computation approach to biology: Modelling cellular processes and populations of cells with stochastic models of P systems*. Unpublished Doctoral Thesis, University of Trento.

Sell, S. (2013). *Stem cells handbook*. New York: Springer. ISBN 978-1-4614-7695-5.

Şen, Z. (2014). *Philosophical, logical and scientific perspectives in engineering*. Heidelberg: Springer. ISBN 978-3-319-01741-9.

Senkerik, R., Zelinka, I., Davendra, D., & Oplatkova, Z. (2010). Utilization of soma and differential evolution for robust stabilization of chaotic logistic equation. *Computers and Mathematics with Applications, 60*, 1026–1037.

Shann, M. (2008). *Emergent behavior in a simulated robot inspired by the slime mold*. Unpublished Bachelor Thesis, University of Zurich.

Shaw, B., Banerjee, A., Ghoshal, S. P., & Mukherjee, V. (2011). Comparative seeker and bio-inspired fuzzy logic controllers for power system stabilizers. *Electrical Power and Energy Systems, 33*, 1728–1738.

Shettleworth, S. J. (2010). *Cognition, evolution, and behavior*. New York: Oxford University Press. ISBN 978-0-19-531984-2.

Shi, Y. (2011a). Brain storm optimization algorithm. In Y. Tan, Y. Shi & G. Wang (Eds.), *ICSI 2011, Pat I, LNCS* (Vol. 6728, pp. 303–309). Berlin: Springer.

Shi, Y. (2011b). An optimization algorithm based on brainstorming process. *International Journal of Swarm Intelligence Research, 2*, 35–62.

Shlesinger, M. F., Klafter, J., & Zumofen, G. (1999). Above, below and beyond Brownian motion. *American Journal of Physics, 67*, 1253–1259.

Silva, D. J. A. D., Teixeira, O. N., & Oliveira, R. C. L. D. (2012). Performance study of cultural algorithm based on genetic algorithm with single and multi population for the MKP. In S. Gao (Ed.), *Bio-inspired computational algorithms and their applications*. Rijeka: InTech.
Sizer, F. S., & Whitney, E. (2014). *Nutrition: Concepts and controversies*. Belmont: Cengage Learning. ISBN 978-1-133-60318-4.
Smolin, L. A., & Grosvenor, M. B. (2010). *Healthy eathing_a guide to nutrition: Nutrition for sports and exercise*. New York: Infobase Publishing. ISBN 978-1-60413-804-7.
Song, M. X., Chen, K., He, Z. Y., & Zhang, X. (2013). Bionic optimization for micro-siting of wind farm on complex terrain. *Renewable Energy, 50*, 551–557.
Srinivasan, S., & Ramakrishnan, S. (2012). Nugget discovery with a multi-objective cultural algorithm. *Computer Science and Engineering: An International Journal, 2*, 11–25.
Steinbuch, R. (2011). Bionic optimisation of the earthquake resistance of high buildings by tuned mass dampers. *Journal of Bionic Engineering, 8*, 335–344.
Steinitz, M. (2014). *Human monoclonal antibodies: Methods and protocols*. New York: Springer. ISBN 978-1-62703-585-9.
Stradner, J., Thenius, R., Zahadat, P., Hamann, H., Crailsheim, K., & Schmickl, T. (2013). *Algorithmic requirements for swarm intelligence in differently coupled collective systems*. Chaos: Solitons and Fractals. 50.
Stukas, A. A., & Clary, E. G. (2012). Altruism and helping behavior. In V. S. Ramachandran, (Ed.), *Encyclopedia of human behavior* (2nd ed.). London: Elsevier, Inc. ISBN 978-0-12-375000-6.
Su, M.-C., Su, S.-Y., & Zhao, Y.-X. (2009). A swarm-inspired projection algorithm. *Pattern Recognition, 42*, 2764–2786.
Subbaiah, K. V., Rao, M. N., & Rao, K. N. (2009). Scheduling of AGVs and machines in FMS with makespan criteria using sheep flock heredity algorithm. *International Journal of Physical Sciences, 4*, 139–148.
Sueur, C., Deneubourg, J.-L., & Petit, O. (2010). Sequence of quorums during collective decision making in macaques. *Behavioral Ecology and Sociobiology, 64*, 1875–1885.
Sulaiman, M. H. (2013, March 15–17). Differential search algorithm for economic dispatch with valve-point effects. In *2nd International Conference on Engineering and Applied Science (ICEAS)* (pp. 111–117). Tokyo: Toshi Center Hotel.
Sulis, W. (1997). Fundamental concepts of collective intelligence. *Nonlinear Dynamics, Psychology, and Life Sciences, 1*, 35–53.
Sun, J., & Lei, X. (2009). Geese-inspired hybrid particle swarm optimization algorithm. In *International Conference on Artificial Intelligence and Computational Intelligence* (pp. 134–138). IEEE.
Taffe, M. A., & Taffe, W. J. (2011). Rhesus monkeys employ a procedural strategy to reduce working memory load in a self-ordered spatial search task. *Brain Research, 1413*, 43–50.
Taherdangkoo, M., Shirzadi, M. H., & Bagheri, M. H. (2012a). A novel meta-heuristic algorithm for numerical function optimization_blind, naked mole-rats (BNMR) algorithm. *Scientific Research and Essays, 7*, 3566–3583.
Taherdangkoo, M., Yazdi, M., & Bagheri, M. H. (2011). Stem cells optimization algorithm. *LNBI* (Vol. 6840, pp. 394–403). Berlin: Springer.
Taherdangkoo, M., Yazdi, M., & Bagheri, M. H. (2012b). A powerful and efficient evolutionary optimization algorithm based on stem cells algorithm for data clustering. *Central European Journal of Computer Science, 2*, 1–13.
Tan, Y., & Zhu, Y. (2010). Fireworks algorithm for optimization. In Y. Tan, Y. Shi & K. C. Tan (Eds.), *ICSI 2010, Part I, LNCS* (Vol. 6145, pp. 355–364). Berlin: Springer
Taylor, K. (2012). *The brain supremacy: Notes from the frontiers of neuroscience*. Oxford: Oxford University Press. ISBN 978-0-19-960337-4.
Tero, A., Takagi, S., Saigusa, T., Ito, K., Bebber, D. P., Fricker, M. D., et al. (2010). Rules for biologically inspired adaptive network design. *Science, 237*, 439–442.
Thammano, A., & Moolwong, J. (2010). A new computational intelligence technique based on human group formation. *Expert Systems with Applications, 37*, 1628–1634.

Theiner, G., Allen, C., & Goldstone, R. L. (2010). Recognizing group cognition. *Cognitive Systems Research, 11*, 378–395.

Tidball, K. G., & Krasny, M. E. (2014). *Greening in the red zone: Disaster, resilience and community greening*. Heidelberg: Springer. ISBN 978-90-481-9946-4.

Tollefsen, D. P. (2006). From extended mind to collective mind. *Cognitive Systems Research, 7*, 140–150.

Touhara, K. (2013). *Pheromone signaling: Methods and protocols*. New York: Springer. ISBN 978-1-62703-618-4.

Ulutas, B. H., & Kulturel-Konak, S. (2011). A review of clonal selection algorithm and its applications. *Artificial Intelligence Review, 36*, 117–138.

Umedachi, T., Takeda, K., Nakagaki, T., Kobayashi, R., & Ishiguro, A. (2010). Fully decentralized control of a soft-bodied robot inspired by true slime mold. *Biological Cybernetics, 102*, 261–269.

Venkumar, P., & Sekar, K. C. (2012). Design of cellular manufacturing system using non-traditional optimization algorithms. In V. Modrák & R. S. Pandian (Eds.), *Operations management research and cellular manufacturing systems: Innovative methods and approaches, Chap. 6* (pp. 99–139). Hershey: IGI Global.

Verdy, A., & Flierl, G. (2008). Evolution and social behavior in krill. *Deep-Sea Research II, 55*, 472–484.

Vucetich, J. A., Peterson, R. O., & Waite, T. A. (2004). Raven scavenging favours group foraging in wolves. *Animal Behaviour, 67*, 1117–1126.

Wang, G., Guo, L., Gandomi, A. H., Cao, L., Alavi, A. H., Duan, H., et al. (2013). Lévy-flight krill herd algorithm. *Mathematical Problems in Engineering, 2013*, 1–14.

Wang, P., & Cheng, Y. (2010). Relief supplies scheduling based on bean optimization algorithm. *Economic Research Guide, 8*, 252–253.

Wang, S., Dai, D., Hu, H., Chen, Y.-L., & Wu, X. (2011). RBF neural network parameters optimization based on paddy field algorithm. In *International Conference on Information and Automation* (*ICIA*) (pp. 349–353). June, Shenzhen, China. IEEE.

Wang, W., Feng, Q., & Zheng, Y. (2008, November 19–21). A novel particle swarm optimization algorithm with stochastic focusing search for real-parameter optimization. In *11th Singapore International Conference on Communication Systems* (*ICCS*) (pp. 583–587). Guangzhou, China. IEEE.

Wang, X., Gao, X.-Z., & Ovaska, S. J. (2009). Fusion of clonal selection algorithm and harmony search method in optimization of fuzzy classification systems. *International Journal of Bio-Inspired Computation, 1*, 80–88.

Wang, Z.-R., Ma, F., Ju, T., & Liu, C.-M. (2010). A niche genetic algorithm with population migration strategy. In *IEEE 2nd International Conference on Information Science and Engineering* (*ICISE*) (pp. 912–915).

Wei, G. (2011). Optimization of mine ventilation system based on bionics algorithm. *Procedia Engineering, 26*, 1614–1619.

Wei, Z. H., Cui, Z. H., & Zeng, J. C. (2010, September 26–28). Social cognitive optimization algorithm with reactive power optimization of power system. In *2010 International Conference on Computational Aspects of Social Networks* (*CASoN*) (pp. 11–14). Taiyuan, China.

Weigert, G., Horn, S., & Werner, S. (2006). Optimization of manufacturing processes by distributed simulation. *International Journal of Production Research, 44*, 3677–3692.

Whitehouse, M. E. A., & Lubin, Y. (1999). Competitive foraging in the social spider *Stegodyphus dumicola*. *Animal Behaviour, 58*, 677–688.

Whitten, K. W., Davis, R. E., Peck, M. L., & Stanley, G. G. (2014). *Chemistry*. Belmont: Cengage Learning. ISBN 13: 978-1-133-61066-3.

Wilson, C. (2013). *Brainstroming and beyond: a user-centered design method.* Waltham: Morgan Kaufmann, Elsevier Inc. ISBN 978-0-12-407157-5.

Wolpert, D. H., & Macready, W. G. (1997). No free lunch theorems for optimization. *IEEE Transactions on Evolutionary Computation, 1*, 67–82.

Woodward, J. (2008). *Climate change*. New York: Dorling Kindersley Limited. ISBN 978-07566-3771-2.

Woodworth, S. (2007). *Computability limits in membrane computing*. Unpublished Doctoral Thesis, University of California, Santa Barbara.

Wu, J., Cui, Z., & Liu, J. (2011, August 18–20). Using hybrid social emotional optimization algorithm with metropolis rule to solve nonlinear equations. In Y. Wang, A. Celikyilmaz, W. Kinsner, W. Pedrycz, H. Leung & L. A. Zadeh (Eds.), *10th International Conference on Cognitive Informatics and Cognitive Computing* (ICCI & CC) (pp. 405–411). Banff, AB. IEEE.

Xiao, J.-H., Huang, Y.-F., & Cheng, Z. (2013). A bio-inspired algorithm based on membrane computing for engineering design problem. *International Journal of Computer Science Issues, 10*, 580–588.

Xu, Y. C., Cui, Z. H., & Zeng, J. C. (2010). Social emotional optimization algorithm for nonlinear constrained optimization problems. In *1st International Conference on Swarm, Evolutionary and Memetic Computing* (*SEMCCO*) (pp. 583–590).

Xue, J., Wu, Y., Shi, Y., & Cheng, S. (2012). Brain storm optimization algorithm for multi-objective optimization problems. In Y. Tan, Y. Shi & Z. Ji (Eds.), *ICSI 2012, Part I, LNCS* (Vol. 7331, pp. 513–519). Berlin: Springer.

Yampolskiy, R. V., Ashby, L., & Hassan, L. (2012). Wisdom of artificial crowds: A metaheuristic algorithm for optimization. *Journal of Intelligent Learning Systems and Applications, 4*, 98–107.

Yang, C., Tu, X., & Chen, J. (2007). Algorithm of marriage in honey bees optimization based on the wolf pack search. In *IEEE International Conference on Intelligent Pervasive Computing* (*IPC*) (pp. 462–467).

Yang, X.-S. (2005). Biology-derived algorithms in engineering optimization. In S. Olarius & A. Zomaya (Eds.), *Handbook of Bioinspired Algorithms and Applications, Chap. 32* (pp. 585–596). Boca Raton: CRC Press.

Yang, X.-S. (2012). Flower pollination algorithm for global optimization. *Unconventional Computation and Natural Computation, LNCS* (Vol. 7445, pp. 240–249). Berlin: Springer.

Yang, X.-S., & Deb, S. (2010). Eagle strategy using Lévy walk and firefly algorithms for stochastic optimization. In J. R. Gonzalez (Ed.), *Nature Inspired Cooperative Strategies for Optimization* (*NISCO 2010*), *SCI* (Vol. 284, pp. 101–111). Berlin: Springer.

Yang, X.-S., & Deb, S. (2012). Two-stage eagle strategy with differential evolution. *International Journal of Bio-Inspired Computation, 4*, 1–5.

Yang, X.-S., Karamanoglu, M., & He, X. (2013). Multi-objective flower algorithm for optimization. *Procedia Computer Science, 18*, 861–868.

Yeagle, P. L. (Ed.). (2005). *The structure of biological membranes*. Boca Raton: CRC Press. ISBN 0-8493-1403-8.

You, S. K., Kwon, D. H., Park, Y.-I., Kim, S. M., Chung, M.-H., & Kim, C. K. (2009). Collective behaviors of two-component swarms. *Journal of Theoretical Biology, 261*, 494–500.

Zaharie, D., & Ciobanu, G. (2006). Distributed evolutionary algorithms inspired by membranes in solving continuous optimization problems. In H. J. Hoogeboom (Ed.), *WMC 7, LNCS* (Vol. 4361, pp. 536–553). Berlin: Springer.

Zang, H., Zhang, S., & Hapeshi, K. (2010). A review of nature-inspired algorithms. *Journal of Bionic Engineering, 7*, S232–S237.

Zelinka, I., & Lampinen, J. (2000). Soma: Self-organizing migrating algorithm. In *The 6th International Conference on Soft Computing, Brno, Czech Republic*.

Zelinka, I., Senkerik, R., & Navratil, E. (2009). Investigation on evolutionary optimization of chaos control. *Chaos, Solitons and Fractals, 40*, 111–129.

Zhan, Z.-H., Zhang, J., Shi, Y.-H., & Liu, H.-L. (2012, June 10–15) A modified brain storm optimization. In *World Congress on Computational Intelligence* (*WCCI*) (pp. 1–8). Brisbane, Australia. IEEE.

Zhang, G., Cheng, J., & Gheorghe, M. (2011). A membrane-inspired approximate algorithm for traveling salesman problems. *Romanian Journal of Information Science and Technology, 14*, 3–19.

Zhang, G., Yang, H., & Liu, Z. (2007). Using watering algorithm to find the optimal paths of a maze. *Computer, 24*, 171–173.

Zhang, W., Luo, Q. & Zhou, Y. (2009). A method for training RBF neural networks based on population migration algorithm. In *International Conference on Artificial Intelligence and Computational Intelligence* (*AICI*) (pp. 165–169). IEEE.

Zhang, W., & Zhou, Y. (2009). Description population migration algorithm based on framework of swarm intelligence. In *IEEE WASE International Conference on Information Engineering* (*ICIE*) (pp. 281–284).

Zhang, X., Chen, W., & Dai, C. (2008a, April 6–9) Application of oriented search algorithm in reactive power optimization of power system. *DRPT 2008* (pp. 2856–2861). Nanjing, China. DRPT.

Zhang, X., Sun, B., Mei, T., & Wang, R. (2010, November 28–30). Post-disaster restoration based on fuzzy preference relation and bean optimization algorithm. In *Youth Conference on Information Computing and Telecommunications* (*YC-ICT*) (pp. 271–274). IEEE.

Zhang, X., Jiang, K., Wang, H., Li, W., & Sun, B. (2012a). An improved bean optimization algorithm for solving TSP. In Y. Tan, Y. Shi & Z. Ji (Eds.), *ICSI 2012, Part I, LNCS* (Vol. 7331, pp. 261–267). Berlin: Springer.

Zhang, X., Huang, S., Hu, Y., Zhang, Y., Mahadevan, S., & Deng, Y. (2013a). Solving 0-1 knapsack problems based on amoeboid organism algorithm. *Applied Mathematics and Computation, 219*, 9959–9970.

Zhang, X., Sun, B., Mei, T., & Wang, R. (2013b). A novel evolutionary algorithm inspired by beans dispersal. *International Journal of Computational Intelligence Systems, 6*, 79–86.

Zhang, X., Wang, H., Sun, B., Li, W., & Wang, R. (2013c). The Markov model of bean optimization algorithm and its convergence analysis. *International Journal of Computational Intelligence Systems, 6*, 609–615.

Zhang, X., Wang, R., & Song, L. (2008b). A novel evolutionary algorithm: Seed optimization algorithm. *Pattern Recognition and Artificial Intelligence, 21*, 677–681.

Zhang, Z.-W., Zhang, H., & Li, Y.-B. (2012b). Biologically inspired collective construction with visual landmarks. *Journal of Zhejiang University-SCIENCE C (Computers and Electronics), 13*, 315–327.

Zhao, Q., & Liu, X. (2011). An improved multi-objective population migration optimization algorithm. In *2nd International Symposium on Intelligence Information Processing and Trusted Computing* (*IPTC*) (pp. 143–146). IEEE.

Zheng, Y., Chen, W., Dai, C., & Wang, W. (2009). Stochastic focusing search: A novel optimization algorithm for real-parameter optimization. *Journal of Systems Engineering and Electronics, 20*, 869–876.

Zhou, D., Shi, Y., & Cheng, S. (2012). Brain storm optimization algorithm with modified step-size and individual generation. In Y. Tan, Y. Shi & Z. Ji (Eds.), *ICSI 2012, Part I, LNCS* (Vol. 7331, pp. 243–252). Berlin: Springer.

Zhou, Y., & Liu, B. (2009). Two novel swarm intelligence clustering analysis methods. In *IEEE Fifth International Conference on Natural Computation* (*ICNC*) (pp. 497–501).

Zhou, Y., & Mao, Z. (2003). A new search algorithm for global optimization: Population migration algorithm. *Journal of South China University of Technology, A31*, 1–5.

Zungeru, A. M., Ang, L.-M., & Seng, K. P. (2012). Termite-hill: Performance optimized swarm intelligence based routing algorithm for wireless sensor networks. *Journal of Network and Computer Applications, 35*, 1901–1917.

Part III
Physics-based CI Algorithms

Chapter 18
Big Bang–Big Crunch Algorithm

Abstract In this chapter, the big bang–big crunch (BB–BC), a global optimization method inspired from one of the cosmological theories known as closed universe, is introduced. We first, in Sect. 18.1, describe the background knowledge regarding the big bang and big crunch. Then, Sect. 18.2 details the fundamentals of BB–BC, the selected variants of BB–BC, and the representative BB–BC application, respectively. Finally, Sect. 18.3 draws the conclusions of this chapter.

18.1 Introduction

Cosmological theory is an exciting subject, because it shows how the universe happens, moves, and revolutions. One of the fascinating topics is where all of the stars and galaxies came from (and how, and why)? This question has long been explored by the physics. For example, two famous physics, i.e., Sir Isaac Newton and Albert Einstein, believed that the universe is unchanging and introduced a term, called cosmological constant. However, this would prove to be a mistake. In 1929, astronomer Edwin Hubble discovered that the universe was expanding. He found that in the early universe, gravity was very strong, as a result of the concentration of matter in a very small space—so small, in fact, that it was compressed down to a single point. Thus, it would suffer an incredible pressure and has expanded ever since, known as the big bang. This event was controversial until 1965, when an accidental discovery supported the theory. Today, the most advanced astronomical observations show that the big bang theory is likely true.

Scientists were originally very upset by the big bang theory, because they believed in an eternal universe (i.e., the universe does not change over time). However, they concerned soon with another question of what is the ultimate fate of the universe? One idea that was popular was that the universe would expand until gravity began to pull it back, resulting in a big crunch, where all matter returned to a single unified point—and then the cycle of expansion would start all over again. This hypothesis is known as closed universe. What happens after that? We cannot exactly tell for now.

B. Xing and W.-J. Gao, *Innovative Computational Intelligence: A Rough Guide to 134 Clever Algorithms*, Intelligent Systems Reference Library 62,
DOI: 10.1007/978-3-319-03404-1_18, © Springer International Publishing Switzerland 2014

18.1.1 Big Bang

Literally, every bit of matter and energy in our universe was created through a singular event (i.e., the unimaginable crucible of heat and light) that we call the big bang (Bauer and Westfall 2011). Just for the record, it was neither big (in fact, it was very small and fit onto the head of a pin), nor was there a bang. This event happened $(13.73 \pm 0.12) \times 10^9$ years ago. Although this theory is not perfect, and over time physics made efforts in order to make it more consistent. One thing is true that the universe was well on its way to becoming what we observe today: filled with galaxies, stars, planets, and all other sorts of strange and exotic things.

18.1.2 Big Crunch

One hypothesis that the future of the universe is called big crunch model, which means that the universe contracts back into a point of mass. This will proceed almost exactly like the big bang, except in reverse. Whether the expansion of the universe will take place forever or will stop some day, it depends on the quantity of matter it has (Scalzi 2008).

18.2 Big Bang–Big Crunch Algorithm

18.2.1 Fundamentals of the Big Bang–Big Crunch Algorithm

Inspired from one of the cosmological theories that universe was "born" in the big bang and might "die" in the big crunch, Erol and Eksin (2006) proposed a new algorithm, namely, the big bang–big crunch (BB–BC) algorithm. In general, the proposed algorithm includes two successive phases. In the big bang phase (corresponding to the disorder caused by the energy dissipation in nature), random points are generated; whereas, in the big crunch phase (corresponding to the order due to gravitational attraction), those points shrank to a single representative point via a centre of mass or minimal cost approach.

18.2.1.1 Big Bang Phase

Just like the expansion of the universe, the main purpose of the big bang phase in BB–BC is to create initial populations. The initial position of each input is generated randomly over the entire search space. Once the population pool is created, fitness values of the individuals are calculated (Genç et al. 2010).

18.2.1.2 Big Crunch Phase

If enough mass and energy (i.e., inputs) is in the universe (i.e., search space), then that mass and energy may cause enough attraction to halt the expansion of the universe and reverse it—bringing the entire universe back to a single point. Similarly, in the big crunch phase, a contraction procedure is applied to form a centre or a representative point for further big bang operations. In other words, the big crunch phase is a convergence operator that has many inputs but only one output, which called "centre of mass". Here, the term mass refers to the inverse of the fitness function value (f^i). The point representing the centre of mass that is denoted by x^c and calculated according to Eq. 18.1 (Erol and Eksin 2006):

$$\vec{x}^c = \frac{\sum_{i=1}^{N} \frac{1}{f^i} \vec{x}^i}{\sum_{i=1}^{N} \frac{1}{f^i}}, \tag{18.1}$$

where x^i is a point within an n-dimensional search space generated, f^i is fitness function value of this point (such as cost function), N is the population size in big bang phase.

Instead of the centre of mass, the best fit individual (i.e., the lowest f^i value) can also be chosen as the starting point in the big bang phase.

The new generation for the next iteration in the big bang phase is normally distributed around the centre of mass, using Eq. 18.2 (Erol and Eksin 2006):

$$x^{new} = x^c + \frac{lr}{k}, \tag{18.2}$$

where x^c stands for centre of mass, l is the upper limit of the parameter, r [or $N(0, 1)$] is a normal random number generated according to a standard normal distribution with mean (μ) zero and standard deviation (σ) equal to one, and k is the iteration step. Then new point (x^{new}) is upper and lower bounded.

Summarizing the steps in the standard BB–BC algorithm yields to Erol and Eksin (2006):

- Step 1: Initiation population of N candidate solution is randomly generated all over the search space.
- Step 2: The fitness function value (f^i) corresponding to each candidate solution is calculated.
- Step 3: The N candidate solutions are contracted into the centre of mass (x^c), either by using the Eq. (18.1) or by choosing the point that has lowest value after the calculation in Step 2.
- Step 4: New population of solutions is generated around x^c by adding or subtracting a random number whose value decreases by increasing the iterations elapsed.
- Step 5: Check if maximum iteration is reached; go to Step 2 for new beginning. If a specified termination criteria is satisfied stop and return the best solution.

18.2.2 Performance of BB–BC

To evaluate the performance of the BB–BC algorithm, (Erol and Eksin 2006) proposed six test functions, namely, Sphere function, Rosenbrock function, Step function, Ellipsoid function, Rastrigin function, and Ackley function. Compared with combat genetic algorithm (CGA), the BB–BC algorithm presented a better results of finding the global best solution.

18.2.3 Selected BB–BC Variants

Although BB–BC algorithm is a new member of computational intelligence (CI) family, a number of BB–BC variations have been proposed in the literature for the purpose of further improving the performance of BB–BC. This section gives an overview to some of these BB–BC variants which have been demonstrated to be very efficient and robust.

18.2.3.1 Hybrid BB–BC Algorithm

In Kaveh and Talatahari (2009, 2010b), the authors developed one of the first BB–BC hybrids, called hybrid BB–BC (HBB–BC). Overall, the HBB–BC introduced two improvements: using the particle swarm optimization (PSO) capacities to improve the exploration ability of BB–BC algorithm, and using sub-optimization mechanism (SOM) to update the search-space of BB–BC algorithm. Compared with the standard BB–BC and other conventional CI optimization methods such as genetic algorithm (GA), ant colony optimization (ACO), PSO, and harmony search (HS), the HBB–BC performed better.

In general, there are also two phases involved in HBB–BC: a big bang phase where candidate solutions are randomly distributed over the search space, and a big crunch phase working as a convergence operator where the centre of mass is generated. Compared with standard BB–BC algorithm, the main difference is that the HBB–BC employed the PSO capacities to improve the exploration ability. Kaveh and Talatahari (2009) pointed out that the reason to select PSO as the first reformation due to at each iteration, the particle moves towards both local best (i.e., a direction computed from the best visited position), and global best (i.e., the best visited position of all particles in its neighbourhood). Inspired by that, the HBB–BC approach not only uses the centre of mass but also utilizes the best position of each candidate and the best global position to generate a new solution. The calculation formulas in big crunch phase are as follows:

- The centre of mass can be computed via Eq. 18.3 (Kaveh and Talatahari 2009):

$$x_i^{c(k)} = \frac{\sum_{j=1}^{N} \frac{1}{f^i} x_i^{(k,j)}}{\sum_{j=1}^{N} \frac{1}{f^i}}, \quad i = 1, 2, \ldots, ng, \tag{18.3}$$

  where $x_i^{c(k)}$ is the ith component of the jth solution generated in the kth iteration; f^i is fitness function value of this point (such as cost function); N is the population size in big bang phase.
- The new generation for the next iteration in the big bang phase is normally distributed around $x_i^{c(k)}$ and can be computed via Eq. 18.4 (Kaveh and Talatahari 2009):

$$x_i^{(k+1,j)} = \alpha_2 x_i^{c(k)} + (1-\alpha_2)\left(\alpha_3 x_i^{gbest(k)} + (1-\alpha_3) x_i^{lbest(k,j)}\right) + \frac{r_j \alpha_1 (x_{\max} - x_{\min})}{k+1}, \quad i = 1, 2, \ldots, ng \quad j = 1, 2, \ldots, N, \tag{18.4}$$

  where r_j is a random number from a standard normal distribution which changes for each candidate; $x_{\max}$ and $x_{\min}$ are the upper and lower limits; α_1 is a parameter for limiting the size of the search space; $x_i^{lbest(k,j)}$ is the best position of the jth particle up to the iteration k; $x_i^{gbest(k)}$ is the best position among all candidates up to the iteration k; α_2 and α_3 are adjustable parameters controlling the influence of the global best and local best on the new position of the candidates, respectively.

Another reformation in the HBB–BC is that the SOM has been employed as an auxiliary tool to update the search space. Based on the principle of finite element method, SOM was introduced by Kaveh et al. (2008). The work principle of SOM is repetitive dividing the search space into sub-domains and employing optimization process into these sub-domains until a specified termination criteria (such as required accuracy) is satisfied and return the best solution.

The SOM mechanism can be calculated as the repetition of the following steps for definite times, nc, (in the stage k of the repetition):

- Calculating cross-sectional area bounds for each group.

  If $x_i^{gbest(k_{SOM}-1)}$ is the global best solution obtained from the previous stage for design variable i, then we have Eq. 18.5 (Kaveh and Talatahari 2009):

$$\begin{cases} x_{\min,i}^{(k_{SOM})} = x_i^{gbest(k_{SOM}-1)} - \beta_1 \cdot \left(x_{\max,i}^{(k_{SOM}-1)} - x_{\min,i}^{(k_{SOM}-1)}\right) \geq x_{\min,i}^{(k_{SOM}-1)} \\ x_{\max,i}^{(k_{SOM})} = x_i^{gbest(k_{SOM}-1)} + \beta_1 \cdot \left(x_{\max,i}^{(k_{SOM}-1)} - x_{\min,i}^{(k_{SOM}-1)}\right) \leq x_{\max,i}^{(k_{SOM}-1)} \end{cases}, \quad i = 1, 2, \ldots, ng \quad k_{SOM} = 2, \ldots, nc, \tag{18.5}$$

where is an adjustable factor which determines the amount of the remaining search space; and $x_{\min,i}^{(k_{SOM})}$, $x_{\max,i}^{(k_{SOM})}$ are the minimum and the maximum allowable cross-sectional areas at the stage, respectively. In stage 1, the amounts of $x_{\min,i}^{(1)}$ and $x_{\max,i}^{(1)}$ are set according to Eq. 18.6 (Kaveh and Talatahari 2009):

$$x_{\min,i}^{(1)} = x_{\min}, \quad x_{\max,i}^{(1)} = x_{\max}, \quad i = 1, 2, \ldots, ng, \tag{18.6}$$

where $x_{\min}, i^1$ and $x_{\max}, i^1$ are the minimum and the maximum allowable cross-sectional areas at the stage 1.

- Determining the amount of increase in allowable cross-sectional areas via Eq. 18.7 (Kaveh and Talatahari 2009).

$$x_i^{*(k_{SOM})} = \frac{\left(x_{\max,i}^{(k_{SOM})} - x_{\min,i}^{(k_{SOM})}\right)}{\beta_2 - 1}, \quad i = 1, 2, \ldots, ng, \tag{18.7}$$

where $x_i^{*(k_{SOM})}$ is the amount of increase in allowable cross-sectional area; and β_2 is the number of permissible value of each group.

- Creating the series of the allowable cross-sectional areas.
  The set of allowable cross-sectional areas for group *i* can be defined as Eq. 18.8 (Kaveh and Talatahari 2009):

$$x_{\min,i}^{(k_{SOM})}, \ x_{\min,i}^{(k_{SOM})} + x_i^{*(k_{SOM})}, \ \ldots, x_{\min,i}^{(k_{SOM})} + (\beta_2 - 1) \cdot x_i^{*(k_{SOM})} = x_{\max,i}^{(k_{SOM})}, \quad i = 1, 2, \ldots, ng. \tag{18.8}$$

- Determining the optimum solution of the stage k_{SOM}.
  This is the last step and the stopping creation for SOM can be defined as Eq. 18.9 (Kaveh and Talatahari 2009):

$$x_i^{*(nc)} \leq x^*, \quad i = 1, 2, \ldots, ng, \tag{18.9}$$

where $x_i^{*(nc)}$ = the amount of accuracy rate of the last stage; and x^* = the amount of accuracy rate of the primary problem.

In addition to HBB–BC, another hybridization between the BB–BC algorithm and simulated annealing (SA) technique was recently proposed in Altomare et al. (2013). In this approach, the value of fitness function is further submitted to a local optimization by SA with a fast annealing schedule. This new hybrid method has been implemented to solve crystal structure problems. Compared with traditional SA algorithm, the hybridized algorithm showed better results in terms of computation time.

18.2.3.2 Improved BB–BC Algorithm

To improve the BB–BC performance, Hasançebi and Azad (2012) proposed two enhanced variants of the BB–BC algorithm, called modified BB–BC (MBB–BC) and exponential BB–BC algorithm (EBB–BC), respectively. In the new formulation, the normal random number (*r*) is changed by using any appropriate statistical distribution in order to eliminate the shortcomings of the standard formulation (e.g., big search dimensionality). Furthermore, to meet the discrete data requirements, the improved BB–BC algorithm employed the way of round-off instead of the real values to nearest integers representing the sequence number. As a result, the new generation for the next iteration in the big bang phase can be formulated as Eq. 18.10 (Hasançebi and Azad 2012):

$$x^{new} = x^c + round\left[\alpha \cdot N(0,1)_i^3 \frac{(x_{\max} - x_{\min})}{k}\right], \tag{18.10}$$

where x^c is the value of discrete design variable, $x_{\max}$ and $x_{\min}$ are its lower and upper bounds, respectively. In addition, the power of random number is set to 3 based on extensive numerical experiments. This reformulation is referred to as MBB–BC.

In a similar vein, Hasançebi and Azad (2012) also proposed an alternative approach called EBB–BC to deal with the discrete design problem where the use of an exponential distribution (*E*) in conjunction with the third power of random number as shown in Eq. 18.11.

$$x^{new} = x^c + round\left[\alpha \cdot E(\lambda = 1)_i^3 \frac{(x_{\max} - x_{\min})}{k}\right]. \tag{18.11}$$

The probability density function for an exponential distribution is given as Eq. 18.12 (Hasançebi and Azad 2012):

$$f(x) = \begin{cases} \lambda e^{-\lambda x} & x \geq 0 \\ 0 & x < 0 \end{cases}, \tag{18.12}$$

where λ is a real, positive constant. The mean and variance of the exponential distribution are given as $1/\lambda$ and $1/\lambda^2$, respectively.

Accordingly, if all the design variables in a new solution remain unchanged after applying Eq. 18.11, i.e., $x^{new} = x^c$, the generation process is iterated in the same way by decreasing the λ parameter of the exponential distribution by half each time, and this is repeated until a different solution is produced, i.e., $x^{new} \neq x^c$.

Hasançebi and Azad (2012) presented two numerical examples to investigate the performance of EBB–BC and MBB–BC. Compared with standard BB–BC, the improved variants gave the better results in terms of balancing between the exploration and exploitation characteristics of the algorithms. More recently, an upper bound strategy (UBS) with MBB–BC and EBB–BC was further integrated in Azad et al. (2013) for optimum design of steel frame structures. Computational results showed that the new effort significantly reduced the number of structural analyses.

Furthermore, to improve the convergence properties of the BB–BC, Alatas (2011) proposed a new methods called uniform big bang–chaotic big crunch (UBB–CBC) algorithm which involves two improved reformulation, i.e., an uniform population method to generate uniformly distributed random points in the big bang phase (called UBB), and the chaotic maps property to rapidly shrink those points to a single representative point in the big crunch phase (called CBC). Compared with benchmark functions, the performance of UBB–CBC showed superiority over the standard BB–BC algorithm.

18.2.3.3 Local Search-Based BB–BC Algorithm

As Kaveh and Talatahari (2009) concluded at the end of their study, the HBB–BC is worse than improved algorithms which have the extra local search ability. To fulfil this gap, Genç et al. (2010) introduced a local search move mechanism to BB–BC algorithm based on defining a possible improving direction to check neighbouring points.

In details, Genç et al. (2010) put the local search methods (i.e., expansion and contraction) between the original "banging" and "crunching" phases. The main objective is to modify the representative point with local directional moves, in order to easily attack the path going to optima and decrease the process time fro reaching the global minima. The direction vector can be formulated as Eq. 18.13 (Genç et al. 2010):

$$IV_1 = P(n) - P(n-1), \tag{18.13}$$

where IV_1 stands for the improvement vector of single step regression BB–BC; $P(n)$ is the current best or fittest point; and $P(n-1)$ is the last stored best or fittest point.

To test the performance of the new algorithm, Genç et al. (2010) implemented it on the target tracking problem. The simulation results showed that the local search-based BB–BC algorithm outperformed the standard BB–BC algorithm in terms of data accuracy.

18.2.4 Representative BB–BC Application

According to the literature review, the main application of the BB–BC algorithm is in structural optimization. In general, there are three main groups of structural optimization applications (Sadollah et al. 2012; Hasançebi and Azad 2012): (1) sizing optimization; (2) shape optimization; and (3) topology optimization. In sizing optimization, it can further be divided into two subcategories: discrete and continuous. Hasançebi and Azad (2012) used MBB–BC and EBB–BC to solve the discrete sizing optimization, whereas Kaveh and Talatahari (2009) and (2010a) proposed the HBB–BC algorithm to solve problems with continuous domains.

18.2.4.1 Truss Optimization

Truss optimization is one of the most active branches of the continuous sizing optimization. The main objective for designing truss structures is to determine the optimum values for member cross-sectional areas (A_i) in order to minimize the structural weight (W), meanwhile, satisfy the inequality constraints that limit design variable sizes and structural responses.

In Kaveh and Talatahari (2009), the authors employed the HBB–BC method to address the above mentioned truss optimization problem. In their work, five truss structures optimization examples were presented, namely, a 25-bar spatial truss structure, a 72-bar spatial truss structure, a 120-bar dome shaped truss, a square on diagonal double-layer grid, and a 26-story-tower spatial truss. Compared with other CI techniques (e.g., GA, ACO, and PSO), the HBB–BC performed well in large size structures characterized by converging difficulty or easily getting trapped at a local optimum.

18.3 Conclusions

In summary, the BB–BC algorithm is a population-based CI algorithm that shares some similarities with evolutionary algorithms (Erol and Eksin 2006), such as randomly selected initialization and refinement of the value of fitness function according to the best fitted answers of the previous loop or loops (Kaveh and Farhoudi 2011). The core working principle of BB–BC is to transform a convergent solution to a chaotic state which is a new set of solutions (Erol and Eksin 2006). The leading advantages of BB–BC are its high convergence speed and the low computation time, together with its simplicity and capability of easy-to-implement (Desai and Prasad 2013).

With the rapid spreading of BB–BC, in addition to the representative applications detailed in this chapter, the BB–BC has also been successfully applied to a variety of optimization problems as outlined below:

- Automatic target tracking (Genç et al. 2010; Genç and Hocaoğlu 2008).
- Fuzzy system control (Kumbasar et al. 2008, 2011; Aliasghary et al. 2011).
- Layout optimization (Kaveh and Farhoudi 2011).
- Linear time invariant systems (Desai and Prasad 2013).
- Course timetabling (Jaradat and Ayob 2010).
- Power system (Sedighizadeh and Arzaghi-Haris 2011; Dincel and Genc 2012; Kucuktezcan and Gen 2012; Zandi et al. 2012).
- Structural engineering (Altomare et al. 2013; Azad et al. 2013; Tang et al. 2010; Camp 2007; Camp and Huq 2013).

Interested readers are referred to them as a starting point for a further exploration and exploitation of the BB–BC algorithm.

References

Alatas, B. (2011). Uniform big bang–chaotic big crunch optimization. *Communications in Nonlinear Science and Numerical Simulation, 16*, 3696–3703.

Aliasghary, M., Eksin, I., & Guzelkaya, M. (2011). Fuzzy-sliding model reference learning control of inverted pendulum with big bang–big crunch optimization method. In *11th International Conference on Intelligent Systems Design and Applications (ISDA)* (pp. 380–384). IEEE.

Altomare, A., Corriero, N., Cuocci, C., Moliterni, A., & Rizzi, R. (2013). The hybrid big bang–big crunch method for solving crystal structure from powder diffraction data. *Journal of Applied Crystallography, 46*, 779–787.

Azad, S. K., Hasançebi, O., & Azad, S. K. (2013). Upper bound strategy for metaheuristic based design optimization of steel frames. *Advances in Engineering Software, 57*, 19–32.

Bauer, W., & Westfall, G. D. (2011). *University physics with modern physics*. New York, USA: McGraw-Hill. ISBN 978-0-07-285736-8.

Camp, C. V. (2007). Design of space trusses using big bang–big crunch optimization. *Journal of Structural Engineering, 133*, 999–1008.

Camp, C. V., & Huq, F. (2013). CO_2 and cost optimization of reinforced concrete frames using a big bang–big crunch algorithm. *Engineering Structures, 48*, 363–372.

Desai, S. R., & Prasad, R. (2013). A novel order diminution of LTI systems using big bang–big crunch optimization and routh approximation. *Applied Mathematical Modelling, 37*, 8016–8028. http://dx.doi.org/10.1016/j.apm.2013.02.052.

Dincel, E., & Genc, V. M. I. (2012, November 23–25). A power system stabilizer design by big bang–big crunch algorithm. In *IEEE International Conference on Control System, Computing and Engineering*, Penang, Malaysia (pp. 307–312). IEEE.

Erol, O. K., & Eksin, I. (2006). A new optimization method: Big bang–big crunch. *Advances in Engineering Software, 37*, 106–111.

Genç, H. M., & Hocaoğlu, A. K. (2008). Bearing-only target tracking based on big bang–big crunch algorithm. In *The Third International Multi-Conference on Information Technology* (pp. 229–233). IEEE.

Genç, H. M., Eksin, İ., & Erol, O. K. (2010, October 10–13). Big bang–big crunch optimization algorithm hybridized with local directional moves and application to target motion analysis problem. In *IEEE International Conference on Systems, Man, and Cybernetics (SMC)*, Istanbul, Turkey (pp. 881–887). IEEE.

Hasançebi, O., & Azad, S. K. (2012). An exponential big bang–big crunch algorithm for discrete design optimization of steel frames. *Computers and Structures, 110–111*, 167–179.

Jaradat, G. M., & Ayob, M. (2010). Big bang–big crunch optimization algorithm to solve the course timetabling problem. In *10th International Conference on Intelligent Systems Design and Applications (ISDA)* (pp. 1448–1452). IEEE.

Kaveh, A., & Farhoudi, N. (2011). A unified approach to parameter selection in meta-heuristic algorithms for layout optimization. *Journal of Constructional Steel Research, 67*, 1453–1462.

Kaveh, A., & Talatahari, S. (2009). Size optimization of space trusses using big bang–big crunch algorithm. *Computers and Structures, 87*, 1129–1140.

Kaveh, A., & Talatahari, S. (2010a). A discrete big bang–big crunch algorithm for optimal design of skeletal structures. *Asian Journal of Civil Engineering (Building and Housing), 11*, 103–122.

Kaveh, A., & Talatahari, S. (2010b). Optimal design of Schwedler and ribbed domes via hybrid big bang–big crunch algorithm. *Journal of Constructional Steel Research, 66*, 412–419.

Kaveh, A., Farahmand, B. A., & Talatahari, S. (2008). Ant colony optimization for design of space trusses. *International Journal of Space Structure, 23*, 167–181.

Kucuktezcan, C. F., & Genc, V. M. I. (2012). Big bang–big crunch based optimal preventive control action on power systems. In *3rd IEEE PES International Conference and Exhibition on Innovative Smart Grid Technologies (ISGT Europe)*, Berlin, Germany (pp. 1–4). IEEE.

Kumbasar, T., Yeşil, E., Eksin, İ., & Güzelkaya, M. (2008, March 12–14). Inverse fuzzy model control with online adaptation via big bang–big crunch optimization. In *3rd International Symposium on Communications, Control and Signal Processing*, Malta (pp. 697–702). IEEE.

Kumbasar, T., Eksin, I., Guzelkaya, M., & Yesil, E. (2011). Adaptive fuzzy model based inverse controller design using BB–BC optimization algorithm. *Expert Systems with Applications, 38*, 12356–12364.

Sadollah, A., Bahreininejad, A., Eskandar, H., & Hamdi, M. (2012). Mine blast algorithm for optimization of truss structures with discrete variables. *Computers and Structures, 102–103*, 49–63.

Scalzi, J. (2008). *The rough guide to the universe*. New York, USA: Rough Guides Ltd. ISBN 9781-84353-800-4.

Sedighizadeh, M., & Arzaghi-Haris, D. (2011). Optimal allocation and sizing of capacitors to minimize the distribution line loss and to improve the voltage profile using big bang–big crunch optimization. *International Review of Electrical Engineering, 6*, 2013–2019.

Tang, H., Zhou, J., Xue, S., & Xie, L. (2010). Big bang–big crunch optimization for parameter estimation in structural systems. *Mechanical Systems and Signal Processing, 24*, 2888–2897.

Zandi, Z., Afjei, E., & Sedighizadeh, M. (2012, Dec 2–5). Reactive power dispatch using big bang–big crunch optimization algorithm for voltage stability enhancement. In *IEEE International Conference on Power and Energy (PECon)*, Kota Kinabalu Sabah, Malaysia (pp. 239–244). IEEE.

Chapter 19
Central Force Optimization Algorithm

Abstract In this chapter, we introduce a new deterministic multi-dimensional search algorithm called central force optimization (CFO), which is based on the metaphor of gravitational kinematics. We first, in Sect. 19.1, describe the general knowledge of the gravitational force. Then, in Sect. 19.2, the fundamentals and performance of CFO are detailed. Finally, Sect. 19.3 draws the conclusions of this chapter.

19.1 Introduction

Nowadays, there are a lot of interactions between physicists and computer scientists that often provide new insights in both fields. On one hand, some physics phenomena can be used as general purpose optimization tools to solve the combinatorial problems, such as simulated annealing. On the other hand, algorithms in linear, nonlinear, and discrete optimization sometimes have the potential to be useful tools in physics, such as Monte Carlo simulations. In the light of these statements, recently, Formato (2007) proposed a new optimization algorithm called central force optimization (CFO) algorithm which is inspired by the theory of particle kinematics in gravitational field.

19.1.1 Gravitational Force

The gravitational force, often simply called gravity, operates over all distances and is always a force of attraction between objects with mass (Bauer and Westfall 2011). For example, if you hold an object in your hand and let go of it, it falls downward. We know what causes this effect: the gravitational attraction between the earth and the object. Despite that, gravitational force is also responsible for holding the moon in orbit around the earth and the earth in orbit around the sun.

B. Xing and W.-J. Gao, *Innovative Computational Intelligence: A Rough Guide to 134 Clever Algorithms*, Intelligent Systems Reference Library 62,
DOI: 10.1007/978-3-319-03404-1_19,

The magnitude of the gravitational force between two point masses is given via Eq. 19.1 (Bauer and Westfall 2011):

$$F(r) = G\frac{m_1 m_2}{r^2}, \tag{19.1}$$

where m_1 and m_2 are two point masses, the coefficient G is the "universal gravitational constant", and $r = \left|\vec{r}_2 - \vec{r}_1\right|$ is a distance from each other.

Each mass is accelerated toward the other. The vector acceleration experienced by mass m_1 due to mass m_2 is given by Eq. 19.2 (Formato 2007):

$$\vec{a}_1 = -G\frac{m_2 \hat{r}}{r^2}, \tag{19.2}$$

where $\hat{r}$ is a unit vector, and the minus sign is taken into account in the order in which the differences in the acceleration expressions are taken.

The position vector a particle subject to constant acceleration during the interval t to $t + \Delta t$ is given by Eq. 19.3 (Formato 2007):

$$\vec{R}(t + \Delta t) = \vec{R}_0 + \vec{V}_0 \Delta t + \frac{1}{2}\vec{a}\Delta t^2, \tag{19.3}$$

where $\vec{R}_0$ and $\vec{V}_0$ are the position and velocity vectors at time t, respectively.

19.2 Central Force Optimization Algorithm

19.2.1 *Fundamentals of Central Force Optimization Algorithm*

Central force optimization (CFO) algorithm was originally proposed in (Formato 2007, 2009). In CFO, each probe which has a position vector (R), an acceleration vector (A), and a fitness value (M) referred to as population. The working principle is CFO "flies" a set of "probes" through the space over a set of discrete "time" steps. With these idea in mind, two simple equations, one for a probe's "acceleration" ($\vec{a}_j^p$) and another for its position vector ($\vec{R}_j^p$) in CFO space, are proposed.

- The total acceleration experienced by p as it "flies" from position $\vec{R}_{j-1}^p$ to $\vec{R}_j^p = \sum_{k=1}^{N_d} x_k^{p,j}\hat{e}_k$, where $x_k^{p,j}$ are the probe p's coordinates at step j, N_d is decision space's dimensionality, and $\hat{e}_k$ is the unit vector along the x_k axis, is given by summing over all other probes as Eq. 19.4 (Formato 2007):

$$\vec{a}_{j-1}^{p} = G \sum_{k=1,k\neq p}^{N_p} U\left(M_{j-1}^{k} - M_{j-1}^{p}\right) \cdot \left(M_{j-1}^{k} - M_{j-1}^{p}\right)^{\alpha} \cdot \frac{\left(\vec{R}_{j-1}^{k} - \vec{R}_{j-1}^{p}\right)}{\left|\vec{R}_{j-1}^{k} - \vec{R}_{j-1}^{p}\right|^{\beta}}, \quad (19.4)$$

where $N_p(1 \leq p \leq N_p)$ is the total number of probes, $N_t(0 \leq j \leq N_t)$ is the total "time" step (iteration) number, G is the gravitational constant, p is the current probe, j is the current time step, α and β are constants that would be one and three in the real world, M^k is the fitness of probe k, U is unit step function in which $U(z) = \begin{cases} 1, & z \geq 0 \\ 0, & \text{otherwise} \end{cases}$, and $U\left(M_{j-1}^{k} - M_{j-1}^{p}\right) \cdot \left(M_{j-1}^{k} - M_{j-1}^{p}\right)^{\alpha}$ is defined as the "mass", where $M_{j-1}^{p} = f\left(x_1^{p,j-1}, x_2^{p,j-1}, \ldots, x_{N_d}^{p,j-1}\right)$ and every other probe also has a fitness given by M_{j-1}^{k}, where $k = 1, \ldots, p-1, p+1, \ldots, N_p$.

- The new position vector for probe p at time step j is computed via Eq. 19.5 (Formato 2007):

$$\vec{R}_{j}^{p} = \vec{R}_{j-1}^{p} + \frac{1}{2}\vec{a}_{j-1}^{p}\Delta t^2, \quad j \geq 1, \quad (19.5)$$

where Δt is the "time" interval between steps during which the acceleration is constant.

Furthermore, probe retrieval is another interesting topic with regards to CFO due to probes may fly outside the decision space and should be returned. There are many possible probe retrieval methods. A useful one is the reposition factor (F_{rep}), which plays an important role in CFO's convergence. It is described in Eqs. 19.6 and 19.7, respectively (Formato 2007):

$$R(p,i,j) = X_{\min}(i) + F_{rep} \cdot (R(p,i,j-1) - X_{\min}(i)), \quad (19.6)$$

$$R(p,i,j) = X_{\max}(i) - F_{rep} \cdot (X_{\max}(i) - R(p,i,j-1)), \quad (19.7)$$

where i is the current dimension, j is the current time step, p is the current probe number, and F_{rep} is a value chosen by the user, typically 0.5.

CFO uses Eq. 19.6 to reposition dimensions of probes that have exceeded their minimum value, while Eq. 19.7 is used to reposition dimensions of probes that have exceeded their maximum value.

Taking into account the above mentioned phases, the steps of implementing standard CFO algorithm can be summarized as follows (Formato 2007):

- Step 1: Initialize position and acceleration vector of each probe to zero.
- Step 2: Compute initial probe distribution.
- Step 3: Calculate initial fitness values.
- Step 4: Evaluate earlier termination criterion or less than N_t, if reach N_t, return to Step 2; if fulfil stop criterion, return Step 5. This step includes several tasks such as Computing the new probe position vectors, retrieving errant particles,

updating fitness matrix for this time step, calculating accelerations for next time step, and increasing F_{rep}, and repeating from Step 4.

- Step 5: Stop and putout best solution have reached so far.

19.2.2 Performance of CFO

In order to show how the CFO algorithm performs, four benchmark functions, namely, Schwefel function, Rastrigin's function, Ackley's function, and Goldstein-Price function, have been tested in (Formato 2007). Results showed that CFO has been very encouraging.

19.3 Conclusions

In this chapter, a gravitational force inspired algorithm was introduced. Each solution candidate is described as a probe herein and it is particularly sensitive to the initial probe distribution. The main concept behind CFO is the small objects (i.e., probes) in space (i.e., search/or decision space) can be dragged by bigger ones (i.e., global optimum of the problem). Although it is a newly introduced CI method, we have witnessed the following rapid spreading of CFO:

First, several enhanced versions of CFO can be found in the literature as outlined below:

- Hybrid CFO Nelder-Mead algorithm (Mahmoud 2011).
- Improved CFO (Formato 2010a).
- Modified CFO (Qubati and Dib 2010).
- Simplified CFO (Formato 2010b).
- Variable initial probe distribution and decision space adaptation integrated CFO (Formato 2011).
- Extended CFO (Ding et al. 2012).
- Pseodurandomness enhanced CFO (Formato 2013).

Second, the CFO algorithm has also been successfully applied to a variety of optimization problems as listed below:

- Antenna design optimization (Mahmoud 2011; Formato 2010a; Qubati et al. 2010; Qubati and Dib 2010).
- Artificial neural network training (Green et al. 2012).
- Electronic circuit optimization (Roa et al. 2012).
- Graphics processing unit computing (Green et al. 2011).
- Piping system monitoring (Haghighi and Ramos 2012).
- Planar steel frames design optimizations (Toğan 2012).

Interested readers are referred to them as a starting point for a further exploration and exploitation of the CFO algorithm.

References

Bauer, W., & Westfall, G. D. (2011). *University physics with modern physics*. New York: McGraw-Hill. ISBN 978-0-07-285736-8.

Ding, D., Qi, D., Luo, X., Chen, J., Wang, X., & Du, P. (2012). Convergence analysis and performance of an extended central force optimization algorithm. *Applied Mathematics and Computation, 219*, 2246–2259.

Formato, R. A. (2007). Central force optimization: A new metaheuristic with applications in applied electromagnetics. *Progress in Electromagnetics Research, PIER, 77*, 425–491.

Formato, R. A. (2009). Central force optimization: A new deterministic gradient-like optimization metaheuristic. *OPSEARCH, 46*, 25–51.

Formato, R. A. (2010a). Improved CFO algorithm for antenna optimization. *Progress in Electromagnetics Research B, 19*, 405–425.

Formato, R. A. (2010b). Parameter-free deterministic global search with simplified central force optimization. In D.-S. Huang (Ed.), *ICIC 2010, LNCS 6215* (pp. 309–318). Berlin: Springer.

Formato, R. A. (2011). Central force optimization with variable initial probes and adaptive decision space. *Applied Mathematics and Computation, 217*, 8866–8872.

Formato, R. A. (2013). Pseudorandomness in central force optimization. *British Journal of Mathematics and Computer Science, 3*, 241–264.

Green, R. C., Wang, L., Alam, M., & Formato, R. A. (2011). Central force optimization on a GPU: A case study in high performance metaheuristics using multiple topologies. In *IEEE Congress on Evolutionary Computation (CEC)* (pp. 550–557). IEEE.

Green, R. C., Wang, L., & Alam, M. (2012). Training neural networks using central force optimization and particle swarm optimization: Insights and comparisons. *Expert Systems with Applications, 39*, 555–563.

Haghighi, A., & Ramos, H. M. (2012). Detection of leakage freshwater and friction factor calibration in drinking networks using central force optimization. *Water Resources Management, 26*, 2347–2363.

Mahmoud, K. R. (2011). Central force optimization: Nelder-Mead hybrid algorithm for rectangular microstrip antenna design. *Electromagnetics, 31*, 578–592.

Qubati, G. M., & Dib, N. I. (2010). Microstrip patch antenna optimization using modified central force optimization. *Progress in Electromagnetics Research B, 21*, 281–298.

Qubati, G. M., Formato, R. A., & Dib, N. I. (2010). Antenna benchmark performance and array synthesis using central force optimisation. *IET Microwaves, Antennas and Propagation, 4*, 583–892.

Roa, O., Ramírez, F., Amaya, I., & Correa, R. (2012). Solution of nonlinear circuits with the central force optimization algorithm. In *IEEE 4th Colombian Workshop on Circuits and Systems (CWCAS)* (pp. 1–6). IEEE.

Toğan, V. (2012). Design of planar steel frames using teaching–learning based optimization. *Engineering Structures, 34*, 225–232.

Chapter 20
Charged System Search Algorithm

Abstract In this chapter, we introduce a novel algorithm called charged system search (CSS) algorithm which is inspired by the coulomb's law and laws of motion. We fist describe the general knowledge of the coulomb's law and laws of motion in Sect. 20.1. Then, the fundamentals and performance of CSS are introduced in Sect. 20.2. Finally, Sect. 20.3 summarises this chapter.

20.1 Introduction

When we were young, several of us may discovered after rubbing a balloon on your hair on a dry day, the balloon can attract bits of paper. When materials behave in this way, they are said to be electrified or to have become electrically charge (Serway and Jewett 2014). As we grow older, we have learned more and known that this physic property of some object is essential to the development of the theory for magnetism, gravity, and objects' motion in response to forces. Based on those famous laws, recently, Kaveh and Talatahari (2010c) proposed a new method, called charged system search (CSS), to deal with combinatorial optimization problems.

20.1.1 Coulomb's Law

Coulomb's law, developed by Charles-Augustin de Coulomb (1736–1806), is the fundamental law governing the electric force between any two charge particles (Holzner 2011). As a scientist, Coulomb is famous for his work in electrostatics. He measured the magnitudes of the electric forces between charged objects using the torsion balance. Based on the observations, he discovered that the electrostatic attraction and repulsion varied inversely as the square of the distance from the charges. The magnitude of the electric force ($\mathbf{F}_{ij}$) can be defined as Eq. 20.1 (Kaveh and Talatahari 2010c):

B. Xing and W.-J. Gao, *Innovative Computational Intelligence: A Rough Guide to 134 Clever Algorithms*, Intelligent Systems Reference Library 62,
DOI: 10.1007/978-3-319-03404-1_20,

$$\mathbf{F}_{ij} = k_e \frac{q_i q_j}{r_{ij}^2}, \tag{20.1}$$

where k_e is a constant called the Coulomb constant, r_{ij} is the separation of the centre of the sphere and the selected point, q_i and q_j are the charges on the two objects, respectively.

In general, the object with charge (i.e., source charge) establishes an electric field through out space. The electric filed (E_{ij}) for charge (q_i) can be defined as Eq. 20.2 (Kaveh and Talatahari 2010c):

$$E_{ij} = \begin{cases} k_e \frac{q_i}{a^3} r_{ij} & \text{if } r_{ij} < a \\ k_e \frac{q_i}{r_{ij}^2} & \text{if } r_{ij} > a, \end{cases} \tag{20.2}$$

where E_{ij} is the magnitude of the electric field at a point inside and outside a charged insulating solid sphere, respectively, k_e is a constant called the Coulomb constant, r_{ij} is the distance from the ith source charge (q_i) to the centre of the sphere, q_i is the magnitude of the charge, and a is the radius of the charged sphere.

Using the principle of superposition, the resulting electric force ($\mathbf{F}_j$) for multiple charged particles is calculated via Eq. 20.3 (Kaveh and Talatahari 2010c):

$$\mathbf{F}_j = k_e q_j \sum_{i, i \neq j} \left(\frac{q_i}{a^3} r_{ij} \cdot i_1 + \frac{q_i}{r_{ij}^2} \cdot i_2 \right) \frac{\mathbf{r}_i - \mathbf{r}_j}{\left\| \mathbf{r}_i - \mathbf{r}_j \right\|}, \quad \begin{cases} i_1 = 1, \quad i_2 = 0 \quad \Leftrightarrow \quad r_{ij} < a \\ i_1 = 0, \quad i_2 = 1 \quad \Leftrightarrow \quad r_{ij} \geq a \end{cases}, \tag{20.3}$$

where $\mathbf{r}_i$ and $\mathbf{r}_j$ are the positions of the ith and jth objects.

20.1.2 Laws of Motion

The classical mechanics are focusing on the motion of objects. Important contributions in this area were provided by Sir Isaac Newton (1642–1727), who proposed three basic laws of motion that deal with forces and masses.

Imagine an object that can be modelled as a particle. When it is moving along a coordinate line (say an s-axis) so that we know its position (s) on that line as a function of time (t) is given by Eq. 20.4 (Serway and Jewett 2014):

$$s = f(t). \tag{20.4}$$

The displacement of the particle over the time interval form t to Δt is given by Eq. 20.5 (Serway and Jewett 2014):

$$\Delta s = f(t + \Delta t) - f(t). \tag{20.5}$$

where $f(t)$ is the initial position, and $f(t + \Delta t)$ is the final position.

Also, the average velocity of the particle over that time interval is given by Eq. 20.6 (Serway and Jewett 2014):

$$v = \frac{\text{displacement}}{\text{travel time}} = \frac{\Delta s}{\Delta t} = \frac{f(t+\Delta t) - f(t)}{\Delta t}, \tag{20.6}$$

where v is the velocity of the particle.

In addition, if the particle has one or more forces acting on it (such as a charged particle in an electric field) so that there is a net force on the particle, it will accelerate in the direction of the net force. The relationship between the net force (**F**) and the acceleration (a) is given by Eq. 20.7 (Serway and Jewett 2014):

$$\mathbf{F} = m \cdot a, \tag{20.7}$$

where m is the mass of the object.

The acceleration of the particle (a) is defined as Eq. 20.8 (Kaveh and Talatahari 2010c):

$$a = \frac{v_{new} - v_{old}}{\Delta t}, \tag{20.8}$$

where a is the acceleration of the particle.

Combining the above equations, the displacement of any object is defined as Eq. 20.9 (Kaveh and Talatahari 2010c):

$$f(t+\Delta t) = \frac{1}{2} \cdot \frac{\mathbf{F}}{m} \cdot \Delta t^2 + v_{old} \cdot \Delta t + f(t). \tag{20.9}$$

20.2 Charged System Search Algorithm

20.2.1 Fundamentals of Charged System Search Algorithm

Charged system search (CSS) algorithm was originally proposed in Kaveh and Talatahari (2010c) and Kaveh and Talatahari (2011b). The quantity of the resultant force and the quality of the movement of the CSS algorithm are based on two physics laws, i.e., Coulomb's law and laws of motion, respectively. In order to apply CSS, the following rules are developed (Kaveh and Talatahari 2010c):

- Rule 1: In CSS, each solution candidate ($\mathbf{X}_i$) is described as a charged particle (CP), in which a number of decision variables ($\mathbf{X}_i = \{x_{i,j}\}$) are involved. The magnitude of each CP (i.e., q_i) is defined as Eq. 20.10:

$$q_i = \frac{fit(i) - fit_{worst}}{fit_{best} - fit_{worst}}, \quad i = 1, 2, \ldots, N, \tag{20.10}$$

where $fit(i)$ denotes the fitness of agent i, fit_{best} and fit_{worst} are the best and worst fitness of all particles, respectively, and N is the total number of CPs. The separation distance (r_{ij}) between two charged particles is defined as Eq. 20.11 (Kaveh and Talatahari 2010c):

$$r_{ij} = \frac{\left\|\mathbf{X}_i - \mathbf{X}_j\right\|}{\left\|(\mathbf{X}_i + \mathbf{X}_j)/2 - \mathbf{X}_{best}\right\| + \varepsilon}, \tag{20.11}$$

where $\mathbf{X}_i$ and $\mathbf{X}_j$ are the positions of the ith and jth CPs, $\mathbf{X}_{best}$ is the position of the best current CP, and ε is a small positive number to avoid singularities.

- Rule 2: The initial positions and velocities of CPs can be defined as Eqs. 20.12 and 20.13, respectively (Kaveh and Talatahari 2010c):

$$x_{i,j}^{(0)} = x_{i,\min} + rand \cdot \left(x_{i,\max} - x_{i,\min}\right), \quad i = 1, 2, \ldots, n, \tag{20.12}$$

$$v_{i,j}^{(0)} = 0, \quad i = 1, 2, \ldots, n, \tag{20.13}$$

where $x_{i,j}^{(0)}$ denotes the initial value of the ith variable for the jth CP, $x_{i,\min}$ and $x_{i,\max}$ are the minimum and the maximum allowable values for the variable, *rand* is a random number in the interval [0, 1], and n is the number of variables.

- Rule 3: The probabilities of moving each CP toward the others can be defined as Eq. 20.14 (Kaveh and Talatahari 2010c):

$$p_{ij} = \begin{cases} 1 & \frac{fit(i) - fit_{best}}{fit(j) - fit(i)} > rand \vee fit(j) > fit(i) \\ 0 & \text{otherwise} \end{cases}. \tag{20.14}$$

- Rule 4: The resulting electric force for each CP is determined as Eq. 20.15 (Kaveh and Talatahari 2010c):

$$\mathbf{F}_j = q_j \sum_{i, i \neq j} \left(\frac{q_i}{a^3} r_{ij} \cdot i_1 + \frac{q_i}{r_{ij}^2} \cdot i_2 \right) p_{ij} (\mathbf{X}_i - \mathbf{X}_j), \quad \begin{cases} j = 1, 2, \cdots, N \\ i_1 = 1, \; i_2 = 0 \Leftrightarrow r_{ij} < a \\ i_1 = 0, \; i_2 = 1 \Leftrightarrow r_{ij} \geq a \end{cases}. \tag{20.15}$$

where $\mathbf{F}_j$ is the resultant force affecting the jth CP.

- Rule 5: The new position and velocity of each CP are determined as Eqs. 20.16 and 20.17, respectively (Kaveh and Talatahari 2010c):

$$\mathbf{X}_{j,new} = rand_{j1} \cdot k_a \cdot \frac{\mathbf{F}_j}{m_j} \cdot \Delta t^2 + rand_{j2} \cdot k_v \cdot \mathbf{V}_{j,old} \cdot \Delta t + \mathbf{X}_{j,old}, \tag{20.16}$$

$$\mathbf{V}_{j,new} = \frac{\mathbf{X}_{j,new} - \mathbf{X}_{j,old}}{\Delta t}, \tag{20.17}$$

where $rand_{j1}$ and $rand_{j2}$ are two random numbers uniformly distributed in the range of $(0, 1)$, m_j is the mass of the jth CP which is equal to q_j, Δt is the time step and is set to unity, k_a is the acceleration coefficient, and k_v is the velocity coefficient to control the influence of the previous velocity. Both can be defined as Eqs. 20.18 and 20.19, respectively (Kaveh and Talatahari 2010c):

$$k_v = 0.5\left(\frac{1 - iter}{iter_{\max}}\right), \tag{20.18}$$

$$k_a = 0.5\left(\frac{1 + iter}{iter_{\max}}\right), \tag{20.19}$$

where $iter$ is the actual iteration number, and $iter_{\max}$ is the maximum number of iterations.

- Rule 6: Defining the rule for charged memory (CM) which is used to save the best CP.
- Rule 7: Defining the search space for each CP as Eq. 20.20 (Kaveh and Talatahari 2010c):

$$x_{i,j} = \begin{cases} \text{w.p. CMCR} & \Rightarrow \text{select a new value for a variable from CM} \\ & \Rightarrow \text{w.p.}(1 - \text{PAR}) \text{ do nothing} \\ & \Rightarrow \text{w.p. PAR choose a neighbouring value} \\ \text{w.p.}(1 - \text{CMCR}) & \Rightarrow \text{select a new value randomly} \end{cases}, \tag{20.20}$$

where "w.p." is the abbreviation for "with the probability", $x_{i,j}$ is the ith component of the CP j, CMCRis the charged memory considering rate varying between 0 and 1, it denotes the rate of choosing a value in the new vector from the historic values stored in the CM, (1 – CMCR) sets the rate of randomly choosing one value from the possible range of value, (1 – PAR) represents the rate of doing nothing, and PAR sets the rate of choosing a value from neighbouring the best CP.

- Rule 8: The terminating criterion can be defined as Eq. 20.21 (Kaveh and Talatahari 2010c):

$$\text{Terminating Criterion} = \begin{cases} \text{Maximum number of iterations} \\ \text{Number of iterations without improvement} \\ \text{Minimum objective function error} \\ \text{Difference between the best and the worst CPs} \\ \text{Maximum distance of CPs} \end{cases}. \tag{20.21}$$

Taking into account a set of rules described above, the steps of implementing CSS algorithm can be summarized as follows (Kaveh and Talatahari 2010c):

- Level-1, Step 1: Initialization. Initialize CSS algorithm parameters (see Rules 1 and 2).
- Level-1, Step 2: CP ranking. Evaluate the values of the fitness function for the CPs.
- Level-2, Step 1: Attracting force determination (see Rules 3 and 4).
- Level-2, Step 2: Solution construction (see Rule 5).
- Level-2, Step 3: CP position correction (see Rule 7).
- Level-2, Step 4: CP ranking. Evaluate the values of the fitness function for the new CPs.
- Level-2, Step 5: CM updating (see Rule 6).
- Level-3, Repeat search level steps until a terminating criterion is satisfied (see Rule 8).

20.2.2 Performance of CSS

In order to show how the CSS algorithm performs, 18 unimodal and multimodal functions are considered in (Kaveh and Talatahari 2010c). In addition, three well-studied engineering design problems are used to illustrate the CSS algorithm's working principles, i.e., a tension/compression spring design problem, a welded beam design problem, and a pressure vessel design problem. Compared with other evolutionary algorithms, the computational results showed that the proposed algorithm outperforms others.

20.3 Conclusions

In this chapter, we introduced a novel computational intelligence (CI) method, called CSS, in which a number of CPs (i.e., agents) cooperate in finding good solutions to difficult optimization problems. The working principle is that on the one hand, each CP emerges an electrical field around its space and exerts an electrical force (i.e., attraction or repulsion) to the other CPs, while on the other hand, the new position of each CP is affected by its previous position, velocity, and acceleration in the space. Overall, it can say that CSS based on two physics theories, i.e., Coulomb law from electrostatics and the Newtonian law from mechanics. The former is used to determine the quantity of the resultant force, while the latter is the quality determination of moved objects. In addition, it is worth to note that CPs act concurrently and independently. In other words, even though each CP can find a solution to the problem, good-quality solutions can only be find as the result of the cooperation among the CPs. Although it is a newly introduced CI method, we have witnessed the following rapid spreading of CSS:

First, several enhanced versions of CSS can be found in the literature as outlined below:

- Adaptive CSS (Talatahari et al. 2012a; Niknam et al. 2013).
- Chaoic CSS (Talatahari et al. 2011; Nouhi et al. 2013; Talatahari et al. 2012b).
- Discrete CSS (Kaveh and Talatahari 2010b).
- Enhanced CSS (Kaveh and Talatahari 2011a).
- Hybrid CSS (Kaveh and Laknejadi 2011).
- Hybridized CSS with big bang-big crunch (BB–BC) algorithm (Kaveh and Zolghadr 2012).
- Magnetic CSS (Kaveh et al. 2013).

Second, the CSS algorithm has also been successfully applied to a variety of optimization problems as listed below:

- Composite slab design optimization (Kaveh and Behnam 2012).
- Concrete retaining wall design optimization (Kaveh and Behnam 2013).
- Dome structure design optimization (Kaveh and Talatahari 2011c).
- Frame structure design optimization (Kaveh and Talatahari 2012).
- Grillage system design optimization (Kaveh and Talatahari 2010a).
- Parameter identification for Bouc-Wen model (Talatahari et al. 2012a).
- Power system optimization (Niknam et al. 2013).
- Skeletal structures design optimization (Kaveh and Talatahari 2010d).
- Truss optimization with discrete variables (Kaveh and Talatahari 2010b).
- Truss optimization with natural frequency constraints (Kaveh and Zolghadr 2012).

Interested readers are referred to them as a starting point for a further exploration and exploitation of the CSS algorithm.

References

Holzner, S. (2011). *Physics I for dummies*. Hoboken: Wiley Publishing, Inc. ISBN 978-0-470-90324-7.

Kaveh, A. & Zolghadr, A. (2012). Truss optimization with natural frequency constraints using a hybridized CSS-BBBC algorithm with trap recognition capability. *Computers and Structures*. http://dx.doi.org/10.1016/j.compstruc.2012.03.016.

Kaveh, A., & Behnam, A. F. (2013). Charged system search algorithm for the optimum cost design of reinforced concrete cantilever retaining walls. *Arabian Journal for Science and Engineering, 38*, 563–570.

Kaveh, A. & Behnam, A. F. (2012). Cost optimization of a composite floor system, one-way waffle slab, and concrete slab formwork using a charged system search algorithm. *Scientia Iranica A*, doi:10.1016/j.scient.2012.04.001.

Kaveh, A., & Laknejadi, K. (2011). A novel hybrid charge system search and particle swarm optimization method for multi-objective optimization. *Expert Systems with Applications, 38*, 15475–15488.

Kaveh, A., & Talatahari, S. (2010a). Charged system search for optimum grillage system design using the LRFD-AISC code. *Journal of Constructional Steel Research, 66*, 767–771.

Kaveh, A., & Talatahari, S. (2010b). A charged system search with a fly to boundary method for discrete optimum design of truss structures. *Asian Journal of Civil Engineering (Building and Housing), 11*, 277–293.

Kaveh, A., & Talatahari, S. (2010c). A novel heuristic optimization method: charged system search. *Acta Mechanica, 213*, 267–289.

Kaveh, A., & Talatahari, S. (2010d). Optimal design of skeletal structures via the charged system search algorithm. *Structural and Multidisciplinary Optimization, 41*, 893–911.

Kaveh, A., & Talatahari, S. (2011a). An enhanced charged system search for configuration optimization using the concept of fields of forces. *Structural and Multidisciplinary Optimization, 43*, 339–351.

Kaveh, A., & Talatahari, S. (2011b). A general model for meta-heuristic algorithms using the concept of fields of forces. *Acta Mechanica, 221*, 99–118.

Kaveh, A., & Talatahari, S. (2011c). Geometry and topology optimization of geodesic domes using charged system search. *Structural and Multidisciplinary Optimization, 43*, 215–229.

Kaveh, A., & Talatahari, S. (2012). Charged system search for optimal design of frame structures. *Applied Soft Computing, 12*, 382–393.

Kaveh, A., Share, M. A. M., & MoslehI, M. (2013). Magnetic charged system search: A new meta-heuristic algorithm for optimization. *Acta Mechanica, 224*(1), 85–107.

Niknam, T., Golestaneh, F., & Shafiei, M. (2013). Probabilistic energy management of a renewable microgrid with hydrogen storage using self-adaptive charge search algorithm. *Energy, 49*, 252–267.

Nouhi, B., Talatahari, S., Kheiri, H., & Cattani, C. (2013). Chaotic charged system search with a feasible-based method for constraint optimization problems. *Mathematical Problems in Engineering, 2013*, 1–8.

Serway, R. A., & Jewett, J. W. (2014). *Physics for scientists and engineers with modern physics*. Boston: Brooks/Cole CENAGE Learning. ISBN 978-1-133-95405-7.

Talatahari, S., Kaveh, A., & Sheikholeslami, R. (2011). An efficient charged system search using chaps for global optimization problems. *International Journal of Optimization in Civil Engineering, 2*, 305–325.

Talatahari, S., Kaveh, A., & Rahbari, N. M. (2012a). Parameter identification of Bouc–Wen model for MR fluid dampers using adaptive charged system search optimization. *Journal of Mechanical Science and Technology, 26*, 2523–2534.

Talatahari, S., Kaveh, A., & Sheikholeslami, R. (2012b). Engineering design optimization using chaotic enhanced charged system search algorithms. *Acta Mechanica, 223*, 2269–2285.

Chapter 21
Electromagnetism-like Mechanism Algorithm

Abstract In this chapter, we present an electromagnetism-like mechanism (EM) algorithm which is inspired by the theory of electromagnetism. We first describe the general knowledge of the electromagnetism field theory in Sect. 21.1. Then, the fundamentals and performance of EM are introduced in Sect. 21.2. Finally, Sect. 21.3 summarises in this chapter.

21.1 Introduction

Every day modern life is pervaded by electromagnetic phenomena, such as television, radio, internet, microwave oven, and Smartphone. One important aspect of electromagnetism is the distribution of charged particles and the motion of charged particles from one place to another, i.e., attraction–repulsion mechanism. Based on this aspect, Birbil and Fang (2003) proposed a new population based algorithm called electromagnetism-like (EM) method. The principles behind the algorithm are that inferior particles prevent a move in their direction by repelling other particles in the population, and that superior particles facilitate moves in their direction.

21.1.1 Electromagnetism Field Theory

The understanding of electromagnetic phenomena is usually treated by electromagnetic field theory which is one of the four fundamental interactions in nature. The concept of a field is at the heart of the particle in a field analysis model. In other words, a particle resides in a area of space in which a field exists. In the electromagnetic version of the particle in a field model, the property of a particle that results in a electromagnetic force (i.e., Lorentz force) which is a relation between electric and magnetic forces, i.e., the interaction of electrically charged

B. Xing and W.-J. Gao, *Innovative Computational Intelligence: A Rough Guide to 134 Clever Algorithms*, Intelligent Systems Reference Library 62,
DOI: 10.1007/978-3-319-03404-1_21,

particles and the interaction of uncharged magnetic force fields with electrical conductors (Serway and Jewett 2014). Both phenomena were long thought to be separate forces, however, in 1861 the Scottish physicist James Clerk Maxwell (1831–1879) showed that electric and magnetic fields travel together through space as waves of electromagnetic radiation.

21.2 Electromagnetism-like Algorithm

21.2.1 Fundamentals of Electromagnetism-like Algorithm

Electromagnetism-like mechanism (EM) algorithm was originally proposed by Birbil and Fang (2003). Typically, the EM algorithm consists of four phases, namely, population construction, objective function evaluation, total force calculation, and the movement. In addition, the EM algorithm used a local search procedure to find an improved solution within the neighbourhood of the current solution. For the rest of this section, we will explain them in detail.

- Population construction: A population of m points is randomly generated from the feasible set which is a N-dimensional hyper-cube as described in Eq. 21.1 (Birbil and Fang 2003):

$$X = \left\{x \in R^n \middle| x_k^i = l_k + \lambda(u_k - l_k), \quad i = 1, 2, \ldots, m, \; k = 1, 2, \ldots N\right\}, \quad (21.1)$$

where X is the search space, $\lambda \sim Unif(0,1)$ for each coordinates of x_k^i, and u_k and l_k are upper bound and lower bound of each point, respectively.
- Objective function evaluation: After a point is chosen from the space, its objective function $(f\{x^i\})$ is calculated and the best function value is identified as Eq. 21.2 (Birbil and Fang 2003):

$$x^{best} = \arg\min\{f(x^i), \forall i\}. \quad (21.2)$$

- Total force calculation: In this step, the charge of each point (q^i) is calculated via Eq. 21.3 (Birbil and Fang 2003):

$$q^i = \exp\left\{-n\frac{f(x^i) - f(x^{best})}{\sum_{k=1}^{N}[f(x^k) - f(x^{best})]}\right\}, \forall i. \quad (21.3)$$

Then, the total force (F^i) between two points x^j and x^i is given by Eq. 21.4 (Birbil and Fang 2003):

$$F^i = \sum_{j \neq i}^{m}\left\{\begin{array}{ll}(x^j - x^i)\frac{q^i \cdot q^j}{\|x^j - x^i\|^2} & \text{if} f(x^j) < f(x^i)\\ (x^i - x^j)\frac{q^i \cdot q^j}{\|x^j - x^i\|^2} & \text{if} f(x^j) \geq f(x^i)\end{array}\right\}, \forall i. \quad (21.4)$$

- Movement according to the total force: After calculating the total force, the point x^i can move in the direction of the force by a random step length as defined by Eq. 21.5 (Birbil and Fang 2003):

$$x^i = x^i + \lambda \frac{F^i}{\|F^i\|}(RNG), \quad i = 1, 2, \ldots, m, \tag{21.5}$$

where λ denotes the random step length and is assumed to e uniformly distributed $\lambda \sim Unif(0, 1)$, RNG is a vector whose components represent the allowed feasible movement toward the upper bound or the lower bound, respectively, for the corresponding dimension.
- Local search: A local search is used to find high-quality solutions. It is based on the iterative exploration of neighbourhoods of solutions trying to improve the current solution by local changes. The maximum feasible step length is determined by the parameter $\delta \in [0, 1]$ and is given by Eq. 21.6 (Birbil and Fang 2003):

$$s_{\max} = \delta\left(\max_k(u_k - l_k)\right). \tag{21.6}$$

For EM, the local search procedure includes four process:

1. Searching one potential adjacent point that its objective function value is better than original one and storing its value into a temporal point y;
2. Choosing a random number λ_1 as the mechanism to put the candidate points closer to the optimum that described as Eq. 21.7 (Birbil and Fang 2003):

$$\begin{cases} \text{if } \lambda_1 \geq 0.5, & y \text{ is selected} \\ \text{if } \lambda_1 < 0.5, & y \text{ is not selected} \end{cases}. \tag{21.7}$$

3. Calculating the objective function of y;
4. Updating the current best point.

Taking into account the key phases described above, the steps of implementing standard EM algorithm can be summarized as follows (Birbil and Fang 2003):

- Step 1: Randomly selecting m points.
- Step 2: Computing the objective function value of each point.
- Step 3: Implementing the local search procedure.
- Step 4: Calculating the total force (attraction or repulsion) based on the objective function value and summing up them for each point.
- Step 5: Moving each point based on the gotten force.
- Step 6: Check if maximum iteration is reached, go to step 3 for new beginning. If a specified termination criteria is satisfied stop and return the best solution.

21.2.2 Performance of EM

In order to show how the EM algorithm performs, a set of test functions are selected in (Birbil and Fang 2003), namely, Complex function, Davis function, Epistacity (4) function, Epistacity (5) function, Griewank function, Himmelblau function, Kearfott function, Levy function, Rastrigin function, Sine Envelope function, Stenger function, Step funciton, Spiky function, Trid (5) function, and Trid (20) function. Computational results showed that EM rapidly converges to the objective function (i.e., optimal status).

21.3 Conclusions

In this chapter, we introduced an EM algorithm that based on the electromagnetism theory. In EM, each sample point (i.e., a solution) creates an electrical field around its space and exerts an electrical force (i.e., attraction or repulsion) to the other points. The magnitude of attraction or repulsion of each point is determined by the charge of each point. Also, it is the building block of the objective function value, i.e., the better the objective function value, the higher the magnitude of attraction. In this way, we can finally find a direction for each point to move. Although it is a newly introduced computational intelligence (CI) method, we have witnessed the following rapid spreading of EM:

First, several enhanced versions of EM can be found in the literature as outlined below:

- Constrained EM (Rocha and Fernandes 2008b; Ali and Golalikhani 2010).
- Discrete EM (Javadian and Golalikhani 2009; Liu and Gao 2010; Chao and Liao 2012).
- Hybrid data envelopment analysis and EM (Guo et al. 2011).
- Hybrid EM (Yurtkuran and Emel 2010; Latifi and Bonyadi 2009).
- Hybrid EM and chaos optimization algorithm (Wang et al. 2010a).
- Hybrid EM and descent search (Rocha and Fernandes 2009).
- Hybrid EM and genetic algorithm (Javadi et al. 2013; Chang and Lee 2008; Lee and Chang 2010).
- Hybrid EM and great deluge algorithm (Abdullah et al. 2009).
- Hybrid EM and modified Davidon-Fletcher-Power algorithm (Yin et al. 2011).
- Hybrid EM and restarted Arnoldi algorithm (Taheri et al. 2007).
- Hybrid EM and simulated annealing (Jamili et al. 2011; Naderi et al. 2010b).
- Hybrid EM with back-propagation technique (Lee et al. 2012).
- Hybrid EM with genetic operators (Chang et al. 2009).
- Hybrid Hopfield neural networks with EM (Hung et al. 2011).
- Hybrid local search and EM (Gilak and Rashidi 2009; Vahdani et al. 2010).
- Hybrid scatter search and EM (Debels et al. 2006).
- Improved EM (Shang et al. 2010).

- Modified EM (Wu et al. 2007; Yi and Ming-Lu 2011; Rocha and Fernandes 2008c, Han and Han 2010).
- Modified EM based on feasibility and dominance rules (Rocha and Fernandes 2008a; Wu et al. 2013).
- Multiobjective EM (Khalili and Tavakkoli-Moghaddam 2012).
- Quantum-inspired EM (Chou et al. 2010; Chang et al. 2010).
- Revised EM and k-opt method (Wu et al. 2006).
- Species-based improved EM (Lee et al. 2010, 2011).

Second, the EM algorithm has also been successfully applied to a variety of optimization problems as listed below:

- Antenna design optimization (Lee and Jhang 2008; Jhang and Lee 2009).
- Artificial neural network training (Wu et al. 2010; Lee et al. 2011).
- Control optimization (Chang and Lee 2008; Lee and Chang 2010).
- Flexible manufacturing system optimization (Souier and Sari 2011).
- Fuzzy system design optimization (Lee et al. 2010, 2012).
- Image processing (Su and Lin 2011).
- knapsack problem (Chou et al. 2010; Chang et al. 2010; Latifi and Bonyadi 2009).
- Lagrangian algorithm optimization (Rocha and Fernandes 2010).
- Layout design optimization (Guan et al. 2012; Jolai et al. 2012; Javadi et al. 2013).
- Matrix optimization (Taheri et al. 2007).
- Milling process optimization (Wu et al. 2013).
- Mixture design optimization (Chang and Huang 2007).
- Robot control optimization (Wang et al. 2010b; Yin et al. 2011).
- Scheduling optimization (Debels et al. 2006; Chang et al. 2009; Gilak and Rashidi 2009; Javadian and Golalikhani 2009; Liu and Gao 2010; Naderi et al. 2010a, b; Vahdani et al. 2010; Jamili et al. 2011; Guo et al. 2011; Chao and Liao 2012; Khalili and Tavakkoli-Moghaddam 2012).
- Set covering problem (Naji-Azimi et al. 2010).
- Timebabling problem (Abdullah et al. 2009).
- Travelling salesman problem (Wu et al. 2006; Bonyadi et al. 2008).
- Vehicle routing problem (Wu et al. 2007; Yurtkuran and Emel 2010).
- Wireless communication networks optimization (Tsai et al. 2010; Hung et al. 2011).

Interested readers are referred to them as a starting point for a further exploration and exploitation of the EM algorithm.

References

Abdullah, S., Turabieh, H., & Mccollum, B. (2009). A hybridization of electromagnetic-like mechanism and great deluge for examination timetabling problems. *Hybrid Metaheuristics, LNCS 5818*, (pp. 60–72). Berlin: Springer.

Ali, M. M., & Golalikhani, M. (2010). An electromagnetism-like method for nonlinearly constrained global optimization. *Computers and Mathematics with Applications, 60*, 2279–2285.

Birbil, Şİ., & Fang, S.-C. (2003). An electromagnetism-like mechanism for global optimization. *Journal of Global Optimization, 25*, 263–282.

Bonyadi, M. R., Azghadi, M. R., & Shah-Hosseini, H. (2008). Population-based optimization algorithms for solving the travelling salesman problem. In F. Greco (Ed.), *Travelling salesman problem* (Chap. 1, pp. 1–34). Vienna, Austria: In-Tech.

Chang, H.-H., & Huang, T.-Y. (2007). Mixture experiment design using artificial neural networks and electromagnetism-like mechanism algorithm. *Second International Conference on Innovative Computing, Information and Control* (*ICICIC*) (pp. 1–4). IEEE.

Chang, F.-K., & Lee, C.-H. (2008). Design of fractional PID control via hybrid of electromagnetism-like and genetic algorithms. *Eighth International Conference on Intelligent Systems Design and Applications* (*ISDA*) (pp. 525–530). IEEE.

Chang, P.-C., Chen, S.-H., & Fan, C.-Y. (2009). A hybrid electromagnetism-like algorithm for single machine scheduling problem. *Expert Systems with Applications, 36*, 1259–1267.

Chang, C.-C., Chen, C.-Y., Fan, C.-W., Chao, H.-C., & Chou, Y.-H. (2010). Quantum-inspired electromagnetism-like mechanism for solving 0/1 knapsack problem. *Second International Conference on Information Technology Convergence and Services* (*ITCS*) (pp. 1–6). IEEE.

Chao, C.-W., & Liao, C.-J. (2012). A discrete electromagnetism-like mechanism for single machine total weighted tardiness problem with sequence-dependent setup times. *Applied Soft Computing, 12*, 3079–3089. http://dx.doi.org/10.1016/j.asoc.2012.05.017.

Chou, Y.-H., Chang, C.-C., Chiu, C.-H., Lin, F.-J., Yang, Y.-J., & Peng, Z.-V. (2010). Classical and quantum-inspired electromagnetism-like mechanism for solving 0/1 knapsack problems. *IEEE International Conference on Systems Man and Cybernetics* (*SMC*) (pp. 3211–3218). IEEE.

Debels, D., Reyck, B. D., Leus, R., & Vanhoucke, M. (2006). A hybrid scatter search/electromagnetism meta-heuristic for project scheduling. *European Journal of Operational Research, 169*, 638–653.

Gilak, E., & Rashidi, H. (2009). A new hybrid electromagnetism algorithm for job shop scheduling. *Third UKSim European Symposium on Computer Modeling and Simulation* (pp. 327–332). IEEE.

Guan, X., Dai, X., Qiu, B., & Li, J. (2012). A revised electromagnetism-like mechanism for layout design of reconfigurable manufacturing system. *Computers & Industrial Engineering, 63*, 98–108.

Guo, Z., Hang, N., & Wu, J. (2011). DEA and EM based multi-objective short-term hydrothermal economic scheduling. *International Conference on Information Technology, Computer Engineering and Management Sciences* (*ICM*) (pp. 159–162). IEEE.

Han, L., & Han, Z. (2010). Electromagnetism-like method for constrained optimization problems. *International Conference on Measuring Technology and Mechatronics Automation* (*IC-MTMA*) (pp. 87–90). IEEE.

Hung, H.-L., Huang, Y.-F., & Cheng, C.-H. (2011). Performance of hybrid Hopfield neural networks with EM algorithms for multiuser detection in ultra-wide-band communication systems. *IEEE International Conference on Systems, Man, and Cybernetics* (*SMC*) (pp. 1423–1429). IEEE.

Jamili, A., Shafia, M. A., & Tavakkoli-Moghaddam, R. (2011). A hybridization of simulated annealing and electromagnetism-like mechanism for a periodic job shop scheduling problem. *Expert Systems with Applications, 38*, 5895–5901.

Javadi, B., Jolai, F., Slomp, J., Rabbani, M., & Tavakkoli-Moghaddam, R. (2013). A hybrid electromagnetism-like algorithm for dynamic inter/intra cell layout problem. *International Journal of Computer Integrated Manufacturing,* http://dx.doi.org/10.1080/0951192X.2013.814167.

Javadian, N., & Golalikhani, M. (2009). Solving a single machine scheduling problem by a discrete version of electromagnetism-like method. *Journal of Circuits, Systems, and Computers, 18*, 1597–1608.

Jhang, J.-Y., & Lee, K.-C. (2009). Array pattern optimization using electromagnetism-like algorithm. *International Journal of Electronics and Communications, 63*, 491–496.

Jolai, F., Tavakkoli-Moghaddam, R., Golmohammadi, A., & Javadi, B. (2012). An electromagnetism-like algorithm for cell formation and layout problem. *Expert Systems with Applications, 39*, 2172–2182.

Khalili, M., & Tavakkoli-Moghaddam, R. (2012). A multi-objective electromagnetism algorithm for a bi-objective flowshop scheduling problem. *Journal of Manufacturing Systems, 31*, 232–239.

Latifi, O. A., & Bonyadi, M. R. (2009). DEM: A discrete electromagnetism-like mechanism for solving discrete problems. *IEEE International Symposium on Computational Intelligence in Robotics and Automation* (*CIRA*) (pp. 120–125). IEEE.

Lee, C.-H., & Chang, F.-K. (2010). Fractional-order PID controller optimization via improved electromagnetism-like algorithm. *Expert Systems with Applications, 37*, 8871–8878.

Lee, K. C., & Jhang, J. Y. (2008). Application of electromagnetism-like algorithm to phase-only syntheses of antenna arrays. *Progress in Electromagnetics Research, PIER, 83*, 279–291.

Lee, C.-H., Chang, F.-Y., & Lee, C.-T. (2010). Species-based hybrid of electromagnetism-like mechanism and back-propagation algorithms for an interval type-2 fuzzy system design. *International Multi Conference of Engineers and Computer Scientists* (*IMECS*) (Vol. I, pp. 1–6) 17–19 March, Hong Kong. IEEE.

Lee, C.-H., Li, C.-T., & Chang, F.-Y. (2011). A species-based improved electromagnetism-like mechanism algorithm for TSK-type interval-valued neural fuzzy system optimization. *Fuzzy Sets and Systems, 171*, 22–43.

Lee, C.-H., Chang, F.-K., Kuo, C.-T., & Chang, H–. H. (2012). A hybrid of electromagnetism-like mechanism and back-propagation algorithms for recurrent neural fuzzy systems design. *International Journal of Systems Science, 43*, 231–247.

Liu, H., & Gao, L. (2010). A discrete electromagnetism-like mechanism algorithm for solving distributed permutation flowshop scheduling problem. *International Conference on Manufacturing Automation* (*ICMA*) (pp. 156–163). IEEE.

Naderi, B., Jenabi, M., Ghomi, S. M. T. F., & Talebi, D. (2010a). An electromagnetism-like metaheuristic for sequence dependent open shop scheduling. *IEEE Fifth Bio-Inspired Computing: Theories and Applications* (*BIC-TA*) (pp. 489–497). IEEE.

Naderi, B., Tavakkoli-Moghaddam, R., & Khalili, M. (2010b). Electromagnetism-like mechanism and simulated annealing algorithms for flowshop scheduling problems minimizing the total weighted tardiness and makespan. *Knowledge-Based Systems, 23*, 77–85.

Naji-Azimi, Z., Toth, P., & Galli, L. (2010). An electromagnetism metaheuristic for the unicost set covering problem. *European Journal of Operational Research, 205*, 290–300.

Rocha, A. M. A. C., & Fernandes, E. M. G. P. (2008a). Feasibility and dominance rules in the electromagnetism-like algorithm for constrained global optimization. *Second International Conference on Signals, Systems and Automation* (*ICSSA*) (pp 1–10).

Rocha, A. M. A. C., & Fernandes, E. M. G. P. (2008b). Implementation of the electromagnetism-like algorithm with a constraint-handling technique for engineering optimization problems. *Eighth International Conference on Hybrid Intelligent Systems* (*HIS*) (pp. 690–695). IEEE.

Rocha, A. M. A. C., & Fernandes, E. M. G. P. (2008c). On charge effects to the electromagnetism-like algorithm. *The 20th International Conference EURO Mini Conference Continuous Optimization and Knowledge-Based Technologies* (*EurOPT-2008*) (pp. 198–203). Vilnius Gediminas Technical University Publishing House, Technika.

Rocha, A. M. A. C., & Fernandes, E. M. G. P. (2009). Hybridizing the electromagnetism-like algorithm with descent search for solving engineering design problems. *International journal of Computer Mathematics, 86*, 1932–1946.

Rocha, A. M. A. C., & Fernandes, E. M. G. P. (2010). A stochastic augmented Lagrangian equality constrained-based algorithm for global optimization. In T. E. Simos, G. Psihoyios & C. Tsitouras (Eds.), *Proceedings of the International Conference on Numerical Analysis and Applied Mathematics* (*ICNAAM*) (Vol. II, pp. 967–970). American Institute of Physics.

Serway, R. A., & Jewett, J. W. (2014). *Physics for scientists and engineers with modern physics*. Boston: Brooks/Cole CENAGE Learning. ISBN 978-1-133-95405-7.

Shang, Y., Chen, J., & Wang, Q. (2010). Improved electromagnetism-like mechanism algorithm for constrained optimization problem. *International Conference on Computational Intelligence and Security* (*CIS*) (pp. 165–169). IEEE.

Souier, M., & Sari, Z. (2011). A software tool for performance metaheuristics evaluation in real time alternative routing selection in random FMSs. *International Conference on Communications, Computing and Control Applications* (*CCCA*) (pp. 1–6). IEEE.

Su, C.-T., & Lin, H.-C. (2011). Applying electromagnetism-like mechanism for feature selection. *Information Sciences, 181*, 972–986.

Taheri, S. H., Ghazvini, H., Saberi-Nadjafi, J., & Biazar, J. (2007). A hybrid of the restarted Arnoldi and electromagnetism meta-heuristic methods for calculating eigenvalues and eigenvectors of a non-symmetric matrix. *Applied Mathematics and Computation, 191*, 79–88.

Tsai, C.-Y., Hung, H.-L., & Lee, S.-H. (2010). Electromagnetism-like method based blind multiuser detection for MC-CDMA interference suppression over multipath fading channel. *International Symposium on Computer Communication Control and Automation* (*3CA*) (pp. 470–475). IEEE.

Vahdani, B., Soltani, R., & Zandieh, M. (2010). Scheduling the truck holdover recurrent dock cross-dock problem using robust meta-heuristics. *International Journal of Advanced Manufacturing Technology, 46*, 769–783.

Wang, Q., Zeng, J., & Song, W. (2010a) A new electromagnetism-like algorithm with chaos optimization. *International Conference on Computational Aspects of Social Networks* (*CASoN*) (pp. 535–538). IEEE.

Wang, Y., Yang, Y., Yuan, X., Yin, F., & Wei, S. (2010b) A model predictive control strategy for path-tracking of autonomous mobile robot using electromagnetism-like mechanism. *International Conference on Electrical and Control Engineering* (*ICECE*) (pp. 96–100). IEEE.

Wu, P., Yang, K.-J., & Fang, H.-C. (2006). A revised EM-like algorithm + K-OPT method for solving the traveling salesman problem. *First International Conference on Innovative Computing, Information and Control* (*ICICIC*) (pp. 546–549). IEEE.

Wu, P., Yang, K.-J., & Huang, B.-Y. (2007). A revised EM-like mechanism for solving the vehicle routing problems. *Proceedings of the Second International Conference on Innovative Computing, Information and Control* (*ICICIC*) (pp. 1–4). IEEE.

Wu, Q., Zhang, C.-J., Gao, L., & Li, X. (2010). Training neural networks by electromagnetism-like mechanism algorithm for tourism arrivals forecasting. *IEEE Fifth International Conference on Bio-Inspired Computing: Theories and Applications* (*BIC-TA*) (pp. 679–688). IEEE.

Wu, Q., Gao, L., Li, X., Zhang, C., & Rong, Y. (2013). Applying an electromagnetism-like mechanism algorithm on parameter optimisation of a multi-pass milling process. *International Journal of Production Research, 51*, 1777–1788.

Yi, X., & Ming-Lu, J. (2011). Electromagnetism-like algorithm without local search on PAPR reduction for OFDM. *Third International Conference on Awareness Science and Technology* (*iCAST*) (pp. 51–56). IEEE.

Yin, F., Wang, Y.-N., & Wei, S.-I. (2011). Inverse kinematic solution for robot manipulator based on electromagnetism-like and modified DFP algorithms. *ACTA Automatica Sinica, 37*, 74–82.

Yurtkuran, A., & Emel, E. (2010). A new hybrid electromagnetism-like algorithm for capacitated vehicle routing problems. *Expert Systems with Applications, 37*, 3427–3433.

Chapter 22
Gravitational Search Algorithm

Abstract In this chapter, we present a gravitational search algorithm (GSA) which is based on the low of gravity. We first describe the general information of the science of gravity and the definition of mass in Sect. 22.1, respectively. Then, the fundamentals and performance of GSA are introduced in Sect. 22.2. Finally, Sect. 22.3 summarises in this chapter.

22.1 Introduction

Physics is present in every action around you. As we were young, you may have heard a famous story about Sir Isaac Newton watched apples which drop off trees. That is the most widely recognized example about gravity. In fact, gravity is one of the nature forces that operates everywhere. As we grow older, this fundamental theory may become less urgent for most of us. However, there are a considerable number of researchers have an interest in it. Recently, based on the law of gravity, Rashedi et al. (2009) developed a new algorithm, called gravitational search algorithm (GSA).

22.1.1 The Science of Gravity

As we know, the central theme of the story of the universe turns out to be gravity. It can be defined as the tendency of masses to accelerate toward each other (Schutz 2003; Ricci 1998). This famous physics fundamental theory can be go right back to Galileo (1564–1642), who founded the science of gravity. In addition, to explain gravity required two other greatest scientists: Isaac Newton and Albert Einstein. Newton's gravity theory discovered the laws of motion, while Einstein's theory of gravitation led to the erratic motion, known today as Brownian motion.

B. Xing and W.-J. Gao, *Innovative Computational Intelligence: A Rough Guide to 134 Clever Algorithms*, Intelligent Systems Reference Library 62,
DOI: 10.1007/978-3-319-03404-1_22,

The Newton's law of universal gravitation states as follows (Serway and Jewett 2014):

> every particle in the universe attracts every other particle with a force that is directly proportional to the product of their masses and inversely proportional to the square of the distance between them.

If the particles have masses m_1 and m_2, they are separated by a distance r, then the magnitude of this gravitational force is defined by Eq. 22.1 (Serway and Jewett 2014):

$$F_g = G\frac{m_1 m_2}{r^2}, \tag{22.1}$$

where G is a constant, called the universal gravitational constant and its value is $G = 6.674 \times 10^{-11}\ \mathrm{N} \times \mathrm{m}^2/\mathrm{kg}^2$.

In fact, the universal gravitational constant (G) was first evaluated in the late nineteenth century. It depends on the actual age of the universe due to the effect of decreasing gravity as defined by Eq. 22.2 (Rashedi et al. 2009):

$$G(t) = G(t_0) \times \left(\frac{t_0}{t}\right)^{\beta},\ \beta < 1, \tag{22.2}$$

where $G(t)$ is the value of the gravitational constant at time t, is the value of the gravitational constant at the first cosmic quantum-interval of time t_0.

In addition, Newton made an important law (i.e., Newton's second law) when combined with Galileo's discovery. It says that (Schutz 2003):

> when a force is applied to a body, the resulting acceleration depends only on the force and on the mass of the body: the larger the force, the larger the acceleration; and the larger the mass of a body, the smaller its acceleration.

The acceleration of an object (a) depends on the force (F) and its mass (M) is given by Eq. 22.3 (Schutz 2003):

$$a = \frac{F}{M}. \tag{22.3}$$

22.1.2 The Definition of Mass

In physics, mass is a amount of matter in an object, giving rise to the phenomena of the object's resistance of being accelerated by a force and the strength of its mutual gravitational attraction with other objects (Serway and Jewett 2014).

In general, there are three kinds of mass that are defined in classical physics (Schutz 2003):

- Active gravitational mass (M_a): It is a static measurement that is proportional to the magnitude of the gravitational force which is exerted by an object.

- Passive gravitational mass (M_p): It is a static measurement that is proportional to the magnitude of the gravitational force which is experienced by an object when interacting with a second object.
- Inertial mass (M_i): It is a dynamic measurement of how much inertia must be accelerated.

Both active and passive gravitational mass are defined by the force of gravitation, which states there is a gravitational force between any pair of objects, while the inertial mass is mainly defined by Newton's law, which states that when a force is applied to an object, it will accelerate proportionally, and the constant of proportion is the mass of that object. Based on those properties, the Eqs. 22.1 and 22.3 can be rewrote as Eqs. 22.4 and 22.5, respectively (Serway and Jewett 2014):

$$F_{ij} = G\frac{m_{aj} \cdot m_{pi}}{R^2}, \tag{22.4}$$

$$a_i = \frac{F_{ij}}{M_{ii}}, \tag{22.5}$$

where F_{ij} denotes the gravitational force, M_{aj} is the active gravitational mass of particle j, M_{pi} is the passive gravitational mass of particle i, and M_{ii} represents the inertia mass of particle i.

22.2 Gravitational Search Algorithm

22.2.1 Fundamentals of Gravitational Search Algorithm

Gravitational search algorithm (GSA) was originally proposed by Rashedi et al. (2009). In GSA, all the individuals can be mimicked as objects with masses. Based on the Newton's law of universal gravitation, the objects attract each other by the gravity force, and the force makes all of them move towards the ones with heavier masses. In addition, each mass of GSA has four characteristics: position, inertial mass, active gravitational mass, and passive gravitational mass. The first one corresponds to a solution of the problem, while the other three are determined by fitness function. The details of GSA can be summarized as follows.

- First, considering a system with N masses (agents) where the ith mass's position is defined by Eq. 22.6 (Rashedi et al. 2009):

$$X_i = \left(x_i^1, \ldots, x_i^d, \ldots, x_i^n\right), \quad i = 1, 2, \ldots, N, \tag{22.6}$$

  where x_i^d is the position of the ith agent in the dth dimension, and n is the search space's dimension.
- Second, the gravitational force ($F_{ij}^d(t)$) that acting on mass i from mass j at time t can be defined by Eq. 22.7 (Rashedi et al. 2009):

$$F_{ij}^d(t) = G(t)\frac{M_{pi}(t) \cdot M_{aj}(t)}{R_{ij}(t) + \varepsilon}\left(x_j^d(t) - x_i^d(t)\right), \tag{22.7}$$

where M_{aj} is the active gravitational mass related to agent j, M_{pi} is the passive gravitational mass related to agent i, $G(t)$ is gravitational constant at time t, ε is a small constant, and $R_{ij}(t)$ is the Euclidian distance between two agents i and j defined by Eq. 22.8 (Rashedi et al. 2009):

$$R_{ij}(t) = \left\|X_i(t), X_j(t)\right\|_2. \tag{22.8}$$

- Third, for the purpose of computing acceleration of an agent i, total forces (from a group of heavier masses) can be defined by Eq. 22.9 (Rashedi et al. 2009):

$$F_i^d(t) = \sum_{j=1, j \neq i}^{N} rand_j F_{ij}^d(t), \tag{22.9}$$

where $rand_j$ is random number in the interval $[0, 1]$.
- Fourth, based on the total forces, the acceleration of the agent i at time t, and in direction dth, is given by Eq. 22.10 (Rashedi et al. 2009):

$$a_i^d(t) = \frac{F_i^d(t)}{M_{ii}(t)}, \tag{22.10}$$

where M_{ii} is the inertial mass of ith agent.
- Fifth, an agent's next velocity can be computed as a fraction of its present velocity added to its acceleration. Both agent i's new velocity and position are given by Eqs. 22.11 and 22.12, respectively (Rashedi et al. 2009):

$$v_i^d(t+1) = rand_i \times v_i^d(t) + a_i^d(t), \tag{22.11}$$

$$x_i^d(t+1) = x_i^d(t) + v_i^d(t+1), \tag{22.12}$$

where $v_i^d(t)$ and $x_i^d(t)$ are the velocity and the position in dth dimension of agent i at time t, respectively, and $rand_i$ is an uniform random variable in the interval $[0, 1]$ which adds a randomized characteristic to the search.
- Finally, after computing current population's fitness, the gravitational and inertial masses can be updated via Eqs. 22.13 and 22.14, respectively (Rashedi et al. 2009):

$$m_i(t) = \frac{fit_i(t) - worst(t)}{best(t) - worst(t)}, \tag{22.13}$$

$$M_i(t) = \frac{m_i(t)}{\sum_{j=1}^{N} m_j(t)}, \tag{22.14}$$

where $fit_i(t)$ denotes the fitness value of the agent i at time t, $best(t)$ and $worst(t)$ are defined by Eqs. 22.15 and 22.16, respectively (Rashedi et al. 2009):

$$\text{For a minimization problem:} \begin{cases} best(t) = \min\limits_{j \in \{1,\ldots,N\}} fit_j(t) \\ worst(t) = \max\limits_{j \in \{1,\ldots,N\}} fit_j(t) \end{cases}. \tag{22.15}$$

$$\text{For a maximization problem:} \begin{cases} best(t) = \max\limits_{j \in \{1,\ldots,N\}} fit_j(t) \\ worst(t) = \min\limits_{j \in \{1,\ldots,N\}} fit_j(t) \end{cases}. \tag{22.16}$$

Furthermore, to balance the exploration and exploitation of GSA, an agent called k_{best} is employed. It is a function of time which with the initial value k_0 at the beginning and decreasing with time. In GSA, k_0 is normally set to N (i.e., the total number of agents), and k_{best} is decreased linearly. Finally, there will be just one agent applying force to the others. Therefore, the Eq. 22.9 can be modified as Eq. 22.17 (Rashedi et al. 2009):

$$F_i^d(t) = \sum_{j \in k_{best}, j \neq i}^{N} rand_j F_{ij}^d(t), \tag{22.17}$$

where k_{best} is the set of first k agents with the best fitness value and biggest mass.

Meanwhile, an initial value of G_0 is always allocated to the gravitational constant, G, which will be reduced with time as defined by Eq. 22.18 (Rashedi et al. 2009):

$$G(t) = G(G_0,\ t). \tag{22.18}$$

The steps of implementing GSA can be summarized as follows (Rashedi et al. 2009):

- Step 1: Determining the system environment.
- Step 2: Randomized initialization.
- Step 3: Fitness evaluation of agents.
- Step 4: Updating the parameters, i.e., $G(t)$, $best(t)$, $worst(t)$, and $M_i(t)$ for $i = 1, 2, \ldots, N$.
- Step 5: Calculation of the total force in different directions.
- Step 6: Calculation of acceleration and velocity.
- Step 7: Updating the position of agents.
- Step 8: Repeat Steps 3–7 until the stop criteria is reached. If a specified termination criteria is satisfied stop and return the best solution.

22.2.2 Performance of GSA

In order to show how the GSA algorithm performs, 23 standard benchmark functions are tested in (Rashedi et al. 2009). Compare with central force optimization (CFO) algorithm, real genetic algorithm (RGA), and particle swarm optimization (PSO) algorithm, the simulation report obtained by GSA in most cases provide superior results.

22.3 Conclusions

In this chapter, a new optimization algorithm called GSA is introduced. It is based on the law of gravity and the gravitational and inertial mass in Newton's principles. The working principle is each agent can be considered as object with different masses. The entire agents move due to the gravitational attraction force acting between them and the progress of the algorithm directs the movements of all agents globally towards the agents with heavier masses. Although the effectiveness of GSA is still under debate (see Gauci et al. (2012) for details), we have witnessed the following rapid spreading of GSA:

First, several enhanced versions of GSA can also be found in the literature as outlined below:

- Binary GSA (Rashedi et al. 2010).
- Chaotic GSA (Li et al. 2012, 2013; Ju and Hong 2013).
- Discrete local search operator enhanced GSA (Doraghinejad et al. 2013).
- Disruption operator enhanced GSA (Sarafrazi et al. 2011).
- Fuzzy GSA (Ghasemi et al. 2013).
- Hybrid GSA with clonal selection algorithm (Gao et al. 2013).
- Hybrid GSA with K-means (Hatamlou et al. 2011).
- Hybrid genetic algorithm with GSA (Seljanko 2011).
- Hybrid K-harmonic means with improved GSA (Yin et al. 2011).
- Hybrid neural network and GSA (Ghalambaz et al. 2011).
- Hybrid particle swarm optimization and GSA (Mallick et al. 2013).
- Hybrid random-key GSA (Chen et al. 2011).
- Improved GSA (Li and Zhou 2011; Li and Duan 2012).
- Modified GSA (Khajehzadeh et al. 2012; Han and Chang 2012a).
- Multiobjective GSA (lez-Álvarez et al. 2013).
- Non-dominated sorting GSA (Nobahari et al. 2011).
- Opposition-based GSA (Shaw et al. 2012).

Second, the GSA has also been successfully applied to a variety of optimization problems as listed below:

- Antenna design optimization (Chatterjee et al. 2010).
- Chaotic system parameters identification (Li et al. 2012).

- Communication system security optimization (Han and Chang 2012a, b).
- Control system optimization (David et al. 2013; Precup et al. 2013).
- Data clustering (Yin et al. 2011; Papa et al. 2011; Hatamlou et al. 2011, 2012; Bahrololoum et al. 2012).
- Drug design optimization (Bababdani and Mousavi 2013).
- Image processing (Zhao 2011).
- Motif discovery problem (lez-Álvarez et al. 2013).
- Oil demand prediction (Behrang et al. 2011).
- Path planning for uninhabited aerial vehicle (Li and Duan 2012).
- Power system optimization (Güvenç et al. 2012; Eslami et al. 2012a, b; Duman et al. 2012; Niknam et al. 2012; Ghasemi et al. 2013; Roy 2013; Chatterjee et al. 2012; Mallick et al. 2013; Zhang et al. 2013; Roy et al. 2012; Li and Zhou 2011; Shaw et al. 2012; Ju and Hong 2013; Li et al. 2013; Mondal et al. 2013; Kumar et al. 2013; Barisal et al. 2012).
- Quality of service optimization (Zibanezhad et al. 2011).
- Retaining wall structure optimization (Khajehzadeh and Eslami 2012).
- Robot control optimization (Seljanko 2011).
- Signal filter optimization (Rashedi et al. 2011).
- Slope stability analysis (Khajehzadeh et al. 2012).
- Travelling salesman problem (Afaq and Saini 2011; Chen et al. 2011).
- Wireless mesh networks optimization (Doraghinejad et al. 2013).

Interested readers please refer to them as a starting point for a further exploration and exploitation of GSA.

References

Afaq, H., & Saini, S. (2011). On the solutions to the travelling salesman problem using nature inspired computing techniques. *International Journal of Computer Science Issues, 8*, 326–334.

Bababdani, B. M., & Mousavi, M. (2013). Gravitational search algorithm: a new feature selection method for QSAR study of anticancer potency of imidazo[4,5-b]pyridine derivatives. *Chemometrics and Intelligent Laboratory Systems, 122*, 1–11.

Bahrololoum, A., Nezamabadi-Pour, H., Bahrololoum, H., & Saeed, M. (2012). A prototype classifier based on gravitational search algorithm. *Applied Soft Computing, 12*, 819–825.

Barisal, A. K., Sahu, N. C., Prusty, R. C., & Hota, P. K. (2012). Short-term hydrothermal scheduling using gravitational search algorithm. IEEE 2nd International Conference on Power, Control and Embedded Systems, pp. 1–6.

Behrang, M. A., Assareh, E., Ghalambaz, M., Assari, M. R., & Noghrehabadi, A. R. (2011). Forecasting future oil demand in Iran using GSA (gravitational search algorithm). *Energy, 36*, 5649–5654.

Chatterjee, A., Mahanti, G. K., & Pathak, N. (2010). Comparative performance of gravitational search algorithm and modified particle swarm optimization algorithm for synthesis of thinned scanned concentric ring array antenna. *Progress in Electromagnetics Research B, 25*, 331–348.

Chatterjee, A., Ghoshal, S. P., & Mukherjee, V. (2012). A maiden application of gravitational search algorithm with wavelet mutation for the solution of economic load dispatch problems. *International Journal of Bio-Inspired Computation, 4*, 33–46.

Chen, H., Li, S., & Tang, Z. (2011). Hybrid gravitational search algorithm with random-key encoding scheme combined with simulated annealing. *International Journal of Computer Science and Network Security, 11*, 208–217.

David, R.-C., Precup, R.-E., Petriu, E. M., Rădac, M.-B., & Preitl, S. (2013). Gravitational search algorithm-based design of fuzzy control systems with a reduced parametric sensitivity. *Information Sciences, 247*, 154–173. doi:http://dx.doi.org/10.1016/j.ins.2013.05.035.

Doraghinejad, M., Nezamabadi-pour, H., & Mahani, A. (2013). Channel assignment in multi-radio wireless mesh networks using an improved gravitational search algorithm. *Journal of Network and Computer Applications.* doi:http://dx.doi.org/10.1016/j.jnca.2013.04.007.

Duman, S., Güvenç, U., Sönmez, Y., & Yörükeren, N. (2012). Optimal power flow using gravitational search algorithm. *Energy Conversion and Management, 59*, 86–95.

Eslami, M., Shareef, H., Mohamed, A., & Khajehzadeh, M. (2012a). Gravitational search algorithm for coordinated design of PSS and TCSC as damping controller. *Journal of Central South University of Technology, 19*, 923–932.

Eslami, M., Shareef, H., Mohamed, A., & Khajehzadeh, M. (2012b). PSS and TCSC damping controller coordinated design using GSA. *Energy Procedia, 14*, 763–769.

Gao, S., Chai, H., Chen, B., & Yang, G. (2013). Hybrid gravitational search and clonal selection algorithm for global optimization. In Tan, Y., Shi, Y. & Mo, H. (Eds.), *Advances in Swarm Intelligence, LNCS 7929,* (pp. 1–10). Hybrid gravitational search and clonal selection algorithm for global optimization: Springer.

Gauci, M., Dodd, T. J., & Groß, R. (2012). Why 'GSA: a gravitational search algorithm' is not genuinely based on the law of gravity. *Natural Computing, 11*, 719–720. doi:10.1007/s11047-012-9322-0.

Ghalambaz, M., Noghrehabadi, A. R., Behrang, M. A., Assareh, E., Ghanbarzadeh, A., & Hedayat, N. (2011). A hybrid neural network and gravitational search algorithm (HNNGSA) method to solve well known Wessinger's equation. *World Academy of Science, Engineering and Technology, 73*, 803–807.

Ghasemi, A., Shayeghi, H., & Alkhatib, H. (2013). Robust design of multimachine power system stabilizers using fuzzy gravitational search algorithm. *Electrical Power and Energy Systems, 51*, 190–200.

Güvenç, U., Sönmez, Y., Duman, S., & Yörükeren, N. (2012). Combined economic and emission dispatch solution using gravitational search algorithm. *Scientia Iranica D, 19*, 1754–1762.

Han, X., & Chang, X. (2012a). A chaotic digital secure communication based on a modified gravitational search algorithm filter. *Information Sciences, 208*, 14–27.

Han, X., & Chang, X. (2012b). Chaotic secure communication based on a gravitational search algorithm filter. *Engineering Applications of Artificial Intelligence, 25*, 766–774.

Hatamlou, A., Abdullah, S., & Nezamabadi-Pour, H. (2011). Application of gravitational search algorithm on data clustering. *Rough Sets and Knowledge Technology, LNCS 6954,* (pp. 337–346). Berlin: Springer.

Hatamlou, A., Abdullah, S., & Nezamabadi-Pour, H. (2012). A combined approach for clustering based on K-means and gravitational search algorithms. *Swarm and Evolutionary Computation, 6*, 47–52. doi:10.1016/j.swevo.2012.02.003.

Ju, F.-Y., & Hong, W.-C. (2013). Application of seasonal SVR with chaotic gravitational search algorithm in electricity forecasting. *Applied Mathematical Modelling, 37*, p. 23. doi:http://dx.doi.org/10.1016/j.apm.2013.05.016.

Khajehzadeh, M., & Eslami, M. (2012). Gravitational search algorithm for optimization of retaining structures. *Indian Journal of Science and Technology, 5*, 1821–1827.

Khajehzadeh, M., Taha, M. R., El-Shafie, A., & Eslami, M. (2012). A modified gravitational search algorithm for slope stability analysis. *Engineering Applications of Artificial Intelligence, 25*, 1589–1597. doi:10.1016/j.engappai.2012.01.011.

Kumar, J. V., Kumar, D. M. V., & Edukondalu, K. (2013). Strategic bidding using fuzzy adaptive gravitational search algorithm in a pool based electricity market. *Applied Soft Computing, 13*, 2445–2455.

Lez-Álvarez, D. L. G., Vega-Rodríguez, M. A., Gómez-Pulido, J. A., & Sánchez-Pérez, J. M. (2013). Comparing multiobjective swarm intelligence metaheuristics for DNA motif discovery. *Engineering Applications of Artificial Intelligence, 26*, 314–326.

Li, P., & Duan, H. (2012). Path planning of unmanned aerial vehicle based on improved gravitational search algorithm. *Science China Technological Sciences, 55*, 2712–2719. doi:10.1007/s11431-012-4890-x.

Li, C., & Zhou, J. (2011). Parameters identification of hydraulic turbine governing system using improved gravitational search algorithm. *Energy Conversion and Management, 52*, 374–381.

Li, C., Zhou, J., Xiao, J., & Xiao, H. (2012). Parameters identification of chaotic system by chaotic gravitational search algorithm. *Chaos, Solitons & Fractals, 45*, 539–547.

Li, C., Zhou, J., Xiao, J., & Xiao, H. (2013). Hydraulic turbine governing system identification using T–S fuzzy model optimized by chaotic gravitational search algorithm. *Engineering Applications of Artificial Intelligence, 26*, 2073–2082. doi:http://dx.doi.org/10.1016/j.engappai.2013.04.002.

Mallick, S., Ghoshal, S. P., Acharjee, P., & Thakur, S. S. (2013). Optimal static state estimation using improved particle swarm optimization and gravitational search algorithm. *Electrical Power and Energy Systems, 52*, 254–265.

Mondal, S., Bhattacharya, A., & Dey, S. H. N. (2013). Multi-objective economic emission load dispatch solution using gravitational search algorithm and considering wind power penetration. *Electrical Power and Energy Systems, 44*, 282–292.

Niknam, T., Golestaneh, F., & Malekpour, A. (2012). Probabilistic energy and operation management of a microgrid containing wind/photovoltaic/fuel cell generation and energy storage devices based on point estimate method and self-adaptive gravitational search algorithm. *Energy, 43*, 427–437.

Nobahari, H., Nikusokhan, M., & Siarry, P. (2011, June 14–15). Non-dominated sorting gravitational search algorithm. International conference on swarm intelligence (ICSI) (pp. 1–10). Cergy, France.

Papa, J. P., Pagnin, A., Schellini, S. A., Spadotto, A., Guido, R. C., Ponti, M., Chiachia, G., & Falcão, A. X. (2011). Feature selection through gravitational search algorithm. IEEE International Conference on Acoustics, Speech (ICASSP), pp. 2052–2055.

Precup, R.-E., David, R.-C., Petriu, E. M., Rădac, M.-B., Preitl, S., & Fodor, J. (2013). Evolutionary optimization-based tuning of low-cost fuzzy controllers for servo systems. *Knowledge-Based Systems, 38*, 74–84. doi:10.1016/j.knosys.2011.07.006.

Rashedi, E., Nezamabadi-Pour, H., & Saryazdi, S. (2009). GSA: A gravitational search algorithm. *Information Sciences, 179*, 2232–2248.

Rashedi, E., Nezamabadi-Pour, H., & Saryazdi, S. (2010). BGSA: Binary gravitational search algorithm. *Natural Computing, 9*, 727–745.

Rashedi, E., Nezamabadi-Pour, H., & Saryazdi, S. (2011). Filter modeling using gravitational search algorithm. *Engineering Applications of Artificial Intelligence, 24*, 117–122.

Ricci, F. (1998). The search for gravitational waves: an experimental physics challenge. *Contemporary Physics, 39*, 107–135.

Roy, P. K. (2013). Solution of unit commitment problem using gravitational search algorithm. *Electrical Power and Energy Systems, 53*, 85–94.

Roy, P. K., Mandal, B., & Bhattacharya, K. (2012). Gravitational search algorithm based optimal reactive power dispatch for voltage stability enhancement. *Electric Power Components and Systems, 40*, 956–976.

Sarafrazi, S., Nezamabadi-Pour, H., & Saryazdi, S. (2011). Disruption: A new operator in gravitational search algorithm. *Scientia Iranica D, 18*, 539–548.

Schutz, B. (2003). *Gravity from the ground up, The Edinburgh Building, Cambridge CB2 8RU*. UK: Cambridge University Press. ISBN 13 978-0-511-33696-6.

Seljanko, F. (2011, June 20–23). Hexapod walking robot gait generation using genetic-gravitational hybrid algorithm. IEEE 15th International Conference on Advanced Robotics (pp. 253–258), Tallinn University of Technology, Tallinn, Estonia.

Serway, R. A., & Jewett, J. W. (2014). *Physics for scientists and engineers with modern physics*. Boston: Brooks/Cole CENAGE Learning. ISBN 978-1-133-95405-7.

Shaw, B., Mukherjee, V., & Ghoshal, S. P. (2012). A novel opposition-based gravitational search algorithm for combined economic and emission dispatch problems of power systems. *Electrical Power and Energy Systems, 35*, 21–33.

Yin, M., Hu, Y., Yang, F., Li, X., & Gu, W. (2011). A novel hybrid K-harmonic means and gravitational search algorithm approach for clustering. *Expert Systems with Applications, 38*, 9319–9324.

Zhang, W., Niu, P., Li, G., & Li, P. (2013). Forecasting of turbine heat rate with online least squares support vector machine based on gravitational search algorithm. *Knowledge-Based Systems, 39*, 34–44.

Zhao, W. (2011). Adaptive image enhancement based on gravitational search algorithm. *Procedia Engineering, 15*, 3288–3292.

Zibanezhad, B., Yamanifar, K., Sadjady, R. S., & Rastegari, Y. (2011). Applying gravitational search algorithm in the QoS-based Web service selection problem. *Journal of Zhejiang University—Science C (Computers & Electronics), 12*, 730–742.

Chapter 23
Intelligent Water Drops Algorithm

Abstract In this chapter, an intelligent water drops (IWD) algorithm is introduced. We first, in Sect. 23.1, describe the general knowledge of nature water drops and the Newton's law of gravity, respectively. Then, the fundamentals of IWD, the selected variant of IWD, and the representative IWD application are detailed in Sect. 23.2, respectively. Finally, Sect. 23.3 draws the conclusions of this chapter.

23.1 Introduction

The river which flows freely through our land has always played an important part in our lives. Without it we could not make a living. The main building block of the flowing river is the water drops, which is impacted by the gravity. Through its long-rang force, it pulls everything toward the centre of the earth in a straight line. Inspired by this nature phenomenon, Shah-Hosseini (2007) proposed a new problem solving algorithm called intelligent water drops (IWP) algorithm.

23.1.1 Key Characteristics of Nature Water Drops

A river is a natural stream of water drops (such as rainwater or melting snow) with significant volume. It begins high in mountains and flows downhill because of gravity. As the water flows down, it may pick up more water from other small streams, and the whole journey is not smooth, i.e., there are barriers and obstacles (such as rocks, pebbles, and soil) on the path. The speed of water moving mainly depends on the slop gradient, the roughness of the channel, and the tides. In addition, river can move soil directly. This is known as "erosion". Typically, the speed and erosion have a direct relationship. That means, faster water gathers more amount of soil than stationary or slow water and can carry them quite a distance. Over time, as the river flows, it deposit all of the stuff it carries and change the land by carving new paths for themselves.

B. Xing and W.-J. Gao, *Innovative Computational Intelligence: A Rough Guide to 134 Clever Algorithms*, Intelligent Systems Reference Library 62,
DOI: 10.1007/978-3-319-03404-1_23, © Springer International Publishing Switzerland 2014

23.1.2 Newton's Laws of Gravity

As we know, gravity is one of the fundamental force that operates everywhere. For example, together with the density difference between water and air, gravity serves to keep the water molecules in the cup and the air above the water's surface, for all relevant purposes. In the 17th century, Sir Isaac Newton was the first to propose the consequent mathematical expression of gravity. He expressed his model by using three assertions, which have come to be known as Newton's laws (Holzner 2011):

- First law explains what happens with forces and motion. His first law states "An object continues in a state of rest or in a state of motion at a constant velocity along a straight line, unless compelled to change that state by a net force".
- Second law says that an object remains in uniform motion unless acted on by a net force. That means, it details the relationship among net force, the mass, and the acceleration.
- Third law tells us "Whenever one body exerts a force on a second body, the second body exerts an oppositely directed force of equal magnitude on the first body".

Nowadays, observations and theory agree that the gravity explains the working principle of the universe, such as the fluids move. In addition, it even laid the path toward the evolution of life itself, such as how the liquid is changed to the gas.

23.2 Intelligent Water Drops Algorithm

23.2.1 Fundamentals of Intelligent Water Drops Algorithm

Intelligent water drops (IWD) algorithm was originally proposed in (Shah-Hosseini 2007, 2008, 2009a, b). Two attributes are proposed for the IWD algorithm, namely, the amount of soil denoted by *soil* (*IWD*) and the velocity of the IWDs denoted by *velocity* (*IWD*). In the following, we will discuss the algorithm rules in details.

- Path selecting rule: For each IWD, the probability ($p(i,j;IWD)$) of choosing the next location is given by Eqs. 23.1 and 23.2, respectively (Shah-Hosseini 2009b):

$$p(i,j;IWD) = \frac{f(soil(i,j))}{\sum_{k \notin vc(IWD)} f(soil(i,k))}, \tag{23.1}$$

$$f(soil(i,j)) = \frac{1}{\varepsilon_s + g(soil(i,j))}, \tag{23.2}$$

where the set $vc(IWD)$ denotes the locations that the IWD should not visit to keep satisfied the constraints of the problem, ε_s is a small positive number to prevent a possible division by zero in the function $(f(.))$, and $g(soil(i,j))$ is used to shift the $soil(i,j)$ of the path joining location i and j toward positive values and is computed by Eq. 23.3 (Shah-Hosseini 2009b):

$$g(soil(i,j)) = \begin{cases} soil(i,j) & \text{if} \min\limits_{l \notin vc(IWD)} (soil(i,l)) \geq 0 \\ soil(i,j) - \min\limits_{l \notin vc(IWD)} (soil(i,l)) & \text{otherwise} \end{cases}, \tag{23.3}$$

where the function min(.) returns the minimum value of its arguments.

- Velocity updating rule: For each IWD that moves from current location i to next location j, updates its velocity (vel^{IWD}) via Eqs. 23.4 and 23.5, respectively (Shah-Hosseini 2009b):

$$vel^{IWD}(t+1) = vel^{IWD}(t) + \Delta vel^{IWD}(t), \tag{23.4}$$

$$\Delta vel^{IWD}(t) = \frac{a_v}{b_v + c_v \cdot soil^{\beta}(i,j)}, \tag{23.5}$$

where $vel^{IWD}(t+1)$ stands for the updated velocity of an IWD at the node j, $soil(i,j)$ is the soil on the path joining the current location and the new location, a_v, b_v, and c_v are constant velocity parameters which are adjustable according to focal problems, and β is user-selected positive parameters.

- Local soil updating rule: For each IWDs moving, the amount of the soil $(soil(i,j))$ and the soil that each IWD carries $(soil^{IWD})$ are updated via Eqs. 23.6–23.8, respectively (Shah-Hosseini 2009b):

$$soil(i,j) = \rho_0 \cdot soil - \rho_n \cdot \Delta soil(i,j), \tag{23.6}$$

$$soil^{IWD} = soil^{IWD} + \Delta soil(i,j), \tag{23.7}$$

$$\Delta soil(i,j) = \frac{a_s}{b_s + c_s \cdot time^2(i,j;vel^{IWD}(t+1))}, \tag{23.8}$$

where $\Delta soil(i,j)$ is the soil which the vel^{IWD} removes from the path between location i and j, ρ_0 and ρ_n are often positive number generated from the interval [0,1], a_s, b_s, and c_s are user-selected positive numbers that depending on the given problem, and $time(i,j;vel^{IWD})$ is the time value that required for the vel^{IWD} and defined by Eq. 23.9 (Shah-Hosseini 2009b):

$$time(i,j;vel^{IWD}(t+1)) = \frac{HUD(i,j)}{vel^{IWD}(t+1)}, \tag{23.9}$$

where a local heuristic $HUD(.,.)$ denotes the heuristic undesirability of moving between two locations.

- Global soil updating rule: At the end of each iteration, the amount of the soil on the arc of the iteration-best solution (T^{IB}) is reduced based on the quality of the iteration-best solution ($q(T^{IWD})$) as defined by Eq. 23.10 (Shah-Hosseini 2009b):

$$soil(i,j) = (1+\rho_{IWD}) \cdot soil(i,j) - \rho_{IWD} \cdot \frac{soil_{IB}^{IWD}}{(N_{IB}-1)}, \forall (i,j) \in T^{IB}, \tag{23.10}$$

where $soil_{IB}^{IWD}$ denotes the soil of the iterations-best IWD, N_{IB} represents the number of locations in the solution T^{IB}, and ρ_{IWD} is the global soil updating parameter generated from the interval [0,1].

In addition, in IWD, every created IWD is designed to move from its initial location to the next ones until it finds a solution. At the end of each iteration, the best solution (T^{IB}) found by the IWDs within such iteration is obtained via Eq. 23.11 (Shah-Hosseini 2009b):

$$T^{IB} = \arg \max_{\forall T^{IWD}} q(T^{IWD}), \tag{23.11}$$

where $q(.)$ is the objective of quality function, and $q(T^{IWD})$ is the quality of a solution (T^{IWD}) found by the IWD. It means that the iteration-best solution T^{IB} is the dominant solution over all other solutions T^{IWD}.

At the end of each iteration of the IWD algorithm, the current iteration-best solution (T^{IB}) is used to update the total best solution T^{IB} via Eq. 23.12 (Shah-Hosseini 2009b):

$$T^{TB} = \begin{cases} T^{TB} & \text{if } q(T^{TB}) \geq q(T^{IB}) \\ T^{IB} & \text{otherwise} \end{cases}, \tag{23.12}$$

Taking into account four key rules described above, the steps of implementing standard IWD algorithm can be summarized as follows (Shah-Hosseini 2009b):

- Step 1: Initialization of both static and dynamic parameters.
- Step 2: Spread the IWDs randomly on the locations and then update the visited nodes.
- Step 3: Repeat the following processes path selecting process, velocity updating process, and local Soil updating process till stopping criteria met.
- Step 4: Find the iteration-best solution (T^{IB}) from all the solutions (T^{IWD}) found by the IWDs.
- Step 5: Global soil updating process.

- Step 6: Update the total best solution (T^{TB}).
- Step 7: Increment the iteration number by $Iter_{count} = Iter_{count} + 1$. Then, go to Step 2 if $Iter_{count} < Iter_{\max}$.
- Step 8: The algorithm stops such that the best solution is kept in T^{TB}.

23.2.2 Performance of IWD

In order to show how the IWD algorithm performs, several researchers have conducted a set of studies, such as the travelling salesman problem, the n-queen puzzle, the multidimensional k006Eapsack problem, and the image segmentation problem. Experimental results showed that the IWD algorithm performs well in finding optimal or near optimal solutions.

23.2.3 Selected IWD Variant

Although IWD algorithm is a new member of computational intelligence (CI) family, a number of IWD variations have been proposed in the literature for the purpose of further improving the performance of IWD. This section gives an overview to one of these IWD variants which has been demonstrated to be very efficient and robust.

23.2.3.1 IWD for Continuous Optimization

Optimization problems can broadly be described as either continuous or discrete. The main feature of continuous problem is the variables are allowed to take on any values permitted by the constraints. Due to the IWD algorithm is inspired by the nature water drops, it can be easily represented for solving continuous optimization problems. In 2012, Shah-Hosseini (2012a) proposed a variant of IWD called IWD-CO (i.e., the IWD algorithm for continuous optimization). In details, IWD-CO uses binary valued variables to represent information in individuals and puts more emphasis on the mutation operator.

- Edge selecting: Let an IWD is at node i and select the edge $\left(e_{i,i+1}(k)\right)$ to visit the next node $i+1$. The probability $\left(P^{IWD}\left(e_{i,i+1}(k)\right)\right)$ for such a selection is described via Eqs. 23.13–23.15, respectively (Shah-Hosseini 2012a):

$$P^{IWD}\left(e_{i,i+1}(k)\right) = \frac{f\left(soil\left(e_{i,i+1}(k)\right)\right)}{\sum_{l=0}^{1} f\left(soil\left(e_{i,i+1}(l)\right)\right)}, \tag{23.13}$$

$$f\left(soil\left(e_{i,i+1}(k)\right)\right) = \frac{1}{0.0001 + g\left(soil\left(e_{i,i+1}(k)\right)\right)}, \tag{23.14}$$

$$g\left(soil\left(e_{i,i+1}(k)\right)\right) = \begin{cases} soil\left(e_{i,i+1}(k)\right) & \text{if } \min\limits_{l=0.1}\left(soil\left(e_{i,i+1}(l)\right)\right) \geq 0 \\ soil\left(e_{i,i+1}(k)\right) - \min\limits_{l=0.1}\left(soil\left(e_{i,i+1}(l)\right)\right) & \text{otherwise} \end{cases}, \tag{23.15}$$

where $e_{i,i+1}(k)$ is a directed edge that connects node i to node $i+1$.

- Local soil updating: When an IWD leaves node i by using edge $e_{i,i+1}(k)$ to arrive at node $i+1$, the soil of the IWD ($soil^{IWD}$) and the soil of the used edge $\left(soil\left(e_{i,i+1}(k)\right)\right)$ are updated via Eqs. 23.16–23.18, respectively (Shah-Hosseini 2012a):

$$soil\left(e_{i,i+1}(k)\right) = 1.1 \cdot soil\left(e_{i,i+1}(k)\right) - 0.01 \cdot \Delta soil\left(e_{i,i+1}(k)\right), \tag{23.16}$$

$$soil^{IWD} = soil^{IWD} + \Delta soil\left(e_{i,i+1}(k)\right), \tag{23.17}$$

$$\Delta soil\left(e_{i,i+1}(k)\right) = 0.001. \tag{23.18}$$

- Mutation-based local search: This process is repeated until the fitness value of the solution is improved. It is noted that this process is applied to all the solutions created in the current iteration by the IWDs.
- Global soil updating: The soils of the edges with the iteration-best solution are updated via Eqs. 23.19 and 23.20, respectively (Shah-Hosseini 2012a):

$$soil\left(e_{i,i+1}(k)\right) = \min\left\{\max\left[Tempsoil\left(e_{i,i+1}(k)\right), MinSoil\right], MaxSoil\right\}, \quad \forall e_{i,i+1}(k) \in T^{IB} \tag{23.19}$$

$$Tempsoil\left(e_{i,i+1}(k)\right) = 1.1 \cdot soil\left(e_{i,i+1}(k)\right) - 0.01 \cdot \frac{soil_{IB}^{IWD}}{(M \cdot P)}, \forall e_{i,i+1}(k) \in T^{IB} \tag{23.20}$$

where $[MinSoil, MaxSoil]$ is the boundary for the global soil updating, $soil_{IB}^{IWD}$ denote the quality of the IWD with T^{IB}, and $(M \cdot P)$ represents the graph which has $M \times P$ nodes.

23.2.3.2 Performance of IWD-CO

To evaluate the performance of the IWD-CO algorithm, a set of benchmark functions are selected in (Shah-Hosseini 2012a). Computational results showed that the proposed algorithm converges to optimal values of the all six functions.

23.2.4 Representative IWD Application

The first application of the IWD algorithm is tested on the TSP, due to: (1) it is an important NP-hard optimization problem (Johnson and Papadimitriou 1985) that arises in several applications; (2) the IWD algorithm can be easily employed in TSP; (3) it is a standard test bed for new algorithmic idea.

23.2.4.1 Travelling Salesman Problem

The main objective of Travelling Salesman Problem (TSP) is to find the shortest tour through all the cities that a salesman has to visit. For such a TSP with n cities, there is an immense number of possible tours: $\frac{(n-1)!}{2}$. Mathematically we may define the TSP as follow (Bellmore and Nemhauser 1968): suppose we are given a complete digraph $G = (N, E)$, where the cities correspond to the node set $N = \{1, 2, \cdots, n\}$.

The key steps of applying IWD to TSP problem are listed as below (Bonyadi et al. 2008):

- Step 1: Initialization of both static and dynamic parameters.
- Step 2: For every water drop, randomly choosing a city and placing that water drop on the city.
- Step 3: Updating the list of visited cities.
- Step 4: Choosing the next visiting city using probability equation.
- Step 5: Updating water drop's velocity.
- Step 6: Calculating the amount of the soil that a water drop is carrying.
- Step 7: Updating the soil amount of the path flowed by a water drop.
- Step 8: For every water drop, completing its tour by repeatedly using Steps 3–7.

The proposed IWD algorithm was tested on some artificial and benchmark TSP instance. Computational results showed that IWD can obtain a set of promising solutions in terms of fast converging speed.

23.3 Conclusions

In this chapter, we focused on the IWD algorithm, which is inspired by the nature water drops. Based on the observations on the behaviour of water drops, there are three intimations that served as basic principle of the developing of IWD: (1) the velocity enables the water drops to transfer soil from one place to another in the front; (2) a high speed water drop gathers more soil than a slower water drop; and (3) a water drop prefers a path with less soil than a path with more soil.

Interestingly, by taking inspiration from those principles, it is possible to design IWDs that, by moving on a graph modelling in which the water drops flowing process is mimicked, find the easier path between the two locations. For IWD, it is assumed that each water drop carries an amount of soil and every two locations are linked by an arc which also holds an amount of soil. In fact, some amount of soil of the river bed is removed by the water drop and is added to the soil of the water drop. That means, each IWD holds soil in itself and removes soil from its path during movement in the environment. In addition, a water drop has also a velocity that impacts the amount of the soil during the water drops process. Although it is a newly introduced CI method, we have witnessed the following rapid spreading of IWD:

First, in addition to the selected variants detailed in this chapter, several enhanced versions of IWD can also be found in the literature as outlined below:

- Adaptive IWD (Msallam and Hamdan 2011).
- Improved IWD (Duan et al. 2009).
- Modified IWD (Kesavamoorthy et al. 2011).
- Neural IWD (Hendrawan and Murase 2011).

Second, apart from the representative applications, the IWD algorithm has also been successfully applied to a variety of optimization problems as listed below:

- Distributed Denial of Service Attacks Mitigation (Lua and Yow 2011).
- Image processing (Hendrawan and Murase 2011, Shah-Hosseini 2012b).
- Job shop scheduling problem (Niu et al. 2012).
- Multiple knapsack problem (Shah-Hosseini 2008, 2009a).
- Power system optimization (Rayapudi 2011, Nagalakshmi et al. 2011).
- Quality of service optimization (Palanikkumar et al. 2012).
- Robotics path planning (Duan et al. 2008, 2009).
- Single UCAV smooth trajectory planning (Duan et al. 2009).
- Travelling salesman problem (Shah-Hosseini 2009a, b, Afaq and Saini 2011, Bonyadi et al. 2008, Kesavamoorthy et al. 2011, Msallam and Hamdan 2011).
- Vehicle routing problem (Kamkar et al. 2010).

Interested readers are referred to them as a starting point for a further exploration and exploitation of the IWD algorithm.

References

Afaq, H., & Saini, S. (2011). On the solutions to the travelling salesman problem using nature inspired computing techniques. *International Journal of Computer Science Issues, 8*, 326–334.

Bellmore, M., & Nemhauser, G. L. (1968). The traveling salesman problem: A survey. *Operations Research, 16*, 538–558.

Bonyadi, M. R., Azghadi, M. R. & Shah-Hosseini, H. (2008). Population-based optimization algorithms for solving the travelling salesman problem. In F. Greco (Ed.) *Travelling salesman problem*, Chap. 1 (pp. 1–34). Vienna, Austria: In-Tech.

Duan, H., Liu, S. & Lei, X. (2008). Air robot path planning based on intelligent water drops optimization. In *IEEE International Joint Conference on Neural Networks (IJCNN)* (pp. 1397–1401).

Duan, H., Liu, S., & Wu, J. (2009). Novel intelligent water drops optimization approach to single UCAV smooth trajectory planning. *Aerospace Science and Technology, 13*, 442–449.

Hendrawan, Y., & Murase, H. (2011). Neural-intelligent water drops algorithm to select relevant textural features for developing precision irrigation system using machine vision. *Computers and Electronics in Agriculture, 77*, 214–228.

Holzner, S. (2011). *Physics I for dummies*. River Street, Hoboken, NJ, USA: Wiley Publishing, Inc. ISBN 978-0-470-90324-7.

Johnson, D. S. & Papadimitriou, C. H. (1985). Computational complexity. In E. L. Lawer, J. K. Lenstra, A. H. D. R. Kan, & D. B. Shmoys (Eds.), *The Traveling Salesman Problem: A Guided Tour of Combinatorial Optimization*. Wiley.

Kamkar, I., Akbarzadeh-T, M.-R. & Yaghoobi, M. (2010). Intelligent water drops a new optimization algorithm for solving the vehicle routing problem. In *IEEE International Conference on Systems, Man, and Cybernetics (IEEE SMC)*, Istanbul, Turkey, (pp. 4142–4146, October 10–13).

Kesavamoorthy, R., Shunmugam, D. A. & Mariappan, L. T. (2011). Solving traveling salesman problem by modified intelligent water drop algorithm. *International Journal of Computer Applications*. In *International Conference on Emerging Technology Trends (ICETT)* (pp. 18–23).

Lua, R., & Yow, K. C. (2011). Mitigating DDoS attacks with transparent and intelligent fast-flux swarm network. *IEEE Network, 25*, 28–33.

Msallam, M. M., & Hamdan, M. (2011). Improved intelligent water drops algorithm using adaptive schema. *International Journal of Bio-Inspired Computation, 3*, 103–111.

Nagalakshmi, P., Harish, Y., Kumar, R. K. & Jada, C. (2011). Combined economic and emission dispatch using intelligent water drops-continuous optimization algorithm. In *IEEE International Conference on Recent Advancements in Electrical, Electronics and Control Engineering (ICONRAEeCE)* (pp. 168–173).

Niu, S. H., Ong, S. K., & Nee, A. Y. C. (2012). An improved intelligent water drops algorithm for achieving optimal job-shop scheduling solutions. *International Journal of Production Research, 50*, 4192–4205.

Palanikkumar, D., Elangovan, G., Rithu, B., & Anbusel, P. (2012). An intelligent water drops algorithm based service selection and composition in service oriented architecture. *Journal of Theoretical and Applied Information Technology, 39*, 45–51.

Rayapudi, S. R. (2011). An intelligent water drop algorithm for solving economic load dispatch problem. *International Journal of Electrical and Electronics Engineering, 5*, 43–49.

Shah-Hosseini, H. (2007). Problem solving by intelligent water drops. In *IEEE Congress on Evolutionary Computation (CEC)* (pp. 3226–3231, September 25–28).

Shah-Hosseini, H. (2008). Intelligent water drops algorithm: a new optimization method for solving the multiple knapsack problem. *International Journal of Intelligent Computing and Cybernetics, 1*, 193–212.

Shah-Hosseini, H. (2009a). The intelligent water drops algorithm: a nature-inspired swarm-based optimization algorithm. *International Journal of Bio-Inspired Computation, 1*.

Shah-Hosseini, H. (2009b). Optimization with the nature-inspired intelligent water drops algorithm. In W. P. D. Santos (Ed.) *Evolutionary Computation*, Chap. 16 (pp. 297–320). Vienna, Austria.

Shah-Hosseini, H. (2012a). An approach to continuous optimization by the intelligent water drops algorithm. *Procedia—Social and Behavioral Sciences, 32*, 224–229.

Shah-Hosseini, H. (2012b). Intelligent water drops algorithm for automatic multilevel thresholding of grey–level images using a modified Otsu's criterion. *International Journal of Modelling, Identification and Control, 15*, 241–249.

Chapter 24
Emerging Physics-based CI Algorithms

Abstract In this chapter, a set of (more specifically 22 in total) emerging physics-based computational intelligence (CI) algorithms are introduced. We first, in Sect. 24.1, describe the organizational structure of this chapter. Then, from Sects. 24.2 to 24.23, each section is dedicated to a specific algorithm which falls within this category. The fundamentals of each algorithm and their corresponding performances compared with other CI algorithms can be found in each associated section. Finally, the conclusions drawn in Sect. 24.24 closes this chapter.

24.1 Introduction

Several novel physics-based algorithms were detailed in previous chapters. In particular, Chap. 18 detailed the big bang-big crunch algorithm, Chap. 19 was dedicated to central force optimization algorithm, Chap. 20 discussed the charged system search algorithm, Chap. 21 introduced the electromagnetism-like mechanism algorithm, Chap. 22 was devoted to the gravitational search algorithm, and Chap. 23 described the intelligent water drops algorithm. Apart from this quasi-mature physics principles inspired CI methods, there are some emerging algorithms also fall within this category. This chapter collects 22 of them that are currently scattered in the literature and organizes them as follows:

- Section 24.2: Artificial Physics Optimization
- Section 24.3: Atmosphere Clouds Model Optimization
- Section 24.4: Chaos Optimization Algorithm
- Section 24.5: Cloud Model-based Algorithm
- Section 24.6: Extremal Optimization
- Section 24.7: Galaxy-based Search Algorithm
- Section 24.8: Gravitation Field Algorithm
- Section 24.9: Gravitational Clustering Algorithm
- Section 24.10: Gravitational Emulation Local Search

B. Xing and W.-J. Gao, *Innovative Computational Intelligence: A Rough Guide to 134 Clever Algorithms*, Intelligent Systems Reference Library 62,
DOI: 10.1007/978-3-319-03404-1_24,

- Section 24.11: Gravitational Interactions Optimization
- Section 24.12: Hysteretic Optimization
- Section 24.13: Integrated Radiation Optimization
- Section 24.14: Light Ray Optimization
- Section 24.15: Magnetic Optimization Algorithm
- Section 24.16: Particle Collision Algorithm
- Section 24.17: Ray Optimization
- Section 24.18: River Formation Dynamics Algorithm
- Section 24.19: Space Gravitational Optimization
- Section 24.20: Spiral Optimization Algorithm
- Section 24.21: Water Cycle Optimization Algorithm
- Section 24.22: Water Flow Algorithm
- Section 24.23: Water Flow-like Algorithm

The effectiveness of these newly developed algorithms are validated through the testing on a wide range of benchmark functions and engineering design problems, and also a detailed comparison with various traditional performance leading CI algorithms, such as particle swarm optimization (PSO), genetic algorithm (GA), differential evolution (DE), evolutionary algorithm (EA), fuzzy system (FS), ant colony optimization (ACO), and simulated annealing (SA).

24.2 Artificial Physics Optimization Algorithm

In this section, we will introduce an emerging CI algorithm that is based on artificial physics or physicomimetics, a concept which was introduced in Spears et al. (2004a, b); Spears and Gordon (1999); Spears and Spears (2012).

24.2.1 Fundamentals of Artificial Physics Optimization Algorithm

Artificial physics optimization (APO) algorithm was recently proposed in Xie et al. (2009a, b, 2010a, b, 2011a, b), Xie and Zeng (2009a). Several APO applications and variants can also be found in the literature (Gorbenko and Popov 2012, 2013; Xie and Zeng 2009b, 2011; Mo and Zeng 2009; Wang and Zeng 2010a, b; Yang et al. 2010; Yin et al. 2010; Xie et al. 2011c, d; Wang et al. 2011). To implement the APO algorithm, the following steps need to be performed (Xie et al. 2009a; Biswas et al. 2013):

- Initialization step: At this step, a swarm of individuals is randomly generated in the *n*-dimensional decision space.

- Force calculation step: At this step, according to the masses and distances between individual particles, the total force exerted on each particle is computed. In APO, the mass is defined via Eq. 24.1 (Xie et al. 2009a; Biswas et al. 2013):

$$mass_i = e^{\frac{f(x_{best}) - f(x_i)}{f(x_{worst}) - f(x_{best})}}. \tag{24.1}$$

The force is then calculated through Eq. 24.2 (Xie et al. 2009a; Biswas et al. 2013):

$$F_{ij,k} = \begin{cases} G \cdot m_i \cdot m_j (x_{j,k} - x_{i,k}) & \text{if } f(X_j) < f(X_i) \\ -G \cdot m_i \cdot m_j (x_{j,k} - x_{i,k}) & \text{if } f(X_j) \geq f(X_i) \end{cases}, \quad \forall i \neq j \text{ and } i \neq best. \tag{24.2}$$

The kth component of the total force $F_{i,k}$ exerted on individual i by all other individuals is acquired via Eq. 24.3 (Xie et al. 2009a; Biswas et al. 2013):

$$F_{i,k} = \sum_{j=1}^{N_{pop}} F_{ij,k} \forall i \neq best. \tag{24.3}$$

- Motion step: In APO, motion is used to indicate the movement of individuals across the decision space. The velocity and coordinates of an individual i at time $t+1$ are calculated via Eq. 24.3 (Xie et al. 2009a; Biswas et al. 2013):

$$\begin{aligned} v_{i,k}(t+1) &= w v_{i,k}(t) + \lambda \frac{F_{i,k}}{m_i}, \quad \forall i \neq best \\ x_{i,k}(t+1) &= x_{i,k}(t) + v_{i,k}(t+1), \quad \forall i \neq best \end{aligned} \tag{24.4}$$

where the kth components of individual i's velocity and coordinates at t iteration is denoted by $v_{i,k}(t)$ and $x_{i,k}(t)$, respectively.

24.2.2 Performance of APO

In order to show how the APO algorithm performs, Xie et al. (2009a) used 8 benchmark test functions, such as Quadric function, Sphere function, Rastrigin function, and Rosenbrock function. Compared with other CI techniques (e.g., PSO, DE, EA, etc.), the performance of APO is very competitive.

24.3 Atmosphere Clouds Model Optimization Algorithm

In this section, we will introduce an emerging CI algorithm that is based on the behaviour of cloud in nature (Wang 2013).

24.3.1 *Fundamentals of Atmosphere Clouds Model Optimization Algorithm*

Atmosphere clouds model optimization (ACMO) algorithm was originally proposed in Yan and Hao (2012). To implement the ACMO algorithm, the following steps need to be performed (Yan and Hao 2012, 2013):

- Initialization phase: At this stage, the whole search space U is split into M disjoint regions according to Eq. 24.5 (Yan and Hao 2012, 2013):

$$I_i = \frac{u_i - l_i}{M}, \quad i = 1, 2, \ldots, D, \tag{24.5}$$

  where I_i denotes the length of interval in the ith dimension, and the upper- and lower-boundary of the ith dimension is indicated by u_i and l_i, respectively.
- Cloud generation phase: In ACMO, the normal could model is adopted to describe the concept of cloud. In order to generate the cloud, three parameters, namely, region, entropy, and the number of droplets need to be determined by Eqs. 24.6–24.8, respectively (Yan and Hao 2012, 2013):

$$H_t = H_{\min} + \lambda(H_{\max} - H_{\min}), \tag{24.6}$$

$$EnM_i = \frac{I_i/M}{A}, \tag{24.7}$$

$$nMax = N - \sum_{i=1}^{C_m} n_i, \tag{24.8}$$

  where the minimum and maximum humidity values of the whole search space is expressed by $H_{\min}$ and $H_{\max}$, respectively, C_m stands for the existing cloud number, and the number of droplets found in cloud C_i is denoted by n_i.
- Cloud movement phase: The moving speed of cloud is calculate through Eqs. 24.9–24.11, respectively (Yan and Hao 2012, 2013):

$$V_{C_i} = e \cdot 6 \cdot Enc_{C_i}, \tag{24.9}$$

$$e = \frac{(1-\beta) \cdot V_{C_i} + \beta \cdot \left(x_B^* - Centre_{C_i}\right)}{\|(1-\beta) \cdot V_{C_i} + \beta \cdot (x_B^* - Centre_{C_i})\|}, \tag{24.10}$$

$$\beta = \frac{\Delta p}{pMax - pMin}, \tag{24.11}$$

where β stands for the air pressure factor, and the minimum and maximum air pressure differences of the search space are expressed by *pMax* and *pMin*, respectively, and the location with the best fit value in region B is indicated by x_B^*.

- Cloud spreading behaviour: The spreading velocity of cloud is computed via Eqs. 24.12 and 24.13, respectively (Yan and Hao 2012, 2013):

$$En_{C_i} = En_{C_i} + \alpha En_{C_i}, \tag{24.12}$$

$$\alpha = \frac{\Delta p}{pMax}, \tag{24.13}$$

where α stands for a spreading factor.

24.3.2 Performance of ACMO

In order to show how the ACMO algorithm performs (Yan and Hao 2012), used 4 benchmark test functions, namely, Schaffer function, Rastrigin function, and Needle in a haystack function. Compared with other CI techniques (e.g., PSO and GA), the ACMO algorithm can avoid premature convergence and the performance is thus very comparable.

24.4 Chaos Optimization Algorithm

In this section, we will introduce an emerging CI algorithm that is based on some properties of chaos.

24.4.1 Fundamentals of Chaos Optimization Algorithm

Chaos optimization algorithm (ChOA) was originally proposed by Li and Jiang (1998). Several ChOA applications and variants can also be found in the literature (Lu et al. 2006; Tavazoei and Haeri 2007a, b; Yang et al. 2007, 2012; Han and Lu 2008; Cheshomi et al. 2010; Henao 2011; Jiang et al. 2012; Hamaizia et al. 2012; Yuan et al. 2012; Bouras and Syam 2013; Shayeghi et al. 2009). To implement the ChOA algorithm, the following steps need to be performed (Li and Jiang 1998; Cheshomi et al. 2010):

- Step 1: Generation i chaos variables where the i chaotic states can be calculated by Eq. 24.14 (Li and Jiang 1998; Cheshomi et al. 2010):

$$x_{n+1} = 4x_n(1 - x_n). \tag{24.14}$$

- Step 2: By using the carrier wave approach, changing i optimization variables to chaos variables. Then, amplifying the i chaotic variables' the ergodic areas to the variance ranges of optimization variables through Eq. 24.15 (Li and Jiang 1998; Cheshomi et al. 2010):

$$x_i'(n+1) = c_i + d_i x_i(n+1), \tag{24.15}$$

  where $x_i(n+1)$ indicates the i chaotic states obtained through previous step.
- Step 3: Performing rough search and calculate the value of the objective function.
- Step 4: Starting the second round of carrier wave through Eq. 24.16 (Li and Jiang 1998; Cheshomi et al. 2010):

$$x_i''(n+1) = x_i^* + \alpha x_i(n+1), \tag{24.16}$$

  where the best solution found so far is denoted by x_i^*, and $\alpha x_i(n+1)$ is used to generate i chaotic states with small ergodic ranges around x_i^*
- Step 5: Fining search procedure in which let $x_i = x_i''(n+1)$, and calculate the objective function value.
- Step 6: Terminating the search process if the stopping criterion is met, and use x^* and f^* as the best solution.

24.4.2 *Performance of ChOA*

In order to show how the ChOA performs, Li and Jiang (1998) used 5 benchmark test functions, such as Quadric function, Sphere function, Rastrigin function, and Rosenbrock function. Compared with other CI techniques (e.g., SA, GA, etc.), the performance of ChOA is very competitive. Through the use of come innovative concepts such as ergodicity, stochastic properties, and regularity of chaos states, COA is a powerful stochastic optimization algorithm candidate which make it a new and efficient method in dealing with complex optimization problems.

24.5 Cloud Model-based Algorithm

In this section, we will introduce an emerging CI algorithm that is used to simulate the model of could observed in nature (Wang 2013).

24.5.1 Fundamentals of Cloud Model-based Algorithm

Cloud model-based algorithm (CMBA) was recently proposed in Wang et al. (2012); Zhu and Ni (2012); Sun et al. (2012); Zhang et al. (2008). To implement the CMBA algorithm, the following steps need to be performed (Sun et al. 2012; Zhu and Ni 2012):

- Cloud model and cloud drop: In CMBA, the distribution of x in domain is referred to as cloud model (or cloud in short) and each x is thus called a cloud drop which can be expresses via Eq. 24.17 (Sun et al. 2012; Zhu and Ni 2012):

$$\mu : U \to [0,1], \quad \forall x \in U, x \to \mu(x), \tag{24.17}$$

  where U stands for a quantity domain expressed with accurate numbers, x denotes a random realization of the quality concept, and $\mu(x)$ indicates the membership degree of x.
- Cloud model's numerical characteristics: In CMBA, the variables of expectation (E_x), entropy (E_n), and hyper-entropy (H_e) are used to express the cloud model's numerical characteristics. By setting up these variables according to the actual situation, the cloud drops of x and membership are given through Eqs. 24.18–24.20, respectively (Sun et al. 2012; Zhu and Ni 2012):

$$E'_n = G(E_n, H_e), \tag{24.18}$$

$$x_i = G\left(E_x, E'_n\right), \tag{24.19}$$

$$\mu_i = e^{-\frac{(x_i - E_x)^2}{2(E'_{ni})^2}}, \tag{24.20}$$

  where E_x indicates the cloud drops' distribution in domain, E_n denotes not only the fuzziness of the concept but also the randomness and their relationships, and H_e represents the coagulation of uncertainty of all points.
- Cloud generator: In CMBA, two sub-generators, namely, backward cloud generator and forward cloud generator are designed based on the cloud production and direction computing mechanism. According the values of three variables (i.e., E_x, E_n, and H_e), the forward cloud generator can create the cloud drops (x, μ), and the backward cloud generator can convert quantity values to a quality concept.

24.5.2 Performance of CMBA

In order to show how the CMBA performs, Zhu and Ni (2012) hybridized CMBA with DE algorithm (referred to as CMDE) and used 9 benchmark test functions, such as Sphere function, Griewangk's function, Rosenbrock function, and Rastrigin function. Compared with other DE variants (e.g., opposition-based DE, Self-adapting DE, etc.), the performance of CMDE algorithm is very competitive. The results showed that CMDE shows better convergence in terms of the rate and the reliability on both unimodal and multimodal benchmark test functions.

24.6 Extremal Optimization Algorithm

In this section, we will introduce an emerging CI algorithm that is based on Bak-Sneppen model and simulating far-from equilibrium dynamics in statistical physics (Serway and Jewett 2014; Holzner 2010, 2011; Bauer and Westfall 2011).

24.6.1 Fundamentals of Extremal Optimization Algorithm

Extremal optimization (EO) algorithm was originally proposed in Boettcher (2005), Boettcher and Percus (2000, 2004), Hartmann and Rieger (2004). To implement the EO algorithm, the following steps need to be performed (Boettcher 2004, 2005; Boettcher and Percus 2000, 2004; Chen and Lu 2008):

- Creating a solution $S = (x_1, x_2, \ldots, x_n)$ in a random manner and letting the optimal solution $S_{best} = S$.
- For the current solution S: First, evaluating the fitness degree for each component, $i \in (1, 2, \ldots, n)$; Second, ranking all the components according to their fitness values and finding the component with the "worst fitness", i.e., $\lambda_j \leq \lambda_i$ for all i; Third, selecting one solution S' in the neighbourhood of S; Fourth, setting $S = S'$ unconditionally; Fifth, if $C(S) < C(S_{best})$, i.e., the value of current cost function is less than the value of so-far minimum cost function, then let $S_{best} = S$.
- Repeating previous step for user-defined times.
- Outputting S_{best} and $C(S_{best})$, respectively.

24.6.2 Performance of EO

In order to show how the EO algorithm performs, several benchmark test graphs, such as Hammond, Barth5, Brack2, and Ocean were employed in Boettcher and Percus (2000) Compared with other CI algorithms (e.g., SA, GA), the experimental results demonstrated the competitiveness of EO algorithm over a large variety of graphs.

24.7 Galaxy-based Search Algorithm

In this section, we will introduce an emerging CI algorithm that resembles some features of galaxy (e.g., spiral arms) (Vakoch 2014). Informally, the galaxy is an island of stars and gas that swirl in spiral arms around a centre. The galaxy that we are living in is called Milky Way and all stars that can be seen with our naked eyes belong to this galaxy. The universe, as observed via telescopes, is full of galaxies (the estimated number can go up to 200 billion of them) (Sasselov 2012; Brekke 2012).

24.7.1 Fundamentals of Galaxy-based Search Algorithm

Galaxy-based search algorithm (GbSA) was originally proposed by Shah-Hosseini (2011a). To implement the GbSA algorithm, the following steps need to be performed (Shah-Hosseini 2011a, b):

- Generating initial solution: In GbSA, the minimum and the maximum gray-level is assumed to be 0 and 255, respectively. The initial solution can be computed via Eq. 24.21 (Shah-Hosseini 2011a, b):

$$S_i = 1 + \left\lfloor i \cdot \frac{253}{L} \right\rfloor, \quad i = (1, 2, \ldots, L), \tag{24.21}$$

  where the number of thresholds is denoted by L, and S_i represents the value of threshold i.
- Local search procedure: In GbSA, a modified hill-climbing mechanism is introduced to search the space around the given solution S with small step sizes.
- Spiral chaotic move phase: In GbSA, the spiral chaotic move is used to do global searching. At this stage, each component S_i of S is revised by Eq. 24.22 (Shah-Hosseini 2011a, b):

$$SNext_i \leftarrow S_i \pm NextChaos() \cdot r \cdot \cos(\theta_i), \tag{24.22}$$

  where $SNext_i$ stands for the ith component of the next solution $SNext$, which is on the arm of the spiral galaxy having core S, and $NextChaos()$ generates a chaotic number between [0,1], which is obtained through the logistic map according to Eq. 24.23 (Shah-Hosseini 2011a, b):

$$x_{n+1} = \lambda x_n (1 - x_n), \quad n = 0,\ 1,\ 2, \ldots. \tag{24.23}$$

24.7.2 Performance of GbSA

In order to show how the GbSA algorithm performs, several benchmark image processing problems, namely, Lena, peppers, and baboon, were employed in Shah-Hosseini (2011a) Compared with exhaustive search methods, the experimental results demonstrated the competitiveness of GbSA.

24.8 Gravitation Field Algorithm

In this section, we will introduce an emerging CI algorithm that is derived from a famous astronomy theory regarding planet formation (Sasselov 2012; Brekke 2012; Schutz 2003).

24.8.1 Fundamentals of Gravitation Field Algorithm

Gravitation field algorithm (GFA) was recently proposed in Zheng et al. (2010). To implement the GFA algorithm, the following steps need to be performed (Zheng et al. 2010, 2012; Rong et al. 2013):

- First, generating n dusts $d_i\,(i = 1, 2, \ldots, n)$ randomly distributed in the mass function domain $[a, b]$ to establish the initial solution space.
- Second, decomposing the solution space, and each subspace (called group in GFA) containing a centre dust which has the largest mass value, and surrounding dust.
- Third, moving dusts. The movement strategy is determined via Eq. 24.24 (Zheng et al. 2010, 2012; Rong et al. 2013):

$$Pace_i = M \cdot dis_i, \tag{24.24}$$

  where the distance between the surrounding dust and the centre dust is denoted by dis_i, and M stands for the weight value of the distance.
- Fourth, absorbing dusts according to an absorption strategy where the surrounding dust is eliminated from the initial solution space for increasing the GFA's speed.
- Fifth, checking the termination criterion. If the algorithm does not meet the stopping condition, GFA will go to the second step, otherwise, the algorithm stops.

24.8.2 Performance of GFA

In order to show how the GFA algorithm performs, Zheng et al. (2010) used 5 benchmark test functions, such as Ackley function, Griewangk function, and Rastrigin function. Compared with other CI techniques (e.g., SA, GA, etc.), the performance of GFA algorithm is very competitive. The initial application case studies (Zheng et al. 2010, 2012; Rong et al. 2013) demonstrated that GFA can handle unimodal and multimodal functions optimization effectively.

24.9 Gravitational Clustering Algorithm

In this section, we will introduce an emerging CI algorithm that is on the studies related to gravity research (Schutz 2003; Serway and Jewett 2014; Holzner 2010, 2011; Bauer and Westfall 2011).

24.9.1 Fundamentals of Gravitational Clustering Algorithm

The concept of gravitational clustering was originally proposed in Wright (1977) and the gravitational clustering algorithm (GCA) was recently proposed in Kundu (1999). In the literature, there are several variants and applications of GCA can be found (Gomez et al. 2003; Zhang and Qu 2010; Sanchez et al. 2012). To implement GCA, the following steps need to be performed (Kundu 1999):

- Step 1: Initializing the matrix $M = I$ which is an identity matrix.
- Step 2: For each P_i, performing the following tasks. First, finding a point P_j, $j \neq i$, which is located at the closest distance from it; Second, if $d(P_i, P_j) \geq \delta$, then determining the total force $\mathbf{F}_i$ exerted on P_i; Third, setting $M[i,j] = M[j,i] = 1$, given the conditions expressed by Eq. 24.25 hold (Kundu 1999).

$$\begin{cases} P_j \text{ is a nearest neighbour of } P_i, \text{ and } d(P_i, P_j) < \delta \\ \qquad\qquad \text{or} \\ \text{there is crossing between } P_i \text{ and } P_j \end{cases}. \tag{24.25}$$

- Step 3: If $(M \neq I)$, computing the transitive closure M^* and performing merging mechanism according to the equivalence classes of M^* and returning to Step 2.
- Step 4: Otherwise, each point P_i (the resultant of merging two or more points during the iterations of Step 2) corresponding to a new cluster in the cluster hierarchy. In such case, $h(P_i)$ is simply the height of any of the points in the cluster. Calculating η_i for each P_i and η_{curr} based on Eqs. 24.26 and 24.27, respectively (Kundu 1999):

$$\eta_{safe} = \min\{\eta_i\} \text{ (over all } i). \tag{24.26}$$

$$\eta_{curr} = \min\{\eta_{safe}, \eta_{\max}\}. \tag{24.27}$$

- Step 5: For each point P_i, calculating its new location $P_i + \eta_{curr}\mathbf{F}_i$ and setting $h(P_i) = h(P_i) + \eta_{curr}$.

24.9.2 Performance of GCA

In order to show how the GCA performs, Kundu (1999) employed several benchmark test functions. Compared with other CI techniques (e.g., fuzzy C-means), the performance of GCA algorithm is very competitive, in particular, it produced a compete cluster-hierarchy in $O(N^3)$ time for N points.

24.10 Gravitational Emulation Local Search Algorithm

In this section, we will introduce an emerging CI algorithm that is based on the studies related to gravity research (Schutz 2003; Serway and Jewett 2014; Holzner 2010, 2011; Bauer and Westfall 2011).

24.10.1 Fundamentals of Gravitational Emulation Local Search Algorithm

Gravitational emulation local search (GELS) algorithm was originally proposed in Webster (2004). In the literature, there are several variants and applications of GELS can be found (Balachandar and Kannan 2007, 2009, 2010; Barzegar et al. 2009; Pooranian et al. 2011). To implement the GELS algorithm, the following steps need to be performed (Webster 2004; Balachandar and Kannan 2007):

- Defining key parameters used in GELS algorithm: *Max velocity*, *Radius*, *Iterations*, and Pointer.
- Calculating gravitational force: In GELS, the gravitational force between two candidate solution is computed based on Eq. 24.28 (Webster 2004; Balachandar and Kannan 2007):

$$F = \frac{G(CU - CA)}{R^2}, \tag{24.28}$$

  where G is a universal gravitation constant and normally equals to 6.672, CU and CA denotes the current and candidate solutions' objective function value, respectively, and the value of radius parameter is represented by R.
- Webster (2004) also introduced two methods and two stepping modes which are named as GELS 11, GELS 12, GELS 21, and GELS 22, respectively.

24.10.2 Performance of GELS

In order to show how the GELS algorithm performs, Webster (2004) employed several benchmark test problems, such as travelling salesman problem, bin packing problem, and file assignment problem. Compared with other CI techniques (e.g., SA, GA), the performance of the GELS algorithm is very competitive in terms of reliability and usability.

24.11 Gravitational Interactions Optimization Algorithm

In this section, we will introduce an emerging CI algorithm that is derived from Newton's gravity theory (Schutz 2003; Serway and Jewett 2014; Holzner 2010, 2011; Bauer and Westfall 2011).

24.11.1 Fundamentals of Gravitational Interactions Optimization Algorithm

Gravitational interactions optimization (GIO) algorithm was recently proposed in (Flores et al. 2011) in which the interactions exhibited by a set of bodies were used to guide the search for the global optimum in an optimization problem. To implement the GIO algorithm, the following steps need to be performed (Flores et al. 2011):

- In GIO, the fitness function is regarded as a mapping which transforms a vector $X = (x_1, x_2, \ldots, x_n)$ to a scalar $f(X)$. The fitness value $f(X)$ is associated by this mapping to each location $X = (x_1, x_2, \ldots, x_n)$ of the search space. A body B is then allocated to each location X in the search space where an individual of the population is discovered. The attracting force that exists between two bodies with masses is thus calculate through Eq. 24.29 (Flores et al. 2011):

$$F_{ij} = \frac{M(f(B_i)) \cdot M\left(f\left(B_j\right)\right)}{\left|B_i - B_j\right|^2} \hat{B}_{ij}, \tag{24.29}$$

  where B_i stands for the ith body's position, B_j denotes the jth body that contributes causing a force on the mass B_i, and M is the mapping function which can be calculate via Eq. 24.30 (Flores et al. 2011):

$$M(f(B_i)) = \left(\frac{f(B_i) - \min f(B)}{\max f(B) - \min f(B)}(1 - mapMin) + mapMin\right)^2, \tag{24.30}$$

where the minimum and maximum fitness value of the positions of the bodies at present are denoted by $\min f(B)$ and $\max f(B)$, respectively, and *mapMin* represents a small positive constant whose value is near zero.

- One feature of GIO lies in its full interaction mechanism which means each body B_i interacts with every other body B_j through their masses. Bearing this in mind, the resulting force exerted on body B_i by bodies B_j is thus calculated by Eq. 24.31 (Flores et al. 2011):

$$F_i = \sum_{j=1}^{n} \frac{M(f(B_i)) \cdot M\left(f\left(B_j^b\right)\right)}{\left|B_i - B_j^b\right|^2} \vec{B_i B_j^b}, \tag{24.31}$$

where the resulting force of the sum of all vector forces is denoted by F_i, and the Euclidean distance between B_i's current positions and the best position so far of the body B_j is represented by $\left|B_i - B_j^b\right|$. Suppose that we are willing to find a location of the body B with $M(f(B)) = 1$, B can be computed through Eq. 24.32 (Flores et al. 2011):

$$B = \sqrt{\frac{M(f(B_i))}{|F_i|}} \hat{F}_i. \tag{24.32}$$

- To updating the position of the bodies, we can use the Eq. 24.33 (Flores et al. 2011):

$$\begin{aligned} V_{t+1} &= \chi(V + R \cdot C \cdot B), \\ B_{t+1} &= B + V_{t+1}, \end{aligned} \tag{24.33}$$

where the present speed of B_i is denoted by V, R represents a random real number which falls within [0, 1), and C stands for the gravitational interaction coefficient.

24.11.2 Performance of GIO

In order to show how the GIO algorithm performs, Flores et al. (2011) used 3 unimodal and 4 multimodal benchmark test functions, such as Goldstein and Price function, booth function, four-variable Colville function, Deb's function, Himmelblau's function, and six-hump camelback function. Compared with other CI techniques (e.g., PSO, GSA), the performance of GIO is very competitive in terms of reliability and usability.

24.12 Hysteretic Optimization Algorithm

In this section, we will introduce an emerging CI algorithm that is based on the findings derived from the magnetism research (Schutz 2003; Serway and Jewett 2014; Holzner 2011; Bauer and Westfall 2011).

24.12.1 Fundamentals of Hysteretic Optimization Algorithm

Hysteretic optimization (HO) algorithm was originally proposed in Zaránd et al. (2002), Pál (2003, 2004, 2006a, b). To implement the HO algorithm, the following steps need to be performed (Zaránd et al. 2002; Gonçalves and Boettcher 2008; Pál 2004):

- Step 1: Setting $H = H_1$ large enough so that $S_i = \xi_i \forall i$ and letting $E_{\{\min\}} = \mathrm{H}(= \mathrm{H}_{HO}|_{H=0})$.
- Step 2: Reducing H until one spin turns to unstable and then allowing the system to relax. If $\mathrm{H} < E_{\{\min\}}$, let $E_{\{\min\}} = \mathrm{H}$.
- Step 3: In HO, this is an optional step in which when H passes zero, randomizing ξ_i and leaving the current configuration stable.
- Step 4: At each turning point $H = H_n = {}_{-\gamma_{n-1}}H_{n-1}$, for $0 < \gamma_n < 1$, changing the direction of H oppositely.
- Step 5: Terminating the algorithm when the amplitude $|H_n| < H_{\{\min\}}$.
- Step 6: Restarting the algorithm from Step 1 for N_{run} times with a new and stochastic set of ξ_i's.
- Step 7: Outputting the best $E_{\{\min\}}$ over all runs.

24.12.2 Performance of HO

Zaránd et al. (2002) tested the proposed HO algorithm on two benchmark problems, namely, frustrated magnetic models and the travelling salesman problem. In comparison with other CI techniques (e.g., SA), the HO algorithm showed a very promising performance.

24.13 Integrated Radiation Optimization Algorithm

In this section, we will introduce an emerging CI algorithm that is derived from Einstein's general theory of relativity (Schutz 2003; Serway and Jewett 2014; Holzner 2011; Bauer and Westfall 2011; Gourgoulhon 2013).

24.13.1 Fundamentals of Integrated Radiation Optimization Algorithm

Integrated radiation optimization (IRO) algorithm was originally proposed in Chuang and Jiang (2007). To implement the IRO algorithm, the following steps need to be performed (Chuang and Jiang 2007):

- In IRO, the massive binary star systems which shifts moving through the universe is used to model the role of search agents in search space. Based on this model, the solution qualities of a population of search agents can be further accumulated to form a stronger source of gravitational radiation. This procedure can be expressed by Eq. 24.34 (Chuang and Jiang 2007):

$$P_{GR} = \frac{dE}{dt} = -\frac{32}{\pi} \cdot \frac{G^4}{c^5} \cdot \frac{(m_1 m_2)^2 \cdot (m_1 + m_2)}{R^5}, \tag{24.34}$$

  where the power emitted by two nearby massive binary star systems is denoted by P_{GR}, c represents the light speed, R stands for the distance between two closing binary star systems, and the total masses of two binary systems are indicated by m_1 and m_2, respectively.
- Initialization: Suppose that there are m variables need to be optimized in a given optimization problem, a section of memory space is thus has be allocated. The desired size of memory can be computed through Eq. 24.35 (Chuang and Jiang 2007):

$$Memory = size(SS) = \prod_{i=1}^{m} (res_m), \tag{24.35}$$

  where *Memory* denotes the total amount of memory required to be allocated for building the search space which is represented by *SS*, and res_m stands for the grid numbers in mth dimension of the search space. In IRO, the costs of all search agents are treated as the spatial parameters. The search agents are then planted into the corresponding location in search space according to Eq. 24.36 (Chuang and Jiang 2007):

$$SS(idx(P_i)) = F(P_i), \tag{24.36}$$

  where F is used to refer the cost function of a target problem, P_i denotes the position of a search agent, and such location in the search space is represented by $idx(P_i)$.
- Reprocessing: The geometry of gravity field and the strength of gravitational radiation are estimated at this stage according to Eq. 24.37 (Chuang and Jiang 2007):

$$GR = SS \cdot Gauss$$
$$= -\sum_{k_1=-\infty}^{\infty} \cdots \sum_{k_m=-\infty}^{\infty} Gauss[(x_1 - k_1), \ldots, (x_m - k_m)] \cdot SS(k_1, \ldots, k_m) \tag{24.37}$$

where GR stands for the approximated power of gravitational radiation, and $Gauss$ represent a Gaussian distribution which can be calculated via Eq. 24.38 (Chuang and Jiang 2007):

$$Gauss(x_1, \ldots, x_m) = A \cdot e^{-\left(\sum_{i=1}^{m} \left(\frac{x_i - x_0}{\sigma}\right)^2\right)}, \tag{24.38}$$

where A denotes the amplitude, the centre of the distribution is represented by x_0, and the blob spreading in every dimension is indicated by σ.

- Ranking and movement: In IRO, the percentiles of each search agent in $SS(idx(P_i))$ and $GR(idx(P_i))$ are computed via Eq. 24.39 (Chuang and Jiang 2007):

$$\begin{aligned} {}_{SS}Pct_i &= percentile[SS(idx(P_i))] \\ {}_{GR}Pct_i &= percentile[GR(idx(P_i))], \end{aligned} \tag{24.39}$$

where ${}_{SS}Pct_i$ and ${}_{GR}Pct_i$ are used to denote the percentiles of P_i in $SS(idx(P_i))$ and $GR(idx(P_i))$, respectively. The displacement of a search agent is dominated by the gravitational radiation emitted from other search agents which can be calculated through Eq. 24.40 (Chuang and Jiang 2007):

$$Disp(P_i) = \sum_{\forall j, j \neq i}^{m} \frac{{}_{SS}Pct_j}{\|\mathbf{u_{ij}}\|^2} \cdot \frac{{}_{GR}Pct_j}{\|\mathbf{u_{ij}}\|^2} \cdot (\mathbf{rnd_j} \cdot \vec{u}_{ij}), \tag{24.40}$$

where $Disp(P_i)$ denotes the displacement of the ith search agent during an iteration, $\vec{u}_{ij}$ represents the unit vector of the geometric vector, and $\mathbf{rnd_j}$ indicates a $1 \times j$ random vector with each element usually falling within [0, 1]. At this stage, each search agent renew its location P_i according to Eq. 24.41 (Chuang and Jiang 2007):

$$P_i = P_i + Disp(P_i). \tag{24.41}$$

24.13.2 Performance of IRO

Chuang and Jiang (2007) employed 2D fixed static polynomials function and controller design optimization problem as benchmarks to test the performance of the IRO algorithm. In comparison with other CI techniques (e.g., FS, ACO), the IRO algorithm showed a very promising performance. Overall, through the introduction of random vectors into IRO, a stochastic exploration capability is thus enabled which can help IRO getting out of the trap of local optimum.

24.14 Light Ray Optimization Algorithm

In this section, we will introduce an emerging CI algorithm that is derived from ray optics research, in particular the famous Fermat's principle that is whenever a beam of light travels from one position to another, its actually selected path is the path that needs the shortest time interval (Serway and Jewett 2014; Bauer and Westfall 2011).

24.14.1 Fundamentals of Light Ray Optimization Algorithm

Light ray optimization (LRO) algorithm was recently proposed in Shen and Li (2008, 2009, 2010, 2012); Shen et al. (2012). To implement the IRO algorithm, the following steps need to be performed (Shen and Li 2010; Shen et al. 2012):

- Step 1: Dividing the search space by selecting a vertical and a horizontal grid length, respectively.
- Step 2: Setting the objective function value at some point in the division as the speed of light rays that travel through such division.
- Step 3: Initializing the value of initial point $X^{(0)}$ and initial vector $P^{(0)}$.
- Step 4: Calculating the next iteration point.
- Step 5: Evaluating if the termination criterion is met. If yes, goes to Step 7; otherwise, goes to next step, i.e., Step 6.
- Step 6: Computing the refraction and reflection condition according to Eq. 24.42 (Shen and Li 2010; Shen et al. 2012):

$$\begin{cases} \text{Refraction:} \quad \text{if} \quad \frac{v_{i+1}}{v_i} \sin\alpha_i \leq 1 \\ \text{Reflection:} \quad \text{if} \quad \frac{v_{i+1}}{v_i} \sin\alpha_i > 1 \end{cases}, \tag{24.42}$$

  where α_i denotes the angle of incidence in D_i, v_i represents the propagation velocity of light in D_i. If the condition of total reflection is met, calculating the next searching direction based on reflection law; otherwise computing the next searching direction based on refraction law and goes to Step 4.
- Step 7: Terminating the optimal search and generating the extreme value.

24.14.2 Performance of LRO

In order to show how the LRO algorithm performs, Shen et al. (2012) used 7 benchmark test functions, such as Sphere function, Rosenbrock function, Goldstein and Price function, and six-hump camelback function. The experimental results showed that LRO is very effective. The theoretical analysis also proved that it is a competitive global optimization algorithm.

24.15 Magnetic Optimization Algorithm

In this section, we will introduce an emerging CI algorithm that is based on the studies related to magnetic research (Serway and Jewett 2014; Bauer and Westfall 2011; Placidi 2012; Holzner 2010; Arfken et al. 2013).

24.15.1 Fundamentals of Magnetic Optimization Algorithm

Magnetic optimization algorithm (MOA) was originally proposed in Tayarani et al. (2008). To implement the MOA algorithm, the following steps need to be performed (Tayarani et al. 2008):

- Initializing the population of particles according to Eq. 24.43 (Tayarani et al. 2008):

$$x_{ij,k}^{t} = R(l_k, u_k), \tag{24.43}$$

  where i and j denotes the particle location in the lattice, t represents the number of iterations, and the lower bound and the upper bound of the kth dimension of search space are indicated by l_k and u_k, respectively.
- Terminating the while loop when the stopping criterion is met.
- Calculating the objective of each particles x_{ij}^{t} in X^t. The values are then kept in the magnetic field denoted by B_{ij}^{t}.
- Performing normalization mechanism on B^t which is defined by Eq. 24.44 (Tayarani et al. 2008):

$$B_{ij} = \frac{B_{ij} - \min}{\max - \min}, \tag{24.44}$$

  where $\min = \underset{i,j=1}{\overset{S}{\text{minimum}}}\left(B_{ij}^{t}\right)$, and $\max = \underset{i,j=1}{\overset{S}{\text{maximum}}}\left(B_{ij}^{t}\right)$.
- Computing the mass of all particles and keep the values in M^t based on Eq. 24.45 (Tayarani et al. 2008):

$$M_{ij}^{t} = \alpha + \rho \cdot B_{ij}^{t}, \tag{24.45}$$

where α and ρ denotes the constant values.

- Calculating the resultant force of all forces exerted on each particle.
- Setting the initial value of the resultant force that is applied to particle $x_{ij}^t (F_{ij})$ to zero.
- In MOA, each particle interact only with its neighbours in the lattice-like environment. The set of neighbours for particle x_{ij} is thus defined by Eq. 24.46 (Tayarani et al. 2008):

$$N_{ij} = \{x_{i'j}, x_{ij'}, x_{i''j}, x_{ij''}\}. \tag{24.46}$$

- Computing the force that is applied to a particle x_{ij}^t by its neighbour, x_{uv}^t.
- Since the force exerted on x_{ij}^t by x_{uv}^t is related to the distance between two particles, it is thus can be computed according to Eq. 24.47 (Tayarani et al. 2008):

$$F_{ij,k} = \frac{\left(x_{uv,k}^t - x_{ij,k}^t\right) \cdot B_{uv}^t}{D\left(x_{ij,k}^t, x_{uv,k}^t\right)}. \tag{24.47}$$

- Calculating the speed and the movement of each particle through Eqs. 24.48 and 24.49, respectively (Tayarani et al. 2008):

$$v_{ij,k}^{t+1} = \frac{F_{ij,k}}{M_{ij,k}} \cdot R(l_k, u_k), \tag{24.48}$$

$$x_{ij,k}^{t+1} = x_{ij,k}^t + v_{ij,k}^{t+1}. \tag{24.49}$$

- Taking into account of acceleration for particle defined by Eqs. 24.50–24.52, respectively (Tayarani et al. 2008):

$$a_{ij,k}^{t+1} = \frac{F_{ij,k}}{M_{ij,k}} \cdot R(l_k, u_k), \tag{24.50}$$

$$v_{ij,k}^{t+1} = v_{ij,k}^t + a_{ij,k}^{t+1}, \tag{24.51}$$

$$x_{ij,k}^{t+1} = x_{ij,k}^t + v_{ij,k}^{t+1}. \tag{24.52}$$

24.15.2 Performance of MOA

Tayarani et al. (2008) employed a set benchmark functions to test the performance of MOA, such as Schwefel function, Rastrigin function, Michalewicz function, Goldberg function, De Jong function, Rosenbrock function, Kennedy function, Ackley function, and Griewank function. In comparison with other CI techniques (e.g., PSO and GA), MOA showed a better performance. MOA is viewed as similar to PSO, however, Tayarani et al. (2008) pointed out that the main difference between both is the interaction of the particles, i.e., the interaction between the particles in MOA is higher then PSO. In other words, MOA has a better swarm intelligence then PSO.

24.16 Particle Collision Algorithm

In this section, we will introduce an emerging CI algorithm that is on some findings related to nuclear physics research (Serway and Jewett 2014; Bauer and Westfall 2011; Holzner 2010; Particle_Data_Group 1998; Klafter et al. 2012; Bruce et al. 2014; Cacuci 2010; National_Research_Council 2012; Ahmed 2012; Waltar et al. 2012; Tsvetkov 2011; Shifman 2012; Bes 2007; Philips 2003).

24.16.1 Fundamentals of Particle Collision Algorithm

Particle collision algorithm (PCA) was recently proposed in Sacco and Oliveira (2005). In the literature, there are several variants and applications of PCA can be found (Luz et al. 2008, 2011; Abuhamdah and Ayob 2009a, b, 2011; Sacco et al. 2006). To implement the PCA algorithm, the following steps need to be performed (Sacco and Oliveira 2005):

- First, generating an initial solution *Old_Config* and creating a random perturbation of the solution according to Eq. 24.53 (Sacco and Oliveira 2005):

$$\begin{array}{c} \text{if } fitness(New\_Config) > fitness(Old\_Config) \\ \text{then } Old\_Config := New\_Config, \text{ and exploring()} \\ \text{otherwise scattering()} \end{array}. \tag{24.53}$$

- Second, performing exploring function exploring(). Generating a small random perturbation of the solution according to Eq. 24.54 (Sacco and Oliveira 2005):

$$\begin{array}{c} \text{if } fitness(New\_Config) > fitness(Old\_Config) \\ \text{then } Old\_Config := New\_Config \end{array}. \tag{24.54}$$

- Third, performing scattering function scattering() based on Eq. 24.55 (Sacco and Oliveira 2005):

$$\begin{array}{c} p_{\text{scattering}} = 1 - \dfrac{fitness(New\_Config)}{best\ fitness} \\ \text{if } p_{\text{scattering}} > random(0,1) \\ \text{then } Old\_Config := \text{random solution} \\ \text{otherwise exploring ()} \end{array}, \tag{24.55}$$

where the scattering probability, $p_{\text{scattering}}$, is inversely proportional to it quality.

24.16.2 Performance of PCA

In order to test the performance of PCA, some benchmark test functions, such as Easom function, Rosenbrock function, and De Jong function were employed in Sacco and Oliveira (2005). Furthermore, Sacco and Oliveira (2005) also applied PCA to nuclear reactor design optimization problem. In comparison with other CI techniques (e.g., GA), the results obtained by PCA is very promising. One of the unique characteristics of PCA lies in that it does not require user-determined variables, similarly to the annealing scheduling in SA, which make it very suitable for solving continuous or discrete optimization problems.

24.17 Ray Optimization Algorithm

In this section, we will introduce an emerging CI algorithm that is also derived from ray optics research, in particular the famous Snell's law of refraction that is the light of different wavelengths is refracted at different angles when incident on a material (Serway and Jewett 2014; Bauer and Westfall 2011; Holzner 2010).

24.17.1 Fundamentals of Ray Optimization Algorithm

Ray optimization (RO) algorithm was recently proposed in (Kaveh and Khayatazad 2012). To implement the RO algorithm, the following steps need to be performed (Kaveh and Khayatazad 2012):

- Step 1: scattering and evaluating. At this step, the agents have to be randomly distributed in the search space according to Eq. 24.56 (Kaveh and Khayatazad 2012):

$$\mathbf{X}_{ij} = \mathbf{X}_{j,\min} + rand \cdot \left(\mathbf{X}_{j,\max} - \mathbf{X}_{j,\min}\right), \tag{24.56}$$

  where $\mathbf{X}_{ij}$ denotes the *j*th variable of the *i*th agent, $\mathbf{X}_{j,\min}$ and $\mathbf{X}_{j,\max}$ are used to represent the minimum and maximum limits of the *j*th variable, and *rand* stands for a random number which falls within the range of [0, 1].
- Step 2: movement vector and motion refinement. For each agent in RO, a group of movement vectors will be allocated to it according to its division. In order to deal with the possibility of boundary violation scenario, the motion refinement mechanism is also introduced in RO.
- Step 3: making origin and converging. In RO, a point called origin is defined by Eq. 24.57 (Kaveh and Khayatazad 2012):

$$\mathbf{O}_i^k = \frac{(ite + k) \cdot \mathbf{GB} + (ite - k) \cdot \mathbf{LB}_i}{2 \cdot ite}, \tag{24.57}$$

where the origin of the ith agent for the kth iteration is represented by $\mathbf{O}_i^k$, ite denotes the total iteration number for the optimization process, and the global best and local best of the ith agent is indicated by **GB** and $\mathbf{LB}_i$ respectively.

- Step 4: completing or redoing. In RO, different stopping criterion are considered for ceasing the searching procedure. Some of these criterion include such as maximum iteration number, number of ineffective iteration, and approaching to a minimum goal function error.

24.17.2 Performance of RO

In order to validate the goodness of RO, some benchmark test functions, such as Rastrigin function, De Joung function, Branin function, and Griewank function were employed in Kaveh and Khayatazad (2012). Furthermore, Kaveh and Khayatazad (2012) also included some benchmark engineering design problems in their study. In comparison with other CI techniques (e.g., PSO, GA), the RO algorithm showed a very good efficiency which make it a very competitive optimization algorithm.

24.18 River Formation Dynamics Algorithm

In this section, we will introduce an emerging CI algorithm that is based on the observation on the behaviour of water forming rivers by eroding the ground and depositing sediments (Samuels et al. 2009; Rahman 2011).

24.18.1 Fundamentals of River Formation Dynamics Algorithm

River formation dynamics (RFD) algorithm was originally proposed in Rabanal et al. (2007). Several RFD applications and variants can also be found in the literature (Rabanal and Rodríguez 2011; Rabanal et al. 2007, 2008a, b, 2009, 2010, 2011, 2013; Gupta et al. 2011; Afaq and Saini 2011). To implement the RFD algorithm, the following steps need to be performed (Rabanal et al. 2007, 2011, 2008a, b, 2009, 2010, 2013; Rabanal and Rodríguez 2011):

- At the beginning, a flat environment is provided (i.e., there is the same altitude).

- In the main loop, gradients are modified, which in turn affects movements of subsequent drops and reinforce the best ones. The following transition rule (See Eq. 24.58) defines the probability that a drop k at a node i chooses the node j to move next (Rabanal et al. 2007, 2008a, b, 2009, 2010, 2011, 2013; Rabanal and Rodríguez 2011):

$$P_k(i,j) = \begin{cases} \frac{\text{decreasing}Gradient(i,j)}{\sum_{l \in V_k(i)} \text{decreasing}Gradient(i,l)} & \text{if } j \in V_k(i) \\ 0 & \text{if } j \notin V_k(i) \end{cases}, \quad (24.58)$$

  where $V_k(i)$ is the set of nodes that are neighbours of node i that can be visited by the drop k and have a negative gradient (decreasing*Gradient*) between nodes i and j, which is defined by Eq. 24.59 (Rabanal et al. 2007, 2008a, b, 2009, 2010, 2011, 2013; Rabanal and Rodríguez 2011):

$$decreasingGradient(i,j) = \frac{altitude(i) - altitude(j)}{distance(i,j)}, \quad (24.59)$$

  where $altitude(x)$ stands for the altitude of the node x and distance(i,j) denotes the length of the edge connecting node i and j.
- Finally, after drops transform the landscape by increasing or decreasing the altitude of places; solutions are given in the form of paths of decreasing altitudes (i.e., either all drops find the same solution, or another alternative finishing condition is satisfied).

24.18.2 Performance of RFD

Rabanal et al. (2007) employed travelling salesman problem as a benchmark to test the performance of the RFD algorithm. In comparison with other CI techniques (e.g., ACO), the RFD algorithm showed a better performance. Overall, the main merit of RFD algorithm is threefold (Rabanal et al. 2013): First, to avoid following a local cycle; second, to quick reinforce the new paths; and third, to provide a focused way to punish wrong paths through the sediment process.

24.19 Space Gravitational Optimization Algorithm

In this section, we will introduce an emerging CI algorithm that is also derived from Einstein's general theory of relativity (Schutz 2003; Serway and Jewett 2014; Holzner 2011; Bauer and Westfall 2011).

24.19.1 Fundamentals of Space Gravitational Optimization Algorithm

Space gravitational optimization (SGO) algorithm was recently proposed in Hsiao et al. (2005). To implement the SGO algorithm, the following steps need to be performed (Hsiao et al. 2005):

- Initialization: At this stage, a group of n asteroids was stochastically generated with a position $x[\,]$ and $y[\,]$. The velocities on both x and y axes are denoted by $vx[\,]$ and $vy[\,]$.
- Searching spacetime variation: In SGO, the summation of the variations in geometry of spacetime on directions of x- and y-axis is the acceleration rate for the asteroids which is defined by Eqs. 24.60 and 24.61, respectively (Hsiao et al. 2005):

$$ax[n] = G \cdot \left\{ \begin{array}{c} [f(x[n], y[n]) - f(x[n] + r_d, y[n])] \\ + \\ [f(x[n] - r_d, y[n]) - f(x[n], y[n])] \end{array} \right\}, \tag{24.60}$$

$$ay[n] = G \cdot \left\{ \begin{array}{c} [f(x[n], y[n]) - f(x[n], y[n] + r_d)] \\ + \\ [f(x[n], y[n] - r_d) - f(x[n], y[n])] \end{array} \right\}, \tag{24.61}$$

  where the acceleration rate on x- and y-axis of asteroid n are denoted by $ax[n]$ and $ay[n]$, respectively, and $f(x[n], y[n])$ represents the cost function that is employed to assess the goodness of the solution.
- Speeding up or slowing down: Once the acceleration rate of each axis is obtained, the velocity of the asteroid n can be updated based on Eq. 24.62 (Hsiao et al. 2005):

$$\begin{aligned} vx[n] &= vx[n] + ax[n] \\ vy[n] &= vy[n] + ay[n]. \end{aligned} \tag{24.62}$$

  And the position of the asteroid n in the solution space is then renewed through Eq. 24.63 (Hsiao et al. 2005):

$$\begin{aligned} x[n] &= x[n] + vx[n] \\ y[n] &= y[n] + vy[n]. \end{aligned} \tag{24.63}$$

- Updating optimal solution and checking for termination criterion: In SGO, if the solution acquired by asteroid n is better than the global optimal solution (denoted by G_{best}), the updating G_{best} according to Eq. 24.64 (Hsiao et al. 2005):

$$\text{if } f(x[n], y[n]) < G_{best}, \text{ then } \begin{cases} G_{best} = f(x[n], y[n]) \\ G_{best\_x} = x[n] \\ G_{best\_y} = y[n] \end{cases}, \tag{24.64}$$

where G_{best_x} and G_{best_y} denotes the optimal solution on x- and y-axis at that time, respectively.

24.19.2 Performance of SGO

Hsiao et al. (2005) applied the SGO algorithm to the optimal controller design problem to test its performance. In comparison with other CI techniques (e.g., FS, ACO), the results obtained by SGO algorithm is very competitive Since a simplified model for asteroids shifting in a curved spacetime was introduced in SGO, the computational complexity is considerably small and thus the possibility of a searching agent being trapped in a local minimal is largely reduced.

24.20 Spiral Optimization Algorithm

In this section, we will introduce an emerging CI algorithm that is based on spiral dynamics phenomenon found in nature (Serway and Jewett 2014; Bauer and Westfall 2011; Reece et al. 2011; Abel 2013).

24.20.1 Fundamentals of Spiral Optimization Algorithm

Spiral optimization algorithm (SpOA) was recently proposed in Jin and Tran (2010); Tamura and Yasuda (2011a, b, c). To implement the SpOA algorithm, the following steps need to be performed (Jin and Tran 2010, Tamura and Yasuda 2011a, b, c):

- Step 0: Preparation. Selecting the number of searching point $m > 2$, setting the parameters of α_i and β_i for $A_{\text{spiral}i}$ as $\sqrt{\alpha_i^2 + \beta_i^2} < 1$, and defining the maximum number of iterations $T_{\max}$.
- Step 1: Initialization. Randomly setting the original points $x_i^0 \in \mathbb{R}^2$ $(i = 1, 2, \ldots, m)$ in the feasibility region and letting the centre x^* as defined by Eq. 24.65 (Tamura and Yasuda 2011a):

$$x^* = x_{i_g}^0, \quad i_g = \arg\ \min_i f\left(x_i^0\right). \tag{24.65}$$

- Step 2: Updating x_i according to Eq. 24.66 (Tamura and Yasuda 2011a):

$$x_i^{k+1} = A_{\text{spiral}i} x_i^k - \left(A_{\text{spiral}i} - I_2\right)x^*, \quad i = 1, 2, \ldots, m. \tag{24.66}$$

- Step 3: Updating x^* based on Eq. 24.67 (Tamura and Yasuda 2011a):

$$x^* = x_{i_g}^{k+1}, \quad i_g = \arg\ \min_i f\left(x_i^{k+1}\right). \tag{24.67}$$

- Step 4: Checking whether the stopping criterion is met. If yes (i.e., $k = T_{\max}$), the algorithm stops; otherwise, let $k = k + 1$ and go back to Step 2.

24.20.2 Performance of SpOA

Tamura and Yasuda (2011a) employed 3 benchmark functions (including Rosenbrock function, 2^n minima function, and Rastrigin function) to test the performance of SpOA. In comparison with other CI techniques (e.g., PSO), the results obtained by SpOA is very competitive At the end of their study, Tamura and Yasuda (2011a) suggested that although SpOA was proposed to solve only two-dimensional problems, it could be improved from many perspectives such as x^*, α_i, and β_i tuning, introducing randomness, and extension to deal with n-dimensional problems.

24.21 Water Cycle Optimization Algorithm

In this section, we will introduce an emerging CI algorithm that is based on the observation of water cycle process in nature (Carey et al. 2008; Samuels et al. 2009; Rahman 2011; Davis 2010; Day 2007; Reece et al. 2011; Whitten et al. 2014; Zumdahl and Zumdahl 2014).

24.21.1 Fundamentals of Water Cycle Optimization Algorithm

Water cycle optimization algorithm (WCOA) was originally proposed in Eskandar et al. (2012). To implement the WCOA algorithm, the following steps need to be performed (Eskandar et al. 2012):

- Step 1: Selecting the initial parameters of WCOA: N_{sr}, $d_{\max}$, N_{pop}, max_iteration.
- Step 2: Creating initial population in a random manner and forming the initial streams, rivers, and sea using the Eqs. 24.68–24.70, respectively (Eskandar et al. 2012):

$$\text{Raindrops population} = \begin{bmatrix} \text{Raindrop}_1 \\ \text{Raindrop}_2 \\ \text{Raindrop}_3 \\ \vdots \\ \text{Raindrop}_{N_{pop}} \end{bmatrix} = \begin{bmatrix} x_1^1 & x_2^1 & x_3^1 & \cdots & x_{N_{\text{var}}}^1 \\ x_1^2 & x_2^2 & x_3^2 & \cdots & x_{N_{\text{var}}}^2 \\ \vdots & \vdots & \vdots & \ddots & \vdots \\ x_1^{N_{pop}} & x_2^{N_{pop}} & x_3^{N_{pop}} & \cdots & x_{N_{\text{var}}}^{N_{pop}} \end{bmatrix}, \tag{24.68}$$

$$N_{sr} = \text{Number of rivers} + \underbrace{1}_{\text{Sea}}, \tag{24.69}$$

$$N_{\text{Raindrops}} = N_{pop} - N_{sr}, \tag{24.70}$$

where Raindrop represents a single solution. In a N_{var} dimensional search space, an raindrop is denoted by an array of $1 \cdot N_{\text{var}}$ which can be defined as Raindrop $= [x_1, x_2, x_3, \ldots, x_N]$. N_{sr} stands for the summation of Number of rivers.

- Step 3: Computing the cost of each raindrops via Eq. 24.71 (Eskandar et al. 2012):

$$C_i = Cost_i = f\left(x_1^i, x_2^i, \ldots, x_{N_{\text{var}}}^i\right), \quad i = 1, 2, \ldots, N_{pop}, \tag{24.71}$$

where N_{pop} indicates the number of raindrops.

- Step 4: Determining the flow intensity for rivers and sea through Eq. 24.72 (Eskandar et al. 2012):

$$NS_{\text{n}} = round\left\{\left|\frac{Cost_n}{\sum_{i=1}^{N_{sr}} Cost_i}\right| \cdot N_{\text{Raindrops}}\right\}, \quad i = 1, 2, \ldots, N_{sr}, \tag{24.72}$$

where NS_{n} denotes the number of streams that flowing into the specific rivers or sea.

- Step 5: The streams flow into the rivers according to Eq. 24.73 (Eskandar et al. 2012):

$$X_{Stream}^{i+1} = X_{Stream}^i + rand \cdot C \cdot \left(X_{River}^i - X_{Stream}^i\right), \tag{24.73}$$

where *rand* denotes a uniformly distributed random number which falls within the range of [0, 1].

- Step 6: The rivers flow into the sea according to Eq. 24.74 (Eskandar et al. 2012):

$$X_{River}^{i+1} = X_{River}^i + rand \cdot C \cdot \left(X_{Sea}^i - X_{River}^i\right), \tag{24.74}$$

where *rand* denotes a uniformly distributed random number which falls within the range of [0, 1].

- Step 7: Exchanging positions of river with a stream for the purpose of generating the best solution.

- Step 8: Exchanging the positions of rivers with the sea given that a river gets a better solution than the sea.
- Step 9: Checking the evaporation situation.
- Step 10: The rain process occurs (if the evaporation situation is met) based on Eqs. 24.75 and 24.76, respectively (Eskandar et al. 2012):

$$X_{Stream}^{new} = LB + rand \cdot (UB - LB), \tag{24.75}$$

$$X_{Stream}^{new} = X_{Sea} + \sqrt{\mu} \cdot randn(1, N_{\text{var}}), \tag{24.76}$$

  where the lower and upper boundaries are denoted by *LB* and *UB*, respectively and μ represents a coefficient showing the range of searching region around the sea.
- Step 11: Reducing the value of user determined parameter $d_{\max}$ according to Eq. 24.77 (Eskandar et al. 2012):

$$d_{\max}^{i+1} = d_{\max}^{i} - \frac{d_{\max}^{i}}{\text{max\_iteration}}. \tag{24.77}$$

- Step 12: Checking the converging criterion. If yes, WCOA will be terminated; otherwise, it will go back to Step 5.

24.21.2 Performance of WCOA

Eskandar et al. (2012) employed a set of benchmark functions and engineering design optimization problems to test the performance of WCOA. In comparison with other CI techniques (e.g., PSO, DE), WCOA generally showed a better performance. At the end of their study, Eskandar et al. (2012) suggested that WCOA may be suitable for dealing with real world optimization problems which need significant computational efforts efficiently, and at the same time, with acceptable solutions accuracy degree.

24.22 Water Flow Algorithm

In this section, we will introduce an emerging CI algorithm that is based on the phenomenon of hydrological cycle in meteorology and the erosion result found in nature (Carey et al. 2008; Samuels et al. 2009; Rahman 2011; Davis 2010; Day 2007; Reece et al. 2011; Whitten et al. 2014; Zumdahl and Zumdahl 2014).

24.22.1 Fundamentals of Water Flow Algorithm

Water flow algorithm (WFA) was recently proposed in (Brodić 2011, 2012; Brodić and Milivojević 2010, 2011; Tran and Ng 2011; Basu et al. 2007) for solving text recognition problem. It has been applied to test line segmentation (Brodić 2011, 2012; Brodić and Milivojević 2010, 2011; Basu et al. 2007) and flow shop scheduling (Tran and Ng 2011). The core idea underlying the WFA is hypothetically assuming a flow of water moving towards a particular direction and crossing the image frame in a way that it faces obstruction from the characters of the text lines (Basu et al. 2007). In other words, the artificial water flow across the targeted image frame is anticipated to fill up the gaps existing between consecutive text lines. Accordingly, the un-wetted areas left on the image frame will ideally lie under the text lines. To implement the WFA algorithm, the following components need to be well-designed (Brodić 2011, 2012; Brodić and Milivojević 2010, 2011; Tran and Ng 2011; Basu et al. 2007):

- Component 1: Labelling line spacing. Technically, recognizing the wetted stripes in a document image is not enough to extract the text lines from the same. All un-wetted stripes in the targeted image file have to be labelled distinctly before text line extraction.
- Component 2: Erosion of the dark stripes. In practice, some isolated parts (e.g., dots) from the text written in Roman may sometimes appear outside the white stripes, i.e., in the dark stripes. Such elements are often neglected when text lines are extracted from the document images after marking all white stripes therein separately. In order to prevent such issues, the dark stripes in document images have to be eroded morphologically.
- Component 3: Extraction of the text lines. To do so, one can simply extract all white stripes from the image one by one.
- Component 4: Detecting the skew angle. To filter out certain upper envelope pixels, a difference vector has to be computed. For each column i, d_i is calculated through Eq. 24.78 (Basu et al. 2007):

$$d_i = e_i - e_{i-1}, \quad i = 1, 2, \ldots, n-1. \tag{24.78}$$

Thus, the slope of the upper envelope within the jth interval S_j, is computed via Eq. 24.79 (Basu et al. 2007):

$$S_j = \frac{\sum_i d_i}{\text{width of the } j\text{th interval}}. \tag{24.79}$$

- Component 5: Separating touching text lines. To do this, a straight line making an angle of α_s with a horizontal line across the targeted image frame and passing through either the right most pixel point or the left most pixel point of the upper envelope of the top most touching line will be drawn.

24.22.2 *Performance of WFA*

To verify the proposed WFA, different samples of English and Bengali documents collected from various sources were employed in (Basu et al. 2007). The selected documents are scanned at a resolution of 300 dpi. The experimental results demonstrated that WFA is very promising in dealing with the targeted problems.

24.23 Water Flow-Like Algorithm

In this section, we will introduce an emerging CI algorithm that is also based on the phenomenon of hydrological cycle in meteorology and the erosion result found in nature (Carey et al. 2008; Samuels et al. 2009; Rahman 2011; Davis 2010; Day 2007; Reece et al. 2011; Whitten et al. 2014; Zumdahl and Zumdahl 2014).

24.23.1 *Fundamentals of Water Flow-Like Algorithm*

Water flow-like algorithm (WFlA) was recently proposed in Yang and Wang (2007). To implement the WFA algorithm, the following four major operations need to be performed (Yang and Wang 2007):

- Flow splitting and moving: A main characteristic of WFlA is its solution agents forking mechanism. To fulfil this design, the number of subflow split from main flow i is defined via Eq. 24.80 (Yang and Wang 2007):

$$n_i = \min\left\{ \max\left\{ 1, \text{int}\left(\frac{M_i V_i}{T} \right) \right\}, \bar{n} \right\}, \tag{24.80}$$

  where $\bar{n}$ represents an imposed upper limit for the number of subflows forked from a main flow, and $M_i V_i$ denotes the subflows' momentum. In WFlA, the flow movements from one location to other new locations are not computed from Euclidean distance. Instead, neighbouring locations are assigned to move or split the flow. For instance, two neighbouring solutions obtained from one step movement of coordinate h are defined by Eqs. 24.81 and 24.82, respectively (Yang and Wang 2007):

$$\mathbf{A}_i^{(+)} = \left\{ A_{i1}^{(+)}, A_{i2}^{(+)}, \ldots, A_{iq}^{(+)} \right\}, \tag{24.81}$$

$$\mathbf{A}_i^{(-)} = \left\{ A_{i1}^{(-)}, A_{i2}^{(-)}, \ldots, A_{iq}^{(-)} \right\}. \tag{24.82}$$

Therefore, the set of all one-step neighbouring solutions for flow i is defined by Eq. 24.83 (Yang and Wang 2007):

$$\mathbf{A}_i = \mathbf{A}_i^{(+)} \cup \mathbf{A}_i^{(-)}. \tag{24.83}$$

Meanwhile, the mass of flow i is discriminately distributed among subflows according to their ranks. In MFlA, the mass distributed from M_i to the subflow U_{ik} is expressed by Eq. 24.84 (Yang and Wang 2007):

$$w_{ik} = \left(\frac{n_i + 1 - k}{\sum_{r=1}^{n_i} r} \right), \quad \mathrm{k} = 1, 2, \ldots, n_i. \tag{24.84}$$

And the velocity of subflow U_{ik} split from flow i can be computed according to Eq. 24.85 (Yang and Wang 2007):

$$\mu_{ik} = \begin{cases} \sqrt{V_i^2 + 2g\delta_{ik}} & \text{if } V_i^2 + 2g\delta_{ik} > 0 \\ 0 & \text{otherwise} \end{cases}, \tag{24.85}$$

where the gravitational acceleration is denoted by g.

- Flow merging: In MFlA, water flows will emerge into a single flow when more than two flows of water move to the same position. Therefore, the mass of flow j is added to flow i based on Eq. 24.86 (Yang and Wang 2007):

$$M_i \leftarrow M_i + M_j. \tag{24.86}$$

And the aggregated speed for flow i is obtained through Eq. 24.87 (Yang and Wang 2007):

$$V_i \leftarrow \frac{M_i V_i + M_j V_j}{M_i + M_j}. \tag{24.87}$$

- Water evaporation: A water evaporating operation is also introduced in WFlA for a better simulating the natural water evaporation phenomenon. During the water evaporation process, the masses of all water flows are renewed based on Eq. 24.88 (Yang and Wang 2007):

$$M_i \leftarrow \left(1 - \frac{1}{t}\right)\bar{M}_i, \quad i = 1, 2, \ldots, N, \tag{24.88}$$

where $\bar{M}_i$ denotes the mass when flow i was originally created or merged with others.

- Precipitation: In order to mimic the natural life cycle of water, two kinds of precipitation mechanisms are performed in WFlA, namely, enforced and regular precipitation. In enforced precipitation, each component x'_{ih} is created randomly from the original coordinate x_{ih} based on Eq. 24.89 (Yang and Wang 2007):

$$x'_{ih} = \begin{cases} x_{ih} + d_h^+ & \text{if } \sim U(0,1) > 0.5 \\ x_{ih} - d_h^- & \text{otherwise} \end{cases}. \tag{24.89}$$

And the mass of M_0 is proportionally distributed among the flows according to their original mass. Accordingly, in enforced precipitation, the mass allocated to flow i is calculated via Eq. 24.90 (Yang and Wang 2007):

$$M'_i = \left(\frac{M_i}{\sum_k M_k} \right) M_0, \quad i = 1, 2, \ldots, N. \tag{24.90}$$

On the other hand, regular precipitation mechanism is applied periodically to let the evaporated water return the ground. Similarly, the location $\bar{X}_i$ of a drop down flow i is randomly derived from the ground flow i's location and it is calculated via Eqs. 24.91 and 24.92, respectively (Yang and Wang 2007):

$$\bar{X}_i = \left[\bar{x}_{i1}, \bar{x}_{i2}, \ldots, \bar{x}_{iq} \right], \tag{24.91}$$

$$\bar{x}_{ih} = \begin{cases} x_{ih} + d_h^+ & \text{if } \sim U(0,1) > 0.5 \\ x_{ih} - d_h^- & \text{otherwise} \end{cases}. \tag{24.92}$$

And the masses of the drop down flows are proportionally allocated according to Eq. 24.93 (Yang and Wang 2007):

$$\bar{M}_i = \left(M_0 - \sum_k M_k \right) \left(\frac{M_i}{\sum_k M_k} \right) = \left(\frac{M_0}{\sum_k M_k} - 1 \right), \quad i = 1, 2, \ldots, N. \tag{24.93}$$

24.23.2 Performance of WFlA

The famous bin packing problem was employed in (Yang and Wang 2007) for the purpose of testing the performance of WFlA. In comparison with other CI techniques (e.g., GA, PSO), the WFlA performed well on the target problem. Overall, through adopting several unique characteristics observed from water flows, this novice CI approach showed a steady and persistent solution search performance.

24.24 Conclusions

In this chapter, 22 emerging physics-based CI methodologies are discussed. Although most of them are still in their infancy, their usefulness has been demonstrated throughout the preliminary corresponding studies. Interested readers are referred to them as a starting point for a further exploration and exploitation of these innovative CI algorithms.

References

Abel, P. G. (2013). *Visual lunar and planetary astronomy*. New York: Springer Science + Business Media. ISBN 978-1-4614-7018-2.

Abuhamdah, A., & Ayob, M. (2009a, October 27–28). Hybridization multi-neighbourhood particle collision algorithm and great deluge for solving course timetabling problems. In *IEEE 2nd Conference on Data Mining and Optimization*, Selangor, Malaysia (pp. 108–114).

Abuhamdah, A., & Ayob, M. (2009b, October 27–28). Multi-neighbourhood particle collision algorithm for solving course timetabling problems. In *IEEE 2nd Conference on Data Mining and Optimization*, Selangor, Malaysia (pp. 21–27).

Abuhamdah, A., & Ayob, M. (2011, June 28–29). MPCA-ARDA for solving course timetabling problems. In *IEEE 3rd Conference on Data Mining and Optimization (DMO)*, Selangor, Malaysia (pp. 171–177).

Afaq, H., & Saini, S. (2011). On the solutions to the travelling salesman problem using nature inspired computing techniques. *International Journal of Computer Science Issues, 8*, 326–334.

Ahmed, W. (Ed.). (2012). *Nuclear power: Practical aspects*. Janeza Trdine 9, 51000 Rijeka, Croatia: InTech. ISBN 978-953-51-0778-1.

Arfken, G. B., Weber, H. J., & Harris, F. E. (2013). *Mathematical methods for physicists: a comprehensive guide*. The Boulevard, Langford Lane, Kidlington, Oxford, OX5 1GB, UK: Elsevier Inc. ISBN 978-0-12-384654-9.

Balachandar, S. R., & Kannan, K. (2007). Randomized gravitational emulation search algorithm for symmetric traveling salesman problem. *Applied Mathematics and Computation, 192*, 413–421.

Balachandar, S. R., & Kannan, K. (2009). A meta-heuristic algorithm for vertex covering problem based on gravity. *International Journal of Computational and Mathematical Sciences, 3*, 324–330.

Balachandar, S. R., & Kannan, K. (2010). A meta-heuristic algorithm for set covering problem based on gravity. *International Journal of Computational and Mathematical Sciences, 4*, 223–228.

Barzegar, B., Rahmani, A. M., & Zamanifar, K. (2009). Gravitational emulation local search algorithm for advanced reservation and scheduling in grid systems. In *IEEE 1st Asian Himalayas International Conference on Internet (AH-ICI)* (pp. 1–5).

Basu, S., Chaudhuri, C., Kundu, M., Nasipuri, M., & Basu, D. K. (2007). Text line extraction from multi-skewed handwritten documents. *Pattern Recognition, 40*, 1825–1839.

Bauer, W., & Westfall, G. D. (2011). *University physics with modern physics*. New York, NY, USA: McGraw-Hill. ISBN 978-0-07-285736-8.

Bes, D. R. (2007). *Quantum mechanics: A modern and concise introductory course*. Berlin Heidelberg: Springer. ISBN 978-3-540-46215-6.

Biswas, A., Mishra, K. K., Tiwari, S., & Misra, A. K. (2013). Physics-inspired optimization algorithms: A survey. *Journal of Optimization, 2013*, 1–16.

Boettcher, S. (2004). Extremal optimization. In A. K. Hartmann & H. Rieger (Eds.) *New optimization algorithms in physics*, Chap. 11 (pp. 227–252). Strauss GmbH, Mörlenbach: WILEY-VCH Verlag GmbH & Co. KGaA, Weinheim. ISBN 3-527-40406.

Boettcher, S. (2005). Extremal optimization for Sherrington-Kirkpatrick spin glasses. *The European Physical Journal B, 46*, 501–505.

Boettcher, S., & Percus, A. (2000). Nature's way of optimizing. *Artificial Intelligence, 119*, 275–286.

Boettcher, S., & Percus, A. G. (2004). Extremal optimization at the phase transition of the 3-coloring problem. *Physical Review E, 69*, 066–703.

Bouras, A., & Syam, W. P. (2013). Hybrid chaos optimization and affine scaling search algorithm for solving linear programming problems. *Applied Soft Computing, 13*, 2703–2710.

Brekke, P. (2012). *Our explosive sun: a visual feast of our source of light and life*. New York Dordrecht Heidelberg London: Springer Science + Business Media, LLC. ISBN 978-1-4614-0570-2.

Brodić, D. (2011). Advantages of the extended water flow algorithm for handwritten text segmentation. In *Pattern recognition and machine intelligence*, LNCS 6744 (pp. 418–423). Berlin: Springer.

Brodić, D. (2012). Extended approach to water flow algorithm for text line segmentation. *Journal of Computer Science and Technology, 27*, 187–194.

Brodić, D., & Milivojević, Z. (2010). An approach to modification of water flow algorithm for segmentation and text parameters extraction. In *Emerging trends in technological innovation, IFIP advances in information and communication technology* (pp. 324–331). Berlin: Springer.

Brodić, D., & Milivojević, Z. (2011). A new approach to water flow algorithm for text line segmentation. *Journal of Universal Computer Science, 17*, 30–47.

Bruce, D. W., O'hare, D., & Walton, R. I. (2014). *Local structural characterisation*. West Sussex, PO19 8SQ, UK: Wiley. ISBN 978-1-119-95320-3.

Cacuci, D. G. (2010). *Handbook of nuclear engineering*. New York, NY, USA: Springer Science + BusinessMedia LLC. ISBN 978-0-387-98130-7.

Carey, V. P., Chen, G., Grigoropoulos, C., Kaviany, M., & Majumdar, A. (2008). A review of heat transfer physics. *Nanoscale and Microscale Thermophysical Engineering, 12*, 1–60.

Chen, M.-R., & Lu, Y.-Z. (2008). A novel elitist multiobjective optimization algorithm: Multiobjective extremal optimization. *European Journal of Operational Research, 188*, 637–651.

Cheshomi, S., Rahati-Q, S., & Akbarzadeh-T, M.-R. (2010, August 13–15). Hybrid of chaos optimization and Baum-Welch algorithms for HMM training in continuous speech recognition. In *IEEE International Conference on Intelligent Control and Information Processing*, Dalian, China.

Chuang, C.-L., & Jiang, J.-A. (2007, September 25–28). Integrated radiation optimization: Inspired by the gravitational radiation in the curvature of space-time. In *IEEE Congress on Evolutionary Computation (CEC)*, Singapore (pp. 3157–3164).

Davis, M. L. (2010). *Water and wastewater engineering: Design principles and practice*. New York, USA: The McGraw-Hill Companies, Inc. ISBN 978-0-07-171385-6.

Day, T. (2007). *Water*. London, UK: Dorling Kindersley Limited. ISBN 978-1-40531-874-7.

Eskandar, H., Sadollah, A., Bahreininejad, A., & Hamdi, M. (2012). Water cycle algorithm: A novel meta-heuristic optimization for solving constrained engineering optimization problems. *Computers & Structures, 110–111*, 151–166.

Flores, J. J., López, R., & Barrera, J. (2011). Gravitational interactions optimization. In *Learning and intelligent optimization* (pp. 226–237). Berlin Heidelberg: Springer.

Gomez, J., Dasgupta, D., & Nasraoui, O. (2003, May 1–3). A new gravitational clustering algorithm. In *3rd Siam International Conference on Data Mining*, San Francisco, CA, USA (pp. 83–94).

Gonçalves, B., & Boettcher, S. (2008). Hysteretic optimization for spin glasses. *Journal of Statistical Mechanics: Theory and Experiment*. doi:10.1088/1742-5468/2008/01/P01003.

Gorbenko, A., & Popov, V. (2012). The force law design of artificial physics optimization for robot anticipation of motion. *Advanced Studies in Theoretical Physics, 6*, 625–628.

Gorbenko, A., & Popov, V. (2013). The force law design of APO for starting population selection for GSAT. *Advanced Studies in Theoretical Physics, 7*, 131–134.

Gourgoulhon, É. (2013). *Special relativity in general frames: from particles to astrophysics*. Berlin Heidelberg: Springer. ISBN 978-3-642-37275-9.

Gupta, S., Bhardwaj, S., & Bhatia, P. K. (2011). A reminiscent study of nature inspired computation. *International Journal of Advances in Engineering & Technology, 1*, 117–125.

Hamaizia, T., Lozi, R., & Hamri, N.-E. (2012). Fast chaotic optimization algorithm based on locally averaged strategy and multifold chaotic attractor. *Applied Mathematics and Computation, 219*, 188–196.

Han, F., & Lu, Q.-S. (2008). An improved chaos optimization algorithm and its application in the economic load dispatch problem. *International Journal of Computer Mathematics, 85*, 969–982.

Hartmann, A. K., & Rieger, H. (Eds.). (2004). New optimization algorithms in physics, Strauss GmbH, Mörlenbach: WILEY-VCH Verlag GmbH & Co. KGaA, Weinheim. ISBN 3-527-40406.

Henao, J. D. V. (2011). An introduction to chaos based algorithms for numerical optimization. *Revista Avances en Sistemas e Informática, 8*, 51–60.

Holzner, S. (2011). *Physics I for dummies*. River Street, Hoboken, NJ, USA: Wiley Inc. ISBN 978-0-470-90324-7.

HOLZNER, S. (2010). *Physics II for dummies*. River Street, Hoboken, NJ, USA: Wiley Inc. ISBN 978-0-470-53806-7.

Hsiao, Y.-T., Chuang, C.-L., Jiang, J.-A., & Chien, C.-C. (2005, October 10–12). A novel optimization algorithm: Space gravitational optimization. In *IEEE International Conference on Systems, Man and Cybernetics (SMC)* (pp. 2323–2328).

Jiang, H., Kwong, C. K., Chen, Z., & Ysim, Y. C. (2012). Chaos particle swarm optimization and T-S fuzzy modeling approaches to constrained predictive control. *Expert Systems with Applications, 39*, 194–201.

Jin, G.-G., & Tran, T.-D. (2010, August 18–21). A nature-inspired evolutionary algorithm based on spiral movements. In *IEEE Sice Annual Conference*, The Grand Hotel, Taipei, Taiwan (pp. 1643–1647).

Kaveh, A., & Khayatazad, M. (2012). A new meta-heuristic method: Ray optimization. *Computers & Structures, 112–113*, 283–294.

Klafter, J., Lim, S. C., & Metzler, R. (2012). *Fractional dynamics: Recent advances*. 5 Toh Tuck Link, Singapore: World Scientific Publishing Co. Pte. Ltd. ISBN 978-981-4340-58-8.

Kundu, S. (1999). Gravitational clustering: A new approach based on the spatial distribution of the points. *Pattern Recognition, 32*, 1149–1160.

Li, B., & Jiang, W. (1998). Optimizing complex functions by chaos search. *Cybernetics and Systems: An International, 29*, 409–419.

Lu, H.-J., Zhang, H.-M., & Ma, L.-H. (2006). A new optimization algorithm based on chaos. *Journal of Zhejiang University SCIENCE A, 7*, 539–542.

Luz, E. F. P. D., Becceneri, J. C., & Velho, H. F. D. C. (2008). A new multi-particle collision algorithm for optimization in a high performance environment. *Journal of Computational Interdisciplinary Sciences, 1*, 3–10.

Luz, E. F. P. D., Becceneri, J. C., & Velho, H. F. D. C. (2011). Multiple particle collision algorithm applied to radiative transference and pollutant localization inverse problems. In IEEE International Symposium on Parallel and Distributed Processing Workshops and PhD Forum (IPDPSW) (pp. 347–351).

Mo, S., & Zeng, J. (2009). Performance analysis of the artificial physics optimization algorithm with simple neighborhood topologies. In *IEEE International Conference on Computational Intelligence and Security (CIS)* (pp. 155–160).

National_Research_Council (Ed.). (2012). *Nuclear physics: Exploring the heart of matter*. 500 Fifth Street, NW Washington, DC, USA: The National Academies Press. ISBN 978-0-309-26040-4.

Pál, K. F. (2003). Hysteretic optimization for the traveling salesman problem. *Physica A, 329*, 287–297.

Pál, K. F. (2004). Hysteretic optimization. In A. K. Hartmann & H. Rieger (Eds.) *New optimization algorithms in physics*, Chap. 10 (pp. 205–226). Strauss GmbH, Mörlenbach: WILEY-VCH Verlag GmbH & Co. KGaA, Weinheim. ISBN 3-527-40406.

Pál, K. F. (2006a). Hysteretic optimization for the Sherrington–Kirkpatrick spin glass. *Physica A, 367*, 261–268.

Pál, K. F. (2006b). Hysteretic optimization, faster and simpler. *Physica A, 360*, 525–533.

Particle_Data_Group. (1998). Review of particle physics. *The European Physical Journal C, 3*, 1–794.

Philips, A. C. (2003). *Introduction to quantum mechanics*. The Atrium, Southern Gate, Chichester, West Sussex, PO19 8SQ, England: Wiley. ISBN 0-470-85323-9.

Placidi, G. (2012). *MRI: essentials for innovative technologies*. Broken Sound Parkway NW, Suite 300, Boca Raton, FL, USA: CRC Press, Taylor & Francis Group, LLC. ISBN 978-1-4398-4062-7.

Pooranian, Z., Harounabadi, A., Shojafar, M., & Hedayat, N. (2011). New hybrid algorithm for task scheduling in grid computing to decrease missed task. *World Academy of Science, Engineering and Technology, 79*, 5–9.

Rabanal, P., & Rodríguez, I. (2011). Hybridizing river formation dynamics and ant colony optimization. In *Advances in artificial life, Darwin LNCS 5778* (pp. 424–431). Berlin: Springer.

Rabanal, P., Rodríguez, I., & Rubio, F. (2007). Using river formation dynamics to design heuristic algorithms. In: S. G. Akl, C. S. C., M. J. Dinneen, G. Rozenber, H. T. Wareham (Eds.), *UC 2007*, LNCS (Vol. 4618, pp. 163–177). Heidelberg: Springer.

Rabanal, P., Rodríguez, I., & Rubio, F. (2008a). Finding minimum spanning/distances trees by using river formation dynamics. In M. Dorigo (Ed.), *ANTS 2008, LNCS 5217* (pp. 60–71). Berlin Heidelberg: Springer.

Rabanal, P., Rodríguez, I., & Rubio, F. (2008b). Solving dynamic TSP by using river formation dynamics. In *IEEE 4th International Conference on Natural Computation (ICNC)* (pp. 246–250).

Rabanal, P., Rodríguez, I., & Rubio, F. (2009). Applying river formation dynamics to solve NP-complete problems. In *Nature-inspired algorithms for optimization, studies in computational intelligence* (Vol. 193, pp. 333–368). Berlin Heidelberg: Springer.

Rabanal, P., Rodríguez, I., & Rubio, F. (2010). Applying river formation dynamics to the Steiner tree problem. In F. Sun, Y. Wang, J. Lu, B. Zhang, W. Kinsner & L. A. Zadeh (Eds.), *9th IEEE International Conference on Cognitive Informatics (ICCI)* (pp. 704–711).

Rabanal, P., Rodríguez, I., & Rubio, F. (2011). Studying the application of ant colony optimization and river formation dynamics to the Steiner tree problem. *Evolution Intelligence, 4*, 51–65.

Rabanal, P., Rodríguez, I., & Rubio, F. (2013). Testing restorable systems: Formal definition and heuristic solution based on river formation dynamic. *Formal Aspects of Computing, 25*, 743–768. doi:10.1007/s00165-011-0206-3.

Rahman, M. (2011). *Mechanics of real fluids*. Ashurst Lodge, Ashurst, Southampton, SO40 7AA, UK: WIT Press. ISBN 978-1-84564-502-1.

Reece, J. B., Urry, L. A., Cain, M. L., Wasserman, S. A., Minorsky, P. V., & Jackson, R. B. (2011). *Campbell biology*. Sansome St., San Francisco, CA: Pearson Education, Inc. ISBN 978-0-321-55823-7.

Rong, G., Liu, G., Zheng, M., Sun, A., Tian, Y., & Wang, H. (2013). Parallel gravitation field algorithm based on the CUDA platform. *Journal of Information & Computational Science, 10*, 3635–3644.

Sacco, W. F. & Oliveira, C. R. E. D. (2005). A new stochastic optimization algorithm based on a particle collision meta-heuristic. In *6th World Congresses of Structural and Multidisciplinary Optimization*, Rio de Janeiro, Brazil, 30 May-03 June (pp. 1–6).

Sacco, W. F., Oliveira, C. R. E. D., & Pereira, C. M. N. A. (2006). Two stochastic optimization algorithms applied to nuclear reactor core design. *Progress in Nuclear Energy, 48*, 525–539.

Samuels, P., Huntington, S., Allsop, W., & Harrop, J. (2009). *Flood risk management: research and practice*. London, UK: Taylor & Francis Group. ISBN 978-0-415-48507-4.

Sanchez, M. A., Castillo, O., Castro, J. R., & Rodriguez-Diaz, A. (2012, August 6–8). Fuzzy granular gravitational clustering algorithm. IEEE Annual Meeting of the North American Fuzzy Information Processing Society (NAFIPS), Berkeley, CA, USA (pp. 1–6).

Sasselov, D. (2012). *The life of super-earths: how the hunt for alien worlds and artificial cells will revolutionize life on our planet*. Park Avenue South, New York, NY, USA: Basic Books. ISBN 978-0-465-02193-2.

Schutz, B. (2003). *Gravity: From the ground up*. The Edinburgh Building, Cambridge CB2 8RU, UK: Cambridge University Press. ISBN 978-0-511-33696-6.

Serway, R. A., & Jewett, J. W. (2014). *Physics for scientists and engineers with modern physics*. Boston, MA, USA: Brooks/Cole CENAGE Learning. ISBN 978-1-133-95405-7.

Shah-Hosseini, H. (2011a). Otsu's criterion-based multilevel thresholding by a nature-inspired meta-heuristic called galaxy-based search algorithm. In *IEEE 3rd World Congress on Nature and Biologically Inspired Computing (NaBIC)* (pp. 383–388).

Shah-Hosseini, H. (2011b). Principal components analysis by the galaxy-based search algorithm: a novel metaheuristic for continuous optimisation. *International Journal of Computational Science and Engineering, 6*, 132–140.

Shayeghi, H., Jalilzadeh, S., Shayanfar, H. A., & Safari, A. (2009, May 6–9). Robust PSS design using chaotic optimization algorithm for a multimachine power system. In *IEEE 6th International Conference on Electrical Engineering/Electronics, Computer, Telecommunications and Information Technology (ECTI-CON)*, Pattaya, Chonburi (pp. 40–43).

Shen, J., & Li, Y. (2008). An optimization algorithm based on optical principles. *Advances in Systems Science and Applications, 5*, 1–8.

Shen, J., & Li, Y. (2009, April 24–26). Light ray optimization and its parameter analysis. In *IEEE International Joint Conference on Computational Sciences and Optimization (CSO)*, Sanya, China (pp. 918–922).

Shen, J., & Li, J. (2010, September 13–14). The principle analysis of light ray optimization algorithm. In *IEEE 2nd Second International Conference on Computational Intelligence and Natural Computing (CINC)*, Wuhan, China (pp. 154–157).

Shen, J., & Li, J. (2012). Light ray optimization algorithm and convergence analysis for one dimensional problems. In *ASME International Conference on Electronics, Information and Communication Engineering (EICE)* (pp. 1–4).

Shen, J., Li, J., & Wei, B. (2012). Optimal search mechanism analysis of light ray optimization algorithm. *Journal of Mathematical Research with Applications, 32*, 530–542.

Shifman, M. (2012). *Advanced topics in quantum field theory: A lecture course*. The Edinburgh Building, Cambridge CB2 8RU, UK: Cambridge University Press. ISBN 978-0-521-19084-8.

Spears, W. M., & Gordon, D. F. (1999). Using artificial physics to control agents. In *IEEE International Conference on Information Intelligence and Systems* (pp. 281–288).

Spears, W. M., & Spears, D. F. (2012). *Physicomimetics: Physics-based swarm intelligence*. Berlin: Springer.

Spears, W. M., Spears, D. F., Hamann, J. C., & Heil, R. (2004a). Distributed, physics-based control of swarms of vehicles. *Autonomous Robots, 17*, 137–162.

Spears, W. M., Spears, D. F., Heil, R., Kerr, W., & Hettiarachchi, S. (2004b). *An overview of physicomimetics, LNCS 3324* (pp. 84–97). Berlin: Springer.

Sun, J., Dong, H., & Pan, Y. (2012). A self-adaptive differential evolution algorithm based on cloud model. In *International Conference on Network and Computational Intelligence (ICNCI)* (Vol. 46, pp. 59–64). IACSIT Press.

Tamura, K., & Yasuda, K. (2011a). Primary study of spiral dynamics inspired optimization. *IEEJ Transactions on Electrical and Electronic Engineering, 6*, S98–S100.

Tamura, K., & Yasuda, K. (2011b). Spiral dynamics inspired optimization. *Journal of Advanced Computational Intelligence and Intelligent Informatics, 15*, 1116.

Tamura, K., & Yasuda, K. (2011c). Spiral multipoint search for global optimization. In *IEEE 10th International Conference on Machine Learning and Applications (ICMLA)* (pp. 470–475).

Tavazoei, M. S., & Haeri, M. (2007a). Comparison of different one-dimensional maps as chaotic search pattern in chaos optimization algorithms. *Applied Mathematics and Computation, 187*, 1076–1085.

Tavazoei, M. S., & Haeri, M. (2007b). An optimization algorithm based on chaotic behavior and fractal nature. *Journal of Computational and Applied Mathematics, 206*, 1070–1081.

Tayarani-N, M. H., & Akbarzadeh-T, M. R. (2008). Magnetic optimization algorithms a new synthesis. In *IEEE Congress on Evolutionary Computation (CEC)* (pp. 2659–2664).

Tran, T. H., & Ng, K. M. (2011). A water-flow algorithm for flexible flow shop scheduling with intermediate buffers. *Journal of Scheduling, 14*, 483–500.

Tsvetkov, P. V. (Ed.). (2011). *Nuclear power: Control, reliability and human factors*. Janeza Trdine 9, 51000 Rijeka, Croatia: InTech. ISBN 978-953-307-599-0.

Vakoch, D. A. (Ed.). (2014). *Extraterrestrial altruism: Evolution and ethics in the cosmos*. Berlin Heidelberg: Springer. ISBN 978-3-642-37749-5.

Waltar, A. E., Todd, D. R., & Tsvetkov, P. V. (2012). *Fast spectrum reactors*. New York, Dordrecht, Heidelberg, London: Springer Science + Business Media LLC. ISBN 978-1-4419-9571-1.

Wang, P. K. (2013). *Physics and dynamics of clouds and precipitation*. Cambridge: UK, Cambridge University Press.

Wang, Y., & Zeng, J.-C. (2010a). A constraint multi-objective artificial physics optimization algorithm. In *IEEE Second International Conference on Computational Intelligence and Natural Computing (CINC)* (pp. 107–112).

Wang, Y., & Zeng, J.-C. (2010b). Multi-objective optimization algorithm based on artificial physics optimization (in Chinese). *Control and Decision, 25*, 1040–1044.

Wang, Y., Zeng, J.-C., Cui, Z.-H., & He, X.-J. (2011). A novel constraint multi-objective artificial physics optimisation algorithm and its convergence. *International Journal of Innovative Computing and Applications, 3*, 61–70.

Wang, L., Li, W., Fei, R., & Hei, X. (2012). Cloud droplets evolutionary algorithm on reciprocity mechanism for function optimization. In Y. Tan, Y. Shi & Z. Ji (Eds.), *ICSI 2012, Part I, LNCS 7331* (pp. 268–275). Berlin Heidelberg: Springer.

Webster, B. L. (2004). *Solving combinatorial optimization problems using a new algorithm based on gravitational attraction*. Unpublished doctoral thesis, Florida Institute of Technology.

Whitten, K. W., Davis, R. E., Peck, M. L., & Stanley, G. G. (2014). *Chemistry*. 20 Davis Drive, Belmont, CA 94002-3098USA: Brooks/Cole, Cengage Learning. ISBN 13: 978-1-133-61066-3.

Wright, W. E. (1977). Gravitational clustering. *Pattern Recognition, 9*, 151–166.

Xie, L.-P., & Zeng, J.-C. (2009a, June 12–14). A global optimization based on physicomimetics framework. In *IEEE First ACM/SIGEVO Summit on Genetic and Evolutionary Computation (GEC)*, Shanghai, China (pp. 609–616).

Xie, L., & Zeng, J. (2009b, December 7–9). An extended artificial physics optimization algorithm for global optimization. In *IEEE 4th International Conference on Innovative Computing, Information and Control (ICICIC)*, Kaohsiung Taiwan (pp. 881–884).

Xie, L., & Zeng, J. (2011). A hybrid vector artificial physics optimization for constrained optimization problems. In *IEEE 1st International Conference on Robot, Vision and Signal Processing (RVSP)* (pp. 145–148).

Xie, L., Zeng, J., & Cui, Z. (2009a, December). General framework of artificial physics optimization algorithm. In *IEEE World Congress on Nature and Biologically Inspired Computing (NaBIC)*, Coimbatore, India (pp. 1321–1326).

Xie, L., Zeng, J., & Cui, Z. (2009b). Using artificial physics to solve global optimization problems. In *8th IEEE International Conference on Cognitive Informatics (ICCI)*, June, Hongkong (pp. 502–508).

Xie, L., Tan, Y., & Zeng, J. (2010a, July 17–19). The convergence analysis and parameter selection of artificial physics optimization algorithm. In *IEEE International Conference on Modelling, Identification and Control*, Okayama, Japan (pp. 562–567).

Xie, L., Tan, Y., Zeng, J., & Cui, Z. (2010b). Artificial physics optimisation: A brief survey. *International Journal of Bio-Inspired Computation, 2*, 291–302.

Xie, L., Tan, Y., & Zeng, J. (2011a). A study on the effect of V_{max} in artificial physics optimization algorithm with high dimension. In *IEEE International Conference of Soft Computing and Pattern Recognition (SoCPaR)* (pp. 550–555).

Xie, L., Tan, Y., Zeng, J., & Cui, Z. (2011b). The convergence analysis of artificial physics optimisation algorithm. *International Journal of Intelligent Information and Database Systems, 5*, 536–554.

Xie, L., Zeng, J., & Formato, R. A. (2011c). Convergence analysis and performance of the extended artificial physics optimization algorithm. *Applied Mathematics and Computation, 218*, 4000–4011.

Xie, L., Zeng, J., & Cai, X. (2011c). A hybrid vector artificial physics optimization with multi-dimensional search method. In *IEEE 2nd International Conference on Innovations in Bio-inspired Computing and Applications (IBICA)* (pp. 116–119).

Yan, G.-W., & Hao, Z. (2012, July 7–9). A novel atmosphere clouds model optimization algorithm. In *IEEE International Conference on Computing, Measurement, Control and Sensor Network (CMCSN)*, Taiyuan, China (pp. 217–220).

Yan, G.-W., & Hao, Z. (2013). A novel optimization algorithm based on atmosphere clouds model. *International Journal of Computational Intelligence and Applications, 12*, 1–16.

Yang, F.-C., & Wang, Y.-P. (2007). Water flow-like algorithm for object grouping problems. *Journal of the Chinese Institute of Industrial Engineers, 24*, 475–488.

Yang, D., Li, G., & Cheng, G. (2007). On the efficiency of chaos optimization algorithms for global optimization. *Chaos, Solitons & Fractals, 34*, 1366–1375.

Yang, G., Xie, L., Tan, Y., & Zeng, J. (2010). A hybrid vector artificial physics optimization with one-dimensional search method. In *IEEE International Conference on Computational Aspects of Social Networks (CASoN)* (pp. 19–22).

Yang, Y., Wang, Y., Yuan, X., & Yin, F. (2012). Hybrid chaos optimization algorithm with artificial emotion. *Applied Mathematics and Computation, 218*, 6585–6611.

Yin, J., Xie, L., Zeng, J., & Tan, Y. (2010). Artificial physics optimization algorithm with a feasibility-based rule for constrained optimization problems. In *IEEE International Conference on Intelligent Computing and Intelligent Systems (ICIS)* (pp. 488-492).

Yuan, X., Yang, Y., & Wang, H. (2012). Improved parallel chaos optimization algorithm. *Applied Mathematics and Computation, 219*, 3590–3599.

Zaránd, G., Pázmándi, F., Pál, K. F., & Zimányi, G. T. (2002). Using hysteresis for optimization. *Physical Review Letters, 89*, 150201-1-150201-4.

Zhang, T., & Qu, H. (2010, August 14–15). An improved clustering algorithm. In *3rd International Symposium on Computer Science and Computational Technology (ISCSCT)*, Jiaozuo, China (pp. 112–115).

Zhang, G.-W., He, R., Liu, Y., Li, D.-Y., & Chen, G.-S. (2008). An evolutionary algorithm based on cloud model. *Chinese Journal of Computers, 31*, 1082–1091.

Zheng, M., Liu, G.-X., Zhou, C.-G., Liang, Y.-C., & Wang, Y. (2010). Gravitation field algorithm and its application in gene cluster. *Algorithms for Molecular Biology, 5*, 1–11.

Zheng, M., Sun, Y., Liu, G.-X., Zhou, Y., & Zhou, C.-G. (2012). Improved gravitation field algorithm and its application in hierarchical clustering. *PLoS ONE, 7*, 1–10.

Zhu, C., & Ni, J. (2012, April 21–23). Cloud model-based differential evolution algorithm for optimization problems. In *IEEE 6th International Conference on Internet Computing for Science and Engineering (ICICSE)*, Henan, China (pp. 55–59).

Zumdahl, S. S., & Zumdahl, S. A. (2014). *Chemistry*. 20 Davis Drive, Belmont, CA 94002-3098, USA: Brooks/Cole, Cengage Learning. ISBN 13: 978-1-133-61109-7.

Part IV
Chemistry-based CI Algorithms

Chapter 25
Chemical-Reaction Optimization Algorithm

Abstract In this chapter, we present a novel optimization approach named chemical-reaction optimization (CRO) algorithm. The main idea behind CRO is that a simulation of the molecules' movements and their resultant chemical reactions. We first describe the general knowledge of the chemical-reaction in Sect. 25.1. Then, the fundamentals and performance of CRO are introduced in Sect. 25.2. Next, a selected variation of CRO is explained in Sect. 25.3. Right after this, Sect. 25.4 presents a representative CRO application. Finally, Sect. 25.5 summarises in this chapter.

25.1 Introduction

Chemistry touches almost every aspect of our lives (e.g., the food we eat), our culture (e.g., the beliefs we believe), and our environments (e.g., the fuel supplies we use). From a chemist's viewpoint, the most interesting things is the concepts of chemical change or chemical reactions which including combustion reactions, oxidation-reduction reactions, gas-formation reactions, acid–base reactions etc. One of the main objectives is to identify "equal" at the molecular level in an equilibrium reaction (i.e., how far does the reaction go in order to reach its lowest energy state?) (Whitten et al. 2014).

Recently, realizing that interaction of substances in a chemical reaction were consistent with the objective of combinatorial optimization problems (e.g., both aim to seek the global equilibrium and experience in a stepwise fashion), Lam and Li (2010a) developed a new computation intelligence (CI) method, called chemical reaction optimization (CRO) for the solution of optimization problems. The basic idea is to mimic the interactions of molecules in a chemical reaction in order to reach the equilibrium with their environment (Lam and Li 2010a).

B. Xing and W.-J. Gao, *Innovative Computational Intelligence: A Rough Guide to 134 Clever Algorithms*, Intelligent Systems Reference Library 62, DOI: 10.1007/978-3-319-03404-1_25, © Springer International Publishing Switzerland 2014

25.1.1 Chemical Reaction and Reaction Mechanism

Even before alchemy became a subject of study, many chemical reactions were used and their products applied to daily life. General speaking, a chemical reaction or just reaction is a natural process in which one of more substances are transformed from the unstable substances to the stable ones through the step by step sequence of elementary reactions (Lam and Li 2012). It is usually accompanied by the release of large amounts of energy in the form of heat and light. Based on the experimental studies, Whitten et al. (2014) pointed out that some reactions change in a single step, but most reactions occur in a series of elementary or fundamental steps. The step-by-step pathway by which a reaction occurs is called the reaction mechanism.

25.1.2 Basic Components

To understand the mechanism of chemical reactions, several basic concepts should be known firstly.

25.1.2.1 Molecules (Molecular Structure)

The chemical view of nature is that everything in the world around us is made up of small units called molecules. It is the smallest particle of an element that can have a stable independent existence (Zumdahl and Zumdahl 2014). It is usually composed of more than one kind of atom in a definite ration. For example, a water molecule consists of two atoms of hydrogen and one atom of oxygen. In terms of molecular structure (i.e., the order in which molecules are connected) chemists found that molecules are close together in a solid and a liquid but far apart in a gas.

25.1.2.2 Kinetic and Potential Energy

Energy is defined as the capacity to do work or to transfer heat (Zumdahl and Zumdahl 2014). Some literatures also described the energy as released (exothermic) or absorbed (endothermic) mechanism. We usually classify it into two general types: kinetic and potential.

The term "kinetic" comes from the Greek word *kinein*, meaning "to move" (Whitten et al. 2014). Kinetic energy represents the capacity for doing work directly. On the other hand, potential energy means that the energy an object possesses because of its position, condition, or composition. In nature environments, both energy can be converted. For instance, if we drop a hammer, its potential energy which is stored in the hammer is converted into kinetic energy as it falls, and it could do work on something it hits such as drive a nail or break a piece of glass.

25.1.2.3 Collision Mechanism

The term collision is defined as an event during which two molecules come close to each other and interact by means of forces (Serway and Jewett 2014). In general, collisions involve two broad categories: elastic and inelastic collisions, depending on whether or not kinetic energy is conserved.

25.1.2.4 Chemical Equilibrium

From a chemist's point of view, chemical equilibrium (or stability) means thermodynamic stability of a chemical system (Zumdahl and Zumdahl 2014). It occurs when a system is in its lowest energy state. In addition, this state can be reached by identifying various reaction conditions such as temperature, pressure, and concentrations of reactants used. It plays an important role in the operation of chemical plants due to it aims for optimizing the amount of different substances.

25.1.3 Basic Laws of Thermodynamics

In this section, we direct our attention to the study of two basic laws of thermodynamics, which provides explanations for the molecules' energy transfer processes: (1) the change in energy, and (2) the change in disorder (called entropy change). The former is based on the first law of thermodynamics. It explains that there is no change in the quantity of matter or energy during a chemical reaction; it can only be converted from one form to other. The latter is based on the second law of thermodynamics. It is a thermodynamic state function that plays an important role in the energy transfer.

25.2 Fundamentals of the Chemical-Reaction Optimization Algorithm

Chemical-reaction optimization (CRO) algorithm is a population-based algorithm that inspired by chemical reaction where a swarm of molecules through interacting with each other and finally reach a steady condition. The basic building blocks of CRO is molecules which are composed of several atoms and characterized by bond length, angle, and torsion (Lam and Li 2010a). In addition, each molecule possesses two kinds of energies, i.e., potential energy and kinetic energy. Also (Lam and Li 2010a) assumed that the chemical reaction of molecules is to take place in a closed container. As a result, the chemical change of each molecule is triggered by a collision that either on the walls of the container or with each other.

Furthermore, there are two fundamental assumptions of CRO: first, the conservation of energy, which means the total amount of energy in the system always remains constant as defined by Eq. 25.1 (Lam and Li 2010a):

$$\sum_{i=1}^{PopSize(t)} \left(PE_{\omega_i(t)} + KE_{\omega_i(t)}\right) + Buffer(t) = C, \tag{25.1}$$

where $PE_{\omega_i(t)}$ is the PE of molecule i, $KE_{\omega_i(t)}$ is the KE of molecule i, $PopSize(t)$ is the number of molecule i, $Buffer(t)$ is the energy in the central buffer at time t, and C is a constant.

Second, the conversation of elementary reactions, which means after an elementary reaction, the total energy remains equilibrium as expressed in Eq. 25.2 (Lam and Li 2010a):

$$\begin{aligned} &\sum_{i=1}^{k} (PE_{\omega_i} + KE_{\omega_i}) \geq \sum_{i=1}^{l} PE_{\omega_i'}, \\ &\sum_{i=1}^{k} (PE_{\omega_i} + KE_{\omega_i}) + Buffer \geq \sum_{i=1}^{l} PE_{\omega_i'}, \end{aligned} \tag{25.2}$$

where k and l are the number of molecules involved before and after a particular elementary reaction, ω and ω' are the molecular structures of an existing molecule and the one to be generated from the elementary reaction, respectively.

More precisely, the general attributes that adopted in the CRO algorithm are described as follows:

- Molecular Structure (ω): In CRO, the term "molecular structure" is used to summarize all the characteristics of the molecules (i.e., bond length, angle, and torsion) and it corresponds to a solution in the mathematical domain. The presentation of a molecular structure depends on the problem we are solving. It can be a number, a vector, or even a matrix.
- Potential Energy (PE): In CRO, PE is defined as the objective function value of the corresponding solution (represented by ω) which is expressed as Eq. 25.3 (Lam and Li 2010a):

$$PE_{\omega} = f(\omega), \tag{25.3}$$

  where f denotes the objective function.
- Kinetic Energy (KE): In CRO, KE can be interpreted as a measure of tolerance for the molecule changing to a less favourable structure (i.e., a solution with higher functional value). For example, Number of hits (*NumHit*): The *NumHit* is a record of the total number of hits (i.e., collisions) a molecule has taken. Minimum Structure (*MinStruct*): The *MinStruct* is the ω with the minimum corresponding PE which a molecule has attained so far. Minimum Potential

Energy (*MinPE*): *MinPE* is the lowest value that corresponds to the *MinStruct*. Minimum Hit Number (*MinHit*): In a similar vein, *MinHit* is the number of hits when a molecule realizes *MinStruct*.

- Central Energy Buffer (*Buffer*): In CRO, *Buffer* has two functions: first, it can used to store the lost kinetic energy; second, it can support the decomposition reaction. Through this transformation, Lam and Li (2010a) attempted to push the molecular structures with lower and lower *PE* in the subsequent changes. This phenomenon is the driving force in CRO to ensure convergence to lower energy state.

25.2.1 Elementary Reactions

In CRO, Lam and Li (2010a) defined four types of elementary reactions, namely on-wall ineffective collision, decomposition, inter-molecular ineffective collision, and synthesis. In the following sections, we will discuss those reactions in details.

25.2.1.1 On-Wall Ineffective Collision

As the name implies, this reaction process happens when a molecule bounces back after hitting the wall of the reaction container. As the collision is not so vigorous, the resultant molecular structure should not be too different from the original one, only some molecular attributes has been changed. Support the current molecular structure is ω, and it is changed in his own neighbourhood as defined by Eq. 25.4 (Lam and Li 2010a):

$$\omega \rightarrow \omega', \tag{25.4}$$

where $\omega' = Neighbor(\omega)$ and it is problem-dependent.

The change is allowed only if Eq. 25.5 holds (Lam and Li 2010a):

$$PE_{\omega} + KE_{\omega} \geq PE_{\omega'}, \tag{25.5}$$

where $PE_{\omega'} = f(\omega')$.

If Eq. 25.5 does not hold, the change is prohibited and the molecule retains its original attributes (i.e., ω, *PE*, and *KE*).

After the collision, a certain portion of molecule's *KE* will be extracted by this reaction and stored in the *Buffer* when the transformation is complete. The volume of *KE* loss depends on a random number q. Its *KE* is updated via Eq. 25.6 (Lam and Li 2010a):

$$KE_{\omega'} = (PE_{\omega} + KE_{\omega} - PE_{\omega'}) \times q, \tag{25.6}$$

where $q \in [KELossRate, 1]$, *KELossRate* is a parameter of CRO that limits the maximum percentage of *KE* lost at a time.

In addition, $(1 - q)$ is defined as the fraction of *KE* lost to the environment when the molecule hits the wall. This transformation can improve the molecule's local search ability and enhance the convergence ability. The update of the buffer is defined as Eq. 25.7 (Lam and Li 2010a):

$$Buffer = Buffer + (PE_{\omega} + KE_{\omega} - PE_{\omega'}) \times (1 - q). \tag{25.7}$$

25.2.1.2 Decomposition

In this type of reaction, a molecule will strike the wall and then decompose into small pieces (e.g., two or more). For simplicity, there are only two parts, i.e., ω'_1 and ω'_2 are considered as expressed in Eq. 25.8 (Lam and Li 2010a):

$$\omega \rightarrow \omega'_1 + \omega'_2. \tag{25.8}$$

The newly generated molecular structures will be very much different from their original one. This transformation implies that other regions (i.e., ω'_1 and ω'_2) of the solution space can be explored after enough local search by the ineffective collisions (Lam and Li 2012).

In general, there are two situations should be taken into account for the decomposition reaction:

- The molecules have enough energy (i.e., *PE* and *KE*) to complete the decomposition. They are expressed as Eq. 25.9 (Lam and Li 2010a):

$$\begin{aligned} KE_{\omega'_1} &= \left(PE_{\omega} + KE_{\omega} - PE_{\omega'_1} - PE_{\omega'_2}\right) \times k, \\ KE_{\omega'_2} &= \left(PE_{\omega} + KE_{\omega} - PE_{\omega'_1} - PE_{\omega'_2}\right) \times (1 - k), \end{aligned} \tag{25.9}$$

where k is a random number uniformly generated from the interval $[0, 1]$.

The condition to allow the decomposition is defined by Eq. 25.10 (Lam and Li 2010a):

$$PE_{\omega} + KE_{\omega} \geq PE_{\omega'_1} + PE_{\omega'_2}. \tag{25.10}$$

- The molecules need to get energy from the *Buffer*. They are expressed as Eq. 25.11 (Lam and Li 2010a):

$$\begin{aligned} KE_{\omega'_1} &= \left(PE_{\omega} + KE_{\omega} - PE_{\omega'_1} - PE_{\omega'_2} + Buffer\right) \times m_1 \times m_2, \\ KE_{\omega'_2} &= \left(PE_{\omega} + KE_{\omega} - PE_{\omega'_1} - PE_{\omega'_2} + Buffer - KE_{\omega'_1}\right) \times m_3 \times m_4, \end{aligned} \tag{25.11}$$

where m_1, m_2, m_3, and m_4 are random numbers independently uniformly generated from the interval $[0, 1]$.

The condition to allow the decomposition is defined by Eq. 25.12 (Lam and Li 2010a):

$$PE_{\omega} + KE_{\omega} + Buffer \geq PE_{\omega'_1} + PE_{\omega'_2}. \quad (25.12)$$

If Eq. 25.12 does not hold, the change is prohibited and the molecule retains its original attributes (i.e., ω, PE, and KE).

The update of the buffer is defined by Eq. 25.13 (Lam and Li 2010a):

$$Buffer = Buffer + \left(PE_{\omega} + KE_{\omega} - PE_{\omega'_1} - PE_{\omega'_2}\right) - KE_{\omega'_1} - KE_{\omega'_2}. \quad (25.13)$$

25.2.1.3 Inter-Molecular Ineffective Collision

When two molecules clash together, instead of combining to form new molecule(s), they bounce away from each other. This type of reaction allows the molecular structure to change in a larger extent but no *KE* is drawn to the *Buffer* (Lam and Li 2010a). Furthermore, due to two molecules are involved, the sum of the possessed *KE* is will larger than that of the on-wall ineffective collision.

Suppose ω_1 and ω_2 represent the original two molecular structures, then we obtain two new molecular (i.e., ω'_1 and ω'_2) from the neighbourhoods as defined by Eq. 25.14 (Lam and Li 2010a):

$$\begin{aligned} \omega_1 &\rightarrow \omega'_1, \\ \omega_2 &\rightarrow \omega'_2. \end{aligned} \quad (25.14)$$

The condition to allow the inter-molecular ineffective collision is defined by Eq. 25.15 (Lam and Li 2010a):

$$PE_{\omega_1} + PE_{\omega_2} + KE_{\omega_1} + KE_{\omega_2} \geq PE_{\omega'_1} + PE_{\omega'_2}. \quad (25.15)$$

If Eq. 25.15 does not hold, the change is prohibited and the molecule retains its original attributes (i.e., ω, PE, and KE).

Then, the kinetic energy (*KE*) of the two new molecules are expressed by Eq. 25.16 (Lam and Li 2010a):

$$\begin{aligned} KE_{\omega'_1} &= \left(PE_{\omega_1} + PE_{\omega_2} + KE_{\omega_1} + KE_{\omega_2} - PE_{\omega'_1} - PE_{\omega'_2}\right) \times p, \\ KE_{\omega'_2} &= \left(PE_{\omega_1} + PE_{\omega_2} + KE_{\omega_1} + KE_{\omega_2} - PE_{\omega'_1} - PE_{\omega'_2}\right) \times (1 - p), \end{aligned} \quad (25.16)$$

where p is a random number uniformly generated from the interval $[0, 1]$.

25.2.1.4 Synthesis

Synthesis does the opposite of decomposition. In the process of synthesis reaction, a new molecule (ω') can be generated through the collision. This process implies that the search regions are expanded, i.e., diversification of solutions.

The combination of these two existing molecular structures is represented by Eq. 25.17 (Lam and Li 2010a):

$$\omega_1 + \omega_2 \to \omega'. \tag{25.17}$$

The condition to allow the synthesis collision is defined by Eq. 25.18 (Lam and Li 2010a):

$$PE_{\omega_1} + PE_{\omega_2} + KE_{\omega_1} + KE_{\omega_2} \geq PE_{\omega'}. \tag{25.18}$$

If Eq. 25.18 does not hold, the change is prohibited and the molecule retains its original attributes (i.e., ω, *PE*, and *KE*).

Then, the kinetic energy (*KE*) of the new molecule is defined by Eq. 25.19 (Lam and Li 2010a):

$$KE_{\omega'} = (PE_{\omega_1} + PE_{\omega_2} + KE_{\omega_1} + KE_{\omega_1}) - (PE_{\omega'}). \tag{25.19}$$

25.2.2 Performance of CRO

To evaluate the performance of the CRO algorithm, Lam and Li (2010a) used quadratic assignment problem, which is a fundamental combinatorial problem in operations research, as a running example. Compared with the variants of some popular evolutionary algorithms, CRO resulted a superior performance in many test instances.

25.3 Selected CRO Variant

Although CRO algorithm is a new member of computational intelligence (CI) family, a number of CRO variations have been proposed in the literature for the purpose of further improving the performance of CRO. This section gives an overview to some of these CRO variants which have been demonstrated to be very efficient and robust.

25.3.1 Real-Coded CRO Algorithm

In 2012, Lam et al. (2012) proposed a variant of CRO, called real-coded chemical reaction optimization (RCCRO) algorithm to solve continuous optimization problems. The main difference between CRO and RCCRO is that RCCRO utilized the Gaussian distribution function as the perturbation function and some real-coded-based mechanisms have been developed to implement RCCRO.

More precisely, there are three modifications that adopted in the RCCRO algorithm are described as follows (Lam et al. 2012):

- Solution Representation: Every solution (s) and its molecular structure (ω) in a continuous search space is a real number vector as defined by Eq. 25.20 (Lam et al. 2012):

$$\begin{aligned} s &= [s(1), \ldots, s(i), \ldots, s(n)], \\ \omega &= [\omega(1), \ldots, \omega(i), \ldots, \omega(n)], \end{aligned} \tag{25.20}$$

  where n is the dimension of the problem, $s(i)$ and $\omega(i)$ is usually a floating-point number in the range of $[l(i), u(i)]$, where $l(i)$ and $u(i)$ are the lower and upper bounds of the dimension, respectively.
- Neighbourhood Search Operator: For RCCRO, Lam et al. (2012) utilized the Gaussian distribution function as the perturbation function to search the continuous neighbourhoods as defined by Eq. 25.21 (Lam et al. 2012):

$$\omega'(i) = \omega(i) + N(0, \sigma^2). \tag{25.21}$$

- Boundary Constraint Handling: For RCCRO, Lam et al. (2012) adopted two schemes (i.e., reflecting scheme and hybrid schem) to handle the boundary constraints as expressed by Eqs. 25.22 and 25.23, respectively (Lam et al. 2012):

$$\omega'(i) = \begin{cases} 2 \times l(i) - \omega(i), & \text{if } \omega(i) < l(i) \\ 2 \times u(i) - \omega(i), & \text{if } \omega(i) > u(i) \end{cases}, \tag{25.22}$$

$$\omega'(i) = \begin{cases} l(i), & \text{if } (t \leq 0.5) \text{ and } (\omega(i) < l(i)) \\ u(i), & \text{if } (t \leq 0.5) \text{ and } (\omega(i) > u(i)) \\ 2 \times l(i) - \omega(i), & \text{if } (t > 0.5) \text{ and } (\omega(i) < l(i)) \\ 2 \times u(i) - \omega(i) & \text{if } (t > 0.5) \text{ and } (\omega(i) > u(i)) \end{cases}, \tag{25.23}$$

  where t is a random number drawn from $[0, 1]$, and $l(i)$ and $u(i)$ are the lower and upper bounds of the dimension, respectively.

25.3.2 Performance of RCCRO

To implement the efficient of RCCRO, a large set of standard benchmark functions have been tested and compared with classical evolutionary programming (CFP), conventional evolutionary strategy (CES), covariance matrix adaptation evolution strategy (CMAES), differential evolution (DE), fast evolutionary programming (FEP), fast evolutionary strategy (FES), generalized generation gap model with generic parent-centric recombination operator (G3PCX), group search optimizer (GSO), real-coded biogeography-based optimization (RCBBO), standard genetic algorithm (GA), and particle swarm optimization (PSO). Computational results showed that RCCRO has a higher ability to work well in the continuous domain.

25.4 Representative CRO Application

In this section, we introduced how the CRO algorithm can be adapted to solve quadratic assignment problem.

25.4.1 Quadratic Assignment Problem

Quadratic assignment problem (QAP) is one of the most difficult problems in the NP-hard class that arises in real-world applications, such as facilities location, parallel and distributed computing, and combinatorial data analysis. The main objective of QAP is to find the optimal assignment of n facilities to n locations (Loiola et al. 2007).

Compared with traditional CI methods [such as fast ant system (FAS), an improved simulated annealing (ISA), and Tabu search (TS)], CRO resulted a superior performance in many test instances. In addition, for solving QAP, Xu et al. (2010b) proposed a parallel version of CRO, in which a synchronous communication strategy is integrated. The computational results showed that the parallel CRO method gets better solutions' quality and less running time with those of the sequential version of CRO.

25.5 Conclusions

The CRO algorithm is a newly developed evolutionary CI approach that motivated by the molecules' energy exchange in a chemical reaction. The basic building block of CRO is the molecules. Each of them has a molecular structure (i.e., a solution of a given problem), and possesses two kinds of energies: (1) potential energy (i.e., the objective function value of a given problem), (2) kinetic energy (i.e., the ability of escaping from a local minimum). In addition, four elementary reactions, which are occurred either with the walls of the container or with each other, have been implemented to balance the exploitation and the exploration process, namely on-wall ineffective collision, inter-molecular ineffective collision, decomposition and synthesis. Also, CRO enjoys the advantages of both SA and

GA in terms of the law of energy conservation in SA and the utilization of recombination operator (i.e., crossover) and the mutation operator in GA, respectively. Although CRO is a newly CI methods, we have witnessed the following rapid spreading of CRO:

First, in addition to the selected variant detailed in this chapter, several enhanced version of CRO can also be found in the literature as outlined below:

- Canonical CRO (Xu et al. 2011b).
- CRO with greedy strategy (Truong et al. 2013).
- CRO with Lin-Kernighan local search (Sun et al. 2011).
- Discrete CRO (Li and Pan 2012).
- Double molecular structure-based CRO (Xu et al. 2013).
- Parallel version of CRO (Xu et al. 2010b).
- Simplified version of CRO (Khavari et al. 2011).
- Super molecule-based CRO (Xu et al. 2011b).

Second, apart from the representative application, the CRO algorithm has also been successfully applied to a variety of optimization problems as listed below:

- Artificial neural network training (Yu et al. 2011).
- Channel assignment problem in wireless mesh networks (Lam and Li 2010a).
- Cognitive radio spectrum allocation problem (Lam and Li 2010b; Lam et al. 2013).
- Flexible job-shop scheduling problem (Li and Pan 2012).
- Fuzzy job-shop scheduling problem (Li and Pan 2013).
- Grid scheduling problem (Xu et al. 2010a, 2011a).
- Network coding optimization problem (Pan et al. 2011).
- Population transition problem in peer-to-peer live streaming (Lam et al. 2010).
- Resource-constrained project scheduling problem (Lam and Li 2010a).
- Standard continuous benchmark functions (Lam et al. 2012).
- Stock portfolio selection problem (Xu et al. 2011b).
- Travelling salesman problem (Sun et al. 2011).

Interested readers are referred to them, together with an excellent tutorial regarding CRO introduced by Lam and Li (2012), as a starting point for a further exploration and exploitation of the CRO algorithm.

References

Khavari, F., Naseri, V., & Naghshbandy, A. H. (2011). Optimal PMUs placement for power system observability using grenade explosion algorithm. *International Review of Electrical Engineering, 6*.

Lam, A. Y. S., & Li, V. O. K. (2010a). Chemical-reaction-inspired metaheuristic for optimization. *IEEE Transactions on Evolutionary Computation, 14*, 381–399.

Lam, A. Y. S., & Li, V. O. K. (2010b). Chemical reaction optimization for cognitive radio spectrum allocation. IEEE Global Communication Conference (IEEE GLOBECOM 2010) (pp. 1–5), December. Miami, FL, USA.

Lam, A. Y. S., & Li, V. O. K. (2012). Chemical reaction optimization: a tutorial. *Memetic Computing, 4*, 3–17.

Lam, A. Y. S., Xu, J., & Li, V. O. K. (2010, July 18–23). Chemical reaction optimization for population transition in peer-to-peer live streaming. IEEE World Congress on Computational Intelligence, CCIB (pp. 1429–1436). Barcelona, Spain.

Lam, A. Y. S., Li, V. O. K., & Yu, J. J. Q. (2012). Real-coded chemical reaction optimization. *IEEE Transactions on Evolutionary Computation, 16*, 339–353.

Lam, A. Y. S., Li, V. O. K., & Yu, J. J. Q. (2013). Power-controlled cognitive radio spectrum allocation with chemical reaction optimization. *IEEE Transactions on Wireless Communications, 12*, 3180–3190. doi:10.1109/TWC.2013.061713.120255.

Li, J.-Q., & Pan, Q.-K. (2012). Chemical-reaction optimization for flexible job-shop scheduling problems with maintenance activity. *Applied Soft Computing, 12*, 2896–2912. http://dx.doi.org/10.1016/j.asoc.2012.04.012.

Li, J.-Q., & Pan, Q.-K. (2013). Chemical-reaction optimization for solving fuzzy job-shop scheduling problem with flexible maintenance activities. *International Journal of Production Economics, 145*, 4–17. doi:http://dx.doi.org/10.1016/j.ijpe.2012.11.005.

Loiola, E. M., Abreu, N. M. M. D., Boaventura-Netto, P. O., Hahn, P., & Querido, T. (2007). A survey for the quadratic assignment problem. *European Journal of Operational Research, 176*, 657–690.

Pan, B., Lam, A. Y. S., & Li, V. O. K. (2011). Network coding optimization based on chemical reaction optimization. IEEE Global Communications Conference (GLOBECOM), December. Houston, TX, USA.

Serway, R. A., & Jewett, J. W. (2014). *Physics for scientists and engineers with modern physics*. Boston: MA, USA, Brooks/Cole CENAGE Learning. ISBN 978-1-133-95405-7.

Sun, J., Wang, Y., Li, J., & Gao, K. (2011). Hybrid algorithm based on chemical reaction optimization and Lin-Kernighan local search for the traveling salesman problem. IEEE Seventh International Conference on Natural Computation (ICNC), pp. 1518–1521.

Truong, T. K., Li, K., & Xu, Y. (2013). Chemical reaction optimization with greedy strategy for the 0–1 knapsack problem. *Applied Soft Computing, 13*, 1774–1780.

Whitten, K. W., Davis, R. E., Peck, M. L., & Stanley, G. G. (2014). *Chemistry*. Brooks/Cole, Cengage Learning: USA. ISBN 13: 978-1-133-61066-3.

Xu, J., Lam, A. Y. S., & Li, V. O. K. (2010a). Chemical reaction optimization for the grid scheduling problem. IEEE International Conference on Communications (ICC), May, pp. 1–5.

Xu, J., Lam, A. Y. S., & Li, V. O. K. (2010b). Parallel chemical reaction optimization for the quadratic assignment problem. Internatimonal Conference of Genetic Evolutionary Methods (GEM), part of 2010 World Congress on Computer Science, Computer Engineering, and Applied Computing (WORLDCOMP) (pp. 1–7). July, Las Vegas, NV, USA.

Xu, J., Lam, A. Y. S., & Li, V. O. K. (2011a). Chemical reaction optimization for task scheduling in grid computing. *IEEE Transactions on Parallel and Distributed Systems, 22*, 1624–1631.

Xu, J., Lam, A. Y. S., & Li, V. O. K. (2011b). Stock portfolio selection using chemical reaction optimization. *World Academy of Science, Engineering and Technology, 77*, 458–463.

Xu, Y., Li, K., He, L., & Truong, T. K. (2013). A DAG scheduling scheme on heterogeneous computing systems using double molecular structure-based chemical reaction optimization. *Journal of Parallel and Distributed Computing, 73*, 1306–1322.

Yu, J. J. Q., Lam, A. Y. S., & Li, V. O. K. (2011). Evolutionary artificial neural network based on chemical reaction optimization. Proceedings of the IEEE Congress on Evolutionary Computation (CEC) (pp. 2083–2090), June, New Orleans, LA, USA.

Zumdahl, S. S., & Zumdahl, S. A. (2014). *Chemistry, 20 Davis Drive, Belmont, CA 94002-3098*. Brooks/Cole, Cengage Learning: USA. ISBN 13: 978-1-133-61109-7.

Chapter 26
Emerging Chemistry-based CI Algorithms

Abstract In this chapter, a set of emerging chemistry-based computational intelligence (CI) algorithms are introduced. We first, in Sect. 26.1, describe the organizational structure of this chapter. Then, from Sect. 26.2 to Sect. 26.5, each section is dedicated to a specific algorithm which falls within this category. The fundamentals of each algorithm and their corresponding performances compared with other CI algorithms can be found in each associated section. Finally, the conclusions drawn in Sect. 26.6 closes this chapter.

26.1 Introduction

A novel chemistry-based algorithm (i.e., chemical-reaction optimization algorithm) was detailed in the previous chapter. Apart from this quasi-mature chemistry principle inspired computational intelligence (CI) methods, there are some emerging algorithms also fall within this category. This chapter collects four of them that are currently scattered in the literature and organizes them as follows:

- Section 26.2: Artificial Chemical Process Algorithm.
- Section 26.3: Artificial Chemical Reaction Optimization Algorithm.
- Section 26.4: Chemical Reaction Algorithm.
- Section 26.5: Gases Brownian Motion Optimization Algorithm.

The effectiveness of these newly developed algorithms are validated through the testing on a wide range of benchmark functions and engineering design problems, and also a detailed comparison with various traditional performance leading CI algorithms, such as particle swarm optimization (PSO), genetic algorithm (GA), differential evolution (DE), evolutionary algorithm (EA), fuzzy system (FS), ant colony optimization (ACO), and simulated annealing (SA).

B. Xing and W.-J. Gao, *Innovative Computational Intelligence: A Rough Guide to 134 Clever Algorithms*, Intelligent Systems Reference Library 62,
DOI: 10.1007/978-3-319-03404-1_26,

26.2 Artificial Chemical Process Algorithm

In this section, we will introduce an emerging CI algorithm that is based on an artificial chemical process.

26.2.1 *Fundamentals of Artificial Chemical Process Algorithm*

Artificial chemical process algorithm (ACPA) was proposed in Irizarry (2004, 2005a, b, 2011). In ACPA, each solution is viewed as a vector and encoded into a set of independent discrete variables (x_j, $j = 1, \ldots, V$), called molecules. The major feature of ACPA is the use of multiple encoding and the application of certain kinds of constraints. The main steps of ACPA are outlined as follows (Irizarry 2004, 2005a, b, 2011):

- Initialization: The algorithm starts by initializing x^g (the value of the best solution found so far, $x^g = \left(x_1^g, \ldots, x_V^g\right)$) randomly and placing all variables in L (set $AR = E = S = \emptyset$, $L = \{x_1, \ldots, x_V\}$). Let L (load unit), AR (activation-reactor), E (extraction unit), and S (separation unit) be four disjunctive sets whose elements are the molecule variables.
- Outer loop: Perturbation to form AR.
  First, select a random number $|T_{rx}|(|T_{rx}| \leq |L|)$, from a uniform probability distribution function (PDF) by using Eq. 26.1 (Irizarry 2004, 2005a, b, 2011):

  $$|T_{rx}| = \min\{\text{int}(\rho[V \cdot c_0] + 1), |L|\}, \tag{26.1}$$

  where ρ is uniformly distributed in (0,1), and c_0 is an adjustable parameter used to select the average fraction of elements to be selected from V, which is the total number of molecules representing the possible solutions to the problem.
  Second, select $|T_{rx}|$ elements randomly from L to form the subset $T_{rx}(T_{rx} \subseteq L)$.
  Third, transfer the subset to AR: $L = L \backslash T_{rx}$; $AR = AR \cup T_{rx}$.
  Fourth, select a random new value for each molecule variable in T_{rx}: $x_j^a \neq x_j^g, \forall j \in T_{rx}$.
  Fifth, the new trial vector ($x^t = \left(x_1^t, \ldots, x_V^t\right)$) is generated using Eq. 26.2 (Irizarry 2004, 2005a, b, 2011):

  $$x_j^t = \begin{cases} x_j^g & \text{if } x_j \notin AR \\ x_j^a & \text{if } x_j \in AR \end{cases}. \tag{26.2}$$

  Sixth, if $F(x^t) < F(x^g)$, the trial vector is accepted as a new best solution found, i.e., $x^g = x^t$. In this case, all of the elements in AR are set to the set S: $S = S \cup AR$; $AR = ø$. If the algorithm termination criterion is achieved, exit the algorithm and return x^g as the solution to the optimization problem.

Seventh, if a better solution was found in Sixth, skip the inner loop and perform another outer loop iteration: go to check the algorithm termination criterion; otherwise, continue with Eighth.
Eighth, initialize the parameter RP (reactor performance): $RP = F(x^t)$. This parameter is used and modified in the "goodness" test in the inner loop. Also set $|AR|_0 = |AR|$, which is the initial number of molecules in AR before starting the next inner loop.

- Inner loop: Iterative improvement of AR.
First, select a random number $|E|$, from a prescribed PDF, $|E| \leq |AR|$. It can be defined by Eq. 26.3 (Irizarry 2004, 2005a, b, 2011):

$$|E| = \min\left[\text{int}\left(\rho\left[|AR|_0 \cdot c_i\right] + 1\right), |AR|\right], \tag{26.3}$$

where $|AR|_0$ is defined in the outer loop (i.e., Eighth), and c_i is another adjustable parameter.
Second, select $|E|$ elements randomly from AR to form the subset E.
Third, extract the subset E from $AR : AR = AR \backslash E$.
Fourth, the new trial vector $(x^t = \left(x_1^t, \cdots, x_V^t\right))$ is generated using Eq. 26.2.
Fifth, if $F(x^t) < F(x^g)$, the trial vector is accepted as a new best solution found, i.e., $x^g = x^t$. In this case, all of the elements in AR are set to the set $S : S = S \cup AR;\ AR = \text{ø}$. If the algorithm termination criterion is achieved, exit the algorithm and return x^g as the solution to the optimization problem.
Sixth, if $F(x^t) \leq RP$, the hypothesis is that there is a high probability that most elements in E will prefer to stay in their ground stat $(x_j = x_j^g)$ to generate better solutions. In this case, the elements in E are transferred to $S : S = S \cup E;\ E = \text{ø}$; and the metric RP is updated, $RP = F(x^t)$. If $F(x^t) > RP$, the hypothesis is that there is a high probability that most elements in E will induce a better solution if they are in a different state from their ground state. In this case, a new activated state is generated for all elements in $E\left(x_j = x_j^a \neq x_j^g, \forall j \in E\right)$, and all of the elements in E are transferred back to $AR(AR = AR \cup E; E = \text{ø})$.
- Check conditions to exit or continue the inner loop.
- Check the number of elements in L and AR using Eq. 26.4 (Irizarry 2004, 2005a, b, 2011):

$$\begin{cases} |L| < LT, & L = L \cup S; S = \text{ø} \\ |AR| = V \text{ or } |L| \leq LT, & L = L \cup AR; AR = \text{ø} \end{cases}. \tag{26.4}$$

- Check the termination criteria.

26.2.2 Performance of ACPA

To test the performance of ACPA, a set of benchmark functions are conducted in Irizarry (2004). Compared with other CI algorithms (such as GA), computational results showed that the proposed algorithm outperforms GA in terms of convergence speed and the solutions' quality.

26.3 Artificial Chemical Reaction Optimization Algorithm

In this section, we will introduce an emerging CI algorithm that is based on the studies of chemical reaction (Koretsky 2013; Miessler et al. 2014; Skoog et al. 2014; Whitten et al. 2014; Zumdahl and Zumdahl 2014; Finlayson 2012; Duncan 2009; Waltar et al. 2012; Weinhold and Landis 2012; Borgnakke and Sonntag 2013; Tadmor et al. 2012).

26.3.1 Fundamentals of Artificial Chemical Reaction Optimization Algorithm

Artificial chemical reaction optimization algorithm (ACROA) was recently proposed in Alatas (2011, 2012). The working principles are described as follows: suppose there is a vessel which has a fixed volume and holds a number of N reactants. These chemical substances are spatially and uniformly mixed and interact with each other via particular ways of chemical reaction. Let R_i (for $1 \leq i \leq N$) be the list of chemical substances, and the authors also make an assumption that these reactants an interact through N specified chemical reaction pipelines. Based on the targeted problem, the reactants involved in ACROA can be encoded in different manners, such as binary, real, and string. A new reactant can be created as a result of the interaction between one of more reactants by following different reaction rules which depends on the previously defined encoding schemes.

Typically, ACROA starts with a group of initial populations (i.e., reactants), then it proceeds with the process of these initial substances being consumed and new substances being reproduced, and finally, the ACROA stops at an inert situation, namely, no more chemical reactions could happen. Built on this simple concept, the following steps need to be performed for implementing ACROA (Alatas 2011, 2012; Yang et al. 2011):

- Stage 1: Initializing the parameters of focal problem and algorithm.
- Stage 2: Determining the initial population of reactants and evaluation.

- Stage 3: Performing different chemical reactions mechanisms such as synthesis, decomposition, single/double displacement, combustion, redox, and reversible reactions.
- Stage 4: Updating chemical reaction reactants.
- Stage 5: Checking the stopping criterion.

26.3.2 Performance of ACROA

In order to show how the ACROA performs, Alatas (2012) employed ACROA to find comprehensible IF–THEN rules within two real-world data sets, namely, Zoo and Nursery by following two procedures listed below (Alatas 2012):

- Rule representation: For each focal data set, let us first suppose the number of predictable attributes is denoted by *na*; then, in ACROA, each participated reactant is assumed to have *na* atoms with each atoms corresponding to one potential attribute.
- Rule evaluation: Normally, the objective function for rule extraction is composed of two objectives (i.e., comprehensibility and predictive accuracy). The mathematical expressions of comprehensibility and predictive accuracy can be obtained via Eqs. 26.5 and 26.6, respectively (Alatas 2012):

$$\text{Comprehensibility} = 1 - \frac{\text{No. of attributes in the reactant}}{\text{No. of predicting attribute}}, \tag{26.5}$$

$$\text{Predictive Accuracy} = \frac{|A + C| - \frac{1}{2}}{|A|}, \tag{26.6}$$

where $|A + C|$ denotes the number of cases satisfying both the antecedent and consequent rules, and $|A|$ represents the number of cases satisfying the antecedent rule only. Built on this, the final objective function can be expressed using Eq. 26.7 (Alatas 2012):

$$\begin{aligned}\text{Objective Function} =& w_1 \times (\text{Comprehensibility}) \\ &+ w_2 \times (\text{Predictive Accuracy})\end{aligned}, \tag{26.7}$$

where w_1 and w_2 are weights that usually defined by user.

The number of the initial population was set as 50 and the stopping criterion was defined as when the best value of objective function has no changes after 10 continuous generations. The rules obtained by using ACROA were compared with the other competitor algorithm called genetic algorithm. At the end of the study, Alatas (2012), claimed that, when taking into account of predictive accuracy, the performance of ACROA is slightly better than GA on the chosen data sets. It seems like the ACROA can be treated as a potential effective search approach in dealing with different types optimization problems.

26.4 Chemical Reaction Algorithm

In this section, we will introduce an emerging CI algorithm that is also derived from the observations of chemical reaction process (Koretsky 2013; Miessler et al. 2014; Skoog et al. 2014; Whitten et al. 2014; Zumdahl and Zumdahl 2014; Finlayson 2012; Duncan 2009).

26.4.1 Fundamentals of Chemical Reaction Algorithm

Chemical reaction algorithm (CRA) was recently proposed in Melin et al. (2013). In CRA, each element (or compound) is viewed as a solution. Like other algorithms, the fitness of the elements will be evaluated in accordance with the objective function. The main difference between CRA and other CI algorithms is that no external parameters are taken into account to evaluate the results. As a result, it is a very straight forward methodology that only takes the basic characterises of chemical reactions (i.e., synthesis, decomposition, single-substitution, and double-substitution) into account to find the optimal solutions. The main steps of CRA can be described as follows (Melin et al. 2013):

- Step 1: Defining the optimization problem and initializing the optimization parameters.
- Step 2: Generating the initial population pool randomly.
- Step 3: Evaluate the initial pool.
- Step 4: Identify the best_solution.
- Step 5: Select elements to react. This is an iterative process that includes four parts: (1) Perform the elements reactions, i.e., synthesis, decomposition, single-substitution, and double-substitution; (2) Evaluate those reactions; (3) Apply elitist_reinsertion and get improved_pool; (4) Update the best_solution.
- Step 6: Post process and visualize results.

26.4.2 Performance of CRA

Melin et al. (2013) applied CRA to solve the tracking control problem, specially for the dynamic model of a unicycle mobile robot. Simulation results showed that CRA outperforms the results previously obtained from GA.

26.5 Gases Brownian Motion Optimization Algorithm

In this section, we will introduce an emerging CI algorithm that is based on the behaviour of molecule motion (Tian 2013; Durrett 1984; Shlesinger et al. 1999; Bolstad 2012; Lyshevski 2007; Yin 2013).

26.5.1 Fundamentals of Gases Brownian Motion Optimization Algorithm

Gases Brownian motion optimization (GBMO) algorithm was originally proposed in Abdechiri et al. (2013). In GBMO, the agents are a swarm of molecules and their corresponding positions are normally used to measure each molecule's performance. Since each position is treated as a part of the candidate solution, the GBMO algorithm thus proceeds through an appropriate adjustment of two types of motions, namely, gases Brownian motion and turbulent rotational motion. Typically, the molecule agents involved in GBMO have four specifications (i.e., position, mass, velocity, and turbulent radius). To implement the GBMO algorithm, the following steps need to be performed (Abdechiri et al. 2013):

- Stage 1: Initializing initial population (i.e., number of molecules and their relative locations). In original GBMO, the molecules' positions are randomly distributed in an array manner.
- Stage 2: Taking into account of a random radius of turbulence for each involved molecule and such radius usually falls within the range of [0, 1].
- Stage 3: Assigning a temperature to the system. Since in real-world temperature can affect molecules' moving velocity, the user defined temperature plays an important role in balancing the GBMO's capability of exploitation and exploration which in turn will have a great influence on the converging speed of the algorithm.
- Stage 4: Updating molecules' velocity and position via Eqs. 26.8 and 26.9, respectively (Abdechiri et al. 2013):

$$v_i^d(t+1) = v_i^d(t) + \sqrt{\frac{3kT}{m}}, \tag{26.8}$$

$$x_i^d(t+1) = x_i^d(t) + v_i^d(t), \tag{26.9}$$

  where the ith molecule's velocity and position are denoted by $\mathbf{X}_i = \left(x_i^1, x_i^2, \ldots, x_i^n\right)$ and $\mathbf{V}_i = \left(v_i^1, v_i^2, \ldots, v_i^n\right)$, respectively.
- Stage 5: Evaluating the fitness values of molecules.
- Stage 6: Performing turbulent rotational motion according to Eq. 26.10 (Abdechiri et al. 2013):

$$x_i^d(t+1) = x_i^d(t) + b - \left(\frac{a}{2\pi}\right) \times \sin\left[2\pi x_i^d(t)\right] \times \mathrm{mod}(1), \tag{26.10}$$

  where, under the settings of $a = 0.5$ and $b = 0.2$, a chaotic sequence within (0, 1) could be created, and x_i^d denotes the present location of a molecule during a turbulent rotational motion.
- Stage 7: Comparing the values of the objective function with molecules' new positions.

- Stage 8: Updating the values of mass and temperature. A lighter mass of a molecule normally results in a higher velocity which is often linked with a more efficient searching ability. In GBMO, the mass of a molecule is updated by Eq. 26.11 (Abdechiri et al. 2013):

$$m_i(t) = \frac{\text{fit}_i(t) - \text{worst}(t)}{\text{best}(t) - \text{worst}(t)}, \tag{26.11}$$

where $\text{fit}_i(t)$ stands for the fitness value of the ith molecule at time t. Meanwhile, the temperature parameter is updated through Eq. 26.12 (Abdechiri et al. 2013):

$$T = T - \left(\frac{1}{mean(\text{fit}_i(t))}\right). \tag{26.12}$$

- Stage 9: Checking whether the stopping criterion is met. If yes, terminating the algorithm; otherwise, repeating the iterations of Stages 3–7.

26.5.2 Performance of GBMO

In order to show how the GBMO algorithm performs, Abdechiri et al. (2013) used seven benchmark test functions, such as Sphere function, Rastrigin function, and Rosenbrock function. Compared with other CI techniques (e.g., PSO, GA), The experimental results demonstrated that GBMO outperforms the other methods when dealing with the high dimensionality optimization problems. In terms of computational complexity, the GBMO is $O(n)$ which is less than GA but slightly higher than PSO.

26.6 Conclusions

In this chapter, four emerging chemistry-based CI methodologies are discussed. Although most of them are still in their infancy, their usefulness has been demonstrated throughout the preliminary corresponding studies. Interested readers are referred to them as a starting point for a further exploration and exploitation of these innovative CI algorithms.

References

Abdechiri, M., Meybodi, M.R., Bahrami, H., (2013). Gases Brownian motion optimization: An algorithm for optimization (GBMO). *Applied Soft Computing, 13,* 2932–2946. (http://dx.doi.org/10.1016/j.asoc.2012.03.068).

Alatas, B. (2011). ACROA: Artificial chemical reaction optimization algorithm for global optimization. *Expert Systems with Applications, 38,* 13170–13180.

Alatas, B. (2012). A novel chemistry based metaheuristic optimization method for mining of classification rules. *Expert Systems with Applications, 39*, 11080–11088.

Bolstad, T. M. (2012). Brownian motion. Department of Physics and Technology, University of Bergen.

Borgnakke, C., & Sonntag, R. E. (2013). *Fundamentals of thermodynamics*. Hoboken: Wiley. ISBN 978-1-118-13199-2.

Duncan, A. (2009). *Introduction to chemical engineering processes*. Chandni Chowk: Global Media. ISBN 978-93-80168-32-6.

Durrett, R. (1984). *Brownian motion and martingales in analysis*. Belmont: Wadsworth Advanced Books & Software, A Division of Wadsworth, Inc. ISBN 0-534-03065-3.

Finlayson, B. A. (2012). *Introduction to chemical engineering computing*. Hoboken: Wiley. ISBN 978-0-470-93295-7.

Irizarry, R. (2004). LARES: An artificial chemical process approach for optimization. *Evolutionary Computation Journal, 12*, 1–8.

Irizarry, R. (2005a). Fuzzy classification with an artificial chemical process. *Chemical Engineering Science, 60*, 399–412.

Irizarry, R. (2005b). A generalized framework for solving dynamic optimization problems using the artificial chemical process paradigm: Applications to particulate processes and discrete dynamic systems. *Chemical Engineering Science, 60*, 5663–5681.

Irizarry, R. (2011). Global and dynamic optimization using the artificial chemical process paradigm and fast Monte Carlo methods for the solution of population balance models. In I. Dritsas (Ed.), *Stochastic optimization—seeing the optimal for the uncertain, Chapter 16*. Rijeka: InTech. ISBN 978-953-307-829-8.

Koretsky, M. D. (2013). *Engineering and chemical thermodynamics*. Hoboken: Wiley.

Lyshevski, S. E. (Ed.). (2007). *Nano and molecular electronics handbook*. Boca Raton: CRC Press, Taylor & Francis Group. ISBN 978-0-8493-8528-5.

Melin, P., Astudillo, L., Castillo, O., Valdez, F., & Valdez, F. (2013). Optimal design of type-2 and type-1 fuzzy tracking controllers for autonomous mobile robots under perturbed torques using a new chemical optimization paradigm. *Expert Systems with Applications, 40*, 3185–3195.

Miessler, G. L., Fischer, P. J., & Tarr, D. A. (2014). *Inorganic chemistry*. Upper Saddle River: Pearson Education, Inc. ISBN 978-0-321-81105-9.

Shlesinger, M. F., Klafter, J., & Zumofen, G. (1999). Above, below and beyond Brownian motion. *American Journal of Physics, 67*, 1253–1259.

Skoog, D. A., West, D. M., Holler, F. J., & Crouch, S. R. (2014). *Fundamentals of analytical chemistry*. Belmont: Brooks/Cole, Cengage Learning. ISBN 13: 978-0-495-55828-6.

Tadmor, E. B., Miller, R. E., & Elliott, R. S. (2012). *Continuum mechanics and thermodynamics: From fundamental concepts to governing equations*. Cambridge: Cambridge University Press. ISBN 978-1-107-00826-7.

Tian, J. (Ed.). (2013). *Molecular imaging: Fundamentals and applications*. Hangzhou, Heidelberg and New York: Zhejiang University Press and Springer. ISBN 978-7-308-08271-6; 978-3-642-34302-5.

Waltar, A. E., Todd, D. R., & Tsvetkov, P. V. (Eds.). (2012). *Fast spectrum reactors*. New York and Heidelberg: Springer Science + Business Media LLC. ISBN 978-1-4419-9571-1.

Weinhold, F., & Landis, C. R. (2012). *Discovering chemistry with natural bond orbitals*. Hoboken: Wiley. ISBN 978-1-118-11996-9.

Whitten, K. W., Davis, R. E., Peck, M. L., & Stanley, G. G. (2014). *Chemistry*. Belmont: Brooks/Cole, Cengage Learning. ISBN 13: 978-1-133-61066-3.

Yang, S.-D., YI, Y.-L., & Shan, Z.-Y. (2011). Gbest-guided artificial chemical reaction algorithm for global numerical optimization. *Procedia Engineering, 24*, 197–201.

Yin, Y. (Ed.). (2013). *Responsive photonic nanostructures: Smart nanoscale optical materials*. Cambridge: The Royal Society of Chemistry. ISBN 978-1-84973-653-4.

Zumdahl, SS., Zumdahl, SA. (2014). Chemistry. Belmont: Brooks/Cole, Cengage Learning. 978-1-133-61109-7

Part V
Mathematics-based CI Algorithms

Chapter 27
Base Optimization Algorithm

Abstract In this chapter, the base optimization algorithm (BaOA), a global optimization method inspired from mathematics research, is introduced. We first, in Sect. 27.1, describe the background knowledge of mathematics. Then, the fundamentals and performance of BaOA are detailed in Sect. 27.2. Finally, Sect. 27.3 draws the conclusions of this chapter.

27.1 Introduction

It is hard to imagine our human society today without vast amount of advanced technologies. In particular during the past decades, we have witnessed a proliferation of personal computers, smart phones, high-speed Internet, to name a few. The rapid development of various technologies has reduce the necessity for human beings to perform manual tasks which are either tedious or dangerous in nature, as computers may now accomplish most of them. As one of the most important building blocks, mathematics plays a crucial role in realizing all these technologies. The history of mathematics is no doubt tremendous long. According to (Anglin 1994), Aristotle thought that is the priests in Egypt who actually started mathematics since the priestly class was allowed leisure. Whereas, Herodotus, believed that geometry was created to re-determine land boundaries due to the annual flooding of the Nile. The accurate beginning of mathematics is of course out of the scope of this book. For the rest of this section, we just want to give readers a quick refreshment about several basic arithmetic operators that form the base of mathematics.

27.1.1 Basic Arithmetic Operators

In mathematics, the four basic arithmetic operators are (Bird 2005): addition (+), subtraction (−), multiplication (×), and division (÷).

B. Xing and W.-J. Gao, *Innovative Computational Intelligence: A Rough Guide to 134 Clever Algorithms*, Intelligent Systems Reference Library 62,
DOI: 10.1007/978-3-319-03404-1_27, © Springer International Publishing Switzerland 2014

- For addition and subtraction, when unlike signs are together in a calculation, the overall sign is negative. Therefore, adding minus 5 to 6 is $6 + (-5)$ and turns out to be $6 - 5 = 1$. Like signs together give an overall positive sign. Thus subtracting minus 5 from 6 is $6 - (-5)$ and becomes $6 + 5 = 11$.
- For multiplication and division, when the numbers have unlike signs, the answer will be negative; nevertheless, when the numbers have like signs, the answer is positive. As a result, $6 \times (-5) = -30$, and $(-6) \times (-5) = +30$. Similarly, the following expressions also hold:

$$\frac{-5}{6} = -\frac{5}{6} \quad \text{and} \quad \frac{-5}{6} = +\frac{5}{6}.$$

27.2 Base Optimization Algorithm

27.2.1 Fundamentals of Base Optimization Algorithm

Inspired from mathematical studies, Salem (2012) recently proposed a novel computational intelligence (CI) technique, namely, the base optimization algorithm (BaOA). The main characteristics of BaOA lies in that the standard arithmetic operators are used in combination with a displacement parameter for leading the solutions to reach the optimum position.

Briefly, the BaOA works as follows (Salem 2012):

- First, a set of the solution points $(S_1 \ldots S_n)$ are randomly created, each with d dimensions;
- Second, each solution point is examined in such a way that each dimension is updated whenever an arithmetic operator move the solution point to a new position with a better value. This process is executed on all solution points so that the search for optimum solution can be continued;
- Third, the algorithm terminates when the current solution points do not give better value or no improvement is acquired after a certain round of iterations. Under this situation, it is concluded that the optimum solution has been located.

For the purpose of simplicity, the arithmetic operators used in BaOA are limited to addition (+), subtraction (−), multiplication (×), and division (÷) for now. These operators can, of course, be extended to any other operators according to users' requirements.

Summarizing the steps in BaOA yields to (Salem 2012):

- Step 1: Randomly generate n solution point $(S_1, \ldots, S_n)$, each with d – dimensions.
- Step 2: Initializing the parameters.
- Step 3: For each solution point S_i, calculate its fitness value $f(S_i)$.

- Step 4: For each dimension of S_i, update the jth dimension with basic arithmetic operators via Eq. 27.1 (Salem 2012):

$$\left.\begin{aligned} S_i^+[j] &= S_i[j] + \delta \\ S_i^-[j] &= S_i[j] - \delta \\ S_i^\times[j] &= S_i[j] \times \delta \\ S_i^\%[j] &= S_i[j]/\delta \end{aligned}\right\} R_{\min} \le S_i^*[j] \le R_{\max}\,. \tag{27.1}$$

- Step 5: Re-compute the fitness values f^+,f^-, $f^\times$, and $f^\%$.
- Step 6: Update the fitness value $f(S_i)$ with the best value via Eq. 27.2 (Salem 2012):

$$f(S_i) = \min\{f_{j+}, f_{j-}, f_{j\times}, f_{j\%}\}. \tag{27.2}$$

- Step 7: Check the termination criteria.

27.2.2 *Performance of BaOA*

In (Salem 2012), three benchmark function sets (i.e., unimodal functions, functions with many or few local maxima) were used to test the performance of the proposed BaOA. These functions include sphere model function, generalized Rosenbrock's function, generalized Schwefel's function, generalized Rastrigin's function, Ackley's function, generalized Griewank's function, six-hump camel back function, and the Goldstein-Price function. Compared with convention CI techniques, the BaOA reaches a 100 % success rate along with the minimum number of iterations.

27.3 Conclusions

In summary, the BaOA used the basic arithmetic operators (i.e., +, −, ×, ÷) combining a displacement parameter to guide and redirect the solutions toward the optimum point. One of the main features of BaOA is, unlike the standard particle swarm optimization (PSO), the search process of BaOA can be performed with a single search which has the capability of learning how to find its way to the desired optimum point. Therefore, the BaOA does not necessarily need the cooperation among a group of particles. The leading advantages of BaOA are its conceptual simplicity and the capability of easy-to-implement. Interested readers please refer to (Salem 2012) as a starting point for a further exploration and exploitation of BaOA.

References

Anglin, W. S. (1994). *Mathematics: A concise history and philosophy*. New York: Springer Inc. ISBN 0-378-94280-7.

Bird, J. (2005). *Basic engineering mathematics*. Jordan Hill: Linacre House. ISBN 0-7506-6575-0.

Salem, S. A. (2012, October 10–11). BOA: A novel optimization algorithm. In *International Conference on Engineering and Technology (ICET)* (pp. 1–5). Egypt: IEEE.

Chapter 28
Emerging Mathematics-based CI Algorithms

Abstract In this chapter, an emerging mathematics-based CI category called matheuristics is introduced. We first, in Sect. 28.1, describe the background knowledge regarding the metaheuritics. Then, the fundamentals and representative application of matheuristics are briefed in Sect. 28.2. Finally, Sect. 28.3 draws the conclusions of this chapter.

28.1 Introduction

In this chapter, we will introduce an emerging computational intelligence (CI) category which has its root in mathematics. In the literature, there is a name specially coined for this class of algorithms, i.e., matheuristics. An essential feature for the algorithms belonging to this group is that some parts of the algorithms exploiting the characteristics derived from the fundamental mathematical model of the targeted problems. And thus, we may also find the model-based metaheuristics appearing in the list of topics-of-interest. Accordingly, before we move on to matheuristics, a quick overview of "What is metaheuristics?" is provided in this section.

28.1.1 Metaheuritics

Metaheuristics, a term often mixed used with CI, represent a set of solution methodologies that harmonize an interaction between local enhancement procedures and higher level strategies for the purpose of creating a process which is capable of running away from local optimal and thus completing a robust search throughout the solution space. Slowly, the range of these approaches has also become wide enough to cover any procedures that utilize strategies for conquering the countless traps of local optimality existing in the focal complex solution spaces, in particular those procedures that employ one or more neighbourhood

B. Xing and W.-J. Gao, *Innovative Computational Intelligence: A Rough Guide to 134 Clever Algorithms*, Intelligent Systems Reference Library 62,
DOI: 10.1007/978-3-319-03404-1_28,

structures as a way of defining acceptable movements to shift from one solution to another, or to establish or remove solutions during the constructive and destructive process (Glover and Kochenberger 2003; Talbi 2009; Yang 2010; Gonzalez 2007; Xhafa and Abraham 2008; Alba 2005; Birattari 2009). Throughout this book, we can see that, no matter which name you would like to use, CI or metaheuristics, all tools and mechanisms within this category share one key feature, i.e., they can not guarantee to find the global optimal solutions. Nevertheless, exact approaches, although theoretically being proved to be able to provide such a warranty if allowed to operate long enough, are often fail to find the answers whose quality is competitive compared with those obtained by leading CI algorithms, in particular for various real world problems.

28.2 Matheuristics

28.2.1 Fundamentals of Matheuristics

In general, matheuristics (Maniezzo et al. 2009; Pirkwieser 2012; Archetti et al. 2013) can be viewed as the combination or interoperation of various CI approaches (or metaheuristics) and mathematical programming techniques (e.g., integer programming). Traditionally, the field of CI has always been very hospitable to new proposals about how to construct algorithms to achieve a better solvability for optimization problems which often have remarkably high levels of complexity. Innovation of solution methods has thus long been one of the major features of the CI field. Inspirations from various sources such as animal, plant, human being, physics, and chemistry have made the well-engineered algorithms a great success. In the context of this trend, the paradigm put forth in this chapter, i.e., matheuristics, represents a "back to the roots" fashion for CI: using mathematics.

Since the heart of mathematics is hybridization, an intent to cover all aspects of matheuristics within single chapter would be too ambitious. Here we use an recently proposed corridor method as an example to showcase the power of matheuristics. Corridor method was originally proposed in (Sniedovich and Voß 2005) as a hybrid metaheuristic. The core idea of corridor method is the utilization of a (possibly exact) method to deal with the smaller "instances" of the original optimization problem. In other words, looking for an optimal solution on a limited part of the solution space is the first stage target of corridor approach. In (Caserta and Voß 2012), the authors made an attempt to investigate how the corridor method concept can be used to solve the deoxyribonucleic acid (DNA) sequencing problem. The authors proposed a set of corridor selection strategies as detailed below (Caserta and Voß 2012):

- Fixed corridor. The position of the oligo with the minimum total overlapping degree is denoted by k which can be expressed via Eq. 28.1 (Caserta and Voß 2012):

$$k \leftarrow \underset{w=1,2,\ldots,m}{\arg\min}\ \{\sigma_w\}, \tag{28.1}$$

where the total overlapping degree score of an oligo o_w is defined by Eq. 28.2 (Caserta and Voß 2012):

$$\begin{aligned} k &\leftarrow \underset{w=1,2,\ldots,m}{\arg\min}\ \{\sigma_w\}, \\ \sigma_w &= od(o_{w-1}, o_w) + od(o_w, o_{w+1}). \end{aligned} \tag{28.2}$$

- Dynamic corridor. The dynamic programming scheme is applied over the set of oligos in O_δ. The scheme is then initialized with the following state vector $\mathbf{s} = \{O, f, l, p_i, i\}$.

Bearing the above description in mind, the overall algorithm employed in (Caserta and Voß 2012) can be outlined as below:

- Require: spectrum S, total length n.
- Ensure permutation π^*.
- Initialization of transition probability matrix P_0.
- Corridor phase.
- Dynamic programming phase.
- Adding π^{DP} to quantile of cross entropy population.

28.2.2 *Performance of Matheuristics*

To evaluate the proposed matheuristics approach, Caserta and Voß (2012) tested it on two sets of benchmark sequencing problems in which the first set consists of 320 cases found from the literature and the second set is characterized by its repeated oligonucleotides nature. Through an intensive study, all experimental results demonstrated the efficacy of the proposed matheuristics method, both in terms of the solution quality and the total running time.

28.3 Conclusions

In this chapter, we briefly introduced a new trend emerged in the CI community that is matheuristics. An illustrative study was presented to illustrate the usefulness of mathematics. Interested readers are referred to the literature mentioned in this chapter as a starting point for a further exploration and exploitation of these hybridized CI algorithms.

References

Alba, E. (Ed.). (2005). *Parallel metaheuristics: A new class of algorithms*. New Jeresy: Wiley. ISBN 978-0-471 -67806-9.

Archetti, C., Corberán, Á., Plana, I., Sanchis, J. M., & Speranza, M. G. (2013). *A matheuristic for the team orienteering arc routing problem*. Report No.: WPDEM 2013/9. Italy: Department of Economics and Management, University of Brescia.

BIRATTARI, M. 2009. *Tuning metaheuristics: A machine learning perspective*. Berlin: Springer.

Caserta, M., & Voß, S. (2012). A hybrid algorithm for the DNA sequencing problem. *Discrete Applied Mathematics*. doi:10.1016/j.dam.2012.08.025.

Glover, F., & Kochenberger, G. A. (Eds.). (2003). *Handbook of metaheuristics*. Dordrecht: Kluwer Academic Publishers. ISBN 1-4020-7263-5.

Gonzalez, T. F. (2007). *Handbook of approximation algorithms and metaheuristics*. Boca Raton: Taylor & Francis Group, LLC. ISBN 978-1-58488-550-4.

Maniezzo, V., Stützle, T., & VOß, S. (eds.). (2009). *Matheuristics: Hybridizing metaheuristics and mathematical programming*. New York: Springer Science + Business Media LLC. ISBN 978-1-4419-1305-0.

Pirkwieser, S. (2012). *Hybrid metaheuristics and matheuristics for problems in bioinformatics and transportation*. Unpublished Doctoral Thesis, Vienna University of Technology, Vienna.

Sniedovich, M., & Voß, S. (2005). The corridor method: A dynamic programming inspired metaheuristic. *Control and Cybernetics, 35*, 551–578.

Talbi, E.-G. (2009). *Metaheuristics: From design to implementation*. New Jersey: Wiley. ISBN 978-0-470-27858-1.

Xhafa, F., & Abraham, A. (2008). *Metaheuristics for scheduling in industrial and manufacturing applications*. Berlin: Springer. ISBN 978-3-540-78984-0.

Yang, X.-S. (2010). *Engineering optimization: An introduction with metaheuristic applications*. New Jersey: Wiley. ISBN 978-0-470-58246-6.

Biographies

Bo Xing, is a senior lecturer under the division of Asset Integrity Management Centre at the Department of Mechanical and Aeronautic Engineering, Faculty of Engineering, Built Environment and Information Technology, University of Pretoria, South Africa. Dr. Xing earned his DIng degree (Doctorate in Engineering with a focus on remanufacturing) in the early 2013 from the University of Johannesburg, South Africa. He also obtained his BSc and MSc degree in Mechanical Engineering from the Tianjin University of Science and Technology, P.R. China, and the University of KwaZulu-Natal, South Africa, respectively. He was a scientific researcher at the Council for Scientific and Industrial Research (CSIR), South Africa. He has published more than 50 research papers in books, international journals, and international conference proceedings. His current research interests lie in applying various nature-inspired computational intelligence methodologies towards miniature robot design and analysis, advanced mechatronics system, reconfigurable manufacturing system, e-maintenance, production planning and scheduling, routing and network design in remanufacturing and closed-loop supply chain. Dr. Xing's latest publications include a book entitled **"Computational Intelligence in Remanufacturing"**.

Wen-Jing Gao, is a senior sales representative affiliated to the Department of New Product Development, Mei Yuan Mould Design and Manufacturing Co., Ltd, P.R. China. Mrs. Gao holds a BCom (Honors in Economics) degree from the University of Kassel, Germany. Since 2005, she has been working closely with Dr. Xing in various academic- or industrial-oriented projects. She has published more than 40 technical articles in books, international journals, and international conference proceedings. Her research interests include computational intelligence, new product development, the Internet of things, information management, recommender system design, customer oriented business model, product service system, ambient intelligence, mechatronics, miniature robot design and analysis, remanufacturing, reconfigurable manufacturing system, cellular manufacturing system, flexible manufacturing system, and closed-loop supply chain management. Mrs. Gao has presented her work at various international level of conferences such as IEEE International Conference on Systems, Man, and Cybernetics (IEEE SMC), Annual IEEE International Conference on Fuzzy Systems (FUZZ-IEEE), Annual

B. Xing and W.-J. Gao, *Innovative Computational Intelligence: A Rough Guide to 134 Clever Algorithms*, Intelligent Systems Reference Library 62, DOI: 10.1007/978-3-319-03404-1, © Springer International Publishing Switzerland 2014

IEEE Congress on Evolutionary Computation (IEEE CEC), IEEE World Congress on Computational Intelligence (IEEE WCCI), IEEE Symposium Series on Computational Intelligence (IEEE SSCI), IEEE/ASME International Conference on Mechatronic and Embedded Systems and Application, International Conference on Industrial Engineering and Systems Management (IESM), International Symposium on Neural Networks (ISNN), and International Conference on Swarm Intelligence (ICSI). Mrs. Gao's latest publications include a book entitled "**Computational Intelligence in Remanufacturing**".

Zbigniew Michalewicz, is an internationally renowned new technologies expert. Dr. Zbigniew Michalewicz has published over 260 articles and 37 books on the subject of predictive data mining and logistics optimization. He is Emeritus Professor at the University of Adelaide, Australia. Prof. Michalewicz received his PhD from the Institute of Computer Science, Polish Academy of Sciences, in 1981. He also holds a Doctor of Science degree in Computer Science from the Polish Academy of Science, and in 2002 he received the title of "Professor" from the President of Poland, Mr. Alexander Kwasniewski. Meanwhile, he holds Professor positions at several other institutions such as the Institute of Computer Science, Polish Academy of Sciences, the Polish–Japanese Institute of Information Technology, and the State Key Laboratory of Software Engineering of Wuhan University, China. He is also associated with the Structural Complexity Laboratory at Seoul National University, South Korea. Zbigniew Michalewicz has also served as the Chairman of the IEEE Technical Committee on Evolutionary Computation, and later as the Executive Vice President of the IEEE Neural Network Council. Apart from this, Zbigniew Michalewicz serves as Chief Scientist for SolveIT Software Pty. Ltd. (recently acquired by Schneider Electronic), a company specializing in custom software solutions for demand forecasting, scheduling, supply chain optimization and mine optimization solutions. He has over 25 years of industry experience, and possesses expert knowledge of many artificial intelligence methods and modern heuristics. He has led numerous data mining and optimization projects for major corporations and also for several government agencies in the United States of America and Poland, and his scientific and business achievements have been recognized by publications such as TIME Magazine, Newsweek, The New York Times, Forbes, and the Associated Press. Dr. Michalewicz's latest publications include a book entitled "**A Guide to Teaching Puzzle-Based Learning**".

Xin Yao, is a Chair (Professor) of Computer Science and the Director of CERCIA (the Centre of Excellence for Research in Computational Intelligence and Applications), University of Birmingham, UK. Prof. Yao received his BSc in 1982, MSc in 1985 and PhD in 1990, and worked in Australian National University, Commonwealth Scientific and Industrial Research Organisation (CSIRO) Division of Building, Construction and Engineering, and University College, the University of New South Wales (UNSW), Australian Defence Force Academy (ADFA), before taking up his professorship at Birmingham in April

1999. He is an IEEE Fellow and a Distinguished Lecturer of IEEE Computational Intelligence Society (CIS). He is also a Director of USTC-Birmingham Joint Research Institute in Intelligent Computation and Its Applications (UBRI) at the University of Science and Technology of China (USTC), Hefei, China. His work won the 2001 IEEE Donald G. Fink Prize Paper Award, 2010 IEEE Transactions on Evolutionary Computation Outstanding Paper Award, 2010 BT Gordon Radley Award for Best Author of Innovation (Finalist), 2011 IEEE Transactions on Neural Networks Outstanding Paper Award, and other best paper awards at conferences. He won a prestigious Royal Society Wolfson Research Merit Award in 2012. He was the Editor-in-Chief (2003–2008) of IEEE Transactions on Evolutionary Computation. His major research interests include evolutionary computation and ensemble learning. He has more than 400 refereed publications in international journals and conferences. Prof. Yao recently received the 2013 Evolutionary Computation Pioneer Award for his outstanding contributions to the theory and applications of evolutionary computation. The award was presented during the 2013 IEEE Congress on Evolutionary computation (IEEE CEC), held from 20–23 June 2013 at Canćun, México.

Printed in the United States
By Bookmasters

Printed in the United States
By Bookmasters